About Pearson

Pearson is the world's learning company, with presence across 70 countries worldwide. Our unique insights and world-class expertise comes from a long history of working closely with renowned teachers, authors and thought leaders, as a result of which, we have emerged as the preferred choice for millions of teachers and learners across the world.

We believe learning opens up opportunities, creates fulfilling careers and hence better lives. We hence collaborate with the best of minds to deliver you class-leading products, spread across the Higher Education and Test Preparation spectrum.

Superior learning experience and improved outcomes are at the heart of everything we do. This product is the result of one such effort.

Your feedback plays a critical role in the evolution of our products and you can contact us - reachus@pearson.com. *We look forward to it.*

What Industry and Academic Experts Say about This Book ...

This book provides a comprehensive and modern perspective on Industry 4.0, highlighting its integration with emerging technologies like AI, Blockchain, and cybersecurity measures. The authors offer valuable insights into the transition from traditional lean systems to smart, agile practices enabled by IoT, AI, supporting real-time waste reduction, predictive maintenance, and digital supply chains tailored for customization and adaptability. What sets this book apart is its practical approach, featuring real-world applications and in-depth case studies across industries that show how various technologies can be effectively combined to build smart manufacturing frameworks. Unlike many books that focus on isolated aspects, this work presents the full scope of Industry 4.0, including future trends and selection strategies for different manufacturing contexts. It is a timely and insightful contribution, useful for both industry professionals and students. As a standardization expert, I also see its potential in guiding standardized approaches for smart manufacturing. Congratulations to the authors, Saikat Dutt, Arindam Ray, and Sobhan Sen Sarma, for their thoughtful and thorough work.

—Anupam Kaul
Founder and Lead, Veriti Masters Guild
Chair, BIS, National Standards Committee on Quality Management & Basic Service Standards

I am very happy to learn that the authors have taken the great initiative to introduce and explain the various aspects of Industry 4.0 (the fourth industrial revolution) to the professional and student community. Industry 4.0 optimises the computerisation of Industry 3.0. There is hardly any good literature on this topic that explains the concepts and recommendations lucidly. This book will serve as a strong guide. The book starts with a lucid introduction followed by explaining all nine tenets that include: Industry IoT, Robotics, 3-D Printing, Cloud Computing, Big Data Analytics, Blockchain, Cybersecurity and so on. Each concept is accompanied with illustrations and plausible applications. I congratulate the authors and wish that the community greatly benefits.

—Bimal K. Roy
Padmashree Awardee, Former Director of Indian Statistical Institute,
Former Chairman, National Statistical Commission

This is a must-read for anyone seeking a clear, in-depth view of the Fourth Industrial Revolution. The authors expertly unpack complex technologies like robotics, AI and cyber-physical systems, and connect them to real-world industrial applications. With its balance of technical insight and accessible language, the book is ideal for professionals, leaders and learners alike. Its standout feature is the thoughtful analysis of how robotics is reshaping manufacturing, from adaptive systems to predictive maintenance. A valuable resource for decision-makers, engineers, researchers and students alike.

—Biren Anshu
Chief Human Resource Officer
Novus Hi-Tech Robotic Systemz & The Hi-Tech Robotic Systemz Limited

I am pleased to recommend this current and comprehensive book on Industry 4.0, which offers both practical guidance and theoretical insight into one of the most significant developments in contemporary business. The book skilfully blends basic theories with real-world applications while covering significant technologies like IoT, AI, robotics, additive manufacturing and digital twins. Its systematic approach, supported by case studies and implementation frameworks, is particularly advantageous for practitioners, educators and students. By connecting operational, technological and strategic aspects, the book demystifies the Industry 4.0 environment and equips readers to spearhead digital transformation across industrial domains.

—Bodhibrata Nag
Professor, Operations Management
Indian Institute of Management Calcutta

This book presents a comprehensive treatment of the various facets and subjects related to Industry 4.0 with both theoretical divisions and practical examples. It clearly outlines the revolutionary advancements of Industry 4.0 beyond its predecessors. It is a must-read for engineering and business students to gain some fundamental knowledge of this industry revolution and for industry practitioners who look for opportunities to implement those revolutionary approaches in their organisations. You would not be disappointed.

—Kefeng Xu
Professor, Supply Chain Management
University of North Carolina Greensboro

In a world defined by accelerated digital transformation, Industry 4.0 is not just a buzzword, it's the next evolutionary step in industrialisation. This book delivers a comprehensive yet practical roadmap for navigating the fourth industrial revolution, designed for engineers, students, researchers and industry leaders alike. Each chapter dives deep into these pillars with a structured approach: first introducing the concept, then explaining its architecture, industrial relevance, challenges and finally, its implementation using real-world case studies and applications that increased productivity, improved safety, enhanced quality, reduced operational costs and higher consistency in production across sectors such as automotive, electronics, pharma and logistics.

—Manojkumar Belgaonkar
Head – Regulations, Standards & Quality Management, Siemens Ltd
Senior Assessor – CII-EXIM Bank & FICCI Quality Awards
Member – BIS Committees on Quality, Safety, Automation & Robotics
Member – Indo-German Working Group on Quality Infrastructure

Industry 4.0 is an essential read for anyone looking to understand the transformative power of modern industrial technologies. This book offers a comprehensive introduction to the key concepts and applications of Industry 4.0, making it an invaluable resource for both students and industry professionals. The authors bring together their deep industry expertise and academic excellence to provide over sixty years of industrial experience to the academic perspective on this new industrial paradigm.

The book covers all nine core pillars of Industry 4.0 in a manner that is both informative and readable, including real-world case studies. I find the chapters and the examples provided bridge the gap between theoretical concepts and practical applications. Whether you are an engineering student or a practitioner, Industry 4.0 will equip you with the knowledge and insights needed to navigate the digital transformation and agile manufacturing processes of today's rapidly evolving industrial landscape.

This book is not just a guide; it's a roadmap to the future of smart factories. Highly recommended for anyone eager to stay ahead in the world of Industry 4.0.

—Matthew Mullarkey
Professor of Instruction and Director, DBA Program
University of South Florida

I would like to commend the authors and Pearson for putting together such a comprehensive, insightful and pedagogically structured resource. The text captures the core elements of Industry 4.0, i.e., Digital Transformation, Cyber-Physical Systems and Horizontal and Vertical Integration, in an engaging and application-oriented manner. What I particularly appreciate is the effective use of practical examples, diagrams and clear distinctions between key concepts like digitisation, digitalisation and transformation. The treatment of CPS architecture and the ISA 95 framework further enhances the technical value of the content for students and professionals alike.

I am confident that this book will be a valuable addition to the academic and industrial ecosystem, serving as a relevant reference for engineering, management and applied sciences courses. It will also prove beneficial for practitioners looking to implement Industry 4.0 principles in their operations.

—Naba Kumar Das
Former Deputy General Manager - Operations, Maruti Suzuki India Limited
Assistant Professor, The NorthCap University, Gurugram, Haryana

Authored by experts from top global institutions, *Industry 4.0* blends academic insight with real-world relevance. It explores transformative technologies like cyber-physical systems, IIoT, AI, additive manufacturing and blockchain, showing how their convergence drives smart, autonomous and resilient production systems. The book stands out for connecting theory to practice. It not only introduces technologies but also addresses implementation challenges like legacy integration, workforce readiness and cybersecurity, through case studies and industry examples. Special attention is given to the Indian manufacturing context, highlighting local infrastructure, policy and workforce issues. With a focus on system integration, agile methods and data-driven decision-making, the book encourages a holistic approach to industrial transformation. Whether you're a student, engineer, or leader, this is a practical guide for navigating and succeeding in the Industry 4.0 era.

—Nagahanumaiah
Director of Central Manufacturing Technology Institute, Bengaluru

This book bridges theory and real-world industrial evolution. With rich case studies on IoT, automation, digital twins and cloud ecosystems, it's a definitive guide to Industry 4.0. A must-read for leaders navigating smart manufacturing – academically rigorous yet deeply practical. It will empower future-ready transformation across both factory floors and boardrooms.

—Sumit Malik
Deputy General Manager, Quality Assurance
Maruti Suzuki India Pvt. Ltd, Gurgaon

The book offers a well-structured and insightful overview of key technologies like industrial automation and robotics, cloud computing, and blockchain, clearly explaining core concepts, technological evolution, and practical applications. The automation chapter provides valuable coverage of robotics types, adoption strategies, and market trends. Adding a deeper discussion of implementation challenges could further enrich this section. The cloud computing chapter stands out with its clear explanations of architectures and future directions. Additional details on topics, such as legacy system migration and vendor lock-in, can be included to make the analysis even more comprehensive. Similarly, the blockchain section is highly relevant, particularly in supply chain applications, and could be enhanced by exploring scalability, energy consumption, and emerging solutions in greater depth. Overall, the content is highly informative and thoughtfully presented.

—Somnath Mitra
Digital Transformation Advisor
Former Head - Centre for Digital Transformation, IIM Ahmedabad

INDUSTRY 4.0

Saikat Dutt
Professor of Practice
Indian Institute of Management Calcutta (IIMC)
Institute of Engineering and Management (IEM)
Kolkata
Former Senior Director at Cognizant Technology Solutions

Arindam Ray
Assistant Professor
University of North Carolina, Greensboro, USA

Sobhan Sen Sarma
General Manager
Ericsson India Global Services

Pearson

Senior Manager—Product: Neha Goomer
Senior Editor—Production: C. Purushothaman

ISBN 978-93-671-3174-9

First Impression, 2025
Second Impression, 2026
Third Impression

Published by Pearson India Education Services Pvt. Ltd, CIN: U72200TN2005PTC057128.

Head Office: 1st Floor, Berger Tower, Plot No. C-001A/2, Sector 16B, Noida - 201 301, Uttar Pradesh, India.
Registered Office: 6th Floor, Tower A, Unit A and B International Tech Park, Capita Land Chennai, 200 Feet Radial Road, Zamin Pallavaram, Old Pallavaram Chennai – 600117, Tamil Nadu, India
Website: in.pearson.com; Email: companysecretary.india@pearson.com

Compositor: SRS Global, Puducherry.
Printed in India at Rajkamal Electric Press, Kundli, Haryana-131 028.

This book is dedicated to the people without whom the dream could not have come true—my parents, Tarun Kumar Dutt and Srilekha Dutt; constant inspiration and support from my wife, Adity and unconditional love of my sons Deepro and Devarko.

Saikat Dutt

To Sudhin Sengupta,
Your unwavering belief in my potential, your discipline and your quiet encouragement helped shape who I am today. You inspired me to strive to be the best I can be.

With deep gratitude.

Arindam Ray

This book is dedicated to my parents, Jaya Sen Sarma and Prity Ranjan Sen Sarma, whose values, love and teachings laid the foundation for everything I am; To my wife, Dia, whose unwavering support and belief in me have been my strength in this journey; And to my sons, Jishu and Rishi, for their abundant love and inspiration. May this work inspire you to dream boldly and pursue your own journeys with passion and purpose.

Sobhan Sen Sarma

Contents

Foreword xxv

Preface xxvii

Acknowledgements xxix

About the Authors xxxiii

Model Syllabus for Industry 4.0 xxxv

Lesson Plan: Industry 4.0 – Technologies, Frameworks and Applications xxxix

Chapter 1 Introduction to Industry 4.0 1

1.1 A Factory That Never Sleeps 1

1.2 What is Industry 4.0 2

1.3 Evolution of Industry 4.0 3

1.3.1 Industry 1.0: Mechanisation and Steam Power 3

1.3.2 Industry 2.0: Mass Production and Electricity 4

1.3.3 Industry 3.0: Automation and Computers 5

1.3.4 Industry 4.0: IoT, AI and Data-Driven Decision-Making 6

1.4 Building Blocks of Industry 4.0 8

1.4.1 Foundational Principles of Industry 4.0 8

1.4.2 Operational Methodologies 9

1.4.3 Key Technologies of Industry 4.0 9

1.5 Connectivity for Industry 4.0 and the Importance of 5G Network 13

1.5.1 Foundations Laid by 4G and 4G+ 14

1.5.2 The Leap Forward: 5G as the Communication Backbone for Advanced Industry 4.0 15

1.6 Benefits of Industry 4.0 18

1.7 Use Cases of Industry 4.0 18

1.7.1 Smart Manufacturing and Intelligent Production Systems 19

1.7.2 Supply Chain and Logistics Optimisation 19

1.7.3 Industry 4.0 in Healthcare and Medical Technologies 20

1.7.4 Smart Cities and Infrastructure Development 21

1.7.5 Industry 4.0 and Sustainable Development 22

1.8 The Impact of Industry 4.0 on Workforce and Education 23

1.8.1 Human-Machine Collaboration in Future Workplaces 23

1.8.2 Need for Reskilling and Upskilling in the Workforce 23

1.8.3 Role of Higher Education in Preparing Students for Industry 4.0 24

1.9 Conclusion 24

Summary 25

Exercise 25

Chapter 2 Key Tenets of Industry 4.0 31

2.1 Introduction 31

2.2 Digitisation, Digitalisation and Digital Transformation 33

2.2.1 Digitisation 33

2.2.2 Digitalisation 33

2.2.3 Digital Transformation 34

2.3 Why is Digital Transformation Key to Industry 4.0? 35

2.4 What is a Cyber-physical System? 36

2.5 CPS and IoT 40

2.5.1 What is IoT 40

2.5.2 Comparisons - CPS and IoT 41

2.6 Key Technology Layers of CPS 43

2.6.1 Device Technologies 43

2.6.2 Computing Technologies 44

2.6.3 CPS Communication and Networking Technologies 44

2.7 5C Architecture of CPS 44

2.7.1 Connection 45

2.7.2 Conversion 46

2.7.3 Cyber 46

2.7.4 Cognition 46
2.7.5 Configuration 46
2.8 Domain of Horizontal and Vertical Integration 47
2.9 Horizontal Integration 48
2.9.1 Within a Production Floor 48
2.9.2 Across Multiple Factories 49
2.9.3 Across Supply Chain 49
2.10 Vertical Integration 51
2.10.1 ISA 95 Framework 52
2.11 Conclusion 54
Summary 55
Exercise 57

Chapter 3 Lean and Agile Production Methodologies 61

3.1 Introduction 61
3.2 Traditional Lean Manufacturing Process 62
3.2.1 Key Lean Manufacturing Tools and Techniques 64
3.3 Industry 4.0-enabled Lean Manufacturing Process 65
3.3.1 Case Study on Lean Manufacturing in Industry 4.0 66
3.4 Traditional Agile Manufacturing Process 67
3.5 Industry 4.0-enabled Agile Manufacturing Process 68
3.5.1 Case Study on Agile Manufacturing in Industry 4.0 70
3.6 Other Notable Processes 71
3.7 Conclusion 72
Summary 73
Exercise 74

Chapter 4 Industrial Internet of Things (IIoT) 79

4.1 Introduction 79
4.2 What is IoT and IIoT? 80
4.2.1 IoT 80
4.2.2 IIoT 80
4.3 Reference Architecture of IIoT 82
4.3.1 Key Components of the Architecture 82
4.3.2 Role of Cloud Computing in IIoT Architecture 87
4.3.3 Role of IIoT Platforms 88

4.4 IIoT Implementation Strategy 89
4.4.1 Organisational Setup for Implementing IIoT 89
4.4.2 Technological Setup for Implementing IIoT 90
4.5 Benefits of IIoT 96
4.6 Challenges in IIoT Implementation 97
4.7 Application Domains of IIoT 99
4.7.1 Smart Manufacturing and Supply Chain 99
4.7.2 Healthcare 100
4.7.3 Smart Farming 101
4.7.4 Energy and Utilities 102
4.7.5 Transportation and Logistics 103
4.8 Conclusion 103
Summary 104
Exercise 105

Chapter 5 Industrial Automation and Robotics 111

5.1 Introduction 111
5.1.1 Definition of Industrial Automation 112
5.1.2 Definition of Robots and Robotics 112
5.2 Evolution of Automation and Robotics 114
5.3 Key Building Blocks of Industrial Automation and Robotics 116
5.4 Types of Automation and Robotics 120
5.4.1 Types of Automation 121
5.4.2 Robot Types and Applications 125
5.5 How to Adopt Automation and Robotics in Industry 4.0? - Step-by-Step Approach 133
5.5.1 Step 1: Identify Problem Areas to Solve Through Automation and Robotics 134
5.5.2 Step 2: Define Objectives - What We Want to Achieve from Automation 134
5.5.3 Step 3: Perform Feasibility Study 134
5.5.4 Step 4: Choose the Right Automation and Robotics Technology 135
5.5.5 Step 5: Integrate Hardware and Software 137
5.5.6 Step 6: Prepare Infrastructure and Safety Setup 138
5.5.7 Step 7: Test and Simulate Before Deployment 140
5.5.8 Step 8: Go for Full-Scale Deployment and Monitoring 141

5.6 Case Study of Real-Life Implementation of Automation and Robotics in Amazon Warehouse 143
5.6.1 Introduction 143
5.6.2 History and Early Investments 143
5.6.3 Robotics and Automation Systems 143
5.6.4 Packaging Systems 144
5.6.5 Usage of Soft Robotics 145
5.6.6 Employee Development and Training 145
5.6.7 Usage of Cloud Computing and Robotics Software 146
5.6.8 Investments and Future Trends 146
5.6.9 Key Takeaways 147
5.7 Conclusion 147
Summary 148
Exercise 151

Chapter 6 Additive Manufacturing/3D Printing 157

6.1 Introduction 157
6.2 What is 3D Printing? 158
6.3 History of 3D Printing 159
6.3.1 Early Innovations 159
6.3.2 Prototyping to Production 160
6.3.3 Democratisation of the Technology 160
6.4 How 3D Printing Works? 161
6.4.1 3D Modelling 161
6.4.2 File Conversion 162
6.4.3 Slicing 162
6.4.4 File Transfer 162
6.4.5 Printing 163
6.4.6 Post-processing 163
6.5 3D Printing Technologies 163
6.5.1 Vat Photopolymerisation 163
6.5.2 Material Extrusion - Fused Deposition Modelling (FDM) 166
6.5.3 Powder Bed Fusion 168
6.5.4 Jet Technologies 169
6.5.5 Metal Printing 172
6.6 Overview of 3D Modelling Software 176

6.7 Benefits of 3D Printing 178
6.7.1 Distributed Manufacturing 178
6.7.2 Mass Customisation 178
6.7.3 Complexity 179
6.7.4 Tool-less Production 179
6.7.5 Sustainable / Environmentally Friendly 179
6.8 Applications of 3D Printing 179
6.8.1 Modern Manufacturing 179
6.8.2 Medical and Dental 180
6.8.3 Aerospace and Aviation 180
6.8.4 Automotive 181
6.8.5 Jewellery 181
6.8.6 Fashion 181
6.9 Conclusion 182
Summary 182
Exercise 183

Chapter 7 Augmented Reality, Virtual Reality and Mixed Reality 189

7.1 Introduction 189
7.2 Fundamentals of All the Realities 190
7.2.1 A Brief History of AR/VR 194
7.2.2 How Does Augmented Reality Work? 195
7.2.3 Types of Augmented Realities 195
7.2.4 How Does Virtual Reality Work? 199
7.2.5 Types of Virtual Realities 199
7.3 Key Technologies Behind AR and MR 200
7.3.1 SLAM (Simultaneous Location and Mapping) 200
7.3.2 Depth Tracking 201
7.3.3 Image Processing and Projection 201
7.3.4 3D Modelling 201
7.3.5 Manipulation and Interaction with Objects 202
7.4 Key Technologies Behind VR 202
7.4.1 VR Headsets (HMD) 202
7.4.2 Lenses and Screens 203
7.4.3 Latency, Field of View and Frame Rate 203

7.4.4 Spatial Audio 204
7.4.5 Head, Movement and Position Tracking 204
7.4.6 Virtual Reality Controllers 204
7.4.7 Virtual Reality Software 204

7.5 AR/VR in Smart Factories 205
7.5.1 Warehouse Management 205
7.5.2 Design and Build 206
7.5.3 Maintenance and Assembly Training 207
7.5.4 Factory Floor Planning 208
7.5.5 Improved Worker Safety 208
7.5.6 How a German Automaker Leverages AR/VR in Production 209

7.6 Other Application Areas of AR / VR / MR 209
7.6.1 Gaming 209
7.6.2 Healthcare 210
7.6.3 Education and Training 211
7.6.4 Retail and Marketing 212
7.6.5 Services 213

7.7 Concept of Metaverse 214

7.8 Conclusion 214

Summary 215

Exercise 216

Chapter 8 Cloud Computing 221

8.1 Introduction 221

8.2 What is Cloud Computing 222
8.2.1 Definition and Features 222
8.2.2 Evolution of Cloud Computing 223

8.3 Cloud Service Models 225
8.3.1 Infrastructure as a Service (IaaS) 226
8.3.2 Platform as a Service (PaaS) 226
8.3.3 Software as a Service (SaaS) 227

8.4 Cloud Deployment Models 229
8.4.1 Public Cloud 229
8.4.2 Private Cloud 229

8.4.3 Hybrid Cloud 230

8.4.4 Community Cloud 231

8.5 Key Components of Cloud Architecture 232

8.5.1 Virtualisation 232

8.5.2 Cloud Orchestration and Management 239

8.5.3 Cloud Native Applications 240

8.5.4 Edge and Fog Computing 241

8.5.5 Handling Data of Things in the Cloud 245

8.5.6 Solution Considerations for IoT with Cloud 246

8.6 Cloud Computing for Industry 4.0: Implementation in Manufacturing Systems 247

8.6.1 Management of Manufacturing through Cloud ERP 248

8.6.2 Manufacturing Execution Systems (MES) in Cloud 250

8.6.3 Supervisory Control and Data Acquisition (SCADA) in Cloud 250

8.6.4 Programmable Logic Controllers (PLCs) in Cloud 253

8.6.5 Sensors in Cloud 254

8.7 Cloud Security 256

8.7.1 Cloud Security Vulnerabilities for IoT Systems 256

8.7.2 Cloud Security Services for IoT Applications 257

8.8 Conclusion 259

***Summary* 259**

***Exercise* 262**

Chapter 9 Big Data 267

9.1 Introduction 267

9.1.1 Evolution of Big Data 268

9.1.2 Different Types of Data 270

9.1.3 Big Data Characteristics 274

9.2 Big Data Storage Solutions 277

9.2.1 Relational Databases 277

9.2.2 NoSQL Databases 280

9.2.3 Data Warehouse 283

9.2.4 Data Marts 283

9.2.5 Data Lakes 284

9.3 Big Data Management Techniques 285

9.3.1 Online Transaction Processing (OLTP) 285

9.3.2 Online Analytical Processing (OLAP) 286

9.3.3 Extract Transform Load (ETL) 286

9.3.4 Extract Load Transform (ELT) 287

9.4 Big Data Processing 288

9.4.1 Parallel Data Processing 289

9.4.2 Distributed Data Processing 290

9.4.3 Processing Workloads 291

9.4.4 Batch Processing Using MapReduce 291

9.4.5 Advantages of Using MapReduce 294

9.4.6 Hadoop 294

9.5 Big Data Implementation Approach 296

9.5.1 Business Case Evaluation 296

9.5.2 Data Identification 297

9.5.3 Data Acquisition and Ingestion 297

9.5.4 Data Storage 297

9.5.5 Data Processing and Transformation 298

9.5.6 Descriptive Analytics and Data Visualisation 298

9.5.7 Deployment, Monitoring and Maintenance 298

9.6 Use of Big Data Across Industries 299

9.6.1 Manufacturing 299

9.6.2 Transportation and Logistics 299

9.6.3 Energy and Utilities 299

9.6.4 Retail 300

9.6.5 Healthcare 300

9.6.6 Finance 300

9.6.7 Entertainment and Media 300

9.6.8 Public Sector 301

9.7 Conclusion 301

Summary 302

Exercise 304

Chapter 10 Analytics 309

10.1 Introduction 309
10.1.1 Evolution of Analytics 310
10.1.2 Why Analytics Matters in Industry 4.0? 311
10.2 Key Types of Analytics 312
10.2.1 Descriptive Analytics 312
10.2.2 Predictive Analytics 314
10.2.3 Prescriptive Analytics 323
10.3 Benefits of Analytics in Industry 4.0 325
10.3.1 Decreasing Downtime 325
10.3.2 Improving Productivity and Optimising Cost 328
10.3.3 Optimising Complex Production Networks 328
10.3.4 Other Benefits 329
10.4 Implementation Strategies 331
10.4.1 10 Steps to Implement Analytics for Industry 4.0 331
10.4.2 Established Process Models 336
10.4.3 From Device to Dashboard – Data Flow 340
10.5 Business Use Cases 342
10.6 Conclusion 343
Summary 344
Exercise 345

Chapter 11 Digital Twins 351

11.1 Introduction 351
11.2 What is Digital Twin 352
11.2.1 Definition 352
11.2.2 Evolution of Digital Twin 353
11.2.3 Types of Digital Twins 356
11.2.4 Digital Thread and Its Relationship with Digital Twins 359
11.2.5 Digital Twin vs Simulation 360
11.2.6 Digital Twins vs Augmented Reality 361
11.3 Cognitive Digital Twins 363
11.4 Role of Digital Twins in Product Life Cycle 365
11.4.1 Role of Digital Twins in Design 365

11.4.2 Role of Digital Twins in Production and Testing Phase 367
11.4.3 Role of Digital Twins in the Operational Phase 368

11.5 Use Cases of Digital Twin Applications in Industry 4.0 369

11.5.1 Use of Digital Twins in Manufacturing 369
11.5.2 Use of Digital Twins in Infrastructure 369
11.5.3 Use of Digital Twins in Aerospace 370
11.5.4 Use of Digital Twins in Automotive 372
11.5.5 Use of Digital Twins in Construction 373
11.5.6 Use of Digital Twins in Oil and Gas 373
11.5.7 Use of Digital Twins in Health Care 374

11.6 Key Implementation Considerations 375

11.6.1 Cybersecurity Threats 375
11.6.2 Choosing Right Data 375
11.6.3 Standardisation Issue 376

11.7 Implementing a Digital Twin 376

11.8 Case Study – Digital Twin Implementation for a Smart Hospital Building 378

11.8.1 Introduction 378
11.8.2 Background 379
11.8.3 Innovative Approach 379
11.8.4 Steps for Continuous Life Cycle Integration 379
11.8.5 Digital Twin Development Process 380
11.8.6 Digital Twin Application 381
11.8.7 Key Takeaways 382

11.9 Conclusion 382

Summary 383

Exercise 384

Chapter 12 Blockchain 389

12.1 Introduction 389

12.2 Key Concepts of Blockchain Technology 391

12.2.1 How Do Blockchains Work? 391
12.2.2 Components of Blockchain 393
12.2.3 Structure of a Block 400
12.2.4 Types of Blockchain Networks 403

12.2.5 Blockchain Protocols Relevant for Industry 4.0 403
12.2.6 Blockchain as a Service (BaaS) 406
12.3 Blockchain Use Cases in Industry 4.0 411
12.3.1 Supply Chain Transparency and Traceability 412
12.3.2 Secure and Efficient Tracking of Physical Assets 412
12.3.3 Reform Predictive Maintenance 413
12.3.4 Secure System for Industrial IoT Data 413
12.3.5 Improve Collaboration Between Industry Players 414
12.3.6 Demonstrate Compliance with Social and Environmental Regulations 414
12.4 Blockchain Benefits for Industry 4.0 415
12.5 Implementation Strategy 415
12.6 Challenges and Considerations for Blockchain Adoption 419
12.7 Conclusion 422
Summary 422
Exercise 424

Chapter 13 Cybersecurity 429

13.1 Introduction 429
13.1.1 Evolution of Cybersecurity 431
13.2 Cybersecurity Principles 432
13.2.1 Confidentiality 432
13.2.2 Integrity 432
13.2.3 Availability 432
13.2.4 Balancing the Elements for Security 433
13.3 Types of Cyber Threats 433
13.3.1 Malware 433
13.3.2 Phishing 435
13.3.3 Denial of Service 435
13.3.4 Data Breach 436
13.3.5 Insider Threat 436
13.4 Security at Different Layers 437
13.4.1 System Level Security 437
13.4.2 Network Level Security 437
13.4.3 Application Level Security 437
13.4.4 Transmission Level Security 437

13.4.5 Cloud Security 438

13.4.6 Mobile Security 438

13.5 Cybersecurity in Industry 4.0 438

13.5.1 Connected Devices 439

13.5.2 Smart Factory 440

13.5.3 Digital Supply Chain 441

13.6 Defence in Depth Model 442

13.6.1 ICS Security vs Traditional IS Security 442

13.6.2 DMZ Layers in ISA-95 Framework or Purdue Model 443

13.6.3 Layered Security Approach 445

13.7 Cybersecurity Implementation Framework 447

13.7.1 Identify 447

13.7.2 Protect 448

13.7.3 Detect 449

13.7.4 Respond 449

13.7.5 Recovery 450

13.8 Key Challenges in Implementing Cybersecurity in Smart Factories 450

13.9 Conclusion 451

Summary 451

Exercise 452

Chapter 14 Conclusion: Bringing It All Together and Way Forward 457

14.1 Introduction 457

14.2 SmartAir Aerospace Case Study: A Digital Transformation Journey in Aircraft Manufacturing 460

14.2.1 Design and Engineering Transformation 460

14.2.2 Smart Manufacturing 462

14.2.3 Smart Maintenance 464

14.2.4 Connectivity Using 5G Technology 466

14.2.5 Final Outcomes 466

14.3 Factors Affecting the Adoption of Industry 4.0 467

14.3.1 Data Leverage 467

14.3.2 Organisational Readiness 468

14.3.3 Legal and Regulatory Factors 468

14.3.4 Cyber Security 469
14.3.5 Infrastructure Readiness 469
14.3.6 Technology Integration 470
14.3.7 Financial Considerations 471
14.4 The Way Forward 472
***Summary* 472**
***Exercise* 473**

***Index* 479**

Foreword

In this book, the authors present a deep insight into recent advances in Industry 4.0, manufacturing processes and smart services, as well as the industrial digital transformation. They share their vision on contemporary technologies and how these are likely to influence new ways of working.

All relevant Industry 4.0 topics are discussed coherently, facilitating the understanding of the different areas of interest. The book serves as a reference for engineers by inviting the readers to explore novel technologies like cyber-physical systems, Internet of Things (IoT), 3D printing, virtual reality, AI, ML, blockchain technology and many more. Added value is given by analysing the strategic and organisational shifts which occur when implementing practical solutions for Industry 4.0. The authors channelise a 'passion for technology' by delivering solutions to bridge the gap between theoretical knowledge and practical adaptability. Sections like 'Did you know?' provide further insight, facilitating the learning and deepening the understanding.

The book is valuable for engineers, students, researchers and industry leaders and provides knowledge on how to implement modern automation strategies powered by digital technologies. Based on my experience in implementing Industry 4.0 in various industries across the world, I have seen the challenges that the engineers or industry leaders face if they want to implement or participate in implementing Industry 4.0 in their organisation without the knowledge and experience of why, when and what to change and implement. Hence, an over-encompassing knowledge of all these seemingly discreet technologies and processes and how they interact with each other to make a connected Industry 4.0 setup would be extremely beneficial for the readers.

Dr Daniel Hahn
Senior Expert, Steinbeis R-Tech / World Bank (Germany/India);
Technical Director – Digitalisation, Quantum Servicios Marítimos (Chile);
Former CEO, APS GmbH – European Centre for Mechatronics;
Member: National Occupational Standards Committee and Industry 4.0 Board, ESSCI under Make in India initiative; Panelist: IEEE IECON IES Industry Forum (Melbourne) on Next-Gen Industrial Communication

Preface

As industries across the globe undergo rapid digital transformation, there is an urgent need to understand the technologies driving this change—collectively known as *Industry 4.0*. This book was conceived to address that need. Conversations with students, educators and working professionals revealed a growing curiosity about how traditional manufacturing and service models are evolving with the advent of technologies like IIoT, AI, robotics and cloud computing. Yet, resources that present these ideas in a clear, application-oriented and accessible format remain limited. This book aims to fill that gap by offering a comprehensive guide to the concepts, technologies and strategic implications of Industry 4.0 for both learners and practitioners.

Industry 4.0 brings together a suite of advanced digital technologies such as the Industrial Internet of Things (IIoT), Additive Manufacturing, Automation and Robotics, Cloud Computing, Big Data and Analytics, Extended Reality (AR/VR/MR), Digital Twin and Blockchain integrated through Cyber-Physical Systems and protected by the Cybersecurity protocols. While these concepts are individually known, the way they converge to enable smart, autonomous and connected industrial systems is not always well understood, especially among students and junior managers.

This book is written to demystify Industry 4.0. It is structured to provide readers not just with theoretical knowledge, but also with implementation strategies, case studies and real-world applications. The aim is to empower readers to go beyond buzzwords and understand the deep transformation underway in industrial ecosystems. The book places special emphasis on how traditional methodologies like Lean and Agile evolve into Lean 4.0 and Agile 4.0, aligning efficiency with digital adaptability.

We have taken deliberate care to write in clear, student-friendly language while maintaining academic rigour and professional relevance. The content has been designed to serve as a textbook for graduate and advanced undergraduate students and as a reference guide for industry professionals, project managers and innovation leaders.

WHO SHOULD READ THIS BOOK?

This book is intended for a diverse audience. Engineering and IT students will gain foundational knowledge about Industry 4.0 technologies and their integration. Factory engineers as well as software professionals will find hands-on guidelines and architectural insights into implementing IIoT, automation, cloud solutions and analytics. Technical and business managers will understand how to align digital transformation strategies with business goals, while educators will find it useful as a structured academic resource.

HOW THIS BOOK IS ORGANISED

Each chapter begins with a conceptual overview followed by clearly structured subtopics that explain the core ideas. Real-world examples, practical implementation tips and case studies are interspersed throughout to help readers bridge theory with practice.
At the end of each chapter, you will find:

- Recap summaries to reinforce key points.
- 'Did You Know' and 'Points to Ponder' sections that encourage deeper thinking.
- Sample questions to aid examination preparation or group discussions.

The initial chapters provide the conceptual groundwork by introducing Industry 4.0 and its foundational pillars. This is followed by in-depth exploration of enabling technologies such as IIoT, Automation & Robotics, 3D Printing, Digital Twin and Blockchain. The book also discusses the evolving roles of AR/VR/MR, Cloud Computing, Big Data and Analytics, Cybersecurity and digital transformation frameworks critical to Industry 4.0 readiness.

The final chapters connect these technologies to practical applications and strategic implementation in real-world industrial settings—preparing the reader to lead or contribute meaningfully to Industry 4.0 initiatives.

Happy learning!

Acknowledgements

We are grateful to Pearson Education, which came forward to publish this book. Ms Neha Goomer and Mr Purushothaman Chandrasekaran of Pearson Education have always been kind and understanding. Ms Neha provided immense guidance on making this book market-ready and also provided ideas for popularising the book. Mr. Purushothaman reviewed this book with abundant patience and helped us say what we had wanted to, improvising each and every page of this book with care. Their patience and guidance were invaluable. Thank you very much, Neha and Puru.

All Authors

This book wouldn't have been possible without the constant inspiration and support of my lovely wife, Adity. My parents have always been enthusiastic about this project and provided me continuous guidance at every necessary step. The unconditional love and affection my sons – Deepro and Devarko ushered on me constantly provided me the motivation to work hard on this crucial project.

The journey through my career, both in Industry and Academics has been very rewarding, as it has given me the opportunity to work for some of the best organisations in the world and learn from some of the best minds. Along the way, several individuals have been instrumental in providing me guidance, opportunities and insights. This book is the culmination of all the learning that I gained from many highly reputed professionals in the industry. I was fortunate to work with them and gained knowledge from them which helped me shape my professional career. Prof. Bodhibrata Nag, Operations Management Group and Prof. Piyush Mehta, Dean Faculty and Research, from Indian Institute of Management Calcutta (IIMC), Prof. Bimal K Roy, Padmashree Awardee, Former Director, Indian Statistical Institute (ISI), Prof Satyajit Chakrabarti, Director of Institute of Engineering and Management (IEM) guided me enormously on different aspects of life and career. I am indebted to Mr Krishna Prasad Yerramilli, Vice-president and Mr Ramesh Dhanakoti, Former Senior Vice President of Cognizant Technology Solutions for their inspiration in creating this book. My heartfelt thanks go to all the wonderful people who contributed in many ways to conceptualising this book and wished for the success of this project.

Saikat Dutt

Writing a book is rarely a solitary endeavour and this one has been no exception. It has been shaped, strengthened and sustained by the kindness, wisdom and support of many people to whom I owe a great deal of gratitude.

First, I want to thank my co-authors, Saikat and Sobhan, who brought not just their expertise, but also their camaraderie and good humour to the table. Collaborating with you was an intellectually enriching experience and your insights helped elevate every chapter. You challenged my thinking in the best possible ways and made this journey far more rewarding than I could have imagined. On the lighter side, I'll definitely miss the banter in our weekly meetings and group messages.

Special thanks to the academic communities at the University of South Florida, where this journey began and the University of North Carolina Greensboro, where it is seeing the light of day. These institutions not only provided me with the time and resources to pursue this endeavour, but also a collegial and intellectually vibrant environment that made writing both possible and joyful. I am also indebted to Dr Joy Bhadury, Dean - Bryan School of Business and Economics and Dr Lakshmi Iyer, Department Head - Information Systems and Supply Chain Management in UNCG, whose enthusiastic approval of this book project meant I could pursue it wholeheartedly and without hesitation.

To my family and friends, your support is the invisible thread woven through every word of this book. Thank you for cheering me on through deadlines, drafts and doubts. Especially to my wife, Gargi and my children, Banhisikha and Abhraneel (see, I didn't miss the chance to use your full names!), thank you for bearing with me, especially during the times I missed being fully present. Your patience and occasional playful sarcasm helped keep me sane. To my parents, Amitabha Ray and Gouri Ray, thank you for everything – your sacrifices, love and unyielding faith. This book would not exist without the foundation you laid.

To my teachers, thank you for igniting the spark of curiosity and guiding me along this lifelong journey of learning. And to my students, for keeping me grounded and making sure I never stop learning.

Arindam Ray

This book has been a journey that spanned over a year and a half and built on more than two decades of experience in the IT industry across telecom, cloud and automation. It reflects not only the knowledge I have gained, but also the people and ideas that have inspired and shaped my thinking on Industry 4.0.

I am deeply grateful to my parents, Jaya Sen Sarma and Prity Ranjan Sen Sarma, for instilling in me the values and curiosity that laid the foundation for all I do. My heartfelt thanks to my wife, Dia, for her unwavering support and patience through every phase of this journey. To my sons, Jishu and Rishi, their love and inspiration kept me going, even when I had to sacrifice precious family time for this project. I also thank my co-authors for their collaboration and deep insights, which enriched the book and added greater depth to its content.

This work has benefited greatly from knowledge sharing and idea exchanges with amazing professionals, including Sristi Saha, Researcher in Agricultural Robotics at NARO, Japan; Soumaya Kanta Nayak, Cloud Solution Architect at Ericsson Global Services, Jayanta Kundu, Innovation Lead at Ericsson South East Asia; Santanu Chakraborty, Leadership Coach and former CEO of Reliance Communications; and Prabir Mukherjee, Senior Planning Manager at Anamind and former Supply Chain Leader at Nestlé.

I have drawn inspiration and valuable learnings from the top leaders of Ericsson Global Services, including Dr Sanjoy Sen, Head of Service Delivery and Automation; Ashwin Kalyanpur, Market Area Delivery Cluster Head; Indrajit Sanyal, Head of Service Delivery - Network Management; and Anand Palkiwala, VP and Head of Service Delivery - Business Area Cloud Software and Services.

Also, I am extremely fortunate to witness great leaders of Ericsson like Indranil Das, Former Vice President and Amitabh Ray, Former Managing Director, during my career at Ericsson. Their thought leadership and passion for digital innovation have helped me see the bigger picture in the evolving industrial landscape.

I am indebted to Muthukumaran Arunachalam, Former Head of Service Delivery, Ericsson Global Services and Ramanuj N, Delivery Head – Middle East at Tech Mahindra, for their great inspiration in creating this book.

Lastly, my heartfelt thanks to all who contributed, directly or indirectly and remained a constant motivation while writing this book.

Sobhan Sen Sarma

About the Authors

Saikat Dutt, PMP, PMI-ACP, CSM, is a Professor of Practice at both the Indian Institute of Management Calcutta (IIMC) and the Institute of Engineering and Management (IEM), Kolkata. He is the author of four acclaimed books—Machine Learning, PMI Agile Certified Practitioner – Excel with Ease, Software Engineering and Software Project Management—published by PEARSON. These titles are widely adopted as textbooks in leading academic institutions across India and are regarded as trusted industry references.

With 26 years of experience in the IT industry, Saikat has led large-scale, multi-location and mission-critical projects. He is a former Senior Director at Cognizant Technology Solutions and has held leadership roles in several top multinational companies. His expertise spans Industry 4.0, Artificial Intelligence, Machine Learning, Program Management, Agile and DevOps practices. Saikat is certified as a Project Management Professional (PMP), PMI Agile Certified Practitioner (PMI-ACP) and Certified Scrum Master (CSM). He earned his B.E. degree from Jadavpur University, Kolkata.

An active speaker at industry forums, Saikat frequently presents on topics such as Industry 4.0, Agile, Project Management and Machine Learning. He also collaborates with IIM Calcutta to develop management case studies used in global business schools.

Dr Arindam Ray is an Assistant Professor of Information Systems at the University of North Carolina, Greensboro, USA. His scholarly work focuses on advanced artificial intelligence and machine learning to address critical issues in business and society, with publications in leading journals and conferences in the field of Information Systems. Dr Ray earned his Ph.D. in Business Administration (Information Systems) from the University of South Florida, M.S. in Analytics from the Georgia Institute of Technology, MBA from the Indian Institute of Management Bangalore and Bachelor's degree in Engineering from Jadavpur University.

Before transitioning to academia, he spent over two decades in the tech industry, helping Fortune 500 customers in their digital transformation journey. He is a former Associate Vice President of Tech Mahindra and has held leadership positions in other global organisations.

Sobhan Sen Sarma, PMP, is General Manager at Ericsson India Global Services, with over 26 years of diverse experience in the IT industry. Throughout his career, he has held key delivery leadership roles across several top-tier multinational organisations, driving large-scale digital transformation initiatives with global impact.

Sobhan brings deep expertise in program and project management, telecommunications, cloud technologies, automation, Agile, DevOps and the evolving landscape of Industry 4.0. He has consistently led cross-functional teams and high-value programs, bridging technology and business to deliver innovation and efficiency at scale.

A certified Project Management Professional (PMP), Sobhan holds MSC degree in Economics from Calcutta University and PGDBA in Systems from ICFAI Business School, Calcutta. His work reflects a blend of strategic vision and hands-on execution, developed through decades of experience in dynamic, high-performance environments.

Model Syllabus for Industry 4.0

Course Title: *Industry 4.0: Technologies, Frameworks and Applications*

Credits: 5

Contact Hours per Week: 3 Lectures + 1 Tutorial

Module I: Foundations of Industry 4.0

- **Introduction to Industry 4.0:** Evolution from Industry 1.0 to 4.0; key drivers and scope
- **Industrial Applications:** Smart manufacturing, healthcare, transportation, energy and smart cities
- **Human-Machine Collaboration:** Ethical issues, workforce reskilling and future job roles
- **Implementation Strategy:** Building a future-ready digital enterprise

[4 Lectures]

Module II: Key Tenets of Industry 4.0

Part 1–Digital Transformation and System Integration

- **Digital Transformation Strategy:** From digitisation to digitalisation to transformation
- **Cyber-Physical Systems (CPS):** 5C architecture, OPC-UA, MQTT and RESTful APIs, real-time monitoring and control, predictive maintenance
- **System Integration – Horizontal Integration:** Supply chain and inter-factory collaboration, Vertical Integration: ERP, MES, SCADA, PLC Layer Interoperability
- **Industry Standards:** RAMI 4.0, IIRA, ISA-95 Automation Pyramid and IT/OT Convergence

Part 2–Lean, Agile and Smart Manufacturing Techniques

- **Lean Manufacturing:** Principles, waste elimination, tools (5S, VSM, Kaizen, Kanban, 5S, SMED, TPM, Jidoka)
- **Agile Manufacturing:** Customisation, responsiveness, flexibility
- **Lean 4.0 and Agile 4.0**: Integration of IoT, AI and analytics for intelligent production

- Smart Six Sigma, TOC 4.0 and Servitisation
- **Case Studies:** Implementation of smart production systems (Toyota, Rolls-Royce, Siemens Digital Enterprise)

[6 Lectures]

Module III: Core Technologies in Industry 4.0

- **Industrial Internet of Things (IIoT):** Architecture, IIoT Protocol Stack: MQTT, CoAP, AMQP, DDS, Role of Cloud vs Edge, Developing IIoT Solutions - Organisational setup, Technological setup, Use cases
- **Industrial Automation and Robotics:** Evolution, Building Blocks of Industrial Automation and Robotics – Sensors, Actuators, Controllers, Vision systems, LiDAR, Robotics safety considerations, etc., Types of Automation: Fixed, Programmable, Flexible, Integrated, Type of Robotic Systems: Cartesian Robots, Articulated Robots, SCARA, AGVs, AMRs, Collaborative Robots (COBOTs), Step by Step Automation Implementation Approach, Case Study
- **Additive Manufacturing (3D Printing):** Working principle, Technologies - FDM, SLA, SLS, DMLS, EBM, Binder Jetting, Applications in design, production and prototyping

[10 Lectures]

Module IV: Enabling Technologies in Industry 4.0

- **AR/VR/MR in Industry:** Differences, devices, applications in design, maintenance, training and metaverse vision
- **Digital Twin and Simulation:** Definition, Types – DTP, DTI, DTA, CDT, Role of Digital Twin, Use cases, Implementation considerations, Case study.
- **Blockchain:** Working Principle, Key Component – Node, Block, Hashing, Consensus Mechanism, Cryptography, Smart Contract, Merkel Tree, Types – Ethereum, Hyperledger, Corda, Cordano, Solana, BaaS, Use cases, Implementation strategy

[8 Lectures]

Module V: Data, Cloud and Analytics in Industry 4.0

- **Cloud Computing:** Architecture – Virtualisation, Hypervisor and VM, Containers, Docker, Kubernetes, Service models – IaaS, PaaS, SaaS models, Deployment models – public/private/hybrid clouds, Cloud-native applications, Edge/Fog computing, Data of Things, Cloud implementation in Manufacturing, Cloud security
- **Big Data Ecosystem:** Types of Data, 5Vs, Storage solutions - RDBMS, NoSQL databases, Data Warehouse, Data Mart, Data Lakes, Big Data Management - OLTP, OLAP, ETL, ELT, Hadoop/Spark, Stream vs Batch processing, Big Data project life cycle
- **Industrial Analytics:** Types – Descriptive: Pareto Analysis, EDA, Predictive: Regression Models, Classification, Forecasting, Prescriptive: Optimisation, Simulation,

Reinforcement Learning, KPI, Tools, use cases (predictive maintenance, energy efficiency, supply chain optimisation), Process Models – KDD, SEMMA, CRISP-DM, Implementation strategy

- **Security and Governance:** Principles, Security Layers, Data privacy, Cybersecurity for connected systems, Defense in Depth models, Cybersecurity Program Development Process

[12 Lectures]

Suggested Practical Components / Tutorials

- Case studies on smart factory implementations (e.g., Amazon, Siemens and GE)
- IIoT architecture simulation and data flow demonstration
- Cloud-based analytics dashboard design using sample datasets
- AR/VR applications in maintenance and remote support
- Evaluation of 3D printed objects for real-life use cases

Lesson Plan: Industry 4.0 – Technologies, Frameworks and Applications

Module#	Chapter #	Chapter Name	Detailed Content	Lectures Allotted
1	1	Introduction of Industry 4.0	Introduction to Industry 4.0: Evolution, key drivers and scope	Lecture 1
			Industrial Applications: Smart manufacturing, healthcare, transportation, energy and smart cities	Lecture 2
			Human-Machine Collaboration: Ethics, reskilling, future job roles	Lecture 3
			Implementation Strategy: Digital enterprise transformation	Lecture 4
2	2	Key Tenets of Industry 4.0 - Digital Transformation, Cyber-Physical Systems, Horizontal and Vertical integration	Digital Transformation: Digitisation to Transformation	Lecture 5
			Cyber-Physical Systems: 5C, OPC-UA, MQTT, REST APIs, monitoring	Lecture 6
			Horizontal Integration: Supply chains; Vertical Integration: ERP, MES, SCADA	Lecture 7
			Industry Standards: RAMI 4.0, IIRA, ISA-95	Lecture 8
	3	Lean and Agile production methodologies	Lean Manufacturing: 5S, VSM, Kaizen, Kanban, SMED, TPM, Jidoka	Lecture 9
			Lean 4.0, Agile 4.0: Integration with IoT, AI, analytics. Smart Six Sigma, TOC 4.0, Servitisation, Case studies	Lecture 10

Module#	Chapter #	Chapter Name	Detailed Content	Lectures Allotted
3	4	Industrial Internet of Things (IIoT)	Architecture	Lecture 11
			IIoT Protocol Stack: MQTT, CoAP, AMQP, DDS, Role of Cloud vs Edge	Lecture 12
			Developing IIoT Solutions: Organisational setup, Technological setup, Use cases	Lecture 13
	5	Industrial Automation and Robotics	Evolution, Building Blocks of Industrial Automation and Robotics: Sensors, Actuators, Controllers, Vision systems, LiDAR, Robotics safety considerations, etc.	Lecture 14
			Types of Automation: Fixed, Programmable, Flexible, Integrated	Lecture 15
			Types of Robots: Cartesian Robots, Articulated Robots, SCARA, AGVs, AMRs, Collaborative Robots (COBOTs)	Lecture 16
			Step by Step Automation Implementation Approach, Case Study	Lecture 17
	6	Additive Manufacturing/ 3D printing	Working principle, Technologies: FDM, SLA, SLS, DMLS, EBM, Binder Jetting	Lecture 18–19
			Applications of Additive Manufacturing in Industry: design, production and prototyping	Lecture 20
4	7	Augmented Reality, Virtual Reality and Mixed Reality	Types, Key Technologies and devices	Lecture 21
			Industrial applications, Concept of Metaverse	Lecture 22
	11	Digital twins	Definition, Types: DTP, DTI, DTA, CDT	Lecture 23
			Role of Digital Twin, Use cases, Implementation considerations, Case study	Lecture 24
	12	Blockchain	Key Components: Node, Block, Hashing, Consensus Mechanism, Cryptography, Smart Contract, Merkel Tree	Lecture 25

Module#	Chapter #	Chapter Name	Detailed Content	Lectures Allotted
			Types: Ethereum, Hyperledger, Corda, Cordano, Solana	Lecture 26
			BaaS, Use cases	Lecture 27
			Implementation strategy	Lecture 28
5	8	Cloud Computing	Architecture: Virtualisation, Hypervisor and VM, Containers, Docker, Kubernetes	Lecture 29
			Service models: IaaS, PaaS, SaaS models, Deployment models - public/private/hybrid clouds	Lecture 30
			Cloud-native applications, Edge/Fog computing, Data of Things, Cloud implementation in Manufacturing, Cloud security	Lecture 31
	9	Big Data	Types of Data, 5Vs, Storage solutions: RDBMS, NoSQL databases	Lecture 32
			Data Warehouse, Data Mart, Data Lakes	Lecture 33
			Big Data Management: OLTP, OLAP, ETL, ELT, Hadoop/Spark, Stream vs Batch processing, Big Data project life cycle	Lecture 34
	10	Analytics	Types - Descriptive: Pareto Analysis, EDA, Predictive: Regression Models, Classification, Forecasting, Prescriptive: Optimisation, Simulation, Reinforcement Learning,	Lecture 35
			KPI, Tools, use cases	Lecture 36
			Process Models: KDD, SEMMA, CRISP-DM, Implementation Strategy	Lecture 37
	13	Cyber Security	Principles, Security Layers, Data privacy, Cybersecurity for connected systems, Defence in Depth models, Cybersecurity Program Development Process	Lecture 38
	14	Case Study	SmartAir Aerospace Case Study: A Digital Transformation Journey in Aircraft Manufacturing	Lecture 39–40

Introduction to Industry 4.0

CHAPTER 1

OBJECTIVES

This chapter aims to provide an overview of Industry 4.0 – its core principles, key technologies, benefits and real-world use cases. Industry 4.0 is reshaping industries by merging the physical and digital worlds through advanced technologies like Industrial Internet of Things (IIoT), Cloud, Big Data and Analytics. Drawing from past industrial revolutions, spanning mechanisation, mass production and automation, this latest shift introduces unprecedented connectivity, flexibility and intelligence in manufacturing processes.

By the end of this chapter, you will be able to:

- Understand the concept of Industry 4.0 and its significance in modern industries.
- Recognise the evolution of industrial revolutions and how they have shaped current manufacturing processes.
- Explain the foundational principles of Industry 4.0 – digital transformation, cyber-physical systems, horizontal/vertical integration and operational methodologies such as lean and agile production methodologies.
- Identify key technologies driving Industry 4.0, such as IIoT, automation, additive manufacturing, big data, analytics and cloud computing.
- Explore real-world use cases of Industry 4.0 in sectors like smart manufacturing, healthcare and sustainable development.
- Understand the need for skill development and the role of higher education in preparing students for the demands of Industry 4.0.

1.1 A FACTORY THAT NEVER SLEEPS

At 2:00 AM, in a high-tech automotive plant in Germany, the factory floor hums with activity; yet, remarkably, there are barely any human workers in sight. Instead, a network of intelligent robots, IoT-enabled machines and AI-driven analytics orchestrates the entire operation with unprecedented precision and efficiency.

Suddenly, one of the assembly-line robots slows down. A machine learning (ML) algorithm monitoring real-time sensor data detects anomalous vibration patterns – a telltale sign that a component is nearing failure. Before the issue escalates, the system automatically schedules maintenance, alerts an on-call technician and orders a replacement part from a nearby warehouse, without human intervention.

But this warehouse is no ordinary storage facility. Instead of keeping excess inventory, the factory employs on-demand 3D printing for critical components. Within minutes, an industrial-grade additive manufacturing system fabricates the replacement part using lightweight yet durable materials, ensuring production continues seamlessly.

Meanwhile, across the globe, a senior engineer in Japan, sipping his morning coffee, logs into the factory's cloud-based dashboard. Using augmented reality (AR) and digital twin technology, he inspects the production line remotely, overlaying real-time performance data onto a virtual replica of the factory. With a simple command, he fine-tunes the robotic workflow, optimising assembly speeds without disrupting operations.

At the same time, in the U.S., a customer places a custom order for an electric vehicle via an online platform. They select a unique colon, interior design and battery configuration. Within seconds, the order is transmitted through the cloud to the factory's AI-powered production scheduling system. The system dynamically adjusts manufacturing parameters to accommodate the customisation, ensuring that no two cars rolling off the line are exactly alike.

However, this level of connectivity and automation also comes with risks. In a control room, dashboards monitor network traffic for anomalies. A sudden spike in activity from an unauthorised IP address triggers an alert. Instantly, an AI-driven cybersecurity system isolates the suspicious connection, preventing potential data theft or system sabotage. With cyber-physical security measures in place, the factory remains resilient against cyber threats while maintaining seamless operations.

This is not a glimpse into the distant future – this is Industry 4.0 in action. Companies like Siemens, GE, Toyota, Bosch, Boeing and Volkswagen have already embraced this transformation, leading to increased productivity, real-time adaptability, reduced waste and hyper-personalised manufacturing.

1.2 WHAT IS INDUSTRY 4.0

The world is witnessing a technological revolution that is transforming industries, economies and societies. This shift, dubbed Industry 4.0, represents the fourth industrial revolution – an era characterised by the convergence of physical, digital and even biological systems. Unlike previous revolutions that were driven by mechanisation, electricity and computing, the Industry 4.0 leverages advanced automation, data-driven intelligence and interconnected technologies to create highly efficient and responsive systems.

Industry 4.0 refers to the integration of smart technologies, including Artificial Intelligence (AI), the Internet of Things (IoT), Big Data and cloud computing, into industrial processes. These technologies enable machines, sensors and digital systems to communicate, analyse data and make autonomous decisions in real-time. The primary goal of Industry 4.0 is to enhance productivity, flexibility and efficiency while reducing costs and resource consumption.

The term was first introduced by the German government in 2011 as part of a strategic initiative to promote smart manufacturing. Today, Industry 4.0 extends beyond manufacturing, influencing various sectors such as healthcare, logistics, energy and smart cities.

DID YOU KNOW?

Example of Integration of Physical, Digital and Biological Systems: AI-Powered Bionic Limbs

Consider a bionic prosthetic limb designed for amputees. Traditionally, prosthetic limbs were purely mechanical, but today, Industry 4.0 technologies have transformed them into intelligent, cyber-physical-biological systems.

1. **Digital Integration:** The prosthetic limb is embedded with IoT sensors, AI-driven control systems and real-time cloud connectivity. These components collect and analyse movement patterns, adapting the limb's function to the user's needs.
2. **Physical Engineering:** The limb is created using 3D printing and advanced materials, ensuring a lightweight and customisable fit. Robotics and actuators enable fluid, natural movement.
3. **Biological Convergence:** Using brain-machine interfaces (BMIs) and neural sensors, the prosthetic can respond directly to nerve signals, allowing users to control it with their thoughts. Some prosthetics even provide haptic feedback, enabling users to 'feel' touch or pressure, bridging the gap between biology and technology.

1.3 EVOLUTION OF INDUSTRY 4.0

The journey of industrialisation has been marked by four distinct technological revolutions, each transforming the way goods are produced, economies operate and societies function. Industry 4.0 is the latest chapter in this ongoing evolution, but to fully appreciate its significance, it is essential to understand how we got here. Figure 1.1 gives a quick overview of the journey and we will discuss each phase in detail in the following subsections.

1.3.1 Industry 1.0: Mechanisation and Steam Power (Late 18th to Mid-19th Century)

The First Industrial Revolution, which began in the late 18th century, was driven by mechanisation, water power and steam engines. Prior to this period, most production was manual, relying on human labour and simple tools.

Key Innovations and Impact

- **Steam engines** powered factories, enabling large-scale mechanised production.
- **Textile manufacturing** saw breakthroughs like the spinning jenny and power loom, increasing productivity.

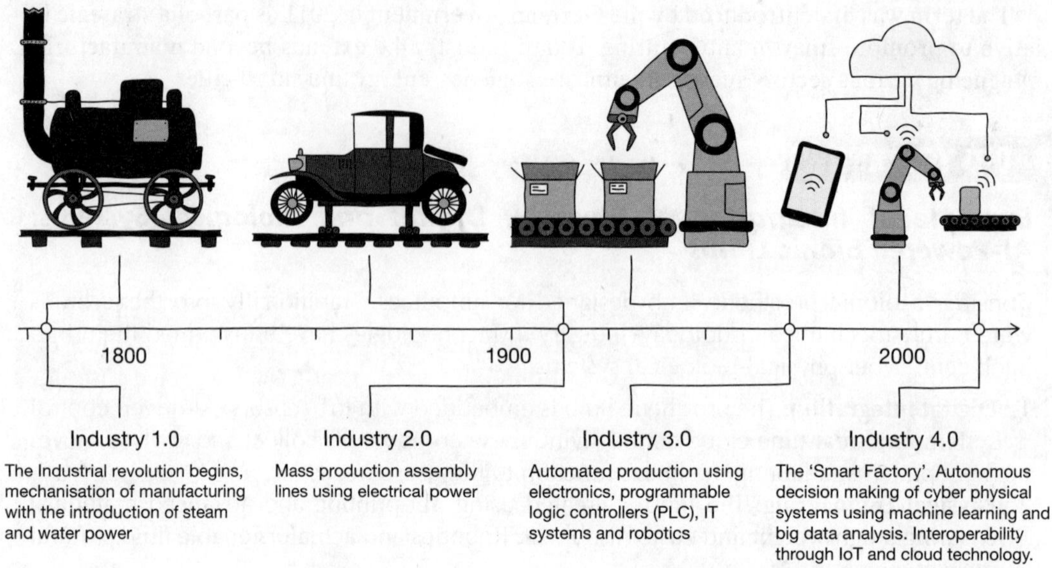

Credit: ChickenDoodleDesigns / Shutterstock

Figure 1.1 History of industrial evolution

- **Iron and coal industries** expanded to support the growing demand for machinery and infrastructure.
- **Transportation revolution:** Steam-powered locomotives and ships improved logistics, facilitating global trade.

The mechanisation of Industry 1.0 marked a shift from agrarian economies to industrial societies, where factories became the backbone of production. However, labour conditions were harsh and manufacturing efficiency was still relatively low compared to later revolutions.

Let's take the automobile industry as an example. While automobiles did not yet exist, this era paved the way for future mobility by advancing metalworking, mechanical engineering and steam-powered locomotion. The steam engine powered trains and early automobiles, such as Nicolas-Joseph Cugnot's steam-powered vehicle (1769), a precursor to modern cars. The assembly of carriages and mechanical components improved, setting the stage for mass vehicle production. Although steam-powered cars never gained widespread use due to inefficiencies, the foundational work on mechanised production and steam energy laid the foundation for the automotive revolution to come.

1.3.2 Industry 2.0: Mass Production and Electricity (Late 19th to Mid-20th Century)

The Second Industrial Revolution, spanning from the late 19th century to the early 20th century, introduced electricity, assembly lines and mass production, making goods cheaper and more accessible.

Key Innovations and Impact

- **Electricity replaced steam power**, allowing greater flexibility and efficiency in factory operations.
- **The assembly line**, pioneered by Henry Ford in 1913, revolutionised manufacturing by introducing standardised, high-speed production.
- **Steel production and chemical industries** advanced, leading to innovations in infrastructure and consumer goods.
- **Telecommunications** emerged, with the invention of the telephone and radio improving long-distance communication.

This era gave rise to urbanisation, global trade expansion and the foundation of modern consumer industries. Mass production made automobiles, appliances and electronics affordable, reshaping everyday life.

Let's continue with our automotive industry example. The invention of the internal combustion engine (late 1800s) made gasoline-powered cars feasible. Karl Benz's Patent Motorwagen (1886) is widely regarded as the first true automobile. Henry Ford's assembly line (1913) revolutionised car manufacturing, reducing production time from 12 and a half hours to just over 90 minutes. Electric starters, headlights and improved steel production enhanced vehicle safety and performance. The rise of consumer cars made automobiles accessible to the middle class, with the Ford Model T (1908) becoming the first mass-produced car. The shift from handcrafted, expensive vehicles to mass production lowered costs, making cars more affordable, reliable and widely available and transforming transportation forever.

1.3.3 Industry 3.0: Automation and Computers (Mid-20th Century to Early 21st Century)

The Third Industrial Revolution began in the mid-20th century with the advent of computers, electronics and early automation technologies. This period saw a shift from purely mechanical and electrical processes to computer-controlled production. This phase is also known as the Digital Revolution.

Key Innovations and Impact

- **Programmable Logic Controllers (PLCs)** and industrial robots automated repetitive tasks, improving precision and efficiency.
- **The rise of personal computers (PCs)** and microprocessors enabled businesses to analyse data, optimise processes and enhance decision-making.
- **The internet and digital communication** transformed global connectivity, leading to e-commerce and remote work.
- **Enterprise Resource Planning (ERP) systems** integrated business operations, improving supply chain efficiency.

This revolution led to lean manufacturing, flexible production lines and early robotics in industries like electronics, automotive and aircraft manufacturing. However, computers were still largely centralised and real-time data processing was limited compared to today's standards.

Robotic automation in manufacturing (e.g., General Motors introduced robotic arms in 1961) allowed for faster, more consistent production with fewer defects. Computer-aided design (CAD) and computer-aided manufacturing (CAM) improved car design and precision. Electronic fuel injection (EFI), anti-lock braking systems (ABS) and early onboard computers made vehicles more efficient and safer. Supply chain optimisation with ERP systems allowed manufacturers to streamline inventory and logistics. Lean manufacturing principles (Toyota Production System, 1970s) reduced waste and improved efficiency. Automakers drastically improved efficiency, vehicle quality and production scalability by leveraging automation, data-driven production and computer-controlled processes.

1.3.4 Industry 4.0: IoT, AI and Data-Driven Decision-Making (21st Century–Present)

We are now in the era of Industry 4.0, characterised by the seamless integration of physical, digital and biological systems. Unlike previous revolutions that focused on mechanisation, mass production, or automation, Industry 4.0 enables smart, interconnected and autonomous industrial systems powered by real-time data.

Key Innovations and Impact

- **Industrial Internet of Things (IIoT):** Smart devices collect and share real-time data across production lines, reducing downtime and improving efficiency.
- **Big Data and Analytics:** Artificial intelligence algorithms analyse vast amounts of data to predict failures, optimise processes and enable autonomous decision-making.
- **Cloud Computing and Edge Computing:** Data storage and processing happen on cloud platforms, allowing remote monitoring, predictive maintenance and faster response times.
- **Digital Twins:** Virtual models of physical assets help in simulation, testing and operational optimisation without disrupting real-world processes.
- **3D Printing (Additive Manufacturing):** Custom parts and products are created on demand, reducing waste and inventory costs.
- **Blockchain:** Secure and transparent transactions improve traceability and trust in manufacturing and logistics.
- **Cybersecurity:** As industries become more connected, cyber threats increase, necessitating robust security frameworks to protect critical infrastructure.

Industry 4.0 transforms entire business models, enabling mass customisation, hyper-efficiency and predictive analytics. Factories no longer just produce – they think, adapt and self-correct in real-time.

Modern car factories use IIoT-enabled machines that communicate in real-time, predicting failures before they happen (predictive maintenance). AI-powered robots assemble vehicles with greater precision, while AI-driven supply chain analytics optimise logistics and demand forecasting. Automakers use cloud-based platforms for vehicle diagnostics, over-the-air (OTA) software updates and remote monitoring. The rise of Tesla, Rivian and legacy automakers' EV lines demonstrates the shift towards battery-powered, eco-friendly transportation. Companies like BMW and Bugatti use 3D printing to produce lightweight, high-performance car parts, reducing waste and production costs. Companies like Waymo and Tesla use AI-driven autonomous driving systems, revolutionising urban mobility. As cars become more software-driven, cybersecurity has become essential to prevent the hacking of vehicle control systems. Automakers create digital replicas of vehicles to simulate real-world conditions, improving crash testing and aerodynamics before manufacturing. In Industry 4.0, cars are no longer just vehicles – they are intelligent, connected and software-driven platforms that continuously evolve with AI-powered updates and new capabilities.

Each industrial revolution has built upon the previous one, transforming production methods, economic structures and human lifestyles. While Industry 1.0 mechanised work, Industry 2.0 scaled production and Industry 3.0 introduced automation, Industry 4.0 creates an intelligent, interconnected and self-optimising industrial ecosystem. Table 1.1 below summarises the evolution stages.

Table 1.1 Evolution of Industrial Revolutions and Their Impact

Industrial Revolution	Key Technologies	General Impact
Industry 1.0 Steam Power and Mechanisation *(Late 18th – Mid-19th Century)*	Steam engines, mechanised textile production, iron working, early factories	• Transition from manual labour to mechanised production. • Increased productivity in agriculture, textiles and mining. • Urbanisation and rise of factory systems.
Industry 2.0 Mass Production and Electricity *(Late 19th – Mid-20th Century)*	Electricity, assembly lines, telecommunication (telegraph, telephone), internal combustion engine, mass steel production	• Introduction of assembly line production across industries. • Faster, cheaper manufacturing; economies of scale. • Electrification transformed industries, homes and transportation.
Industry 3.0 Automation and Computers *(Mid-20th – Early 21st Century)*	Computers, robotics, electronics, early AI, automation, telecommunications (internet, satellites) and semiconductor revolution	• Introduction of computerised control systems. • Widespread automation in manufacturing and logistics. • Early AI and data processing led to digital transformation.
Industry 4.0 Cyber-Physical Systems and AI *(21st Century – Present)*	IoT, AI, big data, cloud computing, 3D printing, blockchain, cybersecurity	• Smart factories with IoT-connected devices enabling real-time monitoring and predictive maintenance. • Data-driven decision-making through AI-powered analytics. • 3D printing enables rapid prototyping and customised manufacturing. • Cybersecurity becomes crucial as industries become digitally interconnected.

1.4 BUILDING BLOCKS OF INDUSTRY 4.0

Industry 4.0 is built upon a set of interconnected technologies and principles. These building blocks work together to create intelligent, data-driven and highly automated industrial environments. This section provides an overview of each component within Industry 4.0 (Refer to Figure 1.2).

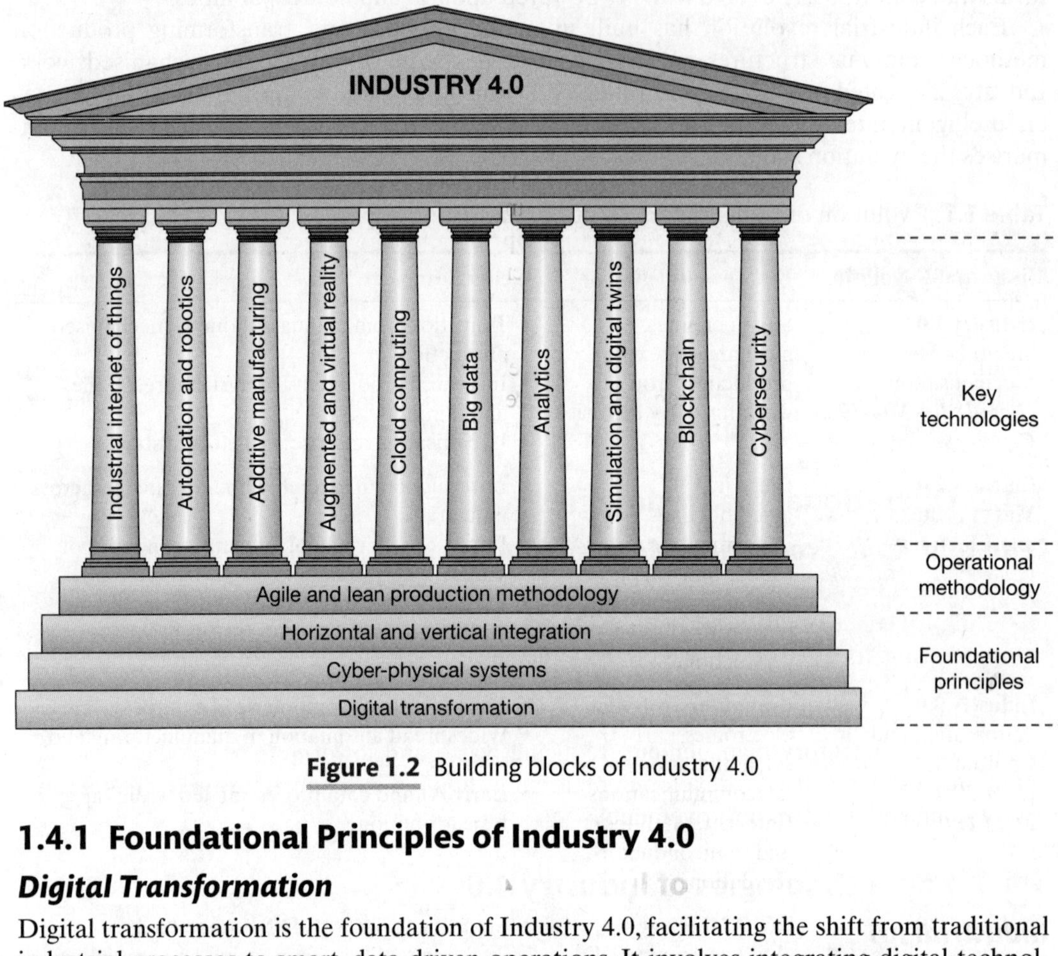

Figure 1.2 Building blocks of Industry 4.0

1.4.1 Foundational Principles of Industry 4.0

Digital Transformation

Digital transformation is the foundation of Industry 4.0, facilitating the shift from traditional industrial processes to smart, data-driven operations. It involves integrating digital technologies across all aspects of an organisation, enhancing efficiency, agility and decision-making capabilities. Businesses that embrace digital transformation benefit from automation, improved data analytics and enhanced customer experiences.

One key aspect of digital transformation is the adoption of cloud computing and big data analytics, which allow businesses to manage and process large amounts of data in real-time. Additionally, AI enables advanced decision-making and predictive capabilities. Companies also leverage digital twins to create virtual models of physical assets, optimising production and maintenance processes.

Cyber-Physical Systems

Cyber-physical systems (CPS) integrate physical machinery with digital computing and communication technologies, creating intelligent and autonomous systems. These systems collect, analyse and respond to data in real-time, improving efficiency and enabling predictive maintenance.

For example, in a smart factory, CPS monitors machine performance and predicts failures before they happen. Sensors installed in machines transmit data to a central analytics platform, where AI-driven insights optimise performance and reduce downtime. CPS also facilitates real-time monitoring and remote control, improving productivity and responsiveness.

Horizontal and Vertical Integration

Industry 4.0 emphasises interconnected supply chains (horizontal integration) and seamless coordination within enterprises (vertical integration). Horizontal integration connects different organisations within a value chain, improving collaboration and efficiency. Vertical integration ensures seamless communication across various levels of an enterprise, from the factory floor to executive management.

In automotive manufacturing, for example, horizontal integration connects suppliers, manufacturers and distributors through a shared digital network, optimising inventory and production schedules. Vertical integration ensures that data flows from production machinery to ERP systems, enabling informed decision-making at all levels.

1.4.2 Operational Methodologies

Lean and Agile Production Methodologies

Lean and agile production methodologies enhance manufacturing efficiency by minimising waste, improving flexibility and enabling rapid responses to fluctuating market demands. Lean manufacturing focuses on eliminating inefficiencies, while agile methodologies prioritise adaptability and customer responsiveness.

For example, Toyota's lean manufacturing system optimises production through just-in-time (JIT) inventory management, reducing waste and lowering costs. Agile principles, commonly used in software development, are now applied in manufacturing to enable rapid prototyping and iterative improvements.

1.4.3 Key Technologies of Industry 4.0

Industrial IoT

The Industrial Internet of Things (IIoT) is a critical technology enabler of Industry 4.0, connecting sensors, devices and machines to create intelligent networks that enhance industrial processes. By embedding smart sensors and actuators into physical assets, IIoT allows manufacturers to collect real-time data, monitor equipment performance and optimise operations.

For instance, in a modern manufacturing facility, IIoT sensors track machine vibrations, temperature and energy consumption. If an anomaly is detected, predictive maintenance algorithms can anticipate equipment failures pre-emptively, reducing downtime and minimising operational costs. This not only increases the lifespan of the equipment but also improves overall efficiency. In supply chain management, IIoT enables real-time tracking of inventory and logistics, ensuring just-in-time deliveries and reducing waste.

IIoT is also transforming industries beyond manufacturing. In agriculture, smart irrigation systems use IIoT sensors to measure soil moisture and weather conditions, optimising water usage and improving crop yields. Similarly, in energy production, smart grids leverage IIoT to monitor electricity demand and automatically adjust supply, improving efficiency and sustainability.

Automation and Robotics

Automation and robotics are at the heart of Industry 4.0, driving efficiency, precision and consistency in industrial processes. Advanced industrial robots like articulated and collaborative robots (cobots), enhance production capabilities by executing repetitive and complex tasks with high accuracy.

In the automotive industry, robotic arms perform intricate assembly operations, such as welding, painting and part installation, with minimal human oversight. These robots can work continuously, improving productivity while reducing errors and material wastage. In contrast, cobots are designed to work alongside operators, augmenting human abilities. Cobots enhance safety and efficiency in tasks like quality inspection and material handling.

Beyond manufacturing, automation is revolutionising logistics and warehousing, as well. Autonomous Mobile Robots (AMRs) and robotic pick-and-place systems are optimising warehouse operations, reducing order fulfilment times and labour costs. In healthcare, robotic-assisted surgeries improve precision and reduce recovery times for patients. Similarly, in agriculture, autonomous tractors and drone-based crop monitoring systems enhance productivity by enabling precise planting, irrigation, and pest control, thereby reducing manual labour and improving yields.

Additive Manufacturing (3D Printing)

Additive manufacturing, commonly referred to as 3D printing, represents a radical shift in production processes. Unlike traditional manufacturing methods that involve cutting or moulding materials, additive manufacturing builds objects layer by layer using digital designs. This approach reduces material waste, shortens production cycles and enables on-demand customisation.

In the aerospace industry, companies like Boeing and Airbus use 3D printing to produce lightweight yet durable aircraft components, improving fuel efficiency and reducing overall manufacturing costs. The automotive sector also benefits from additive manufacturing by enabling rapid prototyping and low-volume production of complex parts. For example, high-performance car manufacturers like Bugatti use 3D-printed titanium brake callipers, reducing weight while maintaining structural integrity.

Beyond manufacturing, 3D printing is revolutionising healthcare. Personalised prosthetics, dental implants and even bio-printed tissues are being developed to enhance patient care. In construction, 3D-printed houses are being built at a fraction of the cost and time required for traditional construction methods, addressing housing shortages in various regions.

Augmented, Virtual and Mixed Reality

Augmented Reality (AR), Virtual Reality (VR) and Mixed Reality (MR) are immersive technologies that enhance human-machine interactions, training and design processes in Industry 4.0. These technologies bridge the gap between digital and physical worlds, providing enhanced visualisation, real-time guidance and interactive simulations.

In manufacturing, AR-enabled smart glasses overlay digital information onto physical objects, assisting workers in assembling complex machinery with step-by-step visual instructions. This reduces human errors and enhances efficiency. VR, on the other hand, is widely used in training programs, allowing workers to simulate hazardous environments without real-world risks. For example, forklift operators can train in a VR environment before handling actual machinery, improving safety.

MR blends AR and VR, enabling real-time collaboration. In product design, engineers in different locations can interact with 3D holographic models, making real-time modifications before production. The automotive industry leverages MR to visualise car interiors and test ergonomic designs before physical prototypes are created.

These technologies also have applications beyond industry. In retail, AR apps allow customers to visualise furniture in their homes before making a purchase. In healthcare, VR is used for pain management and surgical training, improving patient outcomes.

Cloud Computing

Cloud computing is a fundamental enabler of Industry 4.0, providing scalable and on-demand computing power, storage and networking capabilities. By shifting data and applications to the cloud, industries can reduce infrastructure costs, improve collaboration and access real-time insights from anywhere in the world.

Manufacturers leverage cloud platforms to manage and analyse massive datasets collected from IIoT sensors. For instance, predictive maintenance algorithms running in the cloud analyse machine health data from multiple factories, allowing centralised monitoring and remote diagnostics. Cloud-based ERP and Manufacturing Execution Systems (MES) solutions streamline operations, enhancing efficiency across supply chains.

In the automotive sector, cloud computing supports connected vehicles and autonomous driving. Tesla, for example, continuously updates its vehicles with OTA software upgrades via the cloud, improving performance and adding new features. Similarly, cloud-based simulation tools allow car manufacturers to test crash scenarios and aerodynamics before building physical prototypes.

Cloud computing also enhances cybersecurity by offering robust encryption, access control and threat detection mechanisms. However, as industries rely more on cloud-based solutions, ensuring data privacy and regulatory compliance becomes crucial.

Big Data

Big data is a fundamental pillar of Industry 4.0, enabling organisations to process and analyse vast amounts of structured and unstructured data generated from industrial processes. It involves collecting, storing and interpreting data from multiple sources, including sensors, machines, production lines and supply chain networks. The ability to process large datasets in real-time allows businesses to enhance efficiency, optimise decision-making and predict future trends.

In manufacturing, big data plays a key role in predictive maintenance. By continuously monitoring equipment through IIoT sensors, manufacturers can detect anomalies and predict potential failures before they occur, minimising downtime and improving productivity. Additionally, big data is used to optimise supply chain management by analysing demand patterns, inventory levels and logistics data to improve operational efficiency.

Companies such as General Electric (GE) leverage big data analytics in their industrial operations, using data collected from jet engines and power turbines to improve efficiency and reduce operational costs. Similarly, automotive manufacturers utilise big data to analyse customer preferences and enhance vehicle design and production planning.

Analytics

Analytics transforms raw data into actionable insights, allowing businesses to make informed decisions based on real-time and historical data patterns. In Industry 4.0, analytics is applied across multiple domains, including predictive maintenance, process optimisation, quality control and customer behaviour analysis. Advanced analytics, powered by AI and ML, enables organisations to uncover hidden patterns, detect anomalies and enhance automation.

In the manufacturing sector, prescriptive analytics helps improve production efficiency by recommending optimal machine settings, resource allocations and workflow adjustments. For instance, Siemens employs AI-driven analytics to enhance production quality in its smart factories. By analysing sensor data in real-time, it can identify defects and suggest process adjustments to minimise waste and rework.

Predictive analytics is another critical application that allows businesses to foresee potential failures in machinery. Automotive companies like Tesla use data analytics to continuously improve vehicle performance, detect issues remotely and send over-the-air software updates to address concerns before they escalate.

Simulation and Digital Twins

Simulation and digital twin technology enable organisations to create virtual representations of physical assets, processes and systems. A digital twin is a dynamic, real-time virtual model that mimics the behaviour and characteristics of a physical entity, allowing for predictive analysis, performance optimisation and remote monitoring.

In Industry 4.0, digital twins are widely used in manufacturing, supply chain management and equipment maintenance. For example, in aerospace, companies like Boeing utilise digital twins to simulate aircraft performance under different conditions, reducing the need for physical prototypes and minimising development costs.

Within smart factories, simulation enables manufacturers to test different production scenarios before implementing changes. By modelling a factory layout using digital twins, companies can optimise workflows, identify bottlenecks and enhance efficiency without disrupting actual operations. Automotive companies, such as BMW, use digital twins to simulate production lines, ensuring optimal assembly configurations before launching a new model.

Blockchain

Blockchain technology enhances transparency, security and trust in industrial ecosystems by providing a decentralised, tamper-proof ledger for transactions and data exchanges. Unlike traditional centralised databases, blockchain enables multiple stakeholders to verify and record transactions in a secure and immutable manner.

One of the key applications of blockchain in Industry 4.0 is supply chain management. Companies such as Walmart and IBM have implemented blockchain-based solutions to track the movement of goods across supply chains, ensuring product authenticity and preventing fraud. By leveraging blockchain, manufacturers can trace raw materials from their source to the final product, improving compliance with regulatory standards and reducing counterfeiting risks.

Blockchain also enhances cybersecurity by securing IIoT networks. In a smart factory, blockchain can authenticate device identities, prevent unauthorised access and ensure secure data transfers. Additionally, smart contracts, self-executing contracts based on blockchain, can automate business processes such as payments and order fulfilment without the need for intermediaries.

Cybersecurity

As Industry 4.0 advances, cybersecurity becomes a critical concern due to the increasing connectivity of industrial systems. The integration of IoT, cloud computing and AI exposes organisations to potential cyber threats, including data breaches, ransomware attacks and industrial espionage. Effective cybersecurity strategies are essential to protect sensitive data, prevent operational disruptions and ensure system integrity.

One of the biggest cybersecurity challenges in Industry 4.0 is securing industrial control systems (ICS) and operational technology (OT) networks. Traditional manufacturing systems were not designed with cybersecurity in mind, making them vulnerable to cyberattacks. For example, the Stuxnet attack on Iran's nuclear facilities in 2010 demonstrated the potential risks of cyber threats targeting industrial operations.

To mitigate these risks, companies implement multi-layered security measures, including encryption, firewalls, intrusion detection systems and AI-driven threat monitoring. For instance, Siemens has developed an industrial security framework that safeguards connected production environments against cyber threats. Additionally, zero-trust architecture ensures that every device and user is continuously authenticated before accessing critical systems.

Another key aspect of cybersecurity in Industry 4.0 is data privacy. As companies collect vast amounts of data from IIoT sensors and cloud platforms, compliance with data protection regulations such as Global Data Protection Regulation (GDPR) and California Consumer Privacy Act (CCPA) is crucial. Cybersecurity measures help prevent data leaks and unauthorised access to sensitive industrial information.

Though not included within the building blocks, connectivity is a big enabler and distinguishing factor of Industry 4.0. This was made possible by the fifth generation of wireless cellular technology, popularly known as 5G, offering faster speeds, lower latency and increased capacity compared to its predecessor, 4G. In the next section, we discuss 5G as an enabler for connectivity in Industry 4.0.

1.5 CONNECTIVITY FOR INDUSTRY 4.0 AND THE IMPORTANCE OF 5G NETWORK

Industry 4.0 thrives on real-time responsiveness, seamless machine-to-machine (M2M) communication, and intelligent automation – all of which are heavily dependent on high-speed and low-latency network infrastructure. Applications such as predictive maintenance, real-time quality control using vision systems, autonomous robots, digital twins, and remote operation of machinery demand end-to-end (e2e) communication with near-instantaneous feedback. Furthermore, with thousands of sensors, devices, and intelligent machines deployed across a typical factory, high device density and wide coverage are critical. Traditional wired setups or legacy wireless systems cannot meet these modern demands due to bottlenecks in speed, latency and scalability. This is where advanced broadband connectivity emerges as the critical enabler.

1.5.1 Foundations Laid by 4G and 4G+

Fourth-generation (4G) and LTE-Advanced (4G+) technologies played a foundational role in enabling the early stages of Industry 4.0. These communication platforms marked a significant leap over Third-Generation (3G) by offering faster data transfer rates, reduced latency, and more reliable mobile broadband connectivity. While primarily developed to enhance consumer mobile experiences, 4G and 4G+ soon found critical applications in industrial environments, where they supported many of the initial digital transformation initiatives.

4G made it feasible for industries to move away from rigid, wired architectures by introducing mobile and wireless connectivity that could support remote monitoring, mobile access to factory data and real-time dashboards. It enabled operators and maintenance teams to receive alerts and visualise equipment health on handheld devices from virtually anywhere within a facility. This greatly improved responsiveness and supported early forms of predictive maintenance, where historical and sensor data could be used to forecast equipment failure and reduce unplanned downtime.

Another key contribution of 4G technology was its role in integrating basic Industrial Internet of Things (IIoT) systems. Through wireless communication with cloud-based platforms, sensors and programmable logic controllers (PLCs) could transmit operational data for centralised analysis. This supported informed decision-making, condition-based monitoring and the digital integration of production lines. In many cases, private LTE networks were deployed to ensure better data security and local network control within factory environments.

Moreover, 4G provided a stepping stone toward digital cloud integration. Many factories began experimenting with cloud-hosted manufacturing execution systems (MES) and supervisory control and data acquisition (SCADA) tools, using 4G for mobile access to operational insights. This helped bridge the gap between traditional automation systems and modern enterprise IT solutions, leading to more unified and agile production management.

However, despite these advantages, 4G's limitations eventually became apparent. While it offered download speeds of up to 1 Gbps under ideal conditions and latencies ranging from 30 to 50 milliseconds, these parameters fell short for applications requiring deterministic control, ultra-reliability, or dense device connectivity. Industrial environments, particularly those with autonomous robots, machine vision systems, or mission-critical control loops, began demanding more from their communication networks than 4G could deliver. As device density increased, often exceeding 100,000 sensors and actuators/km^2 in large-scale smart factories, the capacity and scalability limits of 4G were tested.

In essence, 4G and its enhanced variants were instrumental in proving the value of wireless communication in industrial settings. They enabled the digitisation of operational processes, helped build confidence in wireless reliability, and supported the development of data-driven decision-making frameworks. These early experiences laid the technological and cultural foundation for industries to embrace more advanced connectivity paradigms like 5G, which not only overcome 4G's constraints but also enable a new class of real-time, autonomous, and interconnected industrial applications.

1.5.2 The Leap Forward: 5G as the Communication Backbone for Advanced Industry 4.0

While 4G laid the groundwork, 5G represents a fundamental shift in connectivity capabilities, tailored specifically for the increasingly demanding requirements of smart factories. 5G introduces ultra-reliable low-latency communication (URLLC), massive device connectivity (over 1 million devices/km^2), network slicing – enabling dedicated virtual networks tailored to specific use cases, Time-Sensitive Networking (TSN) for deterministic communication, seamless edge computing for real-time, localised data processing. These capabilities make 5G not just an enhancement, but a transformational technology for Industry 4.0.

Table 1.2 shows the key differences between 4G and 5G.

Table 1.2 Key differences between 4G and 5G

Parameter	4G	5G
Data Speed	Up to 1 Gbps	Up to 10 Gbps (and beyond in some use cases)
Latency	Around 30–50 milliseconds	As low as 1 millisecond
Device Density	Supports approximately 100,000 devices per square kilometre	Supports over 1 million devices per square kilometre (ideal for dense IIoT environments)
Network Slicing	Not supported	Fully supported – enables virtual networks optimised for specific applications.
Quality of Service (QoS)	Basic QoS based on traffic prioritisation	Enhanced QoS – fine-grained service-level control and reliability across slices
Edge Computing Integration	Limited edge capabilities (relies more on centralised cloud)	Seamless support for edge computing – real-time processing at the network edge
Security	Standard encryption, centralised control	Improved security – includes Zero Trust Architecture, distributed control and slicing-level isolation
Mobility Support	Optimised for mobile broadband	Optimised for both broadband and ultra-reliable low latency communications (URLLC)
Energy Efficiency	Moderate	High – lower energy consumption per bit transmitted
Use Case Flexibility	Primarily designed for consumer and general mobile broadband. Can be used for cloud integration, predictive maintenance and basic IIoT systems.	Designed for diverse use cases, including massive and critical IIoT, AR/VR and autonomous robotics.

5G Use Cases in Industry 4.0

The transformative capabilities of 5G are unlocking a broad spectrum of use cases across smart manufacturing environments. By delivering ultra-low latency, high device density, end-to-end network slicing, and edge computing, 5G empowers industries to transition toward agile, intelligent, and interconnected systems. These applications can be broadly categorised into four major domains:

- **Massive IoT (mIoT):** As the name suggests, the number of connected devices is the key here. In scenarios such as logistics tracking, warehouse automation, inventory management, and environmental monitoring, 5G supports the simultaneous connection of thousands of IoT sensors. This high-density connectivity allows real-time visibility into asset conditions, stock levels, and operational flows, enabling data-driven decisions at every layer of the supply chain.
- **Broadband IoT:** High bandwidth requirement comes from the volume of data to be handled. For data-intensive applications like Augmented Reality (AR) and Virtual Reality (VR), particularly in use cases such as remote diagnostics, hands-free maintenance, and immersive employee training, 5G ensures high throughput with minimal latency. Workers wearing AR headsets can receive live, interactive overlays of equipment data, machine health statistics, or wiring diagrams, even in bandwidth-constrained environments.
- **Critical IoT:** Network speed and latency are very important factors for essential applications and mobile automation systems, including Automated Guided Vehicles (AGVs), Autonomous Mobile Robots (AMRs), and Distributed Control Systems (DCS). These assets require real-time, deterministic communication for safe path planning, obstacle avoidance, and immediate responsiveness. 5G's ultra-reliable low-latency communication (URLLC) enables these devices to operate seamlessly, even in high-traffic or dynamic factory environments.
- **Industrial Automation IoT (Integrated 5G + TSN):** When combined with Time-Sensitive Networking (TSN), 5G offers precise time synchronisation across devices, enabling coordinated operations among CNC machines, robotic arms, and conveyors. TSN ensures deterministic data delivery and quality-of-service (QoS) guarantees, making it ideal for closed-loop control systems, real-time quality inspection, and high-speed actuation where even milliseconds matter. Integration with edge computing further enhances response times and minimises cloud dependency for mission-critical operations. In this case, special mention is needed for brownfield integration.
- **Brownfield Integration:** One of the most vital yet challenging areas in Industry 4.0 adoption is the retrofitting of legacy systems – commonly referred to as brownfield integration. In many factories, legacy PLCs, outdated SCADA systems, and isolated control units still dominate the shop floor. 5G offers a wireless, scalable alternative for connecting these isolated systems to modern digital infrastructure, eliminating the need for expensive rewiring or complete equipment replacement. Through edge gateways and 5G-enabled industrial routers, older machines can now transmit operational data to central analytics platforms or receive control instructions in real-time. Additionally, the use of private 5G networks ensures data privacy, reduces dependency on public infrastructure, and enhances cybersecurity for sensitive legacy equipment. This makes

brownfield integration not just viable, but also cost-effective and scalable for small and mid-sized enterprises.

Together, these use cases demonstrate how 5G, augmented by TSN, Edge Computing, and network slicing, forms the foundational communication backbone of Industry 4.0. It not only enables new digital capabilities but also bridges the technological gap between legacy infrastructure and future-ready smart factories. Figure 1.3 depicts the 5G Use Cases for Industry 4.0 as mentioned above.

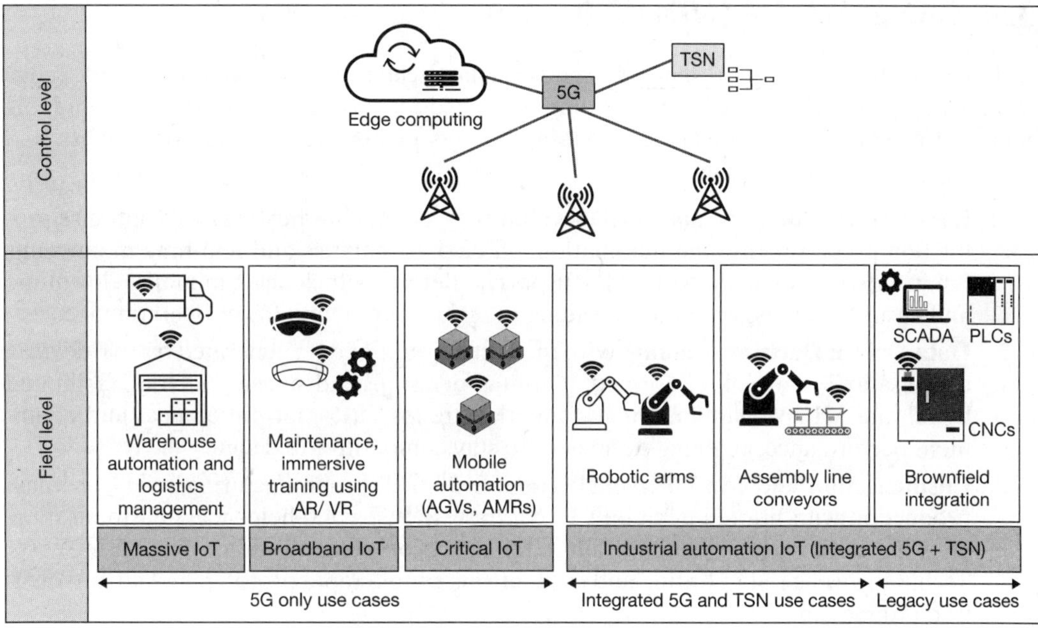

Figure 1.3 Use cases for Industry 4.0

The convergence of these applications enables a smart factory environment where real-time decision-making, automation, and AI-driven control become feasible at scale.

Technical and Process Changes Required for 5G Compatibility

Migrating to 5G is not just a network upgrade. It requires alignment across both the IT and OT domains. Industries need to evaluate and potentially update:

- **Hardware:** Existing PLCs, sensors, and controllers must be 5G-compatible or capable of interfacing with 5G-enabled gateways.
- **Network Architecture:** Legacy SCADA or fieldbus networks may need to integrate with new 5G edge devices and software-defined networks.
- **Cybersecurity:** As data traffic and interconnectivity increase, industries must ensure the implementation of zero-trust architecture, multi-factor authentication, and data encryption at all nodes.
- **Process Standardisation:** To fully leverage features such as network slicing and TSN, industrial processes should be modular, standardised, and digitised to support dynamic reconfiguration and orchestration.

Communication infrastructure is the cornerstone of Industry 4.0. While 4G enabled the initial transition toward connected and mobile systems, 5G dramatically extends these capabilities, supporting the vision of real-time, autonomous, and hyper-connected industrial environments. With features like URLLC, TSN, edge computing, and network slicing, 5G is not merely an upgrade – it is the digital nervous system of the smart factory. As we look ahead, 5G (and soon 6G) will serve as the backbone of future-ready, intelligent, and scalable industrial ecosystems.

1.6 BENEFITS OF INDUSTRY 4.0

As Industry 4.0 continues to reshape the industrial landscape, its adoption brings a wide range of advantages across sectors. By integrating advanced technologies such as IIoT, cloud, big data, and analytics, businesses can unlock new levels of performance, responsiveness, and innovation. The key benefits include:

1. **Increased Productivity and Efficiency:** Industry 4.0 enables businesses to optimise production processes through automation, predictive analytics and real-time monitoring. Smart factories enabled with IoT sensors can detect inefficiencies, minimise downtime and ensure seamless operations, leading to higher output and lower operational costs.
2. **Data-Driven Decision Making:** With the vast amount of data generated by IoT devices, organisations can make informed, real-time decisions. Advanced analytics, facilitated by AI and ML, provide insights into market trends, customer preferences and equipment performance, enabling businesses to stay competitive and agile.
3. **Customisation and Mass Personalisation:** Traditional manufacturing relied on mass production with limited flexibility. In contrast, Industry 4.0 helps mass customisation, allowing companies to produce tailored products on demand without increasing costs. Technologies like 3D printing and digital twins enable rapid prototyping and personalised production.
4. **Enhanced Supply Chain and Logistics:** Industry 4.0 revolutionises supply chains by integrating real-time tracking, predictive analytics and automated logistics. Smart supply chains can dynamically respond to market fluctuations, demand changes and transportation disruptions, ensuring faster and more efficient deliveries.
5. **Improved Workplace Safety and Sustainability:** The use of AI-powered monitoring, robotics and automation reduces human exposure to hazardous environments, enhancing workplace safety. Additionally, Industry 4.0 promotes energy efficiency, waste reduction and sustainable resource utilisation, aligning industries with global sustainability goals.
6. **Transformation of Business Models:** Industry 4.0 is shifting companies from traditional product-based models to service-oriented and data-driven models. Concepts like Product-as-a-Service (PaaS) allow companies to provide subscription-based industrial solutions, leveraging real-time performance data to improve service offerings.

1.7 USE CASES OF INDUSTRY 4.0

Industry 4.0 is revolutionising multiple sectors by integrating digital technologies, automation and data-driven decision-making. While the core principles and technologies provide a foundation, the true impact of Industry 4.0 is best understood through real-world applications. This section explores key use cases where Industry 4.0 has transformed industries, leading to improved efficiency, sustainability and innovation.

1.7.1 Smart Manufacturing and Intelligent Production Systems

Smart manufacturing leverages Industry 4.0 technologies such as IIoT, AI and robotics to create interconnected, data-driven production environments. These systems optimise efficiency, reduce waste and enable real-time decision-making, It also improves worker safety by eliminating the need for human workers in hazardous work zones (Refer to Figure 1.4).

Credit: Pop Nukoonrat / 123RF

Figure 1.4 Engineer checking and controlling welding robotics automatic arms machine in a smart factory

For example, BMW's smart factories utilise IIoT and real-time data analytics to enhance production flexibility. Sensors embedded in assembly lines continuously monitor machinery, predicting maintenance needs before failures occur. AI-powered automation adapts production schedules dynamically, responding to fluctuations in demand and supply chain disruptions.

Additionally, digital twins simulate production processes to test changes before implementation, reducing downtime and costly errors. Collaborative robots (cobots) work alongside human operators, improving productivity while maintaining safety.

Key Benefits

- Enhanced production efficiency and cost savings
- Predictive maintenance reduces unplanned downtime
- Real-time adaptability to market demands

1.7.2 Supply Chain and Logistics Optimisation

Industry 4.0 transforms supply chains by integrating real-time tracking, automation and AI-driven decision-making. With the adoption of IoT sensors, blockchain and advanced analytics, supply chains become more transparent, resilient and efficient.

A notable example is Amazon's fulfilment centres, where AI-driven robotics manage inventory, reducing order processing time. IoT sensors track shipments globally, ensuring real-time visibility (Figure 1.5). Blockchain enhances trust and security by recording transactions and verifying authenticity, particularly in industries like pharmaceuticals and food safety.

Credit: pitinan / 123RF

Figure 1.5 Smart warehouse management system using augmented reality technology to identify package picking and delivery

Autonomous vehicles and drones further streamline logistics, reducing delivery times and transportation costs. Predictive analytics optimise inventory management, minimising stock-outs and overstocking.

Key Benefits

- Improved supply chain visibility and efficiency
- Reduced operational costs and faster deliveries
- Enhanced security and traceability with blockchain

1.7.3 Industry 4.0 in Healthcare and Medical Technologies

The healthcare sector benefits from Industry 4.0 through AI-driven diagnostics, remote patient monitoring and personalised treatments. Technologies like IoT-enabled medical devices, robotic surgeries and 3D printing of prosthetics enhance patient care and medical research.

For example, Siemens Healthineers integrates AI and IoT in MRI scanners to optimise imaging, reducing scan times while improving accuracy. Wearable health devices continuously monitor patients' vitals, enabling early detection of diseases. In surgeries, robotic-assisted systems such as the Da Vinci Surgical System enhance precision, minimising invasiveness and recovery times (Refer to Figure 1.6).

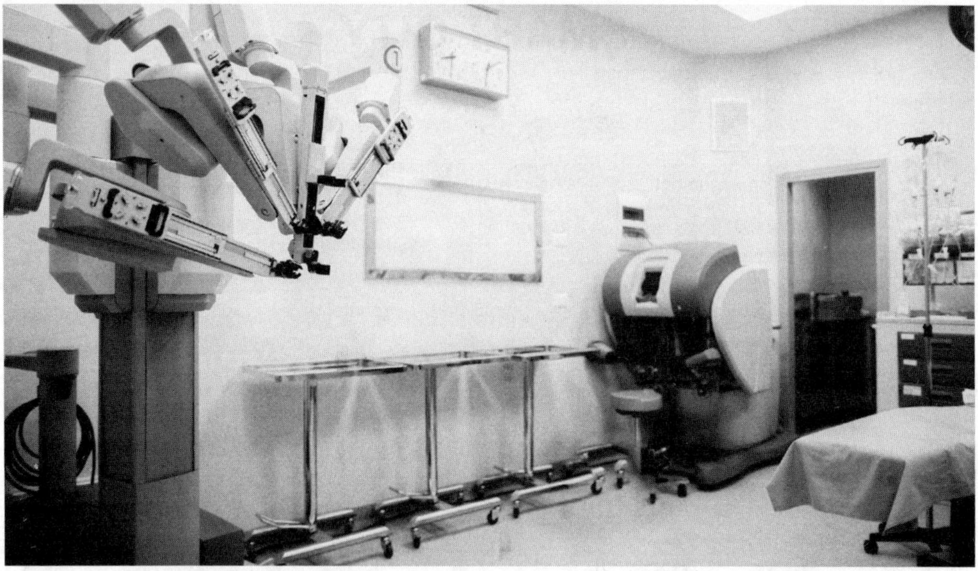

Credit: MAD.vertise / Shutterstock

Figure 1.6 Minimal invasive robot surgical system in hospital

Additionally, 3D printing is revolutionising prosthetics and implants by providing personalised medical solutions tailored to patients' needs.

Key Benefits

- Faster, more accurate diagnoses with AI
- Improved patient monitoring and personalised treatments
- Enhanced surgical precision and reduced recovery times

1.7.4 Smart Cities and Infrastructure Development

Industry 4.0 technologies are instrumental in developing smart cities that enhance urban living through automation, sustainability and connectivity (Refer to Figure 1.7). By integrating IoT, AI and big data, smart cities optimise resource management, traffic flow and public services.

For example, Barcelona has implemented IoT-enabled smart lighting systems that adjust brightness based on real-time traffic and pedestrian activity, reducing energy consumption. AI-driven traffic management systems analyse congestion patterns to optimise signal timings, reducing delays and emissions.

Additionally, predictive maintenance in infrastructure, such as smart grids and water management systems, ensures efficiency and reduces environmental impact.

Key Benefits

- Enhanced sustainability and reduced energy consumption
- Improved urban mobility and reduced congestion
- Better public services through real-time data insights

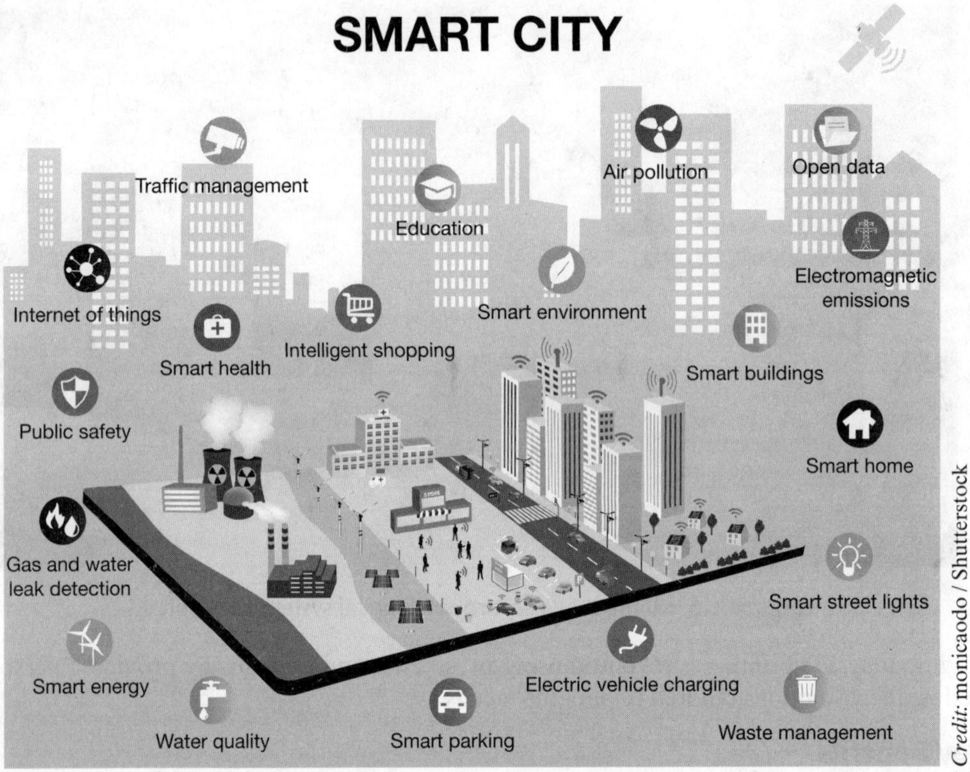

Figure 1.7 Smart city concept design with future technology for living

1.7.5 Industry 4.0 and Sustainable Development

Industry 4.0 plays a crucial role in driving sustainability by reducing resource consumption, minimising waste and enabling circular economy models. Digital technologies optimise energy usage, while AI-driven analytics identify areas for sustainability improvements.

For example, Siemens' smart factories implement AI to optimise energy consumption, significantly reducing carbon footprints. In the fashion industry, Adidas utilises 3D printing and automated production to create sustainable footwear with minimal waste.

Additionally, blockchain technology ensures transparency in supply chains, preventing unethical labour practices and enabling responsible sourcing of materials. Predictive maintenance in industrial settings reduces material waste and prolongs the lifespan of equipment.

Key Benefits

- Reduced environmental impact and improved energy efficiency
- Greater transparency and ethical sourcing in supply chains
- Enabling circular economy models through resource optimisation

1.8 THE IMPACT OF INDUSTRY 4.0 ON WORKFORCE AND EDUCATION

The rapid evolution of Industry 4.0 is reshaping the workforce and redefining the skills required in the modern economy. As automation, artificial intelligence and digitalisation continue to transform industries, there is a growing need for a workforce that can adapt to these changes. This shift has significant implications for education, requiring institutions to evolve their curricula and teaching methods to prepare students for future job roles. This section explores three key aspects of Industry 4.0's impact on the workforce and education: human-machine collaboration in workplaces of the future, the need for reskilling and upskilling and the role of higher education.

1.8.1 Human-Machine Collaboration in Future Workplaces

Industry 4.0 is not just about automation replacing human workers – it is about fostering seamless collaboration between humans and intelligent machines. Future workplaces will be defined by human-machine partnerships, where AI-driven systems enhance human capabilities rather than replace them.

For example, in manufacturing, cobots assist human workers in performing complex assembly tasks, reducing strain and improving efficiency. In healthcare, AI-powered diagnostic tools support doctors in making more accurate medical decisions. Similarly, in finance, ML models help analysts detect fraud and optimise investment strategies.

To prepare for this shift, employees must develop not only technical skills but also soft skills such as problem-solving, adaptability and digital literacy. The ability to work alongside AI-driven tools and interpret complex data insights will be crucial in the future workforce.

Organisations are also focusing on ethical considerations surrounding human-machine collaboration. Ensuring transparency in AI decision-making and maintaining human oversight in automated processes will be essential to building trust in intelligent systems.

1.8.2 Need for Reskilling and Upskilling in the Workforce

The rise of Industry 4.0 has led to the automation of repetitive and labour-intensive tasks, making some traditional job roles obsolete while creating new opportunities in advanced technology fields. As a result, workers must continuously upgrade their skills to remain relevant in the job market.

Reskilling refers to learning new skills to transition into entirely different job roles, while upskilling involves enhancing existing skills to keep up with technological advancements. For example, manufacturing employees who previously operated manual assembly lines may need training in robotics programming or data analytics to work with automated systems. Similarly, logistics professionals must understand AI-powered supply chain optimisation tools to stay competitive.

Governments, businesses and educational institutions are investing in training programs to bridge the skill gap. Initiatives such as online certification programs, corporate training workshops and vocational courses in data science, AI and cybersecurity are becoming increasingly popular. The demand for continuous learning is now a necessity rather than an option, ensuring that employees can adapt to the evolving landscape of Industry 4.0.

1.8.3 Role of Higher Education in Preparing Students for Industry 4.0

Higher education institutions play a critical role in equipping students with the knowledge and skills needed for Industry 4.0 careers. Traditional degree programs are evolving to include courses on automation, IoT, big data analytics and cloud computing, ensuring that graduates are ready to contribute to digitalised industries.

One key trend in higher education is the integration of interdisciplinary learning. As Industry 4.0 requires a blend of technical and business skills, universities are offering programs that combine engineering, computer science and management. For instance, many institutions now offer degrees in Industrial AI, Smart Manufacturing and Digital Supply Chain Management.

Additionally, experiential learning is gaining importance. Universities are increasingly partnering with industries to provide students with hands-on experience through internships, apprenticeships and collaborative research projects. Smart labs and digital twin simulations allow students to experiment with real-world Industry 4.0 applications before entering the workforce. These initiatives ensure that the future workforce is well-equipped to handle the dynamic challenges of Industry 4.0.

The rise of Industry 4.0 presents both challenges and opportunities for education and workforce development. Continuous learning, curriculum transformation and human-machine collaboration will be key to navigating this industrial revolution. By embracing these changes, students, educators and professionals can ensure they remain competitive and prepared for the digital future.

1.9 CONCLUSION

Industry 4.0 represents more than just a technological leap – it is a continuous transformation shaping the future of industries. Its market is experiencing exponential growth, with technologies like AI-driven automation, digital twins and smart manufacturing advancing rapidly. According to the market research report, the global Industry 4.0 market size was valued at USD 52.17 billion in 2023 and is estimated to reach USD 182.01 billion by 2028. Emerging innovations such as 5G, edge computing and advanced AI are improving real-time decision-making and operational efficiency. In parallel, sustainability is taking centre stage, as smart energy systems, additive manufacturing and AI-enabled waste reduction contribute to circular economy goals. The rise of hyperconnected ecosystems is fostering seamless integration across supply chains and platforms, with IoT and blockchain enhancing transparency and security. As automation evolves, the human workforce is shifting toward roles requiring creativity, problem-solving and data expertise; driving demand for skills in AI, robotics, cybersecurity and analytics. Educational institutions must keep pace to prepare future professionals.

The convergence of AI, IoT, big data and robotics will continue to redefine industries, making them smarter, more sustainable and deeply interconnected. Success in this evolving landscape depends on continuous innovation and workforce transformation. Ultimately, the full potential of Industry 4.0 can be realised through close collaboration among industry leaders, governments and academia – ensuring that the digital revolution fosters inclusive, sustainable global growth. To thrive in this evolving landscape, organisations must remain agile and responsive to ongoing change.

SUMMARY

- Industry 4.0 represents a transformative shift by integrating digital and physical systems to enhance efficiency, automation and data-driven decision-making.
- The evolution of industrial revolutions has progressed from mechanisation (Industry 1.0) to mass production (Industry 2.0), automation (Industry 3.0) and now smart, interconnected cyber-physical systems (Industry 4.0).
- Foundational principles such as digital transformation, cyber-physical systems and horizontal and vertical integration are driving industries toward greater connectivity and intelligent operations.
- Key Industry 4.0 technologies, including IIoT, automation, additive manufacturing, AR/VR, cloud computing, big data analytics, blockchain, digital twins and cybersecurity, are revolutionising industrial processes.
- Lean and agile production methodologies are enabling manufacturers to achieve flexibility, reduce waste and respond quickly to changing market demands.
- Real-world use cases showcase the impact of Industry 4.0 in areas like smart manufacturing, supply chain optimisation, healthcare and sustainable development.
- The transition to Industry 4.0 requires a skilled workforce, making reskilling and upskilling essential for future employment.
- Higher education institutions play a critical role in preparing students for Industry 4.0 by equipping them with the necessary technical and analytical skills.
- A collaborative approach among industry, academia and policymakers is vital to maximising the benefits of Industry 4.0 and ensuring sustainable industrial growth.

EXERCISE

Multiple Choice Questions (1 Mark Each)

1. What does Industry 4.0 primarily refer to?

(a) The fourth industrial revolution

(b) A new software update

(c) A type of manufacturing machine

(d) A government policy

2. Which technology is not commonly associated with Industry 4.0?
   (a) Artificial Intelligence
   (b) Blockchain
   (c) Steam engines
   (d) Internet of Things (IoT)
3. Industry 4.0 is characterised by the integration of:
   (a) Manual labour with traditional tools
   (b) Advanced digital technologies in industrial processes
   (c) Agriculture and farming
   (d) The elimination of all industries
4. Which country first introduced the term Industry 4.0?
   (a) United States
   (b) Germany
   (c) China
   (d) Japan
5. One key aspect of Industry 4.0 is:
   (a) Mass production without automation
   (b) Smart factories
   (c) The use of human-only labour
   (d) Complete elimination of robotics
6. The foundation of Industry 4.0 is built on:
   (a) Traditional manufacturing processes
   (b) Digital transformation and automation
   (c) Purely mechanical systems
   (d) Paper-based record keeping
7. The Internet of Things (IoT) in Industry 4.0 enables:
   (a) The connection of devices and machines for data exchange
   (b) The removal of internet from industrial systems
   (c) Manual-only data collection
   (d) The use of traditional telephones for communication
8. One of the major advantages of Industry 4.0 is:
   (a) Decreased efficiency
   (b) Increased production downtime
   (c) Enhanced automation and real-time data monitoring
   (d) Higher reliance on manual work

9. Which of the following technologies plays a key role in predictive maintenance in Industry 4.0?
   (a) Machine learning and data analytics
   (b) Paper logs and manual inspections
   (c) Human intuition
   (d) Traditional calendar-based maintenance

10. Which of the following is a benefit of digital twins in Industry 4.0?
   (a) Physical duplication of machinery
   (b) Virtual replication of systems for monitoring and optimisation
   (c) Elimination of automation
   (d) Slower production processes

11. Which of the following is a key security challenge in Industry 4.0?
   (a) Increased automation efficiency
   (b) Cybersecurity threats due to connected systems
   (c) Use of advanced robotics
   (d) Adoption of 3D printing

12. What does IIoT stand for?
   (a) Industrial Internet of Things
   (b) Internal Information of Technology
   (c) International Institute of Trade
   (d) Independent Industry of Technology

13. Which of the following is NOT an Industry 4.0 technology?
   (a) 3D printing
   (b) Cloud computing
   (c) Horse-drawn cart
   (d) Augmented reality

14. Blockchain in Industry 4.0 is primarily used for:
   (a) Securing data and enhancing traceability
   (b) Replacing all human workers
   (c) Slowing down production processes
   (d) Eliminating the need for cybersecurity

15. A major disadvantage of Industry 4.0 adoption is:
   (a) Higher cybersecurity risks
   (b) Increased operational inefficiency
   (c) Reduced automation
   (d) Poor data collection

16. One challenge of implementing Industry 4.0 in small businesses is:
    (a) High initial investment costs
    (b) Lack of skilled labour
    (c) Data security concerns
    (d) All of the above
17. Which Industry 4.0 technology allows machines to 'learn' from data?
    (a) Analytics
    (b) Steam power
    (c) Assembly lines
    (d) Paper-based record-keeping
18. Additive manufacturing is another term for:
    (a) 3D printing
    (b) Manual crafting
    (c) Subtractive manufacturing
    (d) Traditional machining
19. Which protocol is widely used for IoT communication?
    (a) MQTT
    (b) HTTP only
    (c) Morse code
    (d) Analog signals
20. Industry 4.0 is expected to:
    (a) Transform industries through automation and smart technologies
    (b) Eliminate all traditional manufacturing methods
    (c) Make human labour obsolete
    (d) Reduce efficiency

Short Answer Type Questions (2 Marks Each)

1. What is Industry 4.0 and why is it significant?
2. What are cyber-physical systems (CPS) in Industry 4.0?
3. What role does IoT play in Industry 4.0?
4. What are the key benefits of Industry 4.0?
5. What is the significance of digital twins in Industry 4.0?
6. How does additive manufacturing impact Industry 4.0?
7. Why is cybersecurity crucial in Industry 4.0?
8. How does blockchain contribute to Industry 4.0?

9. How can Industry 4.0 contribute to sustainability?
10. What are some real-world use cases of Industry 4.0?

Long Essay Questions (5 Marks Each)

1. Discuss the evolution of industrial revolutions leading to Industry 4.0.
2. What is the role of IoT in Industry 4.0 and how does it impact industrial operations?
3. How do automation and robotics contribute to Industry 4.0?
4. What role does cloud computing play in Industry 4.0?
5. How does Industry 4.0 impact supply chain and logistics optimisation?
6. Explain the role of 5G technology in Industry 4.0 and how it enhances industrial operations.
7. Discuss the importance of augmented reality (AR), virtual reality (VR) and mixed reality (MR) in Industry 4.0.
8. Discuss the challenges and limitations of implementing 5G in industrial environments.
9. What are the cybersecurity challenges in Industry 4.0 and how can they be addressed?
10. How do lean and agile production methodologies enhance efficiency in Industry 4.0?

CHAPTER 2

Key Tenets of Industry 4.0

Digital Transformation, Cyber-Physical Systems, Vertical and Horizontal Integration

OBJECTIVES

An implementation of Industry 4.0 needs planning and groundwork. It is a transformation journey that starts with the digital enablement of the organisation, as the core concepts of Industry 4.0 are based on digital processing and digital connectivity principles. So, we will discuss the concepts of Digital Transformation in this chapter within the context of Industry 4.0. Integration and interaction between physical and cyber systems are prerequisites for Industry 4.0 implementation. This chapter will also go through the definition and evolution of cyber-physical systems. We will explore why cyber-physical systems and Industry 4.0 are made for each other, the pillars of cyber-physical system, their architecture, features, use cases, security aspects and the challenges and opportunities. As all these systems should interact and collaborate closely, we will also cover the different integration approaches used for Industry 4.0 to create a closely integrated production environment. Integration of factory components at the micro and macro levels drives the idea of a smart factory.

By the end of this chapter, you should be able to:

- understand the concept of digital transformation and its role in preparing an organisation for Industry 4.0 implementation.
- understand how devices' physical and cyber components enable the smart implementation of Industry 4.0.
- how the horizontal and vertical integration approach connects the components within a factory, across factories and within an organisation from the shop floor to the top floor.

2.1 INTRODUCTION

The emergence of Industry 4.0 represents a transformative period in the industrial sector. It involves the seamless integration of advanced technologies into manufacturing and production processes. At the onset, let us understand how an organisation should plan for transformation to adopt Industry 4.0 step by step. This involves a strategic and phased approach, starting with a detailed audit of machines, processes, people and the entire value chain to understand 'why', 'where' and 'what' transformation is needed, together with a clear vision setting and a readiness assessment.

For example, in a traditional steel manufacturing plant, machines might be old and operated manually by operators in challenging conditions. Many tasks are repetitive in nature and labour-intensive. The operators are dependent on manual verification with digital or analog testing equipment. Maintenance of machines is primarily based on limited data based on manual verification and checklists, which can be error-prone. Since the machines are not sending real-time feedback on health status and are not connected, diagnosing failure signals might be wrong, leading to disasters and severe downtime. There are computers for populating data and creating reports for management reporting, but they are not real-time and there are no analytics or insights. There is an inventory management application, but physical inventories are not smart enough to give the real-time feed of inventory status, leading to an under or overstock situation. The suppliers and sellers are not connected, leading to delays in getting market insights.

Here comes the need to adopt digital technologies to fundamentally change a business's operations. This is called digital transformation. The company must now find and rank the pain areas and implement digital transformation in those areas based on priority and Return on Investment (ROI). They must establish a roadmap, initiate pilot projects and invest in employee training and change management.

There can be many decision points and careful considerations for long-term benefits are needed. For example, machines are very old but running. Should we retrofit this with some sensors to get some information (might be limited) or change it to smart and connected machines already equipped with advanced sensors and actuators? Also, can we automate part or full production lines and do the repetitive and hazardous jobs with industrial robots? Once decided, appropriate systems are required to combine physical machinery with digital networks and computational intelligence, enabling real-time monitoring, data collection and autonomous decision-making. This is termed Cyber-Physical Systems (CPS). CPS will help implement the smart and connected machines that can send information and get instructions.

Now that the digital transformation roadmap is ready, machines and shop floors are getting connected. The next step is to uniform management and reporting at the operations level to the executive level and to connect all departments like design, operations, inventory, finance, HR, sales and marketing and, external suppliers and sellers. This can be achieved through horizontal and vertical integration to connect and synchronise different components and levels within an organisation and across the supply chain. Horizontal integration links production stages and external partners for a cohesive value chain, while vertical integration aligns processes from the shop floor to the executive level based on real-time operational data.

Thus, these three tenets – Digital Transformation, Cyber-Physical Systems (CPS) and Horizontal and Vertical Integration, form the backbone of Industry 4.0, driving unparalleled efficiency, innovation and competitive advantage. As we have seen in the above example, these key principles are not isolated advancements but interconnected elements that collectively revolutionise how industries operate, innovate and compete.

This chapter delves into the essence of these concepts, exploring their significance, interconnections and profound impact on modern industrial practices, where industries are connected, smarter, more resilient and can better respond to changing market demands, drive innovation, and create sustainable competitive advantage, by breaking down silos and integrating processes both within and across organisations.

2.2 DIGITISATION, DIGITALISATION AND DIGITAL TRANSFORMATION

The scope of digital transformation is vast and it needs organisation-level change. However, this change is possible only through incremental transformation at the ground level. Digitisation and digitalisation are important pre-steps for preparing an organisation for digital transformation (refer to Figure 2.1).

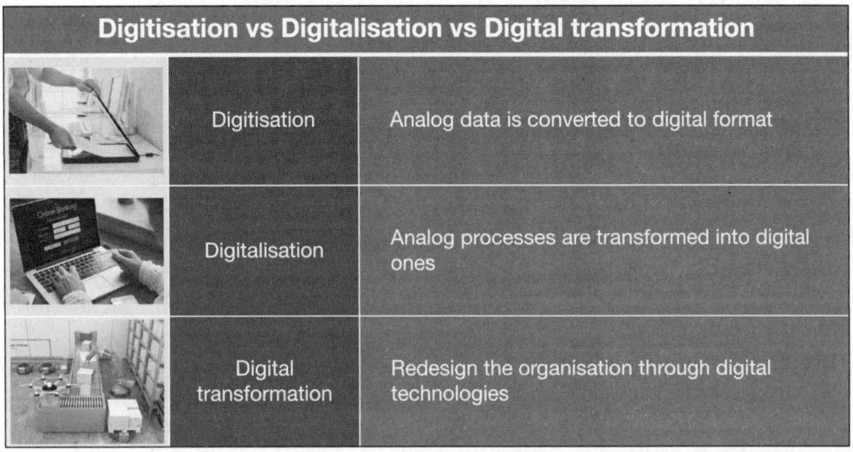

Credit: Song_about_summer / Shutterstock; Rawpixel.com. Shutterstock; Chesky / Shutterstock

Figure 2.1 From digitisation to digital transformation

2.2.1 Digitisation

When we convert a physical object or a non-digital system into a digital representation, it is called digitisation. This can be as simple as scanning a photograph to create a GIF file. Once the physical object is digitised, computer systems can use it for various purposes. We understand that the physical systems or parameters cannot contribute to a digitally connected ecosystem without digitisation. To elaborate on this point, let's take an example of a hygrometer fitted in a data centre. If the hygrometer is an analog one, then someone needs to manually check the readings and make decisions on controlling the humidity as per permissible thresholds. Instead, if a sensor-based digital hygrometer is used, the digital measurement can be fed to a computerised control mechanism for further processing.

So, the first foundation block for digital transformation is digitisation, which initiates consumable data generation by connecting the physical and digital worlds.

2.2.2 Digitalisation

Digitalisation is the next level of maturity after digitisation. When digital technologies and digitised data are leveraged to create more efficient processes, then it is called digitalisation. In our previous example, the hygrometer was transformed into a smart hygrometer that could generate digital signals. A simple PLC can process this signal and generate commands for an actuator to control the humidity. In a large enterprise, a digital process can be more complex, such as an error-monitoring system creating a workflow to generate work orders for plant maintenance. It is important to note that the efficiency of the existing processes only improves through digitalisation, but no process transformation occurs. Mainly, we can interpret the digitalisation process as automating manual processes.

2.2.3 Digital Transformation

Digital Transformation is the transformation in the business processes enabled by digitalisation. The dictionary meaning of transformation is *a marked change in form, nature, or appearance*. So, Digital Transformation is *the integration of computer-based technologies across an organisation's products, processes, and services, aimed at creating marked and beneficial changes for customers, employees and operations*. These changes are not merely incremental changes to improve efficiency but are often large in scope and necessitate assessment and reinvention of multiple facets of an organisation, from customer interactions to supply chains, organisation structures, culture and employee skill sets. Digital Transformation allows an organisation to respond more efficiently to current customer needs and prepares it to effectively meet evolving and future demands.

Let's take the example of Japan-based automobile manufacturer Toyota. The company has invested heavily in digital technologies to transform its manufacturing systems to deal with increasing global competition, price pressure and technological advancements. The transformation is embraced in multiple areas. Key digital technologies like the Internet of Things (IoT), automation and robotics, Artificial Intelligence, and Advanced Analytics are employed to automate and predict maintenance needs in the production lines. Similarly, they leverage CAD, digital twin and 3D printing to expedite the design and rollout of new products. This helped the company to fulfil growing customer expectations and to be more responsive to market changes.

Digital transformation is a long-term strategic objective for any organisation. So, it needs to be planned and executed systematically. Figure 2.2 shows a step-by-step approach to digital transformation for an enterprise.

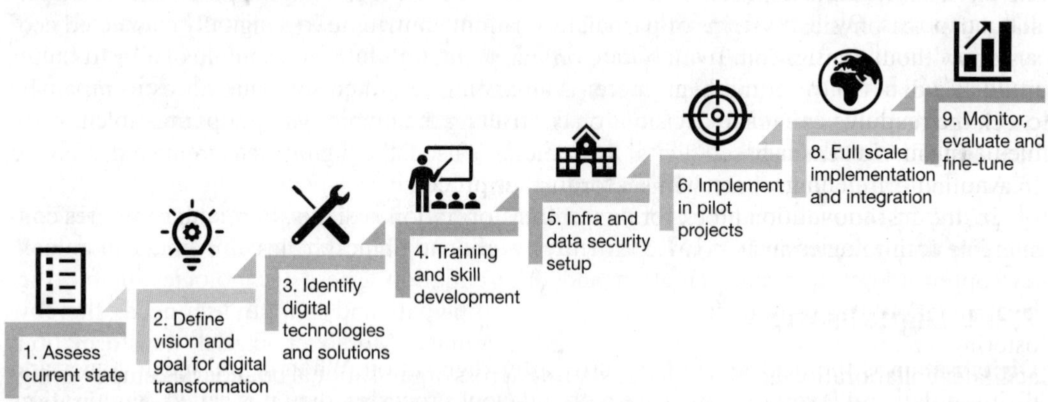

Figure 2.2 Digital transformation steps

DID YOU KNOW?

Digital Transformation has touched the leaders of every business domain. You will get ample examples of large enterprises in each sector, like Manufacturing (GE, Siemens), Healthcare (Mayo Clinic), Telecom (AT&T), Banking and Finance (Goldman Sasch), Retail and Consumer Goods (Nike, Walmart, Starbucks), Travel & Hospitality (Airbnb), Aerospace (Rolls Royce, Boeing), Education (Coursera), etc. who have significantly transformed their business processes with the help of digitalisation.

2.3 WHY IS DIGITAL TRANSFORMATION KEY TO INDUSTRY 4.0?

Digital transformation lies at the heart of Industry 4.0, revolutionising how industries operate and compete in the global landscape. At its core, Industry 4.0 represents the convergence of physical systems with digital technologies, leading to smarter, more connected and more efficient processes across all sectors. In this era of rapid technological advancement, embracing digital transformation isn't merely advantageous; it's imperative for survival and success.

One of the key drivers behind the necessity of digital transformation in Industry 4.0 is the mounting demand for enhanced productivity and efficiency. Figure 2.3 illustrates how digital transformation can impact various aspects of an organisation through the application of diverse technologies. Companies can streamline operations, optimise resource allocation, and minimise downtime by harnessing the power of automation, artificial intelligence and data analytics. Digital technologies enable predictive maintenance, allowing businesses to pre-emptively address equipment failures and prevent costly disruptions to production processes. Moreover, digital transformation facilitates the automation of routine tasks, freeing up human capital to focus on higher-value activities such as innovation and strategic decision-making.

Furthermore, digital transformation empowers companies to meet consumers' evolving expectations in today's interconnected world. With the proliferation of e-commerce platforms and digital marketplaces, customers demand seamless and personalised experiences across all touchpoints. By leveraging data analytics and machine learning algorithms, businesses can gain valuable insights into consumer behaviour and preferences, enabling them to tailor products and services to individual needs. Moreover, digital technologies enable companies to enhance supply chain visibility and agility, ensuring timely delivery and responsiveness to fluctuating market demands.

Another compelling reason why digital transformation is indispensable in Industry 4.0 is its role in driving innovation and fostering a culture of continuous improvement. By embracing emerging technologies such as IoT, augmented reality and blockchain, companies can unlock new opportunities for value creation and differentiation. These technologies enable the development of smart products and services that anticipate and adapt to user needs, thereby fostering greater customer loyalty and brand affinity. Moreover, digital transformation facilitates collaboration and knowledge sharing across organisational boundaries, empowering employees to generate and implement innovative ideas that drive business growth.

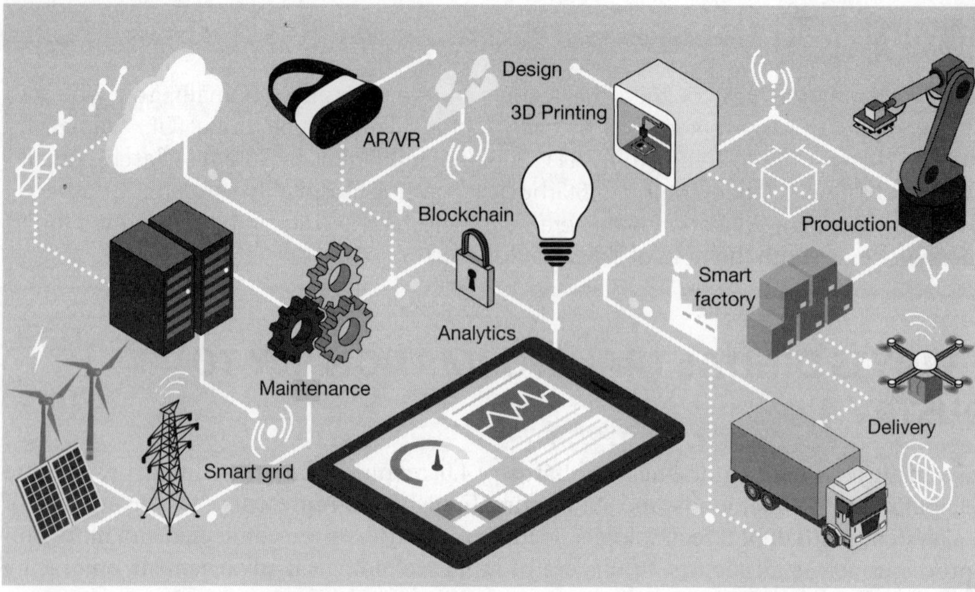

Figure 2.3 Digital transformation in smart factory

In essence, digital transformation is not merely a technological upgrade; it's a strategic imperative for companies seeking to thrive in the era of Industry 4.0. By embracing digital technologies and fostering a culture of innovation, businesses can enhance productivity, meet consumer expectations and unlock new opportunities for growth and competitiveness in an increasingly digital world. Embracing digital transformation isn't just about keeping up with the times; it's about shaping the industry's future and seizing its limitless possibilities.

2.4 WHAT IS A CYBER-PHYSICAL SYSTEM?

Cyber-physical Systems (CPS) is a key tenet of Industry 4.0. The role of CPS is to transform traditional manufacturing processes into intelligent, interconnected systems. According to the National Institute of Standards and Technology (NIST), '*CPSs are smart networked systems with embedded sensors, processors and actuators that are designed to sense and interact with the physical world (including the human users) and support real-time, guaranteed performance in safety-critical applications*'. It integrates the computation layer (cyber) with physical entities like machinery, vehicles, electric supply, etc. These systems' computational and physical components are tightly interconnected and work together effectively, with (full or partial) or without human intervention.

The cyber component acts as the 'central nervous system' of the CPS, providing the intelligence that optimises the physical system's operation. As shown in Figure 2.4, the integration of cyber and physical elements takes place at the cyber-physical bridge. The sensors convert measurable and observable physical quantities into data, forming the *physical-to-cyber* connection. Similarly, the *cyber-to-physical* connection happens at the actuation interface, where processed information leads to decisions that enact physical changes in the system.

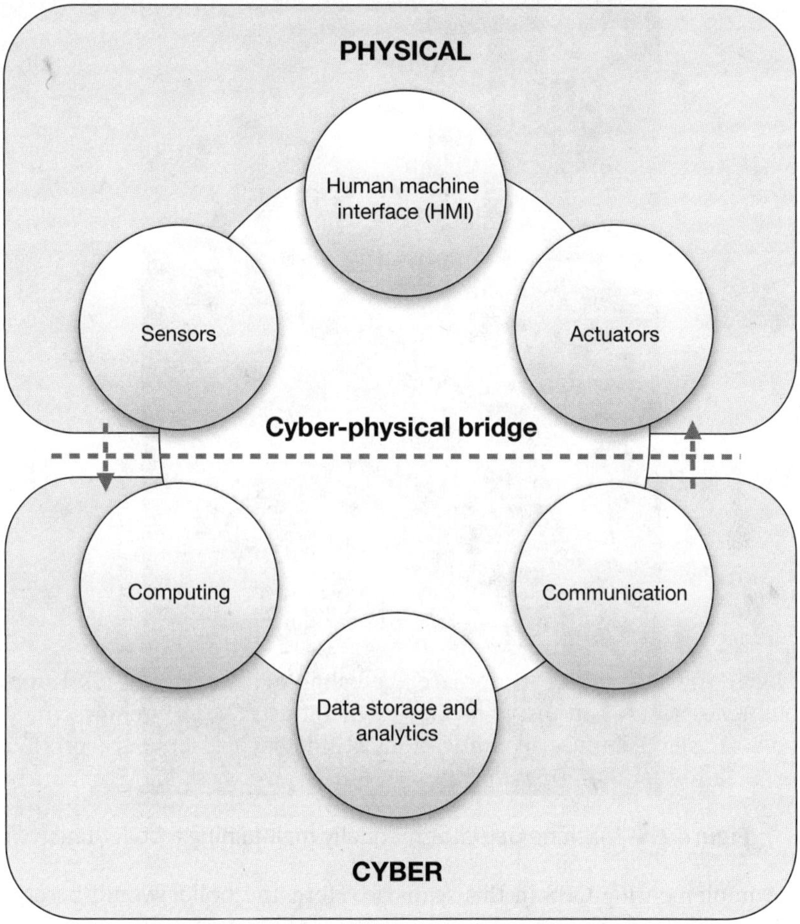

Figure 2.4 Cyber-physical system components

Let us consider a smart machine (physical component) in a smart factory, enabled with heat sensors and actuators, sharing continuous information about the machine's heating to a computational platform with advanced analytics (cyber component). Based on data analysis, this computational framework initiates a corrective mechanism through actuators if the threshold is breached.

For example, imagine the process of boiler temperature control in a power plant. Traditionally, the machine operator would measure the boiler temperature using a mercury-in-glass thermometer inserted into a thermowell or directly into the boiler. Based on experience and a manual checklist, the operator would control the temperature by adjusting the boiler's water supply, fuel supply, air supply, or steam extraction, as shown in Figure 2.5. However, this process is prone to human errors or delays in action that could lead to severe machine downtime.

Credit: Dmitry Kalinovsky / Shutterstock

Figure 2.5 Machine operator manually maintaining a boiler unit

Now, imagine implementing CPS in this scenario. Here, the boiler would be equipped with advanced sensors like infrared and digital temperature sensors and actuators like stepper motors, servo motors, pneumatic actuators and motorised valves. The boiler, sensors and actuators (the physical part of CPS) would be integrated with the cyber parts, such as Programmable Logic Controllers (PLCs). The PLCs are digital control systems that can be further connected to edge or cloud infrastructure to benefit from advanced analytics.

The central digital system would continuously monitor the temperature data sent by the boiler sensors. If thresholds are breached or early warnings are detected, the central system adjusts operations in real-time through the actuators to optimise boiler performance. This system would also proactively notify machine operators and log entries in the trouble ticket system.

We have now discussed the CPS framework for managing a single parameter (heat) for one boiler. In a thermal power plant, there can be thousands of such boilers and other machinery with numerous parameters to monitor. With CPS, all these machines, equipped with sensors and actuators, can be interconnected and linked to a central digital system for effective monitoring, advanced diagnostics and proactive actions autonomously or with human intervention. Hence, CPS is called a 'System of Systems' (SoS). An example diagram of the implementation of CPS in a power plant is shown in Figure 2.6.

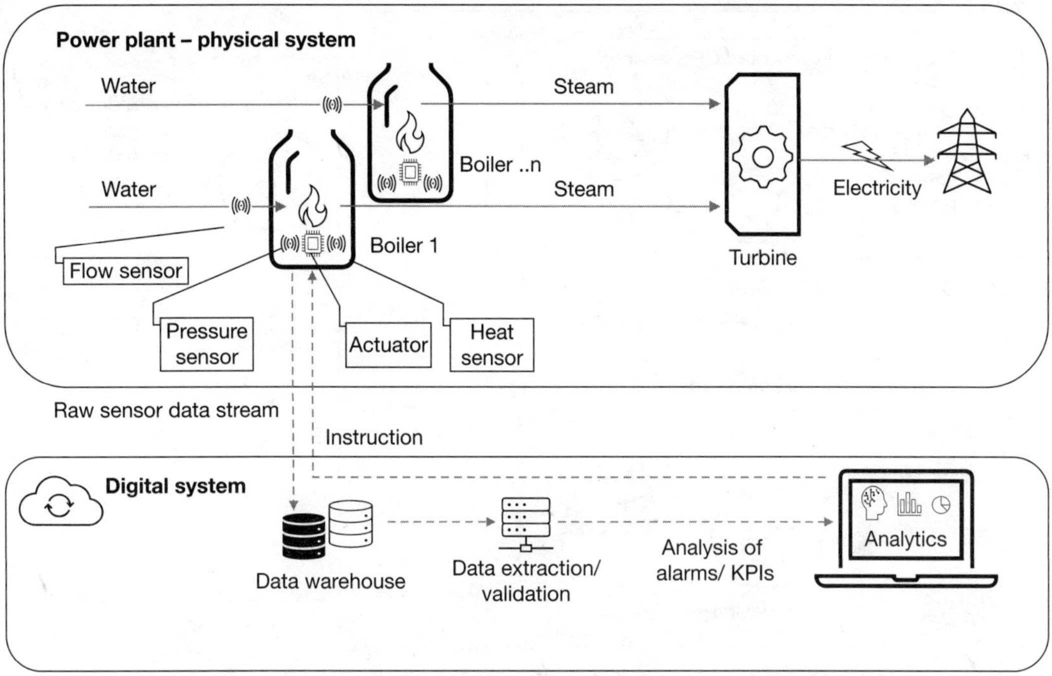

Figure 2.6 Implementation of CPS in a smart thermal power plant

Thus, a CPS solution can vary widely in complexity. As demonstrated in the previous example, it can range from a single device, such as a standalone smart boiler without external connectivity, to a sophisticated interconnected system. This could involve multiple cyber-physical devices or an SOS composed of several such systems. For instance, a network of smart boilers and smart turbines can be interconnected through the internet, all linked to a central digital system. This structure is recursive and can be extended to various scales. A CPS must integrate the decision flow with the information or action flows. The information flow digitally represents the measurement of the physical world's state, while the action flow affects that physical state. Figure 2.7 depicts the concept of SoS.

DID YOU KNOW?

The term 'cyber-physical systems' was coined in 2006 by Helen Gill at the National Science Foundation. Although 'cyberspace' is a related term introduced by William Gibson in his novel *Neuromancer*, both concepts originate from the field of 'cybernetics' (derived from the Greek word 'Kubernetes', meaning pilot), a term coined by American mathematician Norbert Wiener. Wiener developed technology for automatic anti-aircraft gun aiming process during World War II. Though using analog circuits and mechanical parts instead of digital computers, this early work on feedback control systems laid the foundation for modern cybernetics.

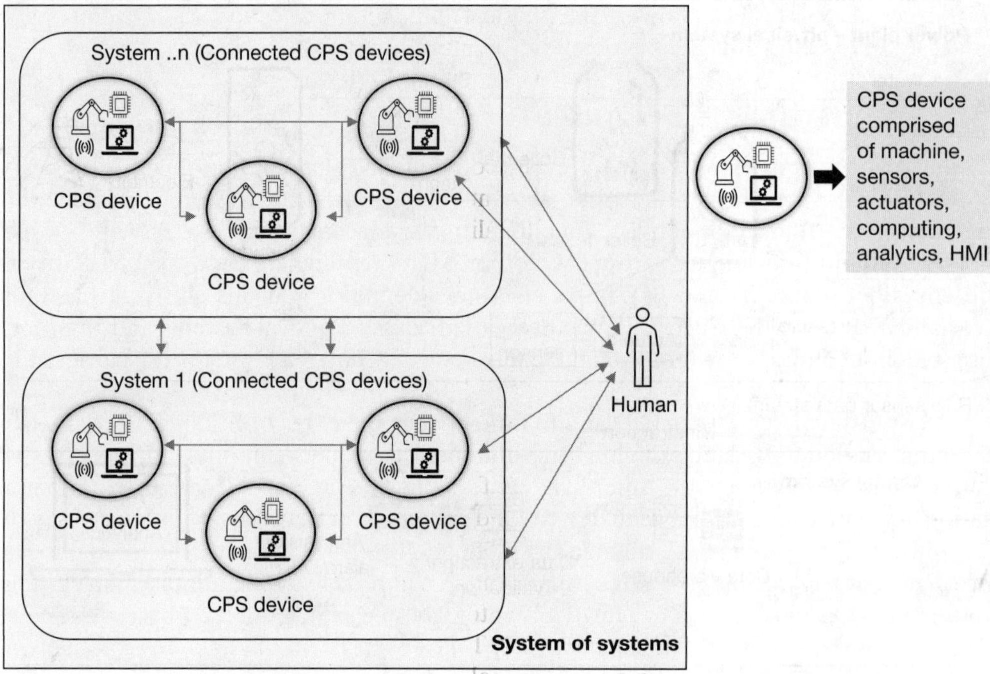

Figure 2.7 CPS as system of systems (SOS)

Are you finding similarities with the concept of the connected world of IoT? We will discuss this in Section 2.5.

2.5 CPS AND IOT

In this section, we will discuss the differences and similarities between CPS and IoT technology. However, before going into that, let us briefly understand IoT.

2.5.1 What is IoT

The IoT refers to a network of uniquely identifiable physical objects equipped with sensors, actuators, software and other technologies that enable them to connect and exchange data with other devices and systems over the Internet. These devices range from everyday household items to sophisticated industrial machines known as Industrial IoT (IIoT).

In recent years, IoT has become one of the most significant technologies of the 21st century. Connecting everything to the internet through embedded devices has enabled seamless communication between people, processes and objects.

By leveraging low-cost computing, cloud, big data, analytics and mobile technologies, physical objects (the 'Things') can now collect and share data with minimal human intervention. In this hyperconnected world, digital systems can record, monitor and adjust interactions between connected devices, allowing the physical and digital worlds to converge and operate seamlessly. IoT will be discussed in detail in Chapter 4.

2.5.2 Comparisons - CPS and IoT

CPS and IoT encompass physical and cyber components, blending the tangible aspects of devices and sensors with the intangible elements of software and data processing to create interconnected systems that enhance functionality and efficiency. Hence, CPS and IoT technologies are hard to distinguish from each other. Many researchers have tried to clarify the differences between CPS and IoT, but a consistent definition remains elusive. Christopher Greer and Team in NIST Special Publication titled 'Cyber-Physical Systems and Internet of Things' identified four schools of thought: equivalency, partial overlap, CPS as a subset of IoT and IoT as a subset of CPS.

The key distinction between CPS and IoT might be in their area of emphasis. CPS discusses integrating embedded systems, highlighting the interaction between physical processes and computational elements. On the other hand, IoT emphasises connectivity, facilitating communication and data exchange between devices and systems over networks. Thus, we can say that CPS concepts are related to the domain of system engineering. Systems Engineering focuses on designing and managing complex systems. CPS is a natural fit within this discipline because it requires a holistic approach to effectively integrate and manage these physical processes, networking and computing. The concept of IoT, however, is closely linked to the Network Engineering domain. Network Engineering deals with designing and managing communication networks, making it a key discipline in developing and implementing IoT solutions.

Greer also identified four key components contributing to the inconsistency in understanding the overlap between CPS and IoT: control, platform, Internet and human. The varying definitions of these components for CPS and IoT and the probable convergence are explained briefly in Table 2.1.

With the above analysis, it can be concluded that no apparent factors distinguish CPS from IoT. According to IEEE Standard 2413, 'An IoT system is a cyber-physical system that interacts with the physical world through sensors and actuators', reflecting the inherent connection between the two concepts. Henceforth, we will maintain CPS and IoT as convergent solutions in this book.

Table 2.1 Convergence of CPS and IoT

Components	CPS	IoT	Convergence
Control	Certain definitions emphasise the control of physical processes	Certain definitions put more emphasis on the information flow of sensors	The difference lies in how the control is defined. If the control means not only information flowing from sensors but also sending alarms to remote monitoring systems, resulting in some actions autonomously or through human intervention, it fits both IoT and CPS.

(*Continued*)

Table 2.1 (*Continued*)

Components	CPS	IoT	Convergence
Platform	Platform choice is a systematic decision that depends on required functionalities such as system requirements, integration capabilities, scalability, safety and compliance.	The IoT can be a platform for or a more straightforward form of CPS to achieve collaboration.	Focusing solely on a platform or application can result in design limitations or oversights that pose safety, security, reliability and other risks in CPS/IoT systems, potentially causing damage or injury. Therefore, IoT and CPS should not be viewed only at the platform level but should always be considered concerning the system's end-to-end functions. When viewed in this broader functional context, platform-based design considerations are similar for both CPS and IoT and do not serve as distinguishing factors between them.
Internet	The primary focus of CPS is often on the tight integration and interaction between computational elements (cyber) and physical processes (physical). The internet might not always be necessary for this interaction. For example, a standalone smart boiler equipped with sensors, actuators and its own PLC can recognise system issues and take action without requiring an internet connection.	Some IoT systems rely heavily on the Internet for their core functionality, using it for data transmission, remote control and cloud-based analytics. Other IoT systems might function primarily on local networks, using the Internet only occasionally or not at all. This variability leads to inconsistencies in how the IoT is perceived and implemented.	CPS and IoT definitions assume end-to-end connectivity involving many standards and protocols across various wireless, backhaul and core networks. However, the inconsistent association of IoT with the Internet, the inclusion of the Internet as an option in CPS and the diversity of protocols used in both CPS and IoT make Internet connectivity an unreliable distinguishing feature.
Human	Some definitions emphasise human interaction.	Some definitions minimise human interaction.	None of the CPS definitions exclude automated system processes; autonomy is a significant goal in CPS research, as seen in areas like autonomous vehicles and robotics. Likewise, IoT definitions do not envision a physical world devoid of human users. Instead, they emphasise the benefits and interactions with human users as key functional goals. Therefore, the roles of humans and automation are not valid factors for distinguishing between IoT and CPS.

2.6 KEY TECHNOLOGY LAYERS OF CPS

CPS is orchestrated by technologies that can be divided into three layers. Figure 2.8 depicts the different technology layers and their interoperability, which are the building blocks for CPS solutions like Smart Cities, Smart Grids, Smart Factories, etc.

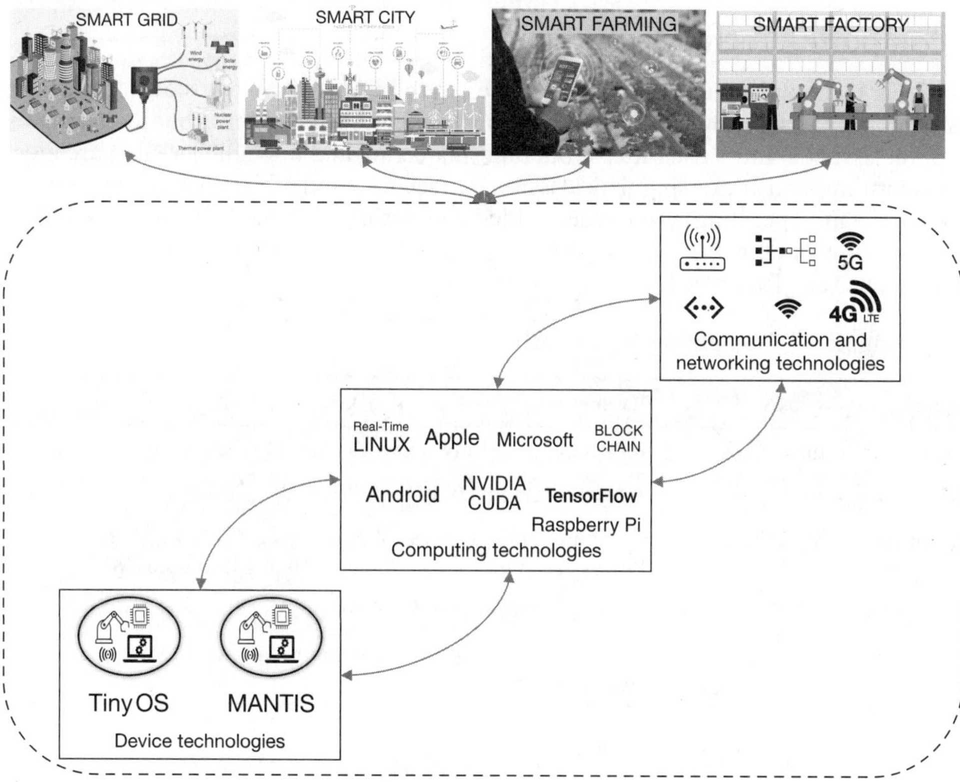

Credit: elenabsl / Shutterstock; Odoroaga Monica / 123RF; BigMouse / Shutterstock; chayanuphol / Shutterstock

Figure 2.8 CPS technology layers

2.6.1 Device Technologies

Device technologies constitute the basic infrastructure of any CPS. These physical and cyber components interact with the surrounding environment to acquire data for computation, analysis and control. Each CPS device technology can be considered an embedded hardware and software system.

Hardware

This part comprises mechanical and electrical components that work together to ensure the smooth operation of the device. Examples include gears, rotors, shafts, integrated circuits (ICs), microcontrollers and microprocessors.

Software

This part includes embedded software and operating systems that enable data abstraction and communication from the hardware components. The software can be directly programmed into the hardware using C or assembly language or operate through embedded operating systems like TinyOS, Mantis, etc.

2.6.2 Computing Technologies

Computational technologies form the second layer of the CPS technology stack, facilitating data abstraction, analysis and processing between CPS devices and other cyber technologies. They form the essential infrastructure for building, managing and upgrading system architectures and algorithms and addressing critical issues like security and privacy. This layer includes various software and operating systems, embedded computing, high-performance computing, data analytics with intelligence and Security and Privacy technologies. Some examples of these technologies are shown in Table 2.2.

Table 2.2 Examples of Computing Technologies

Computing Technologies	Examples
Software and Operating systems	RTOS like RTLinux, VxWorks, JetOS
Embedded Computing	Arduino, Raspberry PI, Beaglebone
High-Performance Computing	Compute Unified Device Architecture (CUDA), Open Multi-Processing (OpenMP), Message Passing Interface (MPI)
Data Analytics with Intelligence	Apache Hadoop, Apache Spark, TensorFlow, Scikit-lean
Security and Privacy Technologies	Blockchain, Encryption Algorithms, Intrusion Detection Systems (IDS)

2.6.3 CPS Communication and Networking Technologies

Communication and Networking technologies like Ethernet, Bluetooth, Wi-Fi, 3G, 4G-LTE and 5G enable seamless data exchange across the various layers and systems within the CPS technology stack. It ensures efficient CPS operation by facilitating interoperability through diverse communication standards and protocols. The advent of IoT-related technologies, cloud and fog/edge computing has enhanced networking and communication within individual CPS systems and among different CPS applications.

Based on these essential technology layers, the CPS reference architecture, the preferred model for a CPS implementation, is developed.

2.7 5C ARCHITECTURE OF CPS

Various references and customised architectures have been developed to support CPS implementations. Notable reference architectures include the Industrial Internet Reference Architecture (IIRA) by the Industrial Internet Consortium, the Reference Architectural

Model Industrie 4.0 (RAMI4.0) by Platform Industrie 4.0 and the Industrial Manufacturing System Architecture (IMSA) by the Chinese Manufacturing Sector. A customised architecture, the 5C architecture developed by Lee et al., complements these reference frameworks with a more detailed, hierarchical and data-centric approach.

The 5C architecture is important for modern industrial applications, emphasising real-time data processing, actionable insights and seamless integration of emerging technologies. It is especially well-suited for manufacturing industry use cases that require detailed, layered analysis and immediate data processing.

The 5C architecture breaks down CPS into layers: Connection, Conversion, Cyber, Cognition and Configuration. This granularity systematically addresses the unique challenges at each CPS development and deployment stage, optimising performance and maintaining flexibility.

The 5C Architecture is depicted in Figure 2.9. Each layer is explained in the subsequent sub-sections.

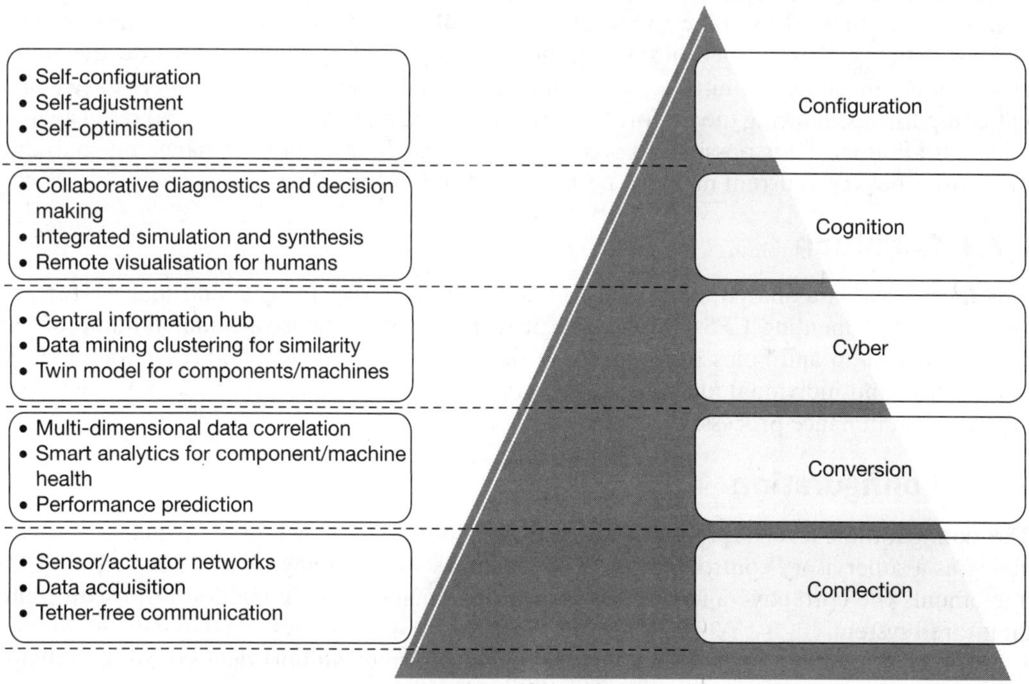

Figure 2.9 CPS 5C architecture

2.7.1 Connection

The first step in developing a CPS for intelligent factories is connecting physical machines and their components to gather accurate and reliable data. Various devices and sensors are used to collect data on voltage, current, temperature, vibration, rotating speed, feed speed, and oil concentration of machines and their components, as well as images and videos. Some sensors also monitor the temperature, humidity, light and atmospheric pressure in the manufacturing environment and warehouse. Additionally, data can be sourced from PLCs or systems like

(Enterprise Resource Planning) ERP, Manufacturing Execution Systems (MES), etc. Two crucial factors should be considered at this level. Firstly, a seamless data-gathering method should be established to capture heterogeneous data from different machines. Secondly, selecting appropriate sensors is also extremely important for this layer.

2.7.2 Conversion

In this layer, data would be converted into meaningful information. Currently, various tools and methodologies are available for converting data into information. In recent years, significant focus has been on developing algorithms specifically for prognostics and industrial health management applications. This second level of the CPS brings self-awareness of machines by calculating metrics such as health value and estimated remaining useful life.

2.7.3 Cyber

The cyber level functions as the **central information hub** in this architecture, collecting data from every connected machine to create a comprehensive network. With the vast amount of information gathered, specialised analytics are employed to gain insights into the status of individual machines within the entire fleet. These analytics enable machines to perform self-comparisons, allowing the performance of each machine to be evaluated and rated against others. Additionally, it is possible to predict future behaviour and performance by analysing similarities between current machine performance and historical data.

2.7.4 Cognition

This layer uses data analytics and machine learning to derive insights and make informed decisions. Implementing CPS at this level provides a comprehensive understanding of the monitored system and helps make accurate decisions. With the availability of comparative information and individual machine status, it becomes possible to prioritise tasks and optimise the maintenance process.

2.7.5 Configuration

The configuration level represents the feedback loop from cyberspace to physical space, acting as a supervisory control that enables machines to self-configure and adapt. This stage implements the corrective and preventive decisions made at the cognition level within the monitored system.

Together, these five layers create a robust architecture to support the dynamic and adaptive nature of Cyber-Physical Systems. By facilitating real-time data collection, sophisticated data processing, precise virtual modelling, intelligent decision-making and efficient system control, the 5C Architecture ensures CPS can address the needs of contemporary industrial, environmental and technological applications. This structured approach boosts CPS's performance and reliability and fosters innovation and efficiency across various domains.

In summary, a digital transformation roadmap creates the blueprint for a smart factory, which is then implemented with a network of cyber-physical systems. However, this network can integrate these systems at different layers. In the next section, we will understand how diverse levels of integration can be leveraged to create globally connected production facilities that embody the true spirit of Industry 4.0.

2.8 DOMAIN OF HORIZONTAL AND VERTICAL INTEGRATION

Horizontal and vertical integration is a key tenet of Industry 4.0, representing different dimensions of how companies organise their operations and supply chains to leverage digital technologies and achieve competitive advantage. In *horizontal integration*, efficiency is achieved at the operational or 'ground level' by tightly integrating processes – on the production floor, across multiple production facilities and throughout the entire supply chain. In *vertical integration*, data flows freely from the shop floor to the top and back, effectively tying an organisation's layers together. This also helps in efficiency in decision-making. Figure 2.10 gives a high-level overview of the horizontal and vertical integration.

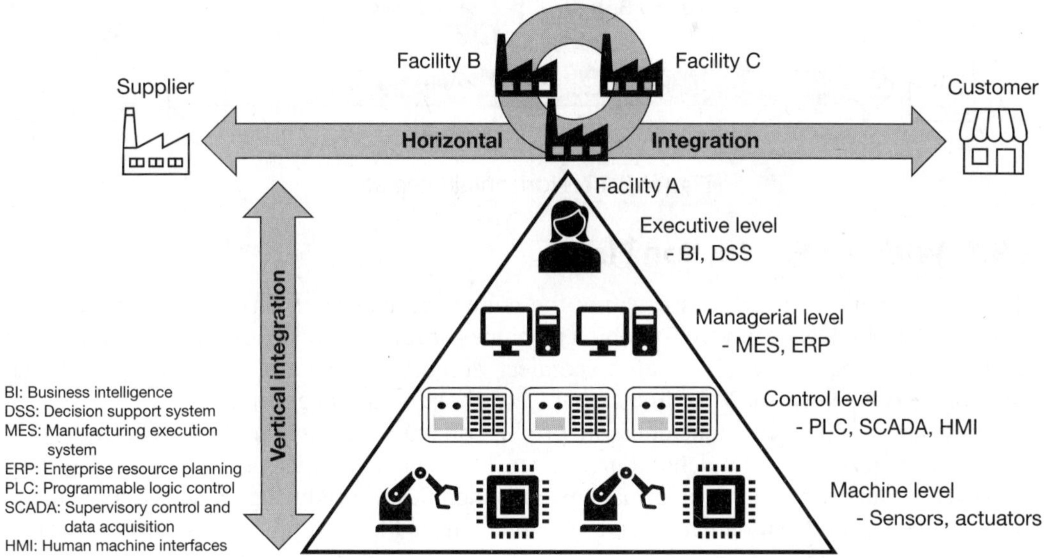

Figure 2.10 Horizontal and vertical integration in industry 4.0

DID YOU KNOW?

This definition of horizontal and vertical integration differs from the horizontal and vertical integration in organisational strategy. There, horizontal integration indicates acquiring organisations with different products in the same market or the same products in different markets. In this way, the acquiring company can diversify their product offerings or increase their market share. A vertical integration strategy entails acquiring companies in the supply chain, either on the supplier or customer side. This way, the acquiring company can secure access to essential supplies and respond faster to new market opportunities, thus extracting more value from handling the more significant portion of the supply chain.

2.9 HORIZONTAL INTEGRATION

Horizontal integration can be categorised into three levels: (i) within a production floor, (ii) across multiple factories, and (iii) across the supply chain. The first level involves integrating multiple machines within the same production facility - for example, Facility A in Figure 2.11. The second level refers to the integration of several manufacturing facilities within a single company, as illustrated by the connection of Facilities A, B, and C in the same figure. The third level extends beyond the organisation, involving integration across different entities in the supply chain. An example of this is the integration of the Supplier and the Customer with the already connected Facilities A, B, and C, as shown in Figure 2.11.

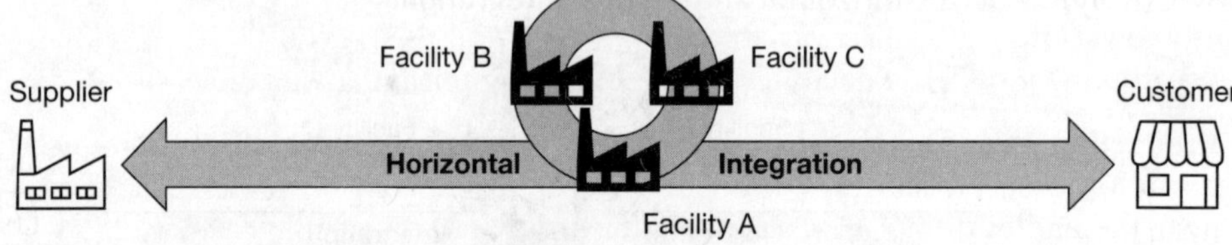

Figure 2.11 Horizontal integration

2.9.1 Within a Production Floor

The starting point in the Industry 4.0 is the connected machines on the production floor. This involves purchasing new IoT-enabled machines or retrofitting the existing machines with sensors and actuators to enable them to connect with other machines. This local Operational Technology (OT) network is the first step towards horizontal integration.

These always-connected machines and production units are integrated into the production network as objects with well-defined properties. They enable predictive maintenance by constantly communicating their performance status, significantly reducing expensive downtime. For instance, sensors can detect when a machine part is likely to fail and prompt a replacement before any disruption occurs.

Furthermore, these intelligent machines can autonomously respond to dynamic production requirements, enabling the production of smaller batch sizes. Smaller batches help lower inventory levels, reduce storage costs and minimise production risks. For example, instead of producing 10,000 identical units and storing them in a warehouse, a manufacturer might produce just 1,000 units based on current demand, which is more efficient and cost-effective.

Smaller batch sizes also allow manufacturers to better respond to changes in demand or customisation requests. This flexibility improves customer satisfaction and loyalty, as businesses can quickly adapt to new trends or individual customer preferences. For instance, a clothing manufacturer can swiftly switch from summer to winter wear as seasons change or create custom designs based on customer specifications.

Traditional production systems focus on standardisation, producing large batch sizes to keep fixed costs low. Factories would make thousands of identical items to benefit from economies of scale. However, this approach often led to overproduction, high inventory costs and less flexibility to meet changing market demands. The shift to smaller, more customisable batches represents a significant evolution in manufacturing, driven by technological advances and the need for greater efficiency and customer-centric production.

The ultimate goal is for intelligent production floors to produce batches of one cost-effectively, meaning each product is personalised for the customer. For example, a customer might order a custom-fit shoe with specific colours and features and the production system can manufacture this unique item just as efficiently as a mass-produced one.

2.9.2 Across Multiple Factories

If an organisation has multiple manufacturing facilities, Industry 4.0 encourages horizontal integration across the plants through advanced MES. Production facility data, such as inventory levels, unexpected delays and equipment status, is seamlessly communicated across the organisation in this scenario. This integration allows production jobs to be automatically redistributed among facilities as needed, enabling the organisation to adapt promptly and efficiently to various production variables.

Redistributing jobs across facilities offers several significant advantages:

- **Production Near Demand Location:** The organisation can significantly reduce transportation costs by shifting production closer to demand locations. Producing goods near the end customer also shortens delivery times, enhancing customer satisfaction. For example, if an auto-parts manufacturing company with factories in the suburbs of Delhi, Mumbai and Chennai receives a large order from an auto manufacturer in Pune, it can allocate production to its Mumbai facility, thereby reducing shipping distances and costs, as well as delivery time.
- **Optimised Resource Utilisation:** Organisations can optimise the use of their resources by redistributing jobs based on facility availability and capability. For example, if one facility has excess capacity while another is overburdened, jobs can be reallocated to balance the workload, preventing overuse of any single plant and reducing overtime costs. This also allows them to compensate for each other during unexpected delays or breakdowns. If a machine in one plant breaks down, production can be swiftly transferred to another, minimising production delay.
- **Customised Production:** Redistributing jobs allows for more tailored production strategies. For example, if different regions have varying preferences or regulatory requirements, production can be adjusted accordingly. An automotive manufacturer might produce electric vehicles in regions with higher demand for eco-friendly transportation while focusing on traditional models in areas where they are more popular.
- **Balanced Inventory Levels:** By monitoring and sharing inventory data across facilities, organisations can maintain balanced inventory levels, reducing the risk of overstocking and stockouts. For example, if one facility has an excess of a particular component, it can transfer it to another facility in need, ensuring efficient use of materials. Efficient job redistribution helps maintain lean inventory levels, reducing the need for extensive storage. For instance, a company can avoid the costs associated with large warehouses by producing goods closer to the point of sale and delivering them just in time.

2.9.3 Across Supply Chain

Industry 4.0 advocates for data openness and high levels of automated collaboration throughout the supply and logistics chain. This includes the upstream supply chain, which sources the raw materials and components to the manufacturing plant and the downstream supply chain, which delivers the final products to the market. For this to be effective, third-party suppliers and service providers must be securely and tightly integrated horizontally into the enterprise's production and logistics control systems. This integration brings numerous benefits and enhances efficiency across the entire supply chain.

- **Improved Forecasting and Demand Planning:** By sharing real-time sales and inventory data with suppliers, manufacturers can improve the accuracy of their demand forecasts. For example, a retailer sharing sales data with its suppliers allows them to better predict which products will be needed and when, leading to more efficient production planning. Collaboration helps mitigate the bullwhip effect in a supply chain disruption. For instance, if a manufacturer has visibility into real-time sales data from retailers, it can adjust production schedules to meet actual demand rather than overproducing based on inaccurate forecasts.
- **Optimised Inventory Management:** With better coordination, companies can implement Just in Time (JIT) inventory systems, reducing the need to hold large stocks of materials and products. For example, an automotive manufacturer can receive components from suppliers exactly when needed for assembly, reducing storage costs and minimising waste. Real-time data sharing allows for precise inventory management. If suppliers know the exact inventory levels at the manufacturer's warehouse, they can ensure timely replenishment, reducing the risk of stockouts or excess inventory. For example, a consumer electronics company can avoid the costly problem of stockouts during peak demand periods by closely coordinating with component suppliers.
- **Streamlined Production Processes:** By integrating suppliers and service providers into production control systems, manufacturers can synchronise production schedules across the supply chain. For example, a fashion retailer can coordinate with fabric suppliers and garment manufacturers to ensure that production timelines align perfectly, reducing lead times, speeding up time-to-market and minimising the production cost for all the participants in the supply chain.
- **Enhanced Supply Chain Visibility:** Integrated systems provide end-to-end visibility into the supply chain, allowing companies to track materials and products from suppliers to customers. For example, a food manufacturer can track the journey of ingredients from farm to factory, ensuring quality and compliance with safety standards. With this end-to-end visibility, companies can identify and address issues before they escalate. For instance, a logistics provider can detect delays in shipment early and reroute deliveries to meet customer deadlines, avoiding costly disruptions.

DID YOU KNOW?

The **bullwhip effect** in supply chain disruption refers to the amplification of demand variability as it moves up the supply chain from the retailer to the manufacturer. Small fluctuations in customer demand can cause larger swings in orders for suppliers, leading to inefficiencies like overproduction or stock shortages. This occurs due to delays in information sharing, overreactions to changes in demand and poor coordination between supply chain partners. In times of disruption, such as during a natural disaster or a global crisis, the bullwhip effect can exacerbate these problems, making recovery and inventory management more challenging.

- **Reduced Operational Costs:** Automation and data integration reduce manual processes, leading to significant cost savings. For example, automated order processing between the manufacturer and a supplier reduces administrative costs and speeds up transactions.
- **Sustainability and Environmental Benefits:** Efficient logistics planning minimises transportation distances and consolidates shipments, reducing the carbon footprint. For instance, a consumer goods company can optimise delivery routes with logistics providers to decrease fuel consumption and emissions.

Large retailers like Walmart exemplify the benefits of Industry 4.0 integration by using advanced data analytics to share real-time sales data with suppliers. This collaboration enhances forecasting and demand planning, allowing suppliers to predict which products are needed and when thus mitigating the bullwhip effect. By implementing Just in Time (JIT) inventory systems, Walmart reduces the need for large stocks of materials, optimising inventory management and lowering storage costs. The integration also streamlines production processes by synchronising schedules across the supply chain, ensuring timely production and delivery. Enhanced supply chain visibility allows Walmart to track products from suppliers to shelves, ensuring quality and compliance while addressing potential issues proactively. Automation and data integration also reduce operational costs through more efficient order processing. Sustainability is also improved as Walmart and its logistics providers optimise delivery routes, reducing fuel consumption and emissions. Through these comprehensive integrations, Walmart effectively minimises the risk of stockouts and overstock situations, demonstrating the multifaceted benefits of Industry 4.0.

2.10 VERTICAL INTEGRATION

The goal of vertical integration is to connect all logical levels of the company, starting from the production floor and working up through departments like R&D, quality control, product management, IT, sales and marketing, up to executive leadership. These layers allow for free and transparent data flow, enabling data-driven tactical and strategic decision-making. Figure 2.12 gives a schematic of the various levels and digital systems used in those levels.

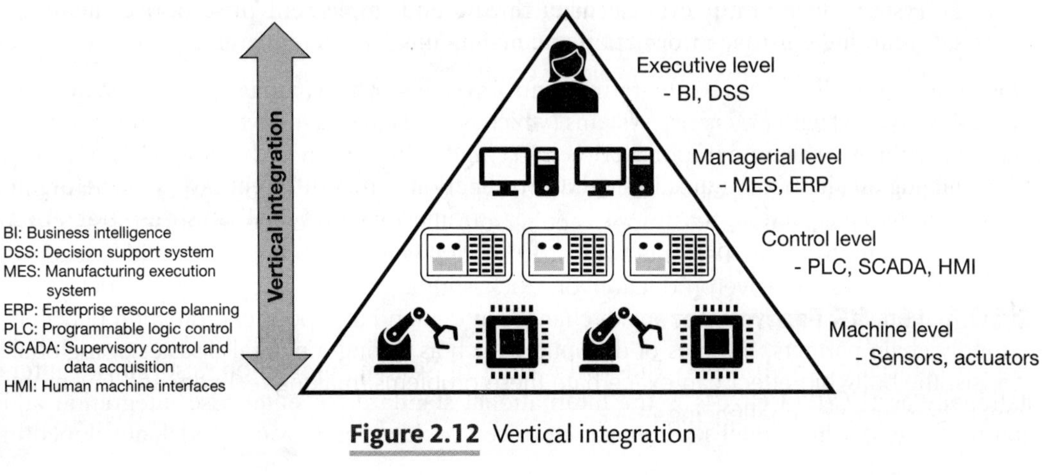

Figure 2.12 Vertical integration

Being able to react quickly to shifting market signals and emerging possibilities gives the vertically integrated Industry 4.0 firm a significant competitive advantage. Some of the advantages are listed below:

- **Enhanced Operational Efficiency:** Integration across departments allows for streamlined workflows and reduced bureaucratic hurdles. For example, a vertically integrated electronics manufacturer can quickly iterate product designs based on real-time feedback from production and quality control teams, reducing time-to-market.
- **Optimised Resource Allocation:** Centralised data flow enables better resource allocation across departments. For instance, a pharmaceutical company integrating R&D with production can prioritise research efforts based on manufacturing capabilities and market demand, ensuring efficient resource use.
- **Collaborative Innovation:** Close integration between R&D, product management and quality control fosters collaborative innovation. For example, an automotive manufacturer can seamlessly integrate feedback from product management into the R&D process, leading to the development of innovative features that meet customer needs and quality standards.
- **Continuous Improvement:** Data transparency allows for continuous quality monitoring and improvement. For instance, a food processing company can leverage real-time data from production and quality control to promptly identify and address quality issues, maintaining high standards across its product line.
- **Responsive Customer Support:** Seamless data flow between customer service and other departments enables faster issue resolution. For instance, a telecommunications provider can integrate customer feedback from support teams into product development processes, improving service reliability and customer retention.
- **Strategic Decision-making:** Comprehensive data flow supports informed decision-making at the executive level. For example, a vertically integrated tech company can leverage data analytics across departments to identify growth opportunities in new markets and allocate resources strategically.
- **Risk Management:** With integrated systems, executive leadership can assess and mitigate risks proactively. For instance, a financial services firm can use real-time data from IT systems to monitor cybersecurity threats and implement pre-emptive measures, safeguarding sensitive information and maintaining customer trust.

One of the major challenges is integrating control systems (which manage production machinery and processes) with enterprise systems (which handle executive-level communication and decision-making). This integration between OT and information technology (IT) is crucial for achieving efficient communication and coordination across different layers of the organisation. In the next section, we discuss ISA 95, a framework that facilitates this integration, bridging the gap between operational and information technology.

2.10.1 ISA 95 Framework

ISA-95 (officially, ANSI/ISA-95 Enterprise-Control System Integration, also known internationally as IEC/ISO 62264) is the international standard for enterprise integration and control systems. This model's origin can be traced back to the Purdue Model, developed in the early 1990s at Purdue University. The Purdue Model defines best practices for the relationship between industrial control systems and business networks. Subsequently, it has been improved to define each layer and add a few more layers.

The framework consists of several layers, each focusing on specific aspects of manufacturing and business processes. Here's a breakdown of the different layers (refer to Figure 2.13):

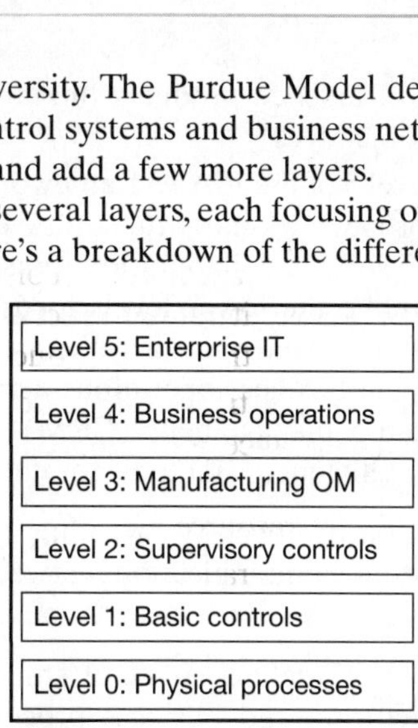

Figure 2.13 Purdue/ISA95 model

- **Level 0 – Physical Processes:** Level 0 represents the physical processes within the manufacturing environment. This includes all the physical devices like sensors, actuators and machinery that interact directly with production processes. For example, temperature sensors in a furnace, motors driving conveyor belts and robotic arms assembling products all operate at this level, capturing real-time data from the production floor.
- **Level 1 – Basic Controls:** Level 1 involves controlling physical processes through sensors and actuators. This level includes the hardware and software used to monitor and manage equipment operations. PLCs are a key component here, collecting data from sensors and sending commands to actuators to control machinery, ensuring that processes run smoothly and efficiently.
- **Level 2 – Supervisory Controls:** Level 2 encompasses systems providing real-time monitoring and supervisory control over production processes. This includes Human-Machine Interfaces (HMIs) and Supervisory Control and Data Acquisition (SCADA) systems. These systems allow operators to oversee production lines and intervene when necessary, maintaining control and ensuring that production meets the required standards and timelines.
- **Level 3 – Manufacturing Operations Management:** Level 3 focuses on managing and optimising manufacturing operations. It includes MES that track and document the transformation of raw materials into finished products. An MES schedules production runs, tracks inventory levels and ensures quality control throughout the manufacturing process, helping to maintain efficiency and product quality.
- **Level 4 – Business Operations:** Level 4 concerns business-related functions such as planning, logistics and enterprise resource management. This level includes Enterprise Resource Planning (ERP) systems that manage supply chain activities, order processing and financials. An ERP system coordinates customer orders, schedules production

and manages the delivery of finished products, linking the manufacturing operations with the broader business objectives.

- **Level 5 – Enterprise IT:** Level 5 of the ISA 95 framework encompasses enterprise-wide IT systems that support broader business functions. These systems include essential infrastructure such as email servers, collaboration tools, corporate networks and enterprise-wide applications. Unlike the lower levels focused on OT, Level 5 IT systems are not directly involved in industrial control systems (ICS) operations but are crucial for supporting and enhancing business operations across the organisation. Examples include customer relationship management (CRM) systems for maintaining client interactions and business intelligence (BI) tools for data analysis and decision-making.

ISA 95 standardises the integration protocol by providing a common language and set of models that help ensure consistent communication between these layers. Here's how it achieves this:

- **Common Terminology:** ISA 95 defines a set of standardised terms and definitions for all aspects of manufacturing and business processes. This common vocabulary ensures that all systems and stakeholders understand each other, reducing the risk of miscommunication.
- **Information Models:** The framework provides detailed models describing the information exchanged between systems at different levels. These models specify what data needs to be shared, such as production schedules, inventory levels and quality metrics.
- **Functional Hierarchies:** ISA 95 outlines functional hierarchies describing the relationships and interactions between systems and layers. These hierarchies help structure the flow of information and ensure that data moves seamlessly from the production floor to executive management.
- **Interface Standards:** By defining standard interfaces for data exchange, ISA 95 enables different systems and software to communicate effectively. This interoperability is crucial for integrating diverse technologies and platforms in manufacturing and enterprise environments.

The ISA 95 framework is crucial for modern manufacturing, where integrating OT and IT is essential for achieving operational excellence. By understanding and implementing ISA 95, organisations can ensure that their production and enterprise systems work together effectively, leading to improved performance and competitiveness in the market.

2.11 CONCLUSION

Though digitisation has roots in the 1940s, most enterprises started leveraging digitally transformed products, services and experiences as business differentiators only around the 2010s. Furthermore, the COVID-19 pandemic expedited digital transformation as it became a compulsory lever for survival. Some of the key transitions that we were mandated during the pandemic, such as automated and low-touch supply chain management, self-service-based customer experience facilities, contactless payment systems, shift in focus from brick-and-mortar to e-commerce platforms, etc., have now become a norm for modern businesses. In this context, tremendous innovations in the generative AI area and its increased adoption have opened up new avenues for digital transformation in the last few years. AI will play a strategic role in improving productivity and gaining a competitive advantage for enterprises and thus will be an integral part of Industry 4.0. As per the prediction of TechTarget's Enterprise Strategy Group, 54% of the organisations it polled will have adopted generative AI within the next 12 months.

Here, the precise implementation of CPS, which is foundational for digital maturity, is crucial because it bridges the physical and digital realms. By integrating physical and virtual systems into intelligent networks, CPS enhances the synergy between humans, machines and information technology. The transformative impact of Industry 4.0 stems from this merger, i.e., combining advanced production methods with digital technologies, creating interconnected enterprises that leverage data to drive smart actions in the physical world. Therefore, achieving the objectives of Industry 4.0 is impossible without proper implementation of CPS. According to the Future Market Insights report, the global CPS market is projected to reach $101.13 billion in 2024. The industry is expected to grow at a compound annual growth rate (CAGR) of 7.8% over the next decade, reaching a valuation of $214.32 billion by 2034. This underscores the increasing importance of CPS in driving the advancements and efficiencies envisioned in Industry 4.0. Once the CPS and the rest of the digital systems are in place, integrating them for seamless data flow across machines, facilities and departments makes an enterprise truly smart.

Vertical and horizontal integration play pivotal roles in the evolution of Industry 4.0, facilitating interconnectedness and efficiency across manufacturing ecosystems. Vertical integration involves the seamless alignment of processes from raw materials to finished products, optimising production and quality control through integrated technologies and data. Meanwhile, horizontal integration focuses on connecting disparate entities within and across industries, enabling collaborative workflows and interoperability of systems and data. According to the MarketsandMarkets research report, the global smart factory market is poised for substantial growth, projected to expand from USD 100.6 billion in 2024 to USD 164.0 billion by 2029 at a CAGR of 10.3%. This growth underscores the increasing adoption of digital technologies and integration strategies in manufacturing, driving operational efficiencies and enhancing competitiveness. Vertical and horizontal integration initiatives are integral to realising the full potential of Industry 4.0, fostering agile, data-driven enterprises capable of smart decision-making and adaptive manufacturing practices.

SUMMARY

- Industry 4.0 represents a transformative phase in the industrial sector through the integration of advanced technologies in manufacturing.
- **Digitisation:** Converting physical objects into digital formats (e.g., scanning a photo).
- **Digitalisation:** Leveraging digital technologies to improve existing processes without fundamental changes.

- **Digital Transformation:** Comprehensive integration of digital technologies to fundamentally change business processes and enhance customer experiences.
- Digital transformation is a key to Industry 4.0 as it is essential for enhancing productivity and operational efficiency, supports predictive maintenance and automation, enables focus on strategic tasks, encourages the adoption of IoT, augmented reality and blockchain for smarter products and facilitates collaboration and knowledge sharing, driving business growth.
- **Cyber-Physical System:** CPS integrates computation with physical processes, enabling real-time interactions and intelligent manufacturing. Sensors gather data from physical systems; actuators implement changes based on computational analysis.
- **CPS and IoT Comparison:**
  - **Control:** CPS emphasises physical process control; IoT highlights information flow.
  - **Platform:** CPS requires systematic decisions for functionality; IoT can be seen as a simpler form of CPS.
  - **Internet Dependence:** CPS may not always need the internet; IoT often relies on it.
  - **Human Interaction:** Both systems acknowledge the role of automation and human users.
- **Technology Layers of CPS**
  - **Device Technologies:** Physical and cyber components, including hardware (mechanical/electrical) and software (embedded systems).
  - **Computing Technologies:** Infrastructure for data processing and analysis, including software, embedded computing and security technologies.
  - **Communication and Networking:** Technologies (Ethernet, Wi-Fi, etc.) enabling data exchange across CPS layers.
- **5C Architecture of CPS is a layered system:**
  - **Connection:** Gathering data from machines and sensors.
  - **Conversion:** Transforming data into meaningful information.
  - **Cyber:** Central hub for data aggregation and analytics.
  - **Cognition:** Utilising analytics and ML for insights and decisions.
  - **Configuration:** Feedback loop enabling self-configuration of machines.
- **Horizontal Integration:** Focuses on integrating processes across production facilities and supply chains for operational efficiency.
- Horizontal Integration Levels
  - **Level (i):** Integration of multiple machines within the same production facility.
  - **Level (ii):** Integration of multiple manufacturing facilities of a single company.
  - **Level (iii):** Integration of different organisations along the supply chain.
- **Vertical Integration:** Ensures free data flow from the shop floor to management for effective decision-making.
- **ISA 95 Framework:** Provides a common language and models for seamless communication across different manufacturing and business levels.

EXERCISE

Multiple Choice Questions (1 Mark)

1. What strategic benefit does digital transformation offer organisations in Industry 4.0?
   (a) Guarantees higher market share.
   (b) Ensures permanent cost reductions.
   (c) Enhances productivity, innovation and competitiveness.
   (d) Eliminates competition.
2. What is the main benefit of predictive maintenance enabled by digital transformation?
   (a) Eliminates the need for maintenance staff.
   (b) Improves equipment lifespan by 50%.
   (c) Prevents costly disruptions to production processes.
   (d) Reduces the cost of digital technologies.
3. What is the key difference between digitisation and digitalisation?
   (a) Digitisation improves efficiency, while digitalisation transforms processes.
   (b) Digitisation involves converting physical objects into digital data, while digitalisation uses that data to improve processes.
   (c) Digitisation automates processes, while digitalisation creates digital objects.
   (d) Digitisation is about customer engagement, while digitalisation focuses on internal processes.
4. A smart hygrometer generating commands for an actuator, is an example of
   (a) Digitisation
   (b) Digitalisation
   (c) Digital transformation
   (d) None of the above
5. A retail company wants to improve customer experience by offering personalised recommendations online. What technology would best support this initiative, as suggested in the text?
   (a) Internet of Things (IoT)
   (b) Predictive maintenance systems
   (c) Data analytics and machine learning algorithms
   (d) Blockchain for secure transactions
6. Your organisation wants to adopt a digital transformation strategy. What initial step should you take to align with the principles discussed in the text?
   (a) Train employees on advanced analytics and artificial intelligence.
   (b) Digitise all physical assets and processes to create a foundation of digital data.
   (c) Begin implementing blockchain technology for all operations.
   (d) Purchase the latest robotics and automation tools immediately.
7. What is the primary role of Cyber-Physical Systems (CPS) in Industry 4.0?
   (a) To replace human operators entirely

(b) To transform traditional manufacturing processes into intelligent, interconnected systems
(c) To minimise data collection to reduce system complexity
(d) To develop advanced robots for industrial applications

8. In the context of IoT and CPS, what is one of the key distinctions mentioned?
(a) IoT focuses on control, whereas CPS focuses on connectivity.
(b) IoT emphasises internet connectivity, while CPS integrates embedded platforms with control mechanisms.
(c) CPS is a subset of IoT, limited to industrial applications.
(d) CPS primarily involves hardware, while IoT is entirely software-based.

9. Which of the following components is NOT part of a typical CPS system?
(a) Heat sensors
(b) Actuators
(c) Mercury-in-glass thermometer
(d) Infrared sensors

10. What does the term 'System of Systems' (SoS) signify in the context of CPS?
(a) A network of standalone devices without interconnection
(b) Integration of multiple CPS devices into a centralised digital system
(c) A method for developing CPS hardware
(d) A focus only on data collection without analysis

11. What does the 5C architecture of CPS NOT focus on?
(a) Real-time data processing
(b) Developing standalone devices
(c) Actionable insights
(d) Seamless integration of technologies

12. What is the primary function of the Cognition layer in the 5C Architecture?
(a) Storing raw data for future use
(b) Making informed decisions using analytical insights
(c) Sending feedback to the machines for self-optimisation
(d) Managing connections between devices

13. A logistics company uses IoT-enabled devices to track shipments in real-time. One of their clients reports that shipment data is not updating correctly. What is the most likely cause of this issue?
(a) Lack of interoperability among IoT tracking devices
(b) Excessive data storage causing delays in updates
(c) Faulty machinery handling the shipments
(d) Over-reliance on manual entry for shipment updates

14. A factory manager notices that unexpected machinery failures are causing significant downtime. Which feature of IoT-enabled Industry 4.0 would best help mitigate this issue?
(a) Implementing a standard maintenance schedule across all machines

(b) Using IoT sensors to monitor machine health and predict failures
(c) Automating all machinery without human supervision
(d) Installing high-capacity machinery to reduce dependency on any single machine

15. Which of the following is NOT a level of horizontal integration?
(a) Integration of multiple machines in the same production facility
(b) Integration of multiple manufacturing facilities in one company
(c) Integration between the production floor and executive leadership
(d) Integration of different organisations along the supply chain

16. Which of the following is NOT a benefit of integrating supply chain partners in Industry 4.0?
(a) Enhanced forecasting and demand planning
(b) Increased bullwhip effect
(c) Streamlined production processes
(d) Reduced operational costs through automation

17. What does vertical integration aim to achieve in Industry 4.0?
(a) Integration of machines across production facilities
(b) Seamless communication between different layers of the organisation
(c) Expansion of product lines through acquisitions
(d) Increased investment in physical infrastructure

18. What is a key advantage of redistributing production jobs across multiple facilities in Industry 4.0?
(a) Increased employee satisfaction
(b) Reduced transportation costs by producing closer-to-demand locations
(c) Increased production of identical units
(d) Reduced reliance on external suppliers

19. A company is working on improving its product offerings by integrating its operational systems across various departments, from production to sales. They want to reduce delays in processing orders and improve overall workflow. What would be a key advantage of this vertical integration?
(a) A more complex risk management framework
(b) Increased reliance on external suppliers for raw materials
(c) Enhanced operational efficiency with streamlined workflows
(d) Greater separation between different organisational layers

20. A company is expanding its operations internationally and decides to integrate its different production sites into a unified system to enhance visibility into production across all sites. What is the most likely result of this improvement in visibility?
(a) Reduced need for coordination between teams
(b) Increased ability to adjust production schedules in real-time based on demand
(c) Greater focus on localised production strategies
(d) Increased complexity in inventory management

Short Answer Type Questions (2 Marks)

1. What is digitisation and why is it essential for digital transformation?
2. How does digitalisation improve processes in an organisation?
3. What distinguishes digital transformation from digitalisation?
4. Why is digital transformation critical for businesses in Industry 4.0?
5. What are the main components of a Cyber-Physical System (CPS)?
6. What role do sensors play in a Cyber-Physical System?
7. Why is real-time data processing critical in Cyber-Physical Systems?
8. How does horizontal integration support global supply chain management in Industry 4.0?
9. In what ways can horizontal integration benefit customer satisfaction in Industry 4.0?
10. Explain the role of vertical integration in creating a smart factory environment in Industry 4.0.

Long Answer Type Questions (5 Marks)

1. Explain the concepts of digitisation and digitalisation with examples. Why are these processes crucial precursors to digital transformation?
2. Define digital transformation and discuss how it differs from digitisation and digitalisation.
3. Why is digital transformation considered a strategic imperative for businesses? Discuss its benefits and challenges.
4. Explain the concept of Cyber-Physical Systems (CPS) and how they differ from traditional computing systems.
5. Discuss the challenges involved in real-time data processing in Cyber-Physical Systems (CPS) to ensure timely and accurate responses.
6. Discuss the role of feedback loops and automation in enhancing operational efficiency using CPS.
8. Discuss the significance of horizontal integration in Industry 4.0 and its impact on inter-organisational collaboration. Provide examples of how it transforms traditional business operations.
9. Explain how vertical integration supports predictive maintenance and operational intelligence in Industry 4.0. Include the technological advancements that enable this integration.
10. Compare and contrast the challenges faced in implementing horizontal and vertical integration in Industry 4.0. What strategies can organisations adopt to overcome these challenges?

Lean and Agile Production Methodologies

CHAPTER

3

OBJECTIVES

Industry 4.0 setup is complex, as multiple technologies, a large set of equipment and an interconnected network create a complex ecosystem. Handling this system requires innovative methodologies. This chapter explains how Lean and Agile methodologies integrate with Industry 4.0 technologies to optimise manufacturing efficiency and flexibility, preparing an organisation to implement these strategies in the modern smart factory.

By the end of this chapter, you should be able to:

- Understand the core principles of traditional Lean and Agile manufacturing methodologies.
- Explain how Industry 4.0 technologies enhance Lean (Lean 4.0) and Agile (Agile 4.0) methodologies.
- Analyse real-world case studies demonstrating the successful integration of Lean 4.0 and Agile 4.0 in smart factories.
- Understand other Industry 4.0-aligned methodologies.

3.1 INTRODUCTION

Why Are Traditional Methodologies Not Effective for Industry 4.0 Factories?

A production methodology is a structured approach that organisations use to improve efficiency, reduce waste and optimise production. Traditional methodologies such as Lean Manufacturing, Six Sigma and Just-in-Time (JIT) were developed to streamline manufacturing processes by eliminating inefficiencies, improving quality and reducing costs. These approaches have been highly effective in conventional production environments, focusing on standardisation, predictable workflows and human-driven decision-making.

Industry 4.0 represents a paradigm shift in manufacturing. Unlike traditional factories, Industry 4.0 factories operate with interconnected devices, real-time data exchange and automated decision-making. We will discuss in this book that the key Industry 4.0 advancements include:

- **Connected Devices and Sensors:** Machines, equipment and products are embedded with IoT sensors, enabling real-time monitoring of operations.
- **Real-Time Data-Driven Decisions:** AI and big data analytics process vast amounts of data instantly, optimising production on the fly.
- **Automation and Robotics:** Smart robots and cobots (collaborative robots) perform repetitive tasks, reducing reliance on manual labour.
- **Predictive and Prescriptive Maintenance:** Instead of scheduled maintenance, AI predicts failures before they occur, minimising downtime.

Traditional manufacturing methodologies were designed for structured, linear production processes and lacked the adaptability to harness the full potential of Industry 4.0 technologies. These older approaches do not effectively utilise real-time data, automation, or AI-driven insights, making them inefficient in modern digital factories.

To remain competitive in the era of Industry 4.0, manufacturers need methodologies that integrate digital transformation principles while preserving the efficiency of traditional approaches. This has led to the evolution of Lean Manufacturing 4.0 and Agile Manufacturing 4.0, which align with Industry 4.0's capabilities.

As we will get to know in the subsequent sections of this chapter, Lean Manufacturing 4.0 builds upon traditional Lean principles by incorporating digital technologies to enhance efficiency, eliminate waste in real-time and improve production flow. IoT sensors, AI-powered analytics and digital twins provide deeper visibility into operations, enabling continuous optimisation.

Similarly, Agile Manufacturing 4.0 emphasises flexibility, rapid customisation and responsiveness to market changes. AI-driven demand forecasting, additive manufacturing (3D printing) and smart supply chains allow manufacturers to adapt production lines dynamically to customer needs.

So, before deep-diving into the Industry 4.0-enabled processes, let's first understand the traditional Lean and Agile manufacturing principles. Then, we will discuss how Industry 4.0 technologies have enhanced these methodologies to suit modern factories.

3.2 TRADITIONAL LEAN MANUFACTURING PROCESS

Lean Manufacturing, originally developed by Toyota through the Toyota Production System (TPS), is a structured approach aimed at maximising customer value while minimising resource use by eliminating non-value-adding activities. Its core principles focus on waste reduction, referred to as *Muda*, to enhance efficiency and productivity. Lean targets the elimination of seven key types of waste in production (refer to Figure 3.1): overproduction, waiting, unnecessary material transport, excessive worker motion, overprocessing, idle inventory and defects.

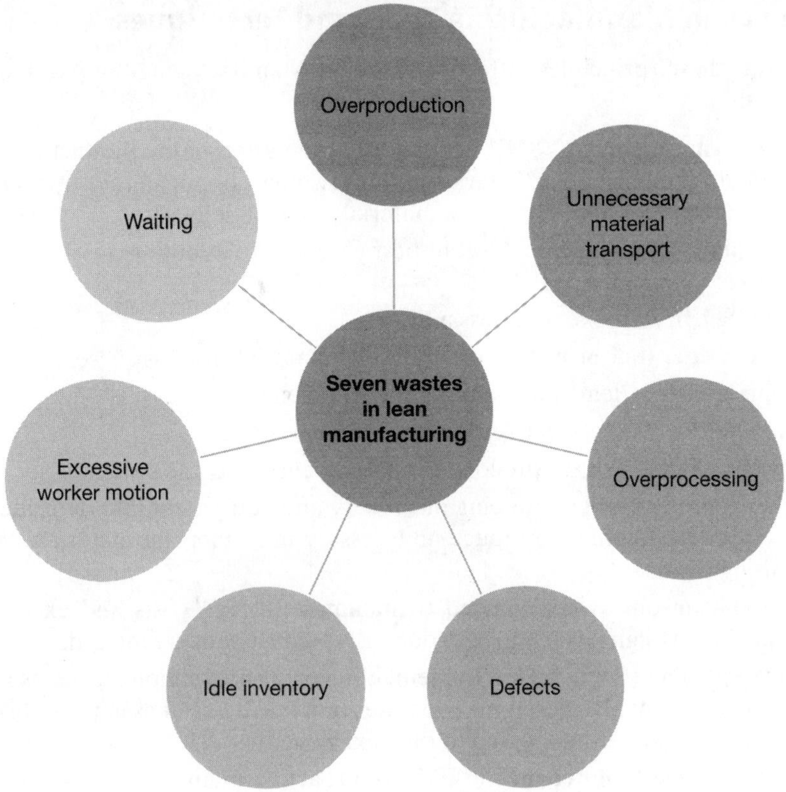

Figure 3.1 Seven wastes that lean manufacturing tries to eliminate

Lean manufacturing is guided by fundamental principles that help organisations systematically eliminate waste while improving productivity and quality:

- **Defining Value from the Customer's Perspective:** Understanding the features or services the customers are paying for clarifies the definition of value from the customer's viewpoint. This ensures that production efforts align with actual demand.
- **Mapping the Value Stream:** Documenting each step in the production process helps distinguish between value-adding and non-value-adding activities. For example, workers waiting for materials to arrive is an instance of waste that can be minimised.
- **Ensuring Continuous Flow:** Designing smooth, uninterrupted workflows reduces bottlenecks. Using smaller batch sizes and organising workstations sequentially improves overall process efficiency.
- **Implementing a Pull-Based System:** Adopting JIT production ensures products are created only when needed, reducing overproduction and excess inventory.
- **Striving for Continuous Improvement (Kaizen):** Promoting a culture of small, incremental changes helps refine processes over time, enhancing efficiency and quality.

3.2.1 Key Lean Manufacturing Tools and Techniques

To implement these principles effectively, Lean Manufacturing employs several key methodologies:

- **Value Stream Mapping (VSM):** A visual representation of the production process that helps identify inefficiencies, delays and improvement opportunities by mapping material and information flows from raw materials to final delivery.
- **5S Workplace Organisation:** A set of best practices to maintain an efficient, organised workspace:
    - **Sort (Seiri):** Remove unnecessary items.
    - **Set in Order (Seiton):** Organise tools and materials for easy access.
    - **Shine (Seiso):** Clean and maintain the workspace.
    - **Standardise (Seiketsu):** Create consistent procedures.
    - **Sustain (Shitsuke):** Ensure long-term adherence to these practices.
- **Cellular Manufacturing:** Grouping machinery and equipment into dedicated 'cells' for specific product families to reduce unnecessary movement of materials and workers, minimising waste.
- **JIT:** Producing only when needed to minimise storage costs and excess inventory. Kanban systems signal when production or replenishment is required.
- **Mistake-Proofing (Poka-Yoke):** Implementing error-prevention mechanisms to reduce defects. This can involve designing processes that make mistakes impossible or ensuring defects are caught before moving to the next stage.
- **Total Productive Maintenance (TPM):** A proactive maintenance strategy that minimises equipment downtime through scheduled inspections, reducing waste from unexpected failures.
- **Standardised Work:** Documenting the most effective way to perform tasks ensures consistency and maximises workforce productivity.
- **Single-Minute Exchange of Die (SMED):** A technique for reducing machine changeover times to enhance production flexibility. It focuses on separating and optimising internal and external setup tasks to minimise downtime and accelerate transitions.
- **Autonomation (Jidoka):** Enabling machines to detect defects and stop production automatically, preventing faulty products from advancing in the process. This ensures quality is built into production rather than relying on inspection at the end.
- **Andon System:** A real-time visual alert mechanism that helps workers and supervisors address production issues immediately, reducing defects and waste.
- **Continuous Improvement (Kaizen):** Involving employees at all levels to contribute small, ongoing process improvements that collectively lead to greater efficiency, quality and operational excellence.

By implementing these tools and practices, a stable and high-volume production line can be run effectively. With the advent of Industry 4.0 technologies, these tools and practices became sharper and more adaptive.

3.3 INDUSTRY 4.0-ENABLED LEAN MANUFACTURING PROCESS

Industry 4.0 transforms an industrial landscape by fully digitalising operations and seamlessly integrating engineering, manufacturing and supply chains with enterprise business systems and processes. This connectivity enables flexible, predictive business operations, with autonomous machines communicating both with each other and with business systems.

Real-time or near-real-time data is continuously gathered from machines, vehicles and personnel using IoT sensors and smart devices. Artificial intelligence enriches this data, integrating it with business information to generate valuable insights across sales, service, engineering, manufacturing and logistics—ultimately driving improved business performance.

As Industry 4.0 technologies evolve, traditional Lean manufacturing principles have also adapts, giving rise to **Lean 4.0**. This next-generation approach maintains the core Lean philosophy while integrating digital tools, enhancing efficiency, reducing waste and maximising the value of advanced automation and data-driven insights. Key enhancements to traditional Lean manufacturing enabled by Industry 4.0 include:

- **Digital Value Stream Mapping (DVSM):** Building upon traditional Value Stream Mapping, DVSM incorporates digital twins, IoT sensors and AI-driven analytics. A digital twin of the production process allows for simulation and optimisation of improvements. IoT sensors provide real-time tracking of material flow, machine performance and bottlenecks. AI-powered analytics further enhance efficiency by predicting inefficiencies before they occur, enabling proactive decision-making and continuous optimisation.
- **Smart JIT and Inventory Control:** Smart JIT enhances traditional JIT practices by integrating Industry 4.0 technologies to improve efficiency and responsiveness. Automated replenishment using RFID tags and smart bins ensures seamless inventory management. AI-driven demand forecasting dynamically adjusts JIT deliveries, while blockchain enhances supply chain transparency, preventing delays. Advanced implementations can even predict component shortages and automatically adjust supplier orders, creating a more adaptive and resilient supply chain.
- **Predictive Maintenance:** An evolution of TPM is Predictive Maintenance. This utilises IoT sensors to monitor equipment in real-time to detect early signs of wear. AI-powered predictive models forecast potential failures before they happen, minimising unplanned downtime. Additionally, Augmented Reality (AR) provides technicians with real-time guidance for repairs, improving maintenance precision and efficiency.
- **Digital Poka-Yoke (Error Prevention):** Traditional mistake-proofing is enhanced through Industry 4.0 technologies that enable real-time error detection and correction. Computer vision and AI-powered inspection systems instantly identify defects, while smart assist systems guide workers to correct errors as they occur. Self-adjusting machines further enhance quality control by automatically compensating for deviations, reducing human errors and ensuring process consistency.
- **Digital Kaizen (Continuous Improvement):** Industry 4.0 enhances Kaizen by integrating AI-driven analytics that continuously identify process optimisation opportunities. Smart wearables, such as AR glasses, provide workers with real-time performance

insights, encouraging a culture of continuous improvement. Automated A/B testing using digital twins allows for safe, data-driven experimentation with process changes, ensuring rapid and effective refinements. You may be aware that A/B testing is the comparison of two versions of the same product to check which one is performing better. So, instead of investing in multiple physical versions of the same product, it is always beneficial to use digital twins to compare performances.

- **Real-Time Andon and Performance Monitoring:** Industry 4.0 upgrades traditional Andon systems with smart dashboards that provide live visibility into Overall Equipment Effectiveness (OEE) and real-time performance tracking. AI-driven alerts notify managers of deviations via mobile apps, enabling quick response. Automated root cause analysis, leveraging historical data and real-time OEE tracking, helps identify inefficiencies and optimise machine utilisation, ultimately driving higher productivity and efficiency.

By integrating these Industry 4.0 technologies with Lean principles, manufacturers can create highly efficient, data-driven and adaptable production environments that minimise waste, enhance flexibility and optimise performance in an increasingly complex industrial landscape.

3.3.1 Case Study on Lean Manufacturing in Industry 4.0

A leading German manufacturer successfully transformed its operations by merging Lean Manufacturing principles with Industry 4.0 technologies, creating a Smart Lean Factory capable of high-mix, low-volume production with minimal waste. Facing challenges in productivity, waste reduction and demand adaptability, the company implemented a fully connected, self-optimising production system with notable results:

- As their traditional VSM lacked real-time adaptability, the company deployed IoT sensors on machines, combined with digital twin simulations to test process changes virtually. This **digital VSM** approach led to a 20% reduction in lead time through dynamic material flow optimisation.
- Earlier manual Kanban systems struggled with variable demand. The organisation enabled RFID-tracked smart bins for automated replenishment and AI-driven demand forecasting for **real-time JIT** adjustments. The result was 30% lower inventory costs while maintaining 99% on-time delivery.
- Unplanned downtime was a major issue faced by the organisation. By installing vibration and thermal sensors on critical machinery and using **AI-based failure prediction**, the factory achieved a 40% reduction in unplanned downtime and extended machine lifespan by 15%.
- Human inspections missed subtle defects, so the company introduced AI-powered vision systems for 100% automated quality checks and self-adjusting CNC machines to correct tool wear in real-time. This **digital Poka-Yoke** method slashed the defect rate from 500 Parts Per Million (PPM) to under 50 PPM.
- Slow manual analysis of improvement ideas was replaced with AI-powered Overall Equipment Effectiveness (OEE) dashboards and AR tablets for worker feedback. Implementation of **Real-Time AI Analytics** prioritised high-impact changes, leading to 50% faster Kaizen implementation and a 10% productivity boost.

Thus, through efficient integration of Lean methodologies with Industry 4.0 innovations, the manufacturer achieved higher efficiency, reliability and quality, proving that the future of smart manufacturing lies in the synergy of human expertise, data-driven insights and automation.

However, you may have observed that Lean practices primarily focus on enhancing efficiency in stable production environments where fluctuations in customer demand or market conditions are minimal. Thus, in dynamic situations where adaptability is crucial, Agile manufacturing is the more suitable approach.

3.4 TRADITIONAL AGILE MANUFACTURING PROCESS

The Agile manufacturing approach differs from traditional mass production by shifting the focus from optimising standardised product output to embedding flexibility throughout the production process to accommodate change. Its core principles revolve around mass customisation, rapid responsiveness and customer-driven production. Unlike lean manufacturing, which primarily targets waste reduction, agile manufacturing prioritises adaptability and responsiveness, with waste reduction occurring as a secondary benefit.

At its core, agile manufacturing emphasises customer-driven production, where decisions are made based on real-time market demands rather than long-term forecasts. It promotes modular and reconfigurable production systems that can swiftly adapt to evolving product requirements or market shifts catering to high-mix and low-volume production (meaning that the batch size can be as small as one with high variance in product specification). A key aspect is decentralised decision-making, empowering frontline teams to make quick adjustments without bureaucratic delays. Additionally, agile manufacturing fosters collaborative supply chains, ensuring transparent information sharing between manufacturers and suppliers to enhance responsiveness.

Another crucial element is the focus on rapid prototyping and iterative development, enabling faster product refinement and reduced time-to-market. This approach requires a cultural shift, encouraging experimentation and adaptability while remaining attuned to market signals. Human expertise and teamwork are highly valued, with cross-functional collaboration playing a vital role in dynamic problem-solving. These principles collectively enable manufacturing systems to remain competitive in volatile markets while maintaining efficiency and quality standards.

Agile manufacturing has gained widespread adoption in unpredictable business environments where increasing customisation demands and supply chain disruptions necessitate shorter product lifecycles and flexible operations. Key tools and techniques used in traditional Agile manufacturing include:

- **Flexible Manufacturing Systems (FMS):** These systems allow a single production line to manufacture multiple products without significant retooling. Technologies such as Computer Numerically Controlled (CNC) machines and other programmable tools enhance FMS flexibility by enabling customisation in product types, volumes and operational sequences. This is also called Programmable Automation, which we will study in Chapter 5 in detail.
- **Postponement or Late-Stage Customisation:** This technique balances bulk production with mass customisation. Products are designed with generic base components that can

be produced in large quantities, while custom features – such as colour, packaging, or software versions are finalised closer to shipment. This improves inventory management, supply chain efficiency and production utilisation.

- **Cross-Functional Teams:** Small teams comprising members from various departments, such as engineering, operations and marketing collaborate to break down silos between design, production and logistics, enabling faster decision-making.
- **Real-Time Production Scheduling:** Unlike traditional static planning, this method dynamically adjusts production workflows based on real-time data from IoT devices, machine sensors and market fluctuations. By leveraging accurate and up-to-date information, manufacturers can optimise resource utilisation, reduce uncertainty and improve operational efficiency.
- **Quick-Change Tooling and SMED:** This technique, also used in Lean manufacturing, minimises changeover time between product switches, enhancing the agility of the production process.

By integrating these strategies, Agile manufacturing enables businesses to remain resilient in fast-changing markets while delivering high-quality, customised products efficiently.

3.5 INDUSTRY 4.0-ENABLED AGILE MANUFACTURING PROCESS

Manufacturing companies face a multitude of challenges, from boosting productivity and maintaining top-tier quality to meeting the rising demand for customised products in an increasingly dynamic market. Fluctuating customer preferences and complex sourcing conditions add further pressure. Additionally, global competition and cost constraints require companies to refine their operations to stay competitive and profitable continuously.

Limited visibility and agility hinder businesses from anticipating, predicting and effectively responding to disruptions. Meanwhile, evolving regulations – particularly in sustainability and life sciences, demand a fundamental rethinking of business processes. Simultaneously, rapid advancements in artificial intelligence, e-mobility, autonomous driving and hyper-automation are making digital transformation a necessity rather than an option.

As discussed earlier, Agile manufacturing prioritises flexibility, rapid adaptation and customer-driven production. Industry 4.0 introduces advanced digital tools that enhance and accelerate this transformation. Together, they enable intelligent, responsive factories that can swiftly adjust to shifting demands, customisation requirements and supply chain disruptions – enhancing overall efficiency, resilience and competitiveness. Let's understand how Industry 4.0 technologies enhanced traditional Agile manufacturing:

- **Smart, Reconfigurable FMS:** Traditional Flexible Manufacturing Systems (FMS) require manual reconfiguration for product variations, with considerable down time of production line. With Industry 4.0, AI-driven robotic arms, smart sensors and collaborative robots (cobots) can be instantly reprogrammed with minimum or zero downtime. Digital twins simulate new setups before implementation, while AI-driven scheduling dynamically adjusts production in real-time, enabling assembly lines to switch between product variations within minutes. This is also called Flexible Automation as we will read in Chapter 5.

- **Mass Customisation with AI and Additive Manufacturing:** AI-powered design tools generate customised product variants, while 3D printing (Additive Manufacturing) enables low-volume, on-demand production. IoT-enabled order tracking ensures real-time personalisation, making mass customisation scalable and efficient.
- **Real-Time Demand Sensing and Dynamic Scheduling:** AI-driven demand forecasting analyses external factors such as social media trends, weather conditions and market shifts to predict demand more accurately. Blockchain-enabled smart contracts automatically adjust supplier orders, while edge computing processes real-time data for instantaneous decision-making.
- **Self-Optimising Factories through Autonomous Cyber-Physical Systems:** Industry 4.0 introduces autonomous cyber-physical systems, autonomous robots and autonomous mobile robots (AMRs), which optimise workflows without human intervention. Machine learning enhances decision-making and IoT-driven closed-loop feedback systems ensure continuous process improvements.
- **Digital Supply Chain and On-Demand Logistics:** Instead of manual supplier coordination, smart supply chains leverage IoT-tracked shipments for real-time visibility. Autonomous drones, AMRs and AGVs enable just-in-time material handling, while digital marketplaces connect manufacturers with available production capacity, improving responsiveness and cost efficiency.
- **Augmented Workforce and Human-Machine Collaboration:** With Industry 4.0, the workforce is enhanced through AR/VR-assisted work instructions, exoskeletons that support physical tasks and cobots that assist in flexible assembly. AI-powered chatbots provide instant troubleshooting, ensuring workers can focus on high-value tasks while machines handle repetitive or complex operations.

When enhanced with Industry 4.0, agile manufacturing transforms into a highly adaptive and intelligent system, seamlessly integrating efficiency, customisation and swift responsiveness to market changes as core capabilities.

Lean and Agile manufacturing process each has their relative strengths and thus can be chosen based on the key focus areas of the organisation. A quick comparison of these two popular manufacturing methodologies is mentioned in Table 3.1.

DID YOU KNOW?

Exoskeletons are wearable robotic systems designed to enhance human capabilities, such as strength, endurance and mobility. Widely used in manufacturing, healthcare and military applications, they help reduce physical strain by supporting tasks like heavy lifting and repetitive movements. In industrial settings, exoskeletons improve worker productivity while lowering the risk of injuries such as back strain. Passive models rely on springs or counterweights for support, whereas active versions use motors and sensors for adaptive assistance. In healthcare, exoskeletons aid mobility for individuals with paralysis or muscle weakness, promoting rehabilitation. As ergonomic workplace solutions, they enhance safety and efficiency, particularly in the automotive and logistics industries. Innovations in lightweight materials and AI are making them more intuitive and accessible.

Table 3.1 Comparison of Agile vs. Lean Manufacturing in Industry 4.0

Aspect	Agile Manufacturing (Industry 4.0)	Lean Manufacturing (Industry 4.0)
Focus	Flexibility, customisation	Waste reduction, efficiency
Key Technologies	Digital twins, 3D printing, AI scheduling	IoT for TPM, AI for waste detection
Production Style	High-mix, low-volume	High-volume, standardised
Supply Chain	Dynamic, on-demand	JIT with smart logistics
Best For	Fast-changing markets (e.g., fashion, semiconductors)	Stable, high-efficiency needs (e.g., automotive)

3.5.1 Case Study on Agile Manufacturing in Industry 4.0

A global leader in industrial automation faced growing challenges due to increasing product customisation, shorter product lifecycles and an unpredictable supply chain. Operating a facility that manufactures over 1,000 variants of industrial controllers and PLCs, the company needed to sustain its exceptional 99.99885% quality rate while significantly enhancing production flexibility. To address these demands, it implemented an Agile 4.0 strategy, blending Agile manufacturing principles with Industry 4.0 technologies and transforming the facility into one of the world's most advanced smart factories.

- The transformation began with an **FMS architecture**, enabling each workstation to be reconfigured, both physically and digitally, within hours instead of days. Cyber-physical systems were introduced, incorporating IoT sensors, machine vision and edge computing for autonomous decision-making. This allowed seamless transitions between product configurations with minimal manual intervention. A continuously updated digital twin provided real-time production insights, allowing engineers to simulate and validate scenarios before implementation, cutting changeover time by 75%.
- A key component of Agile 4.0 was **AI-driven dynamic scheduling**, replacing traditional planning systems with a self-learning algorithm. This system continuously optimises production sequences based on real-time machine availability, material supply and shifting customer priorities. It could automatically reprioritise orders in response to urgent requests or supply chain disruptions, ensuring optimal throughput. Additionally, AR guided operators through changeovers and exception handling, reducing training time for new product variants by 60%.
- **Supplier integration** was crucial in achieving true agility. Blockchain-based smart contracts establish a transparent, responsive supply network. When sensors detected potential material shortages or quality deviations, the system autonomously triggered alternative sourcing strategies or process adjustments. This capability proved invaluable during the semiconductor shortage, allowing the plant to maintain 98% on-time delivery while competitors struggled with severe delays.

The results were transformative. Production flexibility improved by 40%, making batch-size manufacturing as small as one unit profitable. Despite rising product complexity, quality metrics improved, with AI-powered visual inspection reducing escape defects by 90%. Most notably, the time from product design to full-scale production shrank from weeks to days, giving the company a competitive edge in the rapidly evolving industrial automation market.

Agile 4.0 is more than just technological advancement. It requires a fundamental shift in manufacturing paradigms. Success stemmed from combining advanced digital technologies with organisational agility, workforce empowerment and supply chain innovation. The factory continues to push the boundaries of smart manufacturing, recently adopting generative AI for autonomous process optimisation and exploring quantum computing for ultra-complex scheduling.

We have examined how Industry 4.0 can independently enhance Lean and Agile methodologies to transform smart production. Our findings suggest that combining Lean 4.0's precision with Agile 4.0's flexibility creates a manufacturing ecosystem that is both highly efficient and future-proof. Lean 4.0 emphasises data-driven waste elimination, while Agile 4.0 enables dynamic, customer-centric production. This hybrid approach optimises efficiency while ensuring rapid adaptability to market fluctuations. For example, in a digital supply chain, Lean 4.0 focuses on minimising inventory through JIT practices and strong supplier partnerships. Meanwhile, Agile 4.0 enhances this setup with blockchain-tracked on-demand sourcing and distributed manufacturing. While Lean prioritises inventory reduction, Agile ensures swift responses, such as activating alternative suppliers in case of disruptions.

3.6 OTHER NOTABLE PROCESSES

Beyond Lean and Agile, several other methodologies align with Industry 4.0, further enhancing efficiency, flexibility and innovation. These approaches leverage advanced technologies such as AI, IoT and digital twins to optimise manufacturing processes, improve quality and increase resilience in an increasingly complex industrial landscape.

- **Smart Six Sigma (Quality 4.0)** is a data-driven approach to reducing defects through statistical methods, now enhanced with AI and IoT. Industry 4.0 enables AI-powered defect detection using computer vision and vibration analysis, predictive analytics to prevent quality issues before they arise and digital twin simulations for process optimisation. For example, GE Aviation utilises AI-driven Six Sigma to enhance the precision of jet engine manufacturing.
- **Theory of Constraints (TOC)** 4.0 focuses on identifying and eliminating bottlenecks in production. With Industry 4.0, manufacturers can use IoT sensors for real-time bottleneck detection, AI-driven dynamic scheduling to optimise constrained resources and automated buffer management with smart logistics. Intel, for instance, applies TOC alongside real-time data analytics to optimise semiconductor fabrication throughput.
- **Just-in-Sequence (JIS)** Manufacturing builds on JIT by delivering parts in the exact assembly order required. Industry 4.0 enhances JIS with RFID and blockchain-tracked components for accurate sequencing, autonomous mobile robots (AMRs) for precise

material handling and AI-driven sequencing algorithms for mixed-model production. BMW applies JIS with smart logistics to efficiently assemble customised vehicles.

- **Servitisation (Product-as-a-Service)** shifts the focus from selling products to delivering outcomes, such as 'pay-per-use' business models. Industry 4.0 enables this transformation through technologies such as IoT-based performance monitoring, predictive analytics for service optimisation, and blockchain for transparent usage tracking. A well-known example is Rolls-Royce's 'Power-by-the-Hour' program, which provides comprehensive engine and accessory maintenance services at a fixed cost per flying hour. Another example is Philips, which leases medical imaging equipment with IoT-based predictive maintenance.
- **Resilient Manufacturing (Anti-Fragile Operations)** focuses on building systems that adapt and improve under disruption. Industry 4.0 enables resilience through digital twins for risk simulation in supply chains, AI-driven alternative sourcing strategies and cobots (collaborative robots) that allow for rapid reconfiguration during shortages. HP, for example, leverages AI to reroute 3D printer production in response to semiconductor shortages.

Each methodology addresses different aspects of modern manufacturing, making it essential to choose the right approach based on business priorities, as depicted in Table 3.2:

Table 3.2 Mapping of key objectives to the preferred methodologies

Key Objective	Methodology
Waste reduction	Lean 4.0
Customisation	Agile 4.0
Zero defects	Smart Six Sigma
Bottleneck removal	TOC 4.0
Autonomous operations	CPPS
Resilience against disruptions	Resilient Manufacturing

By selectively combining these methodologies with Industry 4.0 technologies, manufacturers can produce customised products at scale, react instantly to market changes and reduce waste while staying adaptable.

3.7 CONCLUSION

The evolution of Agile and Lean 4.0 revolves around the synergy of AI, automation and human expertise. As manufacturing shifts toward a more digital, decentralised and demand-driven landscape, organisations that seamlessly integrate these principles will gain a competitive advantage. Companies that adopt hyper-personalisation, real-time adaptability and predictive intelligence will not only boost efficiency but also safeguard their operations against future uncertainties.

Growing consumer demand for personalised products will push manufacturers to refine batch-size-one production. Leveraging AI-driven demand forecasting, digital twins and flexible manufacturing systems, businesses will achieve mass customisation while upholding Lean efficiency.

With increasing factory complexity, traditional computing will struggle to process vast amounts of data and optimisation challenges. In Agile 4.0 environments, quantum algorithms could revolutionise global supply chain coordination and multi-variable process optimisation, reducing computation times from hours or days to mere seconds.

In the years ahead, smart factories will continue to evolve into self-optimising, ultra-flexible and highly interconnected ecosystems, transforming how products are designed, manufactured and delivered.

SUMMARY

- Traditional Lean methodology aims to eliminate waste (Muda), improve efficiency and enhance value delivery in manufacturing.
- Key Lean principles include continuous improvement (Kaizen), standardised workflows and demand-driven production.
- A few important Lean tools are listed below:
  - **Just-in-Time (JIT) Production:** Reduces inventory and ensures materials arrive exactly when needed.
  - **Value Stream Mapping (VSM):** Identifies inefficiencies in production processes to streamline operations.
  - **Total Productive Maintenance (TPM):** Focuses on proactive maintenance to minimise downtime and improve equipment efficiency.
- Traditional Agile Structure prioritises flexibility, adaptability and rapid response to changing customer demands.
- Key Agile principles are iterative improvements, cross-functional collaboration and real-time decision-making.
- A few important Agile manufacturing tools are as follows:
  - **Flexible Manufacturing Systems (FMS):** Enables production lines to manufacture multiple products with minimal retooling using CNC machines, AGVs and programmable tools for greater customisation.
  - **Real-Time Production Scheduling:** Uses IoT and machine sensor data to dynamically adjust workflows, optimising resource use and improving responsiveness to market changes.
  - **Quick-Change Tooling and SMED:** Reduces changeover time between product variations, enhancing production flexibility and efficiency.
- Lean 4.0 Enhancements include the following:
  - **AI-Driven Predictive Maintenance:** Advanced machine learning algorithms anticipate failures and optimise equipment upkeep.

- **Digital Twins for Lean Optimisation:** Simulates production processes in real-time to prevent inefficiencies.
- **Blockchain in Lean Supply Chains:** Improves transparency, reduces fraud and enhances supplier collaboration through smart contracts.

- Agile 4.0 Enhancements include the following:
  - **Agile 4.0 and Smart Factories:** Uses AI, IoT and automation to enable real-time, autonomous decision-making.
  - **AI-Driven Dynamic Scheduling:** Self-learning systems optimise production sequences based on live data.
  - **Decentralised and Distributed Manufacturing:** Enables rapid response to demand fluctuations and supply chain disruptions.

EXERCISE

Multiple Choice Questions (1 Mark)

1. What is the primary goal of Lean manufacturing?
(a) Maximising production speed
(b) Reducing waste and improving efficiency
(c) Increasing product variety
(d) Enhancing bureaucratic processes

2. Which tool is commonly used in Lean manufacturing to visualise and optimise process flows?
(a) Scrum Board
(b) Value Stream Mapping
(c) Gantt Chart
(d) Waterfall Model

3. What is the key objective of Single-Minute Exchange of Dies (SMED)?
(a) Reducing changeover time between production runs
(b) Increasing work-in-progress inventory
(c) Improving software testing processes
(d) Extending machine maintenance intervals

4. In Agile 4.0, real-time decision-making is enabled through:
(a) Batch processing
(b) Digital Twins and IoT
(c) Increased manual inspections
(d) Standardised rigid processes

5. What role does Blockchain play in Agile 4.0 manufacturing?
   (a) Increasing manual record-keeping
   (b) Improving transparency in supply chains
   (c) Replacing all human decision-making
   (d) Reducing the need for AI systems
6. Which Agile manufacturing tool is primarily used for visualising workflow and limiting work-in-progress?
   (a) Scrum
   (b) Kanban
   (c) JIT
   (d) TPM
7. What is the benefit of cross-functional teams in Agile manufacturing?
   (a) Encourages working in isolated silos
   (b) Enhances collaboration and reduces handover delays
   (c) Eliminates the need for skilled labour
   (d) Increases documentation requirements
8. How does Just-in-Time (JIT) reduce waste in manufacturing?
   (a) By eliminating the need for production scheduling
   (b) By reducing inventory levels and only producing what is needed
   (c) By increasing product variety
   (d) By prioritising large batch production
9. What is the main advantage of using Digital Twins in Agile manufacturing?
   (a) Reducing labour costs by replacing workers
   (b) Enabling real-time simulation and optimisation of production processes
   (c) Increasing reliance on physical prototyping
   (d) Standardising manual quality inspections
10. Which manufacturing approach is best for handling volatile market demand?
   (a) Traditional mass production
   (b) Agile 4.0
   (c) Fixed production lines
   (d) Standardised long-term production planning
11. What is the primary benefit of using AI-driven dynamic scheduling in manufacturing?
   (a) Reducing human involvement in decision-making
   (b) Increasing rigidity in production planning
   (c) Continuously optimising production sequences based on real-time data
   (d) Eliminating the need for automation

12. What is the key difference between Traditional Lean and Lean 4.0?
    - (a) Traditional Lean relies on manual processes, while Lean 4.0 integrates digital technologies
    - (b) Lean 4.0 focuses only on high-volume production
    - (c) Traditional Lean eliminates customer involvement, while Lean 4.0 ignores efficiency
    - (d) Traditional Lean uses AI extensively
13. What is one advantage of using Flexible Manufacturing Systems (FMS) in Agile production?
    - (a) Reducing machine downtime
    - (b) Allowing multiple product types to be produced without significant retooling
    - (c) Standardising large batch production
    - (d) Eliminating automation from production processes
14. Why is supplier integration important in Agile 4.0?
    - (a) It enables real-time collaboration and dynamic supply chain adjustments
    - (b) It increases production lead times
    - (c) It limits manufacturing flexibility
    - (d) It eliminates the need for digital tracking
15. A factory producing industrial controllers needs to handle over 1,000 product variants while maintaining high quality. Which approach would best support this?
    - (a) Traditional mass production
    - (b) Agile 4.0 with AI-driven scheduling and FMS
    - (c) Large batch manufacturing with fixed setups
    - (d) Manual process optimisation
16. A factory wants to reduce changeover time while increasing production flexibility. Which tool should it adopt?
    - (a) SMED
    - (b) Traditional mass production
    - (c) Increased batch sizes
    - (d) Static scheduling
17. A company manufacturing high-end electronics wants to personalise products based on customer preferences while maintaining cost efficiency. Which strategy should they use?
    - (a) Large-scale mass production
    - (b) Postponement or late-stage customisation
    - (c) Fixed production scheduling
    - (d) Increasing safety stock

18. A company is experiencing frequent supply chain disruptions due to unreliable suppliers. What approach can help them mitigate these risks?
    - **(a)** Blockchain-based smart contracts for transparent supplier tracking
    - **(b)** Increasing batch sizes to compensate for shortages
    - **(c)** Reducing supplier diversity
    - **(d)** Eliminating digital tracking systems
19. A manufacturing plant wants to reduce operator errors in assembly processes. Which Agile 4.0 enhancement can improve worker efficiency?
    - **(a)** Augmented Reality (AR) for real-time guidance
    - **(b)** Increasing manual documentation
    - **(c)** Eliminating cross-functional teams
    - **(d)** Reducing automation in workflows
20. A company wants to ensure that its production system remains efficient while also being adaptable to future market changes. What should they implement?
    - **(a)** A hybrid Lean 4.0 and Agile 4.0 approach
    - **(b)** Traditional rigid production lines
    - **(c)** Large-scale inventory buildup
    - **(d)** Increased reliance on fixed schedules

Short Answer Type Questions (2 Marks)

1. What are the key principles of Lean manufacturing?
2. Define Agile 4.0 and explain its significance.
3. How does Kanban improve production workflow?
4. How does AI-driven scheduling improve production efficiency?
5. Describe how Lean 4.0 integrates with Industry 4.0 technologies.
6. What is 'Postponement' in Agile manufacturing and why is it beneficial?
7. How does Just-in-Time (JIT) minimise waste in production?
8. Explain the difference between Traditional Lean and Lean 4.0.
9. How does Agile 4.0 address volatile market demands?
10. What is SMED and how does it benefit manufacturing flexibility?

Long Answer Type Questions (5 Marks)

1. Explain the role of Agile 4.0 in smart factories and its key technological components.
2. Describe how FMS (Flexible Manufacturing System) supports Agile 4.0.

3. How does Agile manufacturing improve customer satisfaction compared to traditional methods?
4. Discuss the benefits of combining Lean 4.0 and Agile 4.0 in manufacturing.
5. How does Agile manufacturing handle supply chain disruptions?
6. What are the future trends in Lean and Agile 4.0 manufacturing?
7. How do Digital Twins transform Lean and Agile manufacturing processes?
8. Discuss the impact of AI-driven dynamic scheduling on modern manufacturing.
9. Explain the role of IoT in Lean 4.0.
10. Explain how Lean 4.0 and Agile 4.0 cater to different focus areas for a manufacturer.

CHAPTER 4

Industrial Internet of Things (IIoT)

OBJECTIVES

We discussed that one of the key tenets of Industry 4.0 is the interconnection of devices. The Industrial Internet of Things (IIoT) serves as the backbone of modern smart manufacturing, enabling digital transformation across industries through seamless integration of physical assets with digital intelligence. In an IIoT ecosystem, the connected industrial edge devices gather data from manufacturing environments, transmit it through network nodes, and process it using cloud-based analytics to generate real-time, actionable insights. This ability to interpret data from diverse industrial assets allows companies to identify inefficiencies, predict maintenance needs, and respond proactively to shifts in demand or production constraints.

By the end of this chapter, you should be able to:

- Understand the architecture of IIoT systems
- Design and setup the IIoT platforms for an enterprise
- Articulate the applications of IIoT in different industry domains

4.1 INTRODUCTION

Smart products and connected products are not new. Since the 1980s when industrial automation came into focus, smart products have evolved. These products have simple embedded devices and can perform various tasks independently. The next level of evolution was the connected products. The connected products can transmit data to each other but do not necessarily work in a symbiotic environment to create intelligence out of the data.

When the products seamlessly connect over a global network to ensure data flow, collaborative work and intelligent decision-making especially in an industrial setup, the concept of the Industrial Internet of Things (IIoT) comes into the picture.

As discussed in Chapter 1, IIoT is a fundamental component of Industry 4.0, to the extent that many believe Industry 4.0 wouldn't exist without it. The wider scope of Industry 4.0 includes IIoT, digitalisation and corporate sustainability. The scope of IIoT extends to data detection, transmission, computation, processing and specialised intelligent applications.

4.2 WHAT IS IoT AND IIoT?

4.2.1 IoT

The term Internet of Things or IoT was invented in 1999 by Kevin Ashton, a British technology pioneer.

The global network infrastructure based on standard and interoperable communication protocol through which the physical and virtual 'things' are interconnected and seamlessly integrated to work symbiotically is called the Internet of Things (IoT).

Any physical or virtual entity that can be *uniquely identified* and is being integrated into communication networks is termed a *'thing'* in the context of IoT. In the IoT environment, the internet connects **things to things** as well as **things to humans** for a regular flow of data that can be analysed to make intelligent decisions.

In today's world, IoT solutions have become ubiquitous and can be found in all application domains like healthcare, manufacturing, transportation and logistics, smart cities, agriculture and many more. Let's take a simple example of the connected cars. Most modern cars are connected over the internet to a central system through which the cars can be tracked in real-time and simple operations like locking the doors or switching off the headlight remotely from home are possible. The application of cutting-edge technologies, such as Artificial Intelligence (AI), Machine Learning (ML), and Big Data Analytics, aims to make IoT devices autonomous, enabling them to process information and data to make real-time decisions without human intervention. IoT goes beyond just data collection, system connectivity, storage and analytics. It plays a crucial role in modernising operations by linking both legacy and smart devices, as well as machines, to the internet. Furthermore, it bridges the gap between Information Technology (IT) and Operational Technology (OT) teams by providing a consolidated view of systems and data.

Two major segments of IoT are as follows:

- **B2C:** Business to Consumer IoT. This segment refers to the IoT solutions that provide service to end-users or individuals directly. A few examples are smart wearables, smart homes, personal tracking devices, etc.
- **B2B:** Business to Business IoT. The domains that constitute this type of IoT are commercial, industrial and infrastructure. The primary customers are commercial establishments or industries. A few examples are smart factories, smart supply chains, smart parking, etc.

Though the underlying principles of IoT are the same for B2C and B2B segments, the scale, complexity and robustness of the architecture are different in these cases.

4.2.2 IIoT

The application, opportunity and impact of IoT in the context of the industry domain are so high that it called for a separate term IIoT to describe the principles. This is a subset of IoT and focuses mainly on automation and efficiency in a connected organisational ecosystem rather than focusing on individual user needs.

The interconnected devices should be able to gather data, transmit it to the central system for analysis, use Artificial Intelligence and Machine Learning on the data to generate actions and send the action signals across other connected devices in the network with an end goal of improving efficiency.

The various components of IIoT like, sensors/actuators, software and applications, middleware and storage systems need to seamlessly connect through networking technologies to form a connected ecosystem (refer to Figure 4.1). This connected and distributed nature of IIoT systems enables enterprises to decentralise their operations and warehouses which can now be set up closer to the markets. Additionally, through the interconnected devices more real-time responses and reactions are feasible. Therefore, this smart ecosystem offers measurable advantages such as enhanced enterprise availability and maintainability, faster time-to-market, increased productivity, greater transparency and maximised operational efficiency. IIoT strives to establish self-sufficient, sturdy and secure links and data transfers between devices and real-world applications. It also aids in achieving Machine-to-Machine (M2M) communication and interconnectivity, as well as incorporating intelligence into devices. As a result, devices are capable of processing information and data and making autonomous, human-like intelligent decisions in real-time without the need for human participation or intervention.

IIoT is transforming every sector of the industrial world by significantly altering the processes at each stage of a product's life cycle, from its design and production to its delivery, sale and maintenance. In the next section, we will see what a reference architecture for the IIoT system should look like.

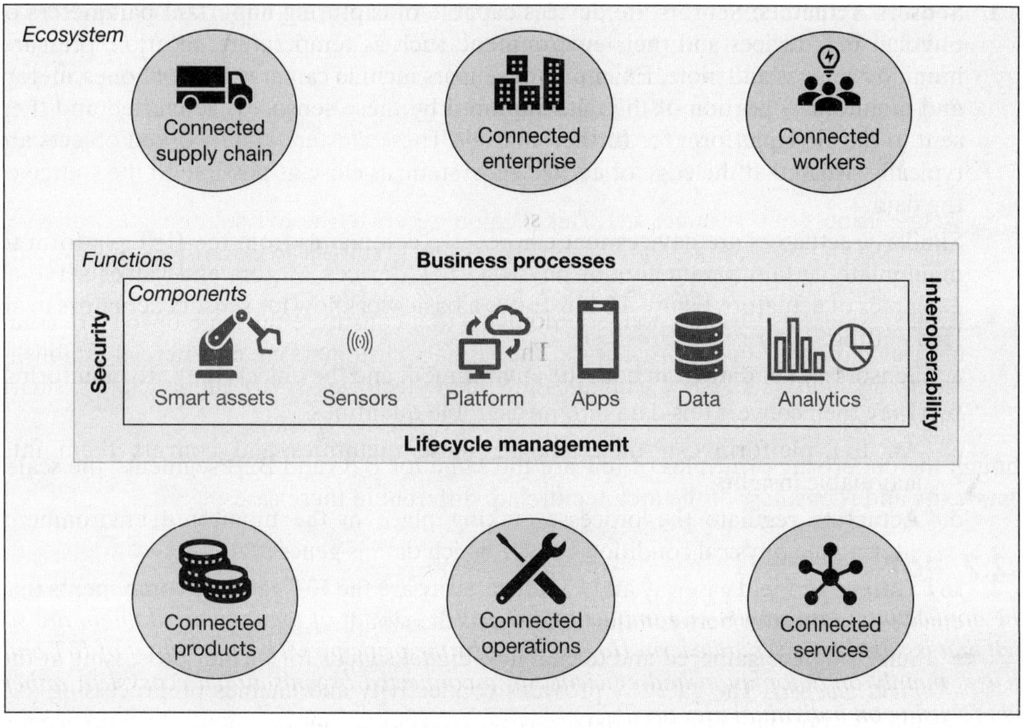

Figure 4.1 Scope of IIoT

4.3 REFERENCE ARCHITECTURE OF IIOT

The Industrial Internet of Things (IIoT) deals with a huge volume of data. The following numbers will provide you with a perspective of the volume: a single jet engine running for just 5 hours produces one terabyte of data. With around 50 billion Industrial IoT devices deployed across the globe, approximately 600 zettabytes of data are expected to be generated per year. An IIoT system gathers data from the manufacturing floor, sends it through connection nodes, processes it on servers and converts this data into actionable insights on a cloud platform. This process enables businesses to make more informed decisions tailored to their specific markets and target demographics. To meet this requirement, the IIoT system links edge devices, such as sensors, actuators, controllers, connection switches, gateways and industrial personal computers (IPC), to the cloud. An effective IIoT architecture should ensure the smooth exchange of data and instructions between devices, along with the ability to extract valuable insights from this data.

4.3.1 Key Components of the Architecture

To understand the architecture of IIoT systems we should understand the underlying components. Figure 4.2 describes different components in an IIoT architecture. Based on the scale of implementation, the number of instances of each component and the complexity of the architecture may change, but the underlying principle of the architecture remains the same.

1. **Sensors/Actuators:** Sensors are devices capable of capturing important parameters of physical IoT devices and their environment, such as temperature, location, pressure, humidity, images and more. Examples of sensors include cameras, microphones, meters and monitors. A portion of the data captured by these sensors is converted and then sent to the IIoT platform for further analysis. These clusters of networked objects are typically situated at the edge of an IoT ecosystem, as close as possible to the source of the data.

   Similarly, actuators are devices that can receive commands from the IIoT platform to manipulate certain parameters of physical IIoT devices. Motors and valves serve as examples of actuators. Figure 4.3 illustrates a basic workflow for sensors/actuators in an IIoT setting:

   a. Sensors gather data from both the environment and the objects they are monitoring.
   b. They then convert this data into measurable quantities.
   c. An IoT platform can then analyse these quantities and convert them into actionable insights.
   d. Actuators regulate the processes taking place in the monitored environment, altering the physical conditions under which data is generated.

2. **IoT Gateways:** Field gateway and Cloud gateway are the IoT gateway components that enable data transfer between the 'things' and the cloud.

   - Sensor data is gathered and turned into digital signals for further processing at the *Field gateway*. The gateway provides connectivity and enables preprocessing and filtering options before the data moves to the cloud gateway. Similarly, control commands get transmitted to actuators from the cloud via this gateway.

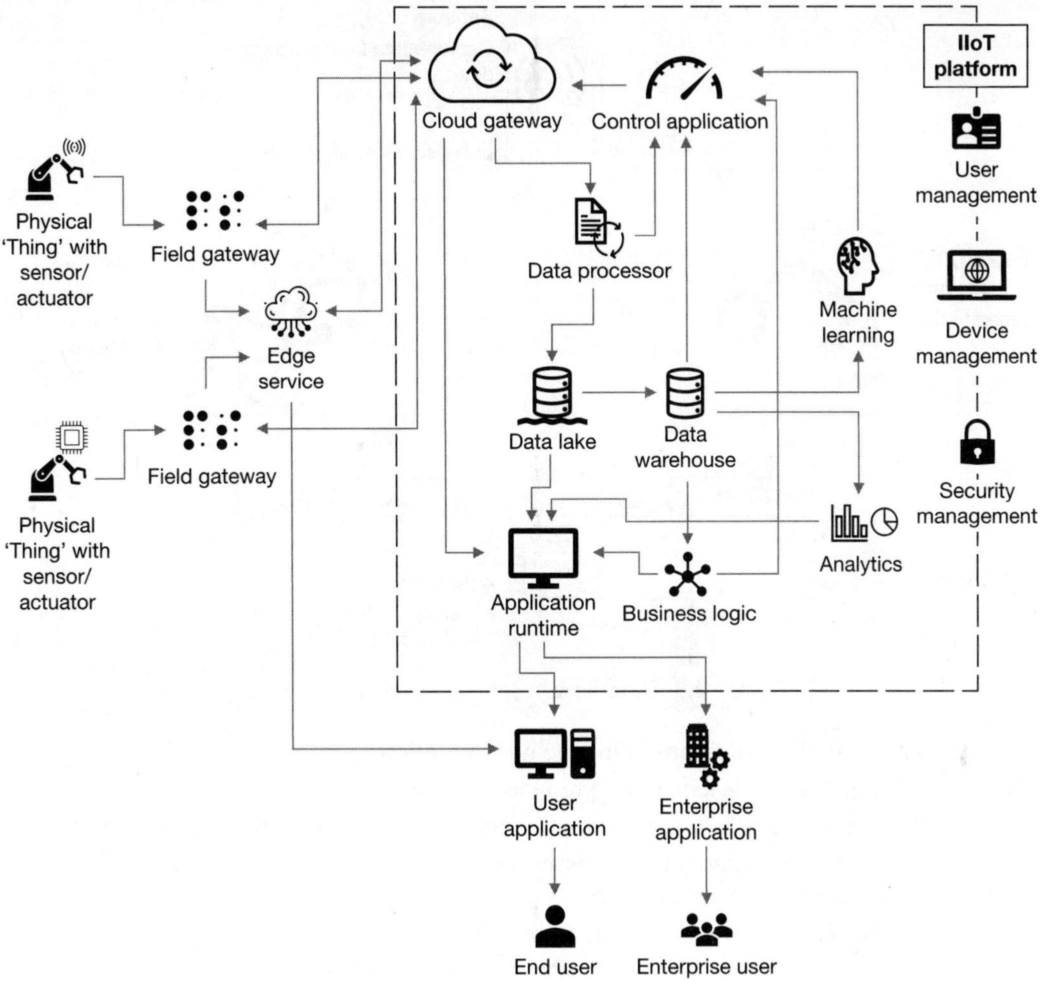

Figure 4.2 IIoT reference architecture

- *Cloud Gateway* enables data compression and secure transmission of data between field gateways and cloud servers. As the field gateways may use different protocols, the cloud gateway provides a seamless connection between the field gateways and the cloud by ensuring the compatibility of protocols.

3. **Edge and Fog Computing:** Edge computing is a concept that involves positioning systems capable of processing and analysing data on-site, or in closer physical proximity to the IIoT network. This approach brings computational power to the 'edge' – the points of data collection or action within the network. This proximity allows for real-time processing of time-sensitive IIoT data, reducing latency and delays. For more in-depth, but less time-critical analysis, IIoT data can be periodically sent to a centralised, AI-powered system. This distributed computing model enables data computation and storage to occur close to the source where it's needed. Azure IoT Edge along with Azure Stream Analytics and AWS Greengrass are examples of a few leading edge-service setups.

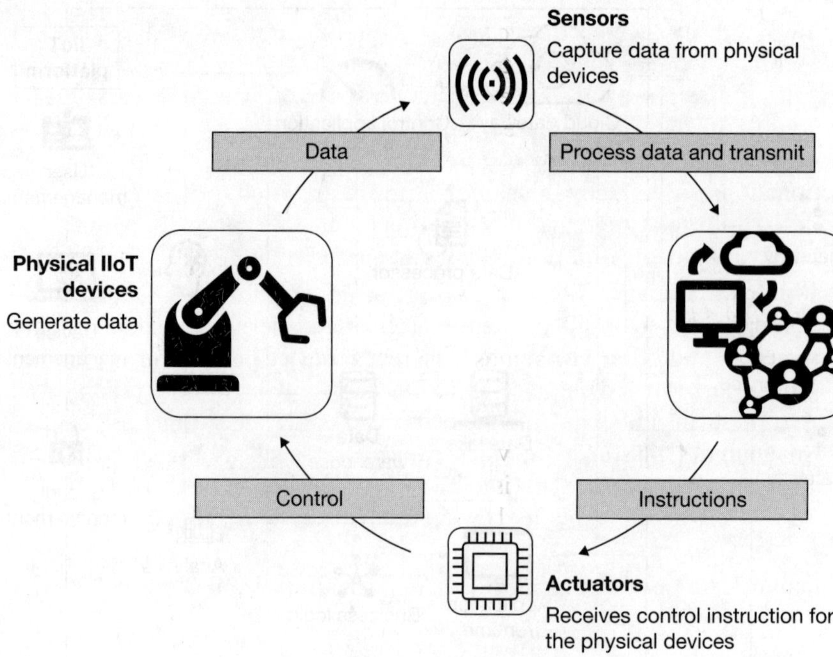

Figure 4.3 Data flow in sensor/actuator devices

The advantages of edge computing include the following:

a. *Improved response time* due to reduced network latency.
b. *Reduced network bandwidth usage* as less data needs to be sent to central processors.
c. *Enhanced data security* as data is stored near the source, reducing the need to send sensitive data over the network.
d. *Increased system scalability* due to the distributed network.
e. *Cost-effectiveness* as costs associated with data storage, network usage, data transmission and data processing are minimised.

As the edge devices are designed to make quick processing at the device end, they have limited processing power. This necessitates that a large volume of data is sent to cloud systems for high-end processing, adding to the network load and latency. A concept called Fog computing has evolved which brings a computing layer between edge and cloud to resolve these issues. One difference between edge and fog computing is their processing location. Edge computing happens directly at the data source as there is a physical connection between the sensor and the processing unit. Fog computing is performed within the IIoT gateway or the fog node with the processor connected through LAN. So, a physical distance from the device is maintained without creating any latency. Moreover, instead of sending all the data to the cloud, a fog computing device sends only the data that needs long-term storage and central processing. The rest of the data is either deleted or stored in the fog device for further processing. Edge and Fog Computing are described in detail in Chapter 8.

4. **Data Storage:** In an industrial setup, the volume of data generated is very high and thus long-term data storage/warehousing and timely retrieval of this 'Big Data' are the crucial responsibilities of the cloud. Once the data processor makes necessary validations and manipulations to the data accumulated on the cloud, those need to be stored. **Data Lakes and Data Warehouses** are used to store data before and after processing. The database functionalities provided by the cloud platforms are often termed *Database as a Service (DBaaS)*. Both relational and non-relational database services are available over the cloud. They provide flexibility, reliability and security along with affordability.
5. **Data Visualisation and Analysis:** The vast amount of information (Big Data) stored in the data warehouse can be used to identify trends and derive valuable insights. Diagrams, infographics and other visual aids help reveal device performance, pinpoint inefficiencies and suggest ways to enhance an IoT system, making it more reliable and customer-centric. These insights can also help in developing algorithms for control applications. An **analytics platform** provides capabilities, such as big data mining, processing and analysis. Advanced visualisation and complex analytics techniques are normally part of this. Figure 4.4 explains how an analytics system should work within the IIoT environment.

   Big data and analytics are discussed in more detail in Chapters 9 and 10, respectively.

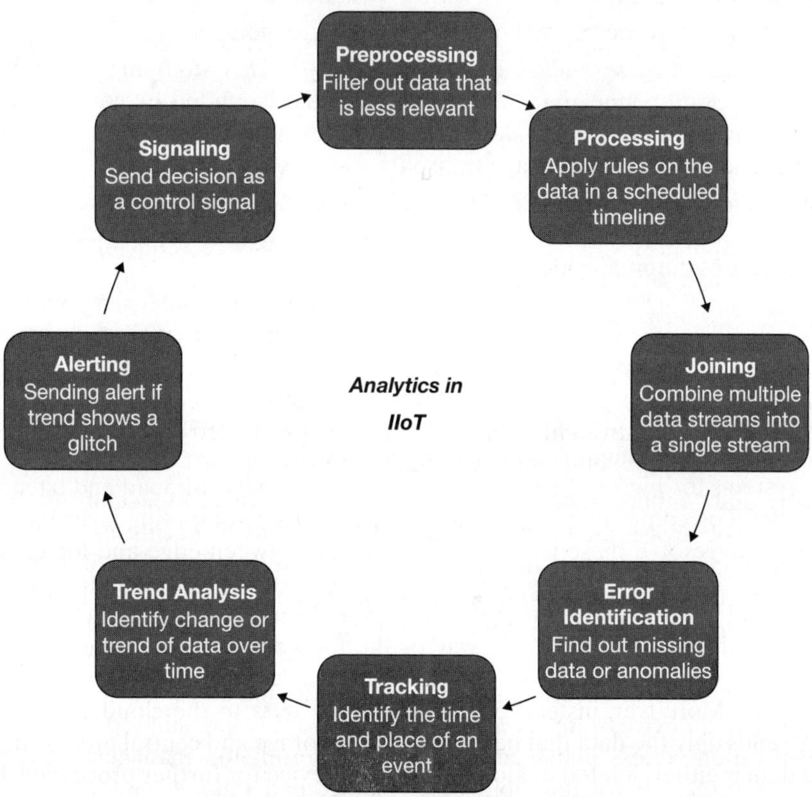

Figure 4.4 Role of analytics in connected IIoT environment

6. **Application Services:** The user applications, control applications, enterprise business applications and rule engines are the decision-making and user interface components in an IIoT architecture. These are commonly hosted on the cloud. Application runtime environment holds the business logic. It processes the necessary data and insights from the Data lake, Data warehouse and Analytics to run the user and enterprise applications. *User applications* enable the connection of end users to the IIoT platform to directly monitor and control smart things. *Control applications* are rule-based or machine learning-based application units that provide automation to the IIoT systems. For example, an AI-driven control application in a data centre can send automated commands to the temperature-control actuator to change the temperature based on the sensor data, machine learning algorithm and historical data stored in the data warehouse. A few examples of high-capability User and *Enterprise application* platforms are SAP (ERP), Salesforce (CRM) and Adobe Commerce (eCommerce) which are hosted on the cloud.
7. **Connectivity Protocols:** IIoT systems need well-defined and secure connectivity protocols standardised across industries so that data transmission across disparate devices can take place. The specifications for protocols might encompass the physical attributes of connections and wiring, the method for setting up a communication channel and the structure of the data transmitted over this channel.

   A few common protocols used in IIoT architecture include:

   - *Advanced Message Queueing Protocol (AMQP):* This was built specifically for IIoT-related cloud connectivity. It is a protocol that is guided by connection, supports two-way communication, allows for multiplexing and employs a compact method of encoding data for message transport.
   - *MQ Telemetry Transport (MQTT):* This compact client-server-based message transport protocol is well-suited for IIoT devices as the message frame sizes are small and it uses minimal code space.
   - *Constrained Application Protocol (CoAP):* This is a condensed version of HTTP which is improvised to cater to IIoT requirements. The IIOT devices such as sensors, actuators, wearables operate in resource constrained environment in terms of processing power availability, memory consumption and network bandwidth utilisation. The lightweight communication protocol used by CoAP helps the IIOT devices to operate efficiently with limited resources.
8. **Security and Management Systems:** Security is a critical aspect of an IIoT system. Thousands of devices are connected via and exposed over the internet. A *Security management system* provides a defence against cybersecurity threats and vulnerabilities through regular monitoring and detection of malware, providing early threat visibility and securing data transfer and encryption. *Device management* systems take control of device identification, configuration, monitoring, diagnostic, upgrades and maintenance activities. They ensure uninterrupted functioning and consistent performance of the connected devices. *Use management systems* are responsible for the identification, authentication, access management, role and permission management of the users. These cyber security related topics are explained in detail in Chapter 13.

4.3.2 Role of Cloud Computing in IIoT Architecture

Cloud computing is a critical component for the smooth operation of IIoT systems, which manage vast quantities of data produced by a variety of sensors and devices connected through the internet. The storage, processing, analytics, applications and communication services that we discussed above can be efficiently handled by cloud computing without needing massive investment in complex on-premise infrastructure. They offer different pay-as-you-use models of service delivery, such as *Software as a Service (SaaS)*, *Platform as a Service (PaaS)* and *Infrastructure as a Service (IaaS)*, which allow enterprises to create and scale their IIoT systems in a fast and flexible manner.

As we have discussed, Edge computing brings some amount of computational power to the physical devices, but these devices have limited capacity for preprocessing. So, to store large volumes of data, perform *in-depth analysis* of it and streamline communication between devices and the cloud, the role of cloud computing becomes critical. Cloud computing also facilitates the *aggregation of data* from multiple sources and provides insights that are not possible at the edge level. Thus, edge computing and cloud computing are complementary architectures. One does not replace the other, but together they create powerful IIoT architecture. Refer to Figure 4.5 for details.

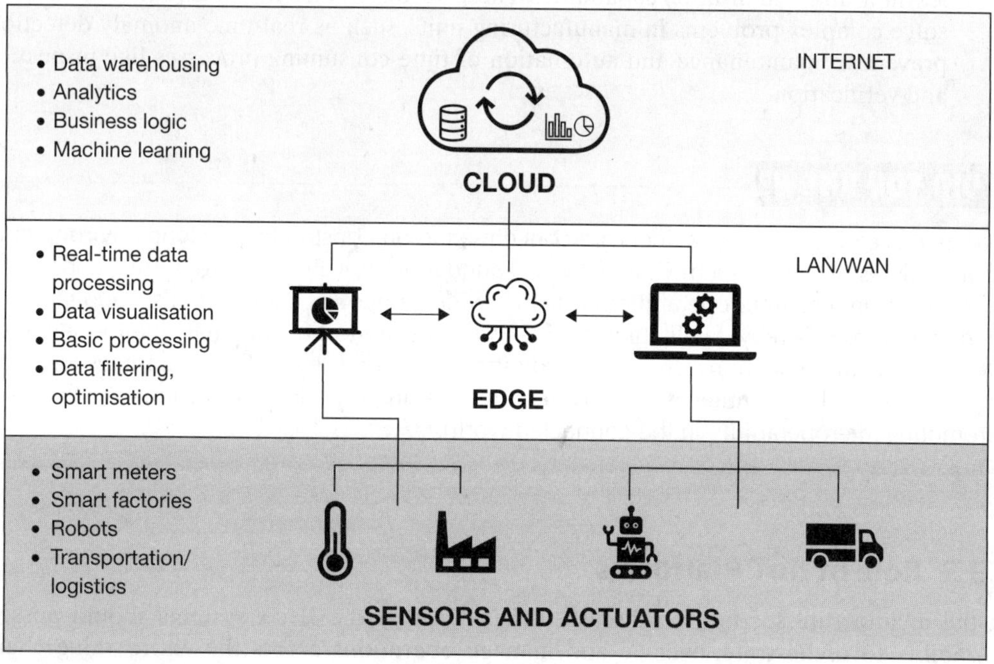

Figure 4.5 Cloud and edge services complement each other to power IIoT

Today, cloud platforms offer built-in capabilities such as machine learning, business intelligence tools and SQL query engines to perform advanced analytics and support business decisions. Some of the leading cloud platforms used for IIoT applications are Azure IoT hub and AWS IoT core. In **Chapter 8,** we will discuss cloud computing in more detail.

Cloud computing brings the following benefits to the IIoT solutions:

a. **Enhanced Storage and Computing Capabilities:** Cloud computing offers extensive storage capacity and computing power at a reduced infrastructure cost through IaaS, along with easy access to business functionalities via SaaS and PaaS.

b. **Heightened Data Security:** The cloud employs advanced encryption algorithms and authentication mechanisms, providing users with robust security.

c. **Simplified Hosting:** The infrastructure of cloud computing enables ready-to-use plug-and-play hosting services, eliminating the need for offline hardware storage device setup.

d. **Optimised Device-to-device Communication:** IoT APIs allow smart devices to connect effortlessly, facilitating swift and easy internal communication between devices.

e. **Lowered Ownership Cost:** Cloud computing reduces expenditure on infrastructure acquisition and maintenance.

f. **Boosted Business Continuity:** Cloud providers manage the infrastructure and disaster recovery systems and data is stored in diverse locations, enhancing business continuity.

g. **Advanced Analytics:** Cloud computing platforms are powered with complex machine learning and visualisation capabilities. These advanced analytics techniques are used to solve complex problems in manufacturing units, such as real-time anomaly detection, preventive maintenance and automation of time-consuming processes like inspection and verification.

DID YOU KNOW?

Numerous cloud service providers are concentrating on constructing a cloud environment that adheres to the Open Connectivity Foundation (OCF) standards. This approach ensures seamless interoperability with a wide range of applications, appliances, and platforms, thereby facilitating device-to-device (D-to-D) and machine-to-machine (M-to-M) communication. The standardisation provided by OCF guarantees secure connection and communication between devices in any cloud environment, thereby promoting interoperability in the connected world.

4.3.3 Role of IIoT Platforms

In the architecture section, we discussed the idea that the IIIoT systems should possess the ability to orchestrate, oversee and manage operations across the entire value chain. To achieve this, all components within the architecture must work seamlessly together to deliver end-to-end functionality. This includes managing device data, performing analytics, visualising data, and executing AI tasks – from edge devices and sensors all the way to the cloud and back.

Since IIoT is a system of systems, organisations may not have in-house expertise in every area. This is where IIoT platforms prove invaluable. These platforms act as the backbone of an IIoT system, connecting various components and ensuring they function as a unified whole.

Key capabilities of IIoT platforms include:

- Connecting to hardware devices
- Managing diverse communication protocols
- Enforcing security and identity verification for devices and users
- Collecting, visualising, and analysing data
- Integrating with external web services

This saves a lot of time and money for an organisation while setting up IIoT systems. These platforms offer a subscription-based model also called IIoT as a Service (IITaaS). A few examples of well-known IIoT platforms are IBM Watson IoT, Oracle IoT cloud, AWS Greengrass IoT, Flutura Cerebra, etc.

4.4 IIoT IMPLEMENTATION STRATEGY

The establishment of a new smart factory or the transformation of a legacy factory into a smart one both need long-term planning and investment. Moreover, effective management guidance and effective change management are key to driving operational efficiencies, cost savings and quality through the deployment of IIoT solutions. The areas of focus for creating an IIoT solution can be generally divided into two categories – Organisational setup and Technology setup.

4.4.1 Organisational Setup for Implementing IIoT

The broad steps include identifying and addressing existing business problems, constructing a business case for investment, being adaptable in adopting new technologies to address multiple use cases, evaluating initial results and implementing the solution throughout the factory.

Before Starting the Transformation Planning

1. Do you know what problem/pain point you are targeting to solve?
2. Did you define the success measures of the project?
3. Do you have the required management sponsorship, with a long-term vision of how the business will utilise the new technologies?
4. Do you have representatives from both the IT and OT teams involved in the transformation journey?
5. Did you plan for the upskilling of support staff and required change management for the new applications?
6. Is there a plan to start with small proof-of-concept projects with a limited number of assets, before rolling out the solution to the whole shop floor?

Current Capabilities and Solutions

Based on the current state of the facilities in the organisation, the need for investment can vary significantly.

1. What is the state of the legacy devices on the shop floor? Do you need additional sensors to be installed? If yes, then what type?
2. What is the current state of connectivity on the shop floor? Is it connected to support the collection of large amounts of data quickly?
3. What connectivity protocol works best for your factory – hard-wired or wireless?
4. What is the current state and capability of the cloud platform in the existing setup? Is it equipped to support IIoT components such as data aggregation, analytics and data visualisation?
5. What is the current state of cybersecurity in the facility? Are hardware, software and data all covered?
6. What enterprise and user applications are in place to enable processing information and to feed results back to the user?

People Readiness

1. Plan proper training for employees on new technologies and processes which will drive awareness and ensure compliance on the shop floor.
2. Change management is a key to any transformation program. A connected workflow brings about a lot of changes in the way the employees work in an enterprise. They should not only accept the changes but should be ready for them.
3. Long-term senior management buy-in is crucial for the sustenance and adoption of the technology and process changes across all facilities.

Robust Data Strategy

1. The monitoring processes should be transitioning from a manual to an automated process of capturing data.
2. The targeted frequency of data collection should ensure a high confidence in the data used.
3. Ensure that the collected data is tagged to the outcome of production. This makes it easier to build analytics which is useful for the business.
4. The processes and steps to ensure that the data collected are consistent and accurate should be planned right from the sensors.
5. The architecture should be designed for scalability so that transitioning from the proof-of-concept stage to the big-bang rollout does not face any architectural roadblocks.

4.4.2 Technological Setup for Implementing IIoT

The concept of IIoT is not based on the invention of any new technologies, but it is based on creating an ecosystem using the existing ones that can make it possible for different things to connect easily and efficiently and exchange data. The resulting *smart* environment reduces production costs and improves market profitability. In this section, let us discuss how to set up a manufacturing environment that can produce, gather, transfer, store and process the data both inside the shop floor and on the cloud seamlessly on a very large scale. While considering a smart environment, the smartness can be classified into four increasing levels:

1. **Monitoring:** It can monitor the product conditions, external operations and usages and can generate alerts and notifications of anomalies or changes.
2. **Controlling:** The embedded software either within the product or in the cloud can control product functions and the user experience through personalisation.
3. **Optimising:** The software system can create algorithms that enhance product performance and improve operational efficiency by enabling predictive diagnostics, service, maintenance and optimisation.
4. **Autonomous:** The highest level of optimisation which can self-coordinate with other systems for diagnosis, maintenance and optimisation through autonomous control.

The model of computer-integrated manufacturing (CIM) which originated in the factories way back in the 1980s continuously improvised to reach its current form in smart manufacturing. Different operational levels and the corresponding components of CIM can be represented like a pyramid structure as shown in Figure 4.6. The five levels are discussed in Table 4.1.

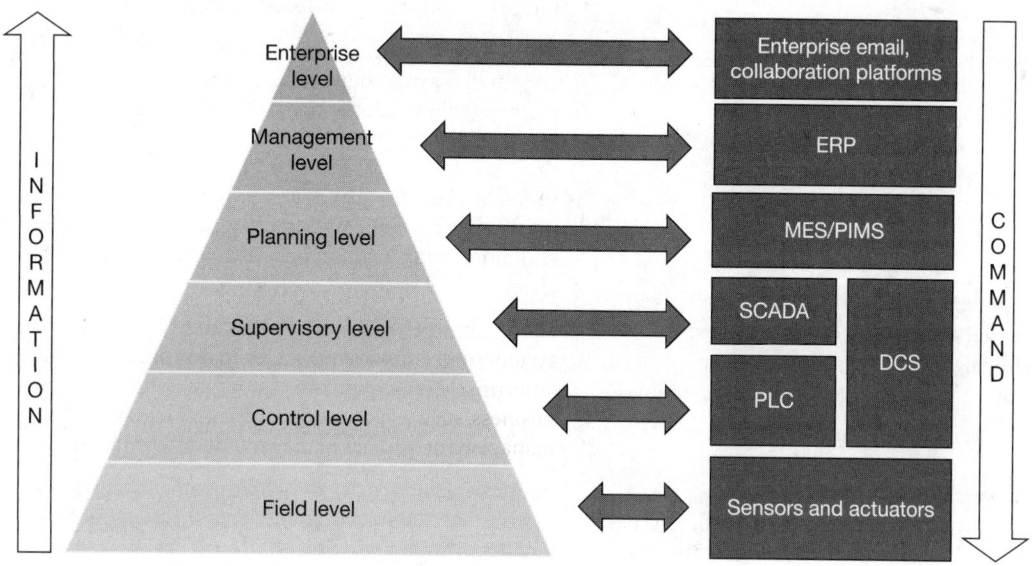

Figure 4.6 CIM operational levels

Table 4.1 Details of the operational levels

Level	Components	Descriptions
Level 0 Field Level	Sensors, actuators and hardware	Sensors normally measure a magnitude and then generate a signal, typically an electrical one. Actuators use incoming signals to set the value of the control variables for the process/ products.
Level 1 Control level	Control systems, such as PLCs and DCSs	PLCs (Programmable Logic Controllers) are industrial controllers that are dedicated to industrial process control. DCSs (Distributed Control Systems) have the combined functionality of control functions available in PLC and the supervision function of the SCADA systems. They are typically used in factories where the processes are continuous, like in refineries, energy production, or chemical factories.
Level 2 Supervisory level	SCADA and Historian systems, time-series databases, production reporting and alarm management	SCADA (Supervisory Control and Data Acquisition) system refers to a set of centralised and computerised applications that are used to control industrial machinery and processes remotely and in real-time. Historian systems gather data from different data sources and have capabilities such as built-in data buffering and store-and-forward mechanisms.
Level 3 Planning level	MES	MES (Manufacturing Execution System) is a software system that synchronises the business management, planning and production systems to optimise processes and resources. It is normally placed between the ERP and the SCADA or PLC systems. Some of the key functionalities are – order management, production planning and tracking, asset management and monitoring, maintenance management, quality checking and so on.
Level 4 Management level	ERP systems	ERP (Enterprise Resources Planning) is a set of systems and software packages that is used by enterprises to manage the regular activities of the business, such as accounting, procurement, inventory management, project management and production.
Level 5 Enterprise level	Enterprise-wide applications such as email, collaboration platforms, etc.	These applications are used across the enterprise to establish communication between different entities and departments.

While implementing these levels and components in an IIoT environment, the key task is to manage the interconnectivity and data flow between these devices/systems. Based on the functionality of these devices/systems the overall data flow can be categorised into four processes: data acquisition, data storage, data processing and execution. The IIoT architecture that we discussed in the previous section can be modified to show these four processes and the underlying data flow. In Figure 4.7, we have shown the generic model of the IIoT data flow diagram and in Figure 4.8 we have explained the implementation detail of the same through AWS. Let us discuss each of the processes and their corresponding components in detail.

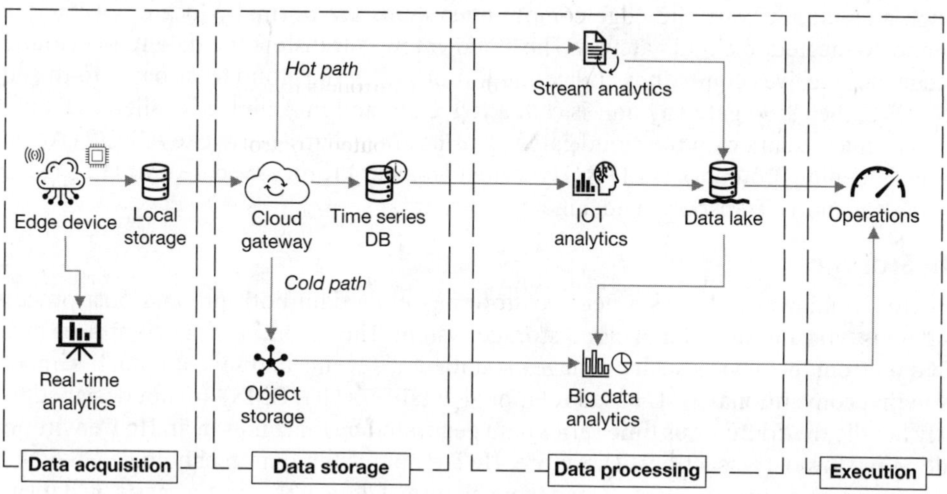

Figure 4.7 IIoT data flow

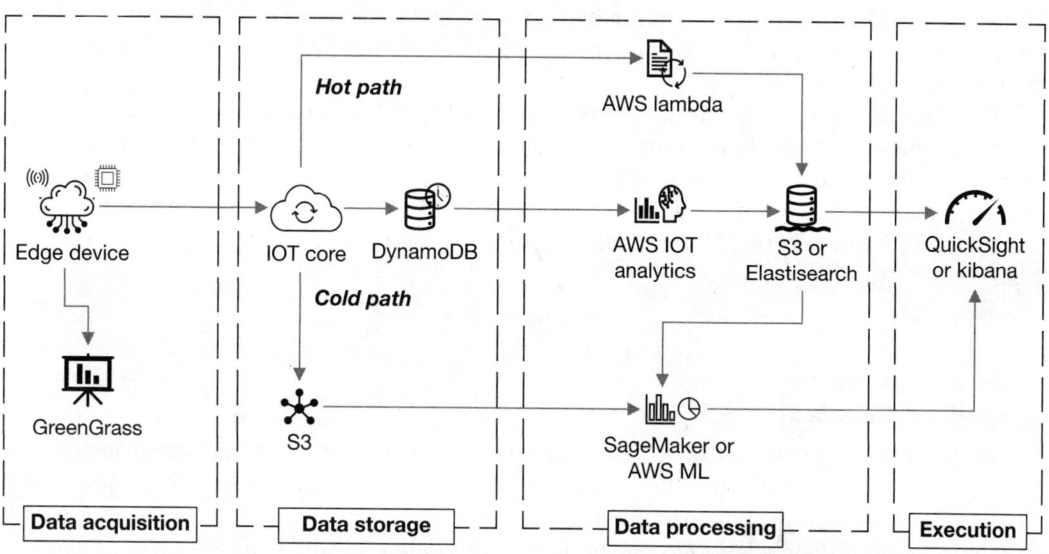

Figure 4.8 AWS implementation of IIoT data flow

Data Acquisition

The sensors are the main components that acquire data from the devices. They either are connected physically to the devices or are close to them to measure the surrounding conditions, like temperature, moisture content, etc., of a data centre. Raw data from the sensors are combined and controlled through the PLC, DCS, SCADA and Historian components. The edge devices are connected on one side to these data sources and on the other side to the cloud through a cloud gateway. The edge computing systems are normally located on the factory floor and connected to cloud gateway. The cloud gateway transfers the data to the cloud from the edge and receives commands and configurations from the cloud to dispatch to the edge.

In AWS, the cloud gateway unit is called IoT Core and in Azure it is called IoT hub. The devices exchange data with the cloud gateway using several protocols, like AMQP (Advanced Message Queuing Protocol), HTTPS (Hypertext Transfer Protocol Secure), MQTTS (Message Queuing Telemetry Transport) and a message broker.

Data Storage

In an IIoT framework, there's a need to store incoming data both pre and post-processing. This data is typically stored in a cloud storage system. The incoming data comprises unstructured data from the device, such as images, sounds, logs, events and sensor data. It's important to note that conventional SQL-based relational databases (RDBMS) are not designed to efficiently handle the continuous time series data generated by machines in an IIoT environment, resulting in slower transactions. Therefore, IIoT employs big data technologies like NoSQL databases, which are more adept at managing historical data with less emphasis on data aggregation and interrelationships. In an industrial context, data often originates from continuous events like alerts, signals and measurements, where the timestamp is crucial. A Time Series Database (TSDB) is the most commonly used historical database. For instance, in an oil rig, the continuous data flow of measuring the temperature of Rig#1 needs to be stored in a database along with the timestamp.

The data is stored in its natural format, like object blobs, images, log files, etc., in a *data lake*. The time-series data like events and measures from the sensors and output of the analytics can also be stored in data lakes. AWS S3 or Hadoop HDFS are the most common data lake implementations. Though AWS doesn't have native support for time-series data, DynamoDB can be used for this purpose. DynamoDB provides key-value-based NoSQL store service with high scalability.

DID YOU KNOW?

Time Series Database (TSDB): A Time-Series Database (TSDB) is a computational system engineered to store and fetch data records that belong to a 'time series'. This refers to a collection of data points linked with timestamps, which offer a crucial context for each data point in terms of their relation to others. Time series data typically comprises a continuous stream of data such as sensor measurements and intraday stock prices. A TSDB enables the storage of substantial volumes of timestamped data in a format that facilitates swift insertion and retrieval, thereby supporting intricate analysis of the data.

Data Processing

Based on the urgency of the data we can either choose to process those immediately or store and process them later along with other data in a batch. Analytics is the key part of data processing which processes the data to extract several types of information, such as anomalies in processes or devices, the fitness of the machines and the efficiency of the shopfloor and also generates key performance indicators (KPIs) to quantitatively drive the insights.

Hot Path Analytics Simple rule engine-based platforms that use data-stream analytics can process data instantaneously based on fixed or intelligent thresholds. This is called hot path analytics. This is generally implemented with serverless Lambda functions. This is a very easy and customisable form of analytics and is well-supported in Python, Node.js, C# and Java. In this context, it's important to understand the concept of AWS Lambda functions. AWS Lambda allows you to run code automatically in response to specific events, known as triggers, without the need to provision or manage servers. AWS handles the underlying infrastructure dynamically based on the trigger. For example, a Lambda function can be set up to execute as soon as a sensor sends data to the cloud—immediately analysing whether the incoming data falls within the permissible range for the equipment. The term serverless refers to a computing model where the user doesn't need to manage or provision the servers or infrastructure, as AWS automatically handles these responsibilities to run the code.

AWS Greengrass is the component that can collect and analyse the data closer to the information source and thus brings cloud capability to the local devices. It can run serverless AWS Lambda functions both on-premises and on the cloud and can deploy it to the devices for local execution. AWS IoT Analytics is also a single service that can handle end-to-end data handling from storage to enrichment and processing.

Cold Path Analytics More advanced analytics involves the processing of a large volume of data from multiple sources like sensors, asset registries, digital twins and historical data. This needs substantial time and computational resources. This advanced analytics is also called cold path analytics and involves machine learning, deep learning and data-driven or physics-based analytics. IIoT can also use big data analytics which uses technologies like data lake and big data platform to process a huge amount of data, such as images or raw data. In an industrial setup, big data describes large sets of diverse data, which can be structured, unstructured and semi-structured and get generated continuously at high speed and in high volumes. The role of big data analytics is to find patterns, trends and relationships in these massive datasets. These analytics processes are normally very complex and thus require specific tools and technologies, high computational power and large data storage that can support the scale.

AWS QuickSight is a serverless, embeddable analytics service that is highly scalable and powered by machine learning. It can help create and publish interactive BI (business intelligence) dashboards that include machine learning-powered insights. Also, AWS SageMaker is a service that can manage the entire machine learning workflow and helps enterprises build, train and deploy their ML models quickly into production.

Execution

Building visualisation from the analytics insights is the final stage of the operations. Calculated data from the analytics can show expected performance, an operational parameter summary, an alarm, a recommendation, or a report. Either these data are used by the operations manager directly or a visualisation layer is built with PowerBI or other advanced BI tools.

4.5 BENEFITS OF IIoT

An IIoT solution is the building block for the overall digital transformation of an enterprise. It provides greater efficiency, competitiveness and resilience to the organisation. The key benefits recognised through a well-architected IIoT system are as follows:

1. **Improved Customer Service:** In an IIoT system, not only the devices and machines are connected, but also the Customer Relationship Management (CRM) applications are integrated within the ecosystem. This provides a better real-time experience to the customers along with benefits like more transparency, faster response to customer feedback and greater ability to cater to individual preferences in manufacturing and production. For example, more transparent and personalised logistics based on customer preferences can be arranged when the systems are seamlessly connected.
2. **Increased Visibility and Reduced Time-to-market:** IIoT devices are constantly capturing and transmitting data for analysis. This continuous flow of information allows for the generation of analytics related to the manufacturing process, workflow and network, thereby enabling the identification of risks and opportunities and facilitating quicker decision-making. Through the IIoT-enabled sensors and monitors, the manufacturers can track real-time updates of the production processes across locations, including remote sites. Not only is visibility established at the device level, but it also generates a comprehensive intelligence of the entire production system. Once it is combined with other data sources like market conditions, weather data and historical enterprise data, it can lead to more informed decisions regarding logistics, inventory, sustenance, quality control and marketing.

   Furthermore, these industrial systems are not just data transmitters, but they can also receive instructions and implement corrective measures to enhance their performance. This two-way communication allows businesses to respond to challenges and opportunities with greater agility, ultimately reducing their time-to-market.
3. **Improved Operational Efficiency:** The convergence of IT and OT systems is a fundamental aspect of IIoT solutions. This integration facilitates automation and optimisation, improves visibility of the supply chain and logistics and enhances efficiency. IIoT provides end-to-end visibility to the operations, including data from remote assets and systems. When these data are analysed and compared with the context of historical data, then powerful and actionable insights about the operational process come out. Moreover, the use of sensors and actuators simplifies the monitoring and control of physical infrastructures within an industrial system, leading to a reduction in unplanned downtime and an optimisation of overall operational efficiency, thereby reducing operational costs. For instance, research indicates that predictive maintenance can decrease machine downtime by 30–50% and increase machine lifespan by 20–40%.
4. **Increased Safety:** An IIoT setup with sensors, Augmented Reality (AR)/Virtual Reality (VR) enabled systems and real-time dashboards of machine parameters combine the precision of smart devices with human interactivity. This enables businesses to reduce the risk of injuries and unexpected strain. Moreover, the connected sensors can penetrate the regions that are hazardous for human interventions. Machine learning-enabled intelligent systems can automate many such activities and in turn, make the industrial setup much safer. For example, in an oil rig, hundreds of sensors are fitted on the driller

that continuously transmits data about the surroundings and the health of the machine which are monitored and the device is controlled remotely through connected systems.

5. **Better Visibility in the Supply Chain:** An IIoT system allows businesses to decentralise their manufacturing units and warehouses, promoting the concept of 'local' production units. With customers increasingly demanding personalised choices and quicker delivery cycles, industries are responding by establishing smaller, geographically dispersed and interconnected production units. The management of a diverse and extensive inventory, the maintenance of a continuous supply chain and the provision of on-demand products necessitate real-time visibility into the location of inventory and warehouses at all times. This level of oversight is only achievable through the interconnected delivery networks provided by IIoT solutions.
6. **Improved Productivity:** Data from sensors and other sources, processed in real-time, aids industrial machinery and infrastructure in their decision-making process. This data-driven approach allows for the generation of insights and the execution of specific actions. The ability of machines to undertake and automate tasks has unlocked immense potential for unparalleled economic growth and efficiency in productivity.

In essence, the appropriate integration of IIoT can enhance asset utilisation, anticipate failure points and even autonomously initiate maintenance processes. By embracing connected and intelligent devices, businesses are empowered to collect and analyse larger volumes of data at higher speeds. This not only boosts scalability and performance but also bridges the gap between production floors and customer-facing units. The integration of IIoT can provide industrial organisations with a more precise understanding of their operational progress, aiding them in making informed business decisions.

4.6 CHALLENGES IN IIoT IMPLEMENTATION

We have discussed how IIoT can revolutionise industries through the connected devices that collaborate over the internet. However, this principle of connectivity and exposure to physical devices over the internet poses new challenges unless the architecture is implemented with great care.

1. **Security Vulnerability:** As we previously discussed, the convergence of IT and OT systems is a key aspect of an IIoT solution. However, there are numerous legacy systems and processes that have been operational for a long time and modernising these systems to adopt new technologies can be challenging. This complexity can hinder the adoption of IIoT and create vulnerabilities in the connected ecosystem.

   While many large enterprises are well-equipped to maintain the safety and upkeep of physical industrial devices, the prominent role of IT in IIoT architecture necessitates that they also manage digital components. Cybersecurity threats such as hacking, malware, system manipulation, data leakage and software vulnerabilities are becoming increasingly prevalent across industries.

   For instance, in 2015, a cyberattack on the power grid infrastructure in Ukraine led to the shutdown of critical systems, impacting thousands of households. Therefore, the integration of IIoT not only involves the adoption of new technologies but also the management of digital security risks.

2. **Privacy:** IIoT relies on the data it generates to derive insights. These devices are engineered to gather sensitive data, which may include personal information. The challenge for IIoT architectures often lies in integrating industrial systems with IT, where both the connection and the exchange of information need to be secure. These systems must adhere to regulations when handling personal data. For instance, the General Data Protection Regulation (GDPR) of the European Union (EU) stipulates that personally identifiable information (PII) should be stored in an encrypted database.
3. **Standardisation:** The successful implementation and expansion of an IIoT architecture depend on the seamless connectivity among the devices involved. As such, interoperability and standards are pivotal in IIoT systems. Interoperability implies that a variety of devices from different manufacturers should be able to integrate effectively and one IIoT system should be able to communicate with another effortlessly. This necessitates standardisation across device manufacturers. Challenges arise when a vendor attempts to impose their proprietary protocols on their devices to establish a monopoly or bypass the standards to expedite market entry.
4. **Complexity:** given that IIoT solutions typically involve a large number of interconnected devices, applications and services, the number of connections between these devices is also substantial. One of the primary sources of complexity is the involvement of multiple technologies and a vast array of components from various vendors in the ecosystem. Designing, deploying and maintaining these components while keeping up with fluctuating user requirements and preferences is a complex task.

   Businesses should consider making substantial investments in cybersecurity to receive early warnings of threats and facilitate swift action. Establishing a Security Operations Centre (SOC) is a crucial step in this direction. A SOC team can centrally monitor and analyse security incidents or unusual activities, enabling issues to be addressed before a full-scale compromise occurs. When designing security systems for IIoT, several key factors are taken into account:
   - The threat defence system should operate end-to-end, from the gateway to the endpoint.
   - Data transfer should be secured.
   - Enhanced visibility should be provided for threat and anomaly detection.
   - Servers and applications across data centres and clouds should be protected.
   - The central control centre should be well-protected, which typically resides on the gateway and sends commands to various devices. This unit could be a prime target for malware and hackers.

DID YOU KNOW?

In 2016 Yahoo publicly revealed that it faced a privacy breach in 2013 where data of 2 billion user accounts got compromised. It faced another attack in 2014 where data from 500 million user accounts was stolen. Sensitive user information like name, email address, phone number, password, and date of birth was stolen in the attack.

4.7 APPLICATION DOMAINS OF IIoT

4.7.1 Smart Manufacturing and Supply Chain

A large part of the term 'Industry' in the Industrial Internet of Things refers to the factories and production units. Smart factories are those where connected devices operate in a highly digitised and connected environment driven by *intelligent automation*. Smart factories are autonomous as the network enables data to flow between the sensors, control units and actuators through which the changes in the operating conditions are detected and more automated, adaptive and flexible responses to those are initiated (refer to Figure 4.9). The advancements in a few key technologies that contributed significantly to this smart integration are wireless sensors, radio frequency identification (RFID) tags, data analytics, cloud computing and machine learning. The point to be noted here is that these technologies are not new. Those have been used for decades in various parts of the operations. But these devices were not connected. The people, assets and data management systems used to work as isolated units in a traditional shop floor and thus needed manual coordination to integrate those. In contrast, the ecosystem in a smart digital factory and supply chain integrates people, machines and big data in a single system. It is driven by the IIoT system that gathers data from the machines, curates and analyses them and then gains insights to forecast trends and recommends process and workflow improvements. Thus, a smart factory learns as it goes to become more productive, resilient and safe.

Smart manufacturing and smart supply chain are the key domains of a smart factory and IIoT addresses the connectivity and collaboration among the components of these domains. It is also evident that smart factories will work best when the encompassing enterprise is equally smart and connected. As shown in Figure 4.9, all the departments should be connected along with the customers and the suppliers to make this model sustainable.

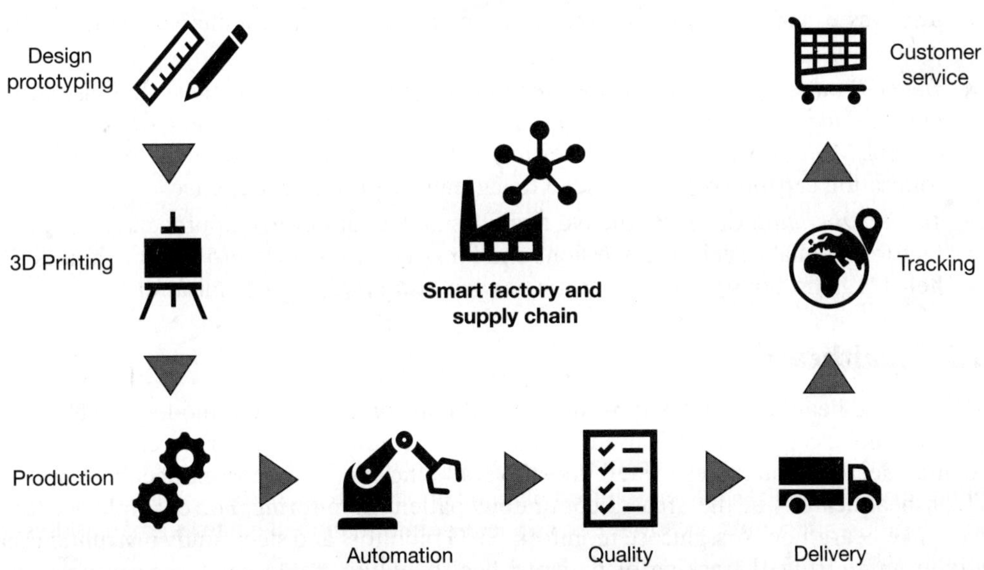

Figure 4.9 Smart factory setup

Four key principles that drive the digitisation of production chains are listed hereunder.

- *Interoperability:* Machines and devices should be able to communicate with one another and with the network seamlessly.
- *Virtualisation:* Smart factories extensively use data from simulation and virtual copies of real assets (like digital twins). It helps them to create more transparent insights into the entire manufacturing processes and operations as data from both physical systems (real-time monitoring) and virtual systems (simulated data) can provide better context-specific information.
- *Decentralisation*: The computing capability is not limited to any central unit, but is distributed closer to the collection and decision points.
- *Real-time Capability*: Smart factories can react to the changes in the surrounding operating conditions and the anomalies very quickly as data is being monitored in real-time. Moreover, many decision points are automated creating the capability to predict and respond to risks as well as improvement opportunities quickly.

Some of the key benefits of smart factories are listed hereunder.

- *Improved Productivity and Efficiency:* Predictive analytics and Big data for process improvement. Optimised supply chain for faster time to market.

  An analysis from a renowned research firm shows that manufacturing output, factory utilisation and labour productivity increase up to 12% in smart factories compared to traditional ones. By 2030, smart factories will surpass traditional factories by 30% in labour productivity.
- *Enhanced Sustainability and Safety*: As a result of improved transparency of the operating parameters, the scope for proactive maintenance increases. Along with that, the use of automation and robots enhances workplace safety. Also, streamlined processes, less waste and intelligent power usage contribute to a sustainable manufacturing process.
- *Better Product Quality and Customer Experience*: Cloud connectivity and visibility to end-to-end production chain enable smart factories to constantly keep track of product quality. Moreover, smart manufacturing techniques like 3D printing help in more customisation capability. All these lead to a greater customer experience.
- *Lower Operating Costs*: Predictive analytics and a data-driven approach enable smart factories to make real-time decisions and long-term process improvement choices. This helps organisations to drive operating expenditure (OPEX) optimisation.

4.7.2 Healthcare

By nature, the healthcare industry is an early adopter of many of the modern technologies and IIoT is no exception. IIoT is used by numerous healthcare providers to enhance patient outcomes, decrease human error and boost overall efficiency. The most noticeable impact of IIoT on healthcare is in the area of continuous patient monitoring. For example, wearable devices like heart monitors, glucose monitors, EKG monitors and sleep study instruments can facilitate uninterrupted tracking of patients' health status, enable healthcare professionals for early identification of potential health problems and promote more timely and suitable medical interventions. Furthermore, IIoT technologies significantly lower the likelihood of medical errors by reducing manual data entry, simplifying patient identity verification and effectively managing patient information and treatment.

Benefits to the Patients

- Continuous monitoring and automatic logging of medical parameters
- Proactive identification and notification of adverse conditions to medical personnel
- Faster and more effective diagnosis and treatment
- Holistic care beyond medical facilities
- Automatic analysis of data collected by devices for predictive and long-term treatments

Benefits to Medical Practitioners

- Continuous access to diagnostic data
- Faster and easy access to patient's medical history
- Machine Learning driven algorithms that can detect anomalies
- Comprehensive patient database through connected medical devices and facilities

4.7.3 Smart Farming

The agriculture sector has turned to IIoT to bridge the gap between global demand and food production. According to an estimation by the United Nations Food and Agriculture Organisation, food production needs to grow by 70% to meet the demand of the growing population. On the other side, production is affected by numerous factors like climate change, inadequate arable land, water scarcity, labour shortage, distress migration, crop failures due to epidemics and other variables. Smart farming uses sensors and intelligent connected equipment to gather data and convert this data into meaningful insights that can help improve productivity and bring down manual efforts. Using the data, it is possible for the farmers to accurately determine and predict rainfall, soil nutrition levels and crop yields and to avoid pest invasions and crop failures. Some of the areas in agriculture where rapid advancement in the use of IIoT is seen are represented in Figure 4.10.

- *Precision Farming:* Use of modern technologies like Robotics (seeding/harvesting, weed control, soil analysis), Global Positioning System (farm planning and mapping, farm vehicle control, crop scouting), Remote sensing (sensors embedded in the soil can produce data to determine optimal time and place to water, spray fertiliser or pesticide), Automated Steering Systems (smart guiding systems deployed on the farm vehicles to assist while planting, spraying or harvesting). A few examples of precision farming platforms are EOSDA Crop Monitoring, GeoPard, Cropin and FarmRoad.
- *Smart Irrigation:* IIoT-based spray irrigation mechanisms like micro-sprinklers, drip lines and central pivots can be controlled remotely and help avoid problems of traditional irrigation systems such as over-watering and water wastage.

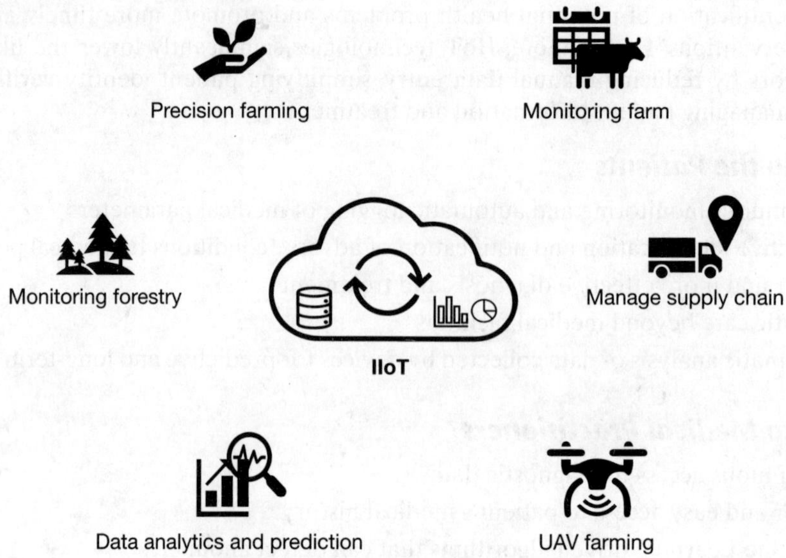

Figure 4.10 IIoT in smart farming

- *Livestock Management:* Using IIoT, farmers can monitor and manage the location, health and safety of their livestock. The sensor data on eating patterns, impending illness, reproductive cycles, calving, etc. can alert the farmers to take proactive actions.
- *UAV Farming:* UAV farming is an innovative use of the IIoT in agriculture, which can assist farmers in fulfilling the growing needs of food production and consumption, while also safeguarding the environment and public health. Drones, or unmanned aerial vehicles (UAVs), are employed to carry out various tasks in agriculture, such as crop monitoring, irrigation, spraying, planting and harvesting. UAV farming can help farmers enhance the efficiency, productivity, quality and sustainability of their operations.

4.7.4 Energy and Utilities

In the energy sector, IIoT applications can significantly enhance efficiency, safety and reliability. These applications not only manage resources but also propose cost-effective strategies. They are reliable and efficient and benefit the organisation by identifying and alerting before a threat arises. There are a multitude of applications of IIoT in this sector:

- *Remote Asset Monitoring/Management:* The use of connected sensors to track parameters such as wear, tear, vibration and temperature helps in assessing the health of assets ranging from turbines to transmission lines. This aids in predicting the 'time to failure' of crucial infrastructure and scheduling maintenance, thereby minimising downtime caused by unexpected maintenance.

- *Automation and Process Optimisation:* The application of real-time data from sensors and cameras enables automation and refinement of plant operations, enhances energy conversion from fuels and curtails maintenance costs. Smart grids enabled by the IIoT utilise sensors, connectivity and data analytics to supervise and regulate the distribution of electricity, thereby optimising energy consumption and minimising waste.
- *Energy System Management:* Smart energy management systems, like smart meters, keep track of a business's energy sources. Techniques like lowering energy bills, reducing carbon footprints and enhancing energy efficiency contribute to the optimisation of energy consumption.
- *Disaster Prevention:* IoT suggests solutions to mitigate risks and enhance safety for both workers and power plants. Sensors installed around the plant detect any leakage or damage and send alerts to a mobile application. For example, the oil and gas industries also make use of IIoT wells with smart sensors. These sensors detect any oil or gas leakage in any of its pipelines and inform the maintenance teams immediately.

4.7.5 Transportation and Logistics

The IIoT is revolutionising the transportation and logistics sector by offering real-time tracking, predictive maintenance and intelligent logistics. Through IIoT devices, businesses can monitor vehicle performance, track shipments in real-time and strategise routes considering traffic and weather conditions. This enhances operational efficiency, reduces expenses and elevates customer service. For example, logistics firms are employing IIoT to monitor their truck fleets in real-time. The data collected helps in predictive maintenance, fuel consumption optimisation and route planning optimisation, which can lead to significant cost savings and better service delivery.

The application of IIoT is numerous across industries and in a wide variety of use cases. With the increased focus on automation, the adoption is bound to rise and IIoT is going to become the new norm in the industrial sector.

4.8 CONCLUSION

We have seen in this chapter how IIoT is essential for Industry 4.0 transformation. Many manufacturers are investing more in IIoT because of its advanced technologies that boost efficiency and productivity. The market also offers more devices and software solutions at lower costs and stronger security protocols that make adoption easier. Furthermore, IIoT helps companies gain a competitive advantage by improving their performance significantly. However, most companies are still in the initial stages of adoption and the market has a lot of potential to grow faster in the future. The IIoT market is projected to reach USD 263 billion by 2027, growing at around 17% annually. Figure 4.11 illustrates the global trend of the IIoT market.

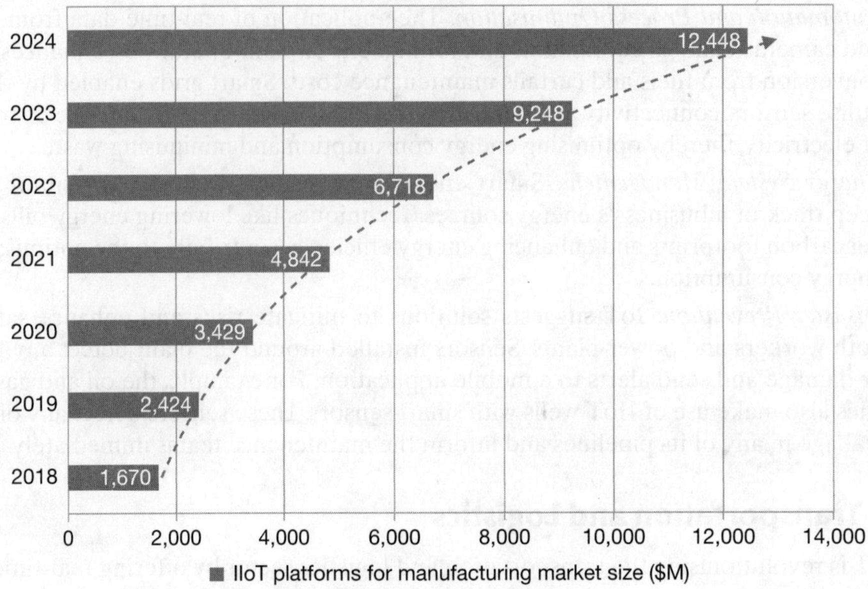

Figure 4.11 Future of IIoT

Manufacturers should realise that becoming smart factories is a gradual process and they need to keep innovating to sustain the positive trend.

SUMMARY

- The global network infrastructure based on standard and interoperable communication protocol through which the physical and virtual 'things' are interconnected and seamlessly integrated to work symbiotically is called the Internet of Things (IoT). Any physical or virtual entity that can be uniquely identified and integrated into communication networks is termed a 'thing' in the context of IoT.
- IoT's application, opportunities and impact in the industrial domain are so significant that a separate term, Industrial Internet of Things (IIoT), was coined to define its principles. This is a subset of IoT and focuses mainly on automation and efficiency in a connected organisational ecosystem rather than focusing on individual user needs.
- Key components of IIoT architecture:
  - Sensors/Actuators
  - IoT Gateways
  - Edge Services
  - Data lake and Big data warehouse
  - Analytics
  - Application services

- Connectivity Protocols
- Security and Management Systems
- The overall data flow can be categorised into four processes:
  - Data Acquisition,
  - Data Storage,
  - Data Processing and
  - Execution
- Benefits of IIoT
  - Improved customer service
  - Increased visibility and reduced time-to-market
  - Improved operational efficiency
  - Increased safety
  - Better visibility in the supply chain
  - Improved productivity
- Challenges in IIoT implementation
  - Security Vulnerability
  - Privacy
  - Standardisation
  - Complexity
- Most companies are still in the initial stages of adoption and the market has a lot of potential to grow faster in the future.
- The IIoT market is expected to grow rapidly, with industries progressively adopting IIoT for greater automation and efficiency. The future holds innovations in smart cities, autonomous systems and more widespread automation in manufacturing and energy sectors.

EXERCISE

Multiple Choice Questions (1 Mark)

1. What is the primary goal of the Industrial Internet of Things (IIoT)?
   (a) To automate all industrial operations
   (b) To increase the volume of data generated
   (c) To connect industrial systems and enable real-time data analysis
   (d) To reduce operational costs in manufacturing
2. A large manufacturing company is planning to implement IIoT.

   Which of the following is NOT a benefit they will get from IIoT integration in manufacturing?
   (a) Increased operational efficiency
   (b) Enhanced data security

(c) Reduced time-to-market
(d) Improved customer service

3. IIoT solutions typically involve the convergence of which two types of systems?
   (a) Information Technology (IT) and Operational Technology (OT)
   (b) Customer Relationship Management (CRM) and Social Media Platforms
   (c) Cloud Computing and Big Data
   (d) Robotics and Artificial Intelligence
4. What does IIoT use to improve visibility in the supply chain?
   (a) Predictive analytics
   (b) Real-time data transmission from interconnected devices
   (c) Satellite technology
   (d) Blockchain
5. Which of the following is a challenge in IIoT implementation?
   (a) Better customer experience
   (b) Standardisation and interoperability of devices
   (c) Increased manufacturing productivity
   (d) Predictive maintenance
6. How can IIoT improve operational safety in industries?
   (a) By increasing production speed
   (b) Through remote monitoring of hazardous areas
   (c) By eliminating all physical devices
   (d) Through increased machine usage
7. What technology is primarily responsible for enabling real-time communication between devices in IIoT systems?
   (a) Virtual Reality (VR)
   (b) Cloud Computing
   (c) Blockchain
   (d) Wireless sensors and communication networks
8. A large agricultural farm has deployed IIoT system. They were able to achieve optimisation in multiple areas.

   Which IIoT application did help them optimise irrigation systems?
   (a) Smart agriculture sensors
   (b) Precision farming technology
   (c) Smart irrigation systems
   (d) Robotics for seeding
9. What kind of devices are used for continuous patient monitoring in healthcare via IIoT?
   (a) Wearable devices like heart and glucose monitors
   (b) Smart robots for surgery

(c) Augmented Reality (AR) glasses
(d) Automated diagnosis machines

10. Which industry benefits from the integration of IIoT for remote asset monitoring and disaster prevention?
(a) Healthcare
(b) Energy and Utilities
(c) Transportation and Logistics
(d) Smart Manufacturing

11. What does predictive maintenance in IIoT systems help reduce?
(a) Machine failure and downtime
(b) The volume of data generated
(c) Operational costs by automating all tasks
(d) The need for human intervention

12. What is one of the benefits of IIoT in smart factories?
(a) Decrease in employee productivity
(b) Automation of manufacturing processes
(c) Reduced connectivity among machines
(d) Increased human error in processes

13. What is the function of digital twins in smart manufacturing?
(a) To provide real-time virtual simulations of assets
(b) To predict customer demands
(c) To control all physical machines
(d) To store employee data

14. Which type of machine learning model is commonly used in IIoT for anomaly detection in manufacturing?
(a) Supervised learning
(b) Reinforcement learning
(c) Unsupervised learning
(d) Deep learning

15. In which sector is UAV (Unmanned Aerial Vehicle) technology utilised as part of IIoT applications?
(a) Healthcare
(b) Agriculture
(c) Energy and Utilities
(d) Transportation and Logistics

16. What does IIoT offer for improving decision-making in smart factories?
(a) Automated robots for every task
(b) Integration of historical and real-time data for insights

(c) Elimination of human operators
(d) Removal of all sensors

17. Which principle is key to the digitisation of production chains in smart factories?
(a) Centralisation
(b) Virtualisation
(c) Manual operations
(d) Limited data accessibility

18. What is the potential result of IIoT-based systems for the energy sector?
(a) Increased energy consumption
(b) Uncontrolled power outages
(c) Optimised energy management and reduced waste
(d) Increased need for manual monitoring

19. Which industry could benefit from IIoT to reduce fuel consumption, predict maintenance needs and track shipments in real-time
(a) Healthcare
(b) Transportation and Logistics
(c) Agriculture
(d) Smart Manufacturing

20. A large factory is planning to implement the IIoT system and is discussing the benefits and threats of the system.

What is a major security concern for IIoT systems?
(a) The cost of implementation
(b) Data encryption and privacy breaches
(c) Lack of available data
(d) Data redundancy

Short Answer Type Questions (2 Marks)

1. What is IIoT (Industrial Internet of Things)?
2. How does IIoT improve operational efficiency?
3. What are the primary components of an IIoT system?
4. What are the key benefits of IIoT in manufacturing?
5. Explain how IIoT helps improve customer service.
6. What is the role of cybersecurity in IIoT?
7. How does IIoT contribute to sustainability in manufacturing?
8. What is 'predictive maintenance' in the context of IIoT?

9. How does IIoT impact healthcare?
10. What challenges does IIoT face during implementation?

Long Answer Type Questions (5 Marks)

1. What is IIoT and how does it differ from traditional industrial systems?
2. Discuss the key benefits of implementing an IIoT system in a manufacturing environment.
3. How does IIoT contribute to the efficiency of predictive maintenance in manufacturing?
4. What are the challenges of integrating IT and OT systems in IIoT implementation?
5. Explain how IIoT enhances safety in hazardous environments.
6. What are the privacy concerns associated with IIoT and how can they be addressed?
7. How does IIoT support the transformation of traditional factories into smart factories?
8. What is the role of data analytics in enhancing productivity within IIoT-enabled factories?
9. How does IIoT contribute to sustainability in industrial operations?
10. Describe the applications of IIoT in the healthcare industry and the benefits it brings.

Industrial Automation and Robotics

CHAPTER 5

OBJECTIVES

In today's rapidly evolving industrial landscape, automation and robotics have emerged as essential tools for enhancing productivity, efficiency and safety across various sectors. As the global marketplace demands higher-quality goods at lower costs, factory floor automation has shifted from isolated machines with manual controls to an integrated manufacturing ecosystem featuring robots and other automation enablers connected to advanced control and data systems. Therefore, it is vital to comprehend industrial automation and robotics technology, along with their use cases, in the context of Industry 4.0.

By the end of this chapter, you should be able to:

- Understand the role of industrial automation and robotics in implementing Industry 4.0.
- Describe the concepts of automation and robotics and their major building blocks.
- Analyse the various types of automation and robots commonly used in industrial settings based on their use cases.
- Understand the step-by-step approach to implementing automation and robotics.
- Recognise the impact of automation and robotics through a real-world industrial case study.

5.1 INTRODUCTION

Automation is a key enabler for Industry 4.0, driving industries toward higher productivity and greater cost-efficiency. To strengthen and sustain their competitive advantage, manufacturers have advanced from basic mechanical systems to intelligent, interconnected technologies powered by AI, IIoT, and real-time data analytics. Companies are now critically analysing their production processes to identify the most effective ways to leverage automation, aiming to increase output without significantly increasing operational costs.

Manufacturers typically adopt one of two strategies to increase production output. The first, while effective, involves adding more production lines, an approach that significantly raises costs. A more efficient and cost-effective alternative is to maximise existing lines' productivity by reducing the cycle time required to manufacture each part or product. This cycle time optimisation is achieved through process improvements and automation using reprogrammable, self-controlled equipment, including robotics. In the following sections, we will explore advancements in industrial automation and robotics, their impact on modern industries and how they drive Industry 4.0-compliant production systems.

5.1.1 Definition of Industrial Automation

Automation involves using mechanical, electronic and computer systems to manage and control manufacturing processes with minimal human intervention. Derived from the Greek words 'auto' (self) and 'matos' (moving), automation enables systems to operate autonomously.

Industrial automation involves computers, robots and control systems to operate and monitor equipment in an industrial environment with limited human intervention. It integrates automated technologies to improve industrial productivity, precision and safety. By automating repetitive and complex tasks, industrial automation lowers costs, reduces human error and facilitates real-time monitoring, ultimately optimising the production process.

In a traditional factory, workers rely on their senses and hands to perform tasks. With automation, these actions are replaced by sensors, controllers and actuators, where sensors detect and transmit information, controllers make decisions and actuators execute the physical tasks. This transition to automated systems allows for continuous, reliable and precise production with minimal human involvement. Now, let us discuss robotics, which can be seen as a key enabler for industrial Automation.

5.1.2 Definition of Robots and Robotics

Automation refers to utilising technology to optimise processes and tasks involving minimal human intervention. This includes the application of robots, but also covers a range of other automatic systems, from simple timers to complex industrial controllers. Therefore, robots can be considered a subset of industrial automation as depicted in Figure 5.1.

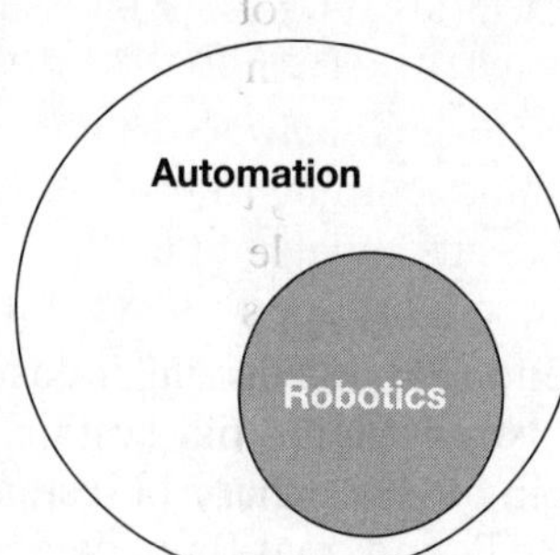

Figure 5.1 Robotics as subset of automation

What are Robots?

A robot is a reprogrammable, multifunction machine that operates automatically, often taking over tasks traditionally performed by humans. While it may or may not resemble a human in appearance or behaviour, a robot is designed to perform specific functions independently or with minimal supervision. Robots can interact with their environment by integrating mechanical components, sensors and computer systems, enabling movement, manipulation, or information processing. They play a critical role across various fields, including industrial automation, healthcare, exploration and entertainment. According to the ISO 8373 definition, a robot is 'an automatically controlled, reprogrammable, multipurpose manipulator, programmable in three or more axes, which may be either fixed in place or mobile for use in industrial automation applications'.

What is Robotics?

Robotics is a multidisciplinary field that connects computer science and engineering to create, design, produce and operate robots. Its primary aim is to develop intelligent machines to assist humans with various tasks. Key disciplines in robotics include mechanical and electrical engineering, mechatronics, computer science, control systems, artificial intelligence, ethics and policy and human-machine interaction.

In this context, it is interesting to know what is not considered as a robot:

For a system to be classified as a robot, it must possess three key elements: sensing, processing and actuation (physical movement). Chatbots, AI, RPA, voice assistants, ATMs and smart appliances utilise automation, AI and intelligent decision-making. Nevertheless, they lack mechanical movement and the ability to engage in autonomous physical interactions with their environment, features that must be considered to meet the above ISO definition of a robot.

- **Chatbots and AI:** These systems process and generate text-based or voice responses using Natural Language Processing (NLP) and machine learning algorithms. Still, they lack physical movement and mechanical actuation. They operate exclusively in the digital realm and do not interact with the physical world.
- **Robotic Process Automation (RPA):** Despite its name containing 'robotic', RPA pertains to software bots that automate repetitive digital tasks, including data entry and workflow automation. Unlike industrial robots, RPAs function entirely in virtual environments and do not have mechanical components or physical actuation.
- **Voice Assistants (e.g., Alexa, Siri, Google Assistant):** These AI-driven systems utilise speech recognition and machine learning to process commands and provide information but lack physical components capable of autonomous movement and task execution. They are software-based, not robotic systems.
- **Automated Teller Machines (ATMs):** While ATMs conduct automated transactions, they are pre-programmed, fixed-function machines that do not display autonomous decision-making or movement. They merely process financial transactions based on user input.

- **Cooking Machines and Smart Washing Machines:** Although these appliances utilise automation, sensors and AI-driven control systems, they function based on predefined settings and cycles. They do not demonstrate autonomous adaptability, environmental interaction, or independent movement, which are essential characteristics of a genuine robotic system.

In this chapter, we will concentrate on our understanding of the robots as categorised by ISO 8373.

5.2 EVOLUTION OF AUTOMATION AND ROBOTICS

The evolution of industrial automation and robotics has been characterised by continuous technological advancements that have reshaped manufacturing, healthcare, exploration and daily life. It began in the late 19th century with the introduction of the electric motor, which revolutionised production by automating processes such as lock manufacturing. Early automation concepts were also influenced by steam-powered mechanisms and gear-driven systems, laying the foundation for mechanised operations.

The 20th century witnessed rapid progress, beginning with Henry Ford's introduction of the conveyor belt assembly line in 1913, dramatically enhancing manufacturing efficiency. Around the same time, the idea of robots originated when Karel Čapek introduced the term in 1920. The first humanoid robot was showcased in London in 1928. Industrial robots designed for specific tasks were created during the 1930s and 1940s, such as DeVilbiss's paint-spraying robot. In 1937, George Devol came up with programmable automation. This led the foundation for the invention of Programmable Logic Controllers (PLCs) in 1968, which transformed industries with flexible automation systems. There were groundbreaking developments in the field of robotics as well. First robotic arm was created in 1963. Advancements in androids and six-axis robots in Japan during the 1970s, enhanced manufacturing flexibility.

During 1980s and 1990s, computers and Internet connectivity were integrated into automation. This enabled sophisticated control systems and web-based manufacturing operations. In robotics also there were revolutions, from NASA's Mars Pathfinder mission to surgical robots like CyberKnife. Honda's research and MIT's development of autonomous multi-legged robots improved the field of humanoid robotics as well. The rise of the Internet also enabled remote automation and globalised industrial operations.

In the 21st century, automation and robotics have evolved into intelligent, interconnected systems, fuelled by AI, IoT and machine learning. By 2000, over 742,000 industrial robots were operational worldwide. Innovations like self-learning robots, AI-powered automation and humanoid robots like Sophia and Nadine have blurred the lines between human and machine capabilities. Robotics has played a crucial role in space exploration, with Robonaut 2 and Mars Rover Curiosity, while exoskeletons and autonomous vehicles have redefined mobility and transportation.

Today, automation and robotics continue to push technological boundaries, with applications ranging from ultra-precise surgical robots to fully autonomous industrial systems. The integration of AI, IoT and smart technologies guarantees greater adaptability, efficiency and autonomy, reinforcing robotics' role in shaping the future of industries and human life. The evolution of industrial automation and robotics is illustrated in Figure 5.2.

1882 -1890	• The first electric motor used in automation, automating lock production.
1913	• Ford introduced the conveyor belt assembly line, revolutionising mass production.
1920	• The term “robot” was coined by Karel Čapek.
1928	• The first humanoid robot capable of moving its hands and head was displayed in London.
1937	• George Devol patented the first programmable device, a precursor to PLCs.
1940s	• DeVilbiss Company introduced the first paint-spraying robot for industrial tasks.
1963	• World’s first robotic arm developed for precise industrial tasks.
1968	• General Motors and Westinghouse introduced the first Programmable Logic Controller (PLC).
1970s	• KUKA introduced a six-axis industrial robot; Japan launched advanced android robots.
1980s	• Computers integrated into automation, enhancing precision and control.
1989	• MIT developed a six-legged robot with advanced controls.
1990s	• Internet-enabled web-based automation; NASA launched Pathfinder robot to Mars; • CyberKnife robotics introduced in healthcare.
2000	• Over 742,000 industrial robots are estimated in global use by the United Nations.
2006	• Cornell's Starfish robot introduced self-learning and adaptive capabilities.
2011	• Robonaut 2 became the first humanoid robot in space.
2015 -2017	• Nadine, a human-like robot, was created; Sophia was granted citizenship; WABOT -1, the first Android Robot that can able to walk, grip, transport objects, sense objects, and communicate, was created in Japan; exoskeletons empowered mobility.
Present	• Innovative technologies like AI and IoT drive the convergence of automation and robotics.

Figure 5.2 Evolution of industrial automation and robotics

5.3 KEY BUILDING BLOCKS OF INDUSTRIAL AUTOMATION AND ROBOTICS

Automation and robotics rely on a combination of advanced technologies to create intelligent, efficient and autonomous systems. These systems integrate various hardware and software components to sense their environment, process information and execute precise actions. In this section, we will discuss these core building blocks.

1. **Sensors for Industrial Automation and Robots:** Sensors are essential to automation, acting as the eyes and ears of robots and industrial systems by gathering real-time environmental data that allows machines to detect objects, analyse conditions and make intelligent decisions. Proximity sensors enable robots to navigate obstacles and interact with materials without physical contact. Vision sensors capture images and process visual data for object recognition, quality inspection and precise robotic guidance. Force and torque sensors gauge the applied force, ensuring precision during delicate material handling and assembly tasks. Temperature sensors monitor heat levels to enhance safety in industrial processes such as welding and metal casting. In agriculture, specialised soil moisture sensors optimise irrigation by managing water usage, while pH and nutrient sensors analyse soil conditions to improve fertiliser application and crop yield. Some notable sensor models include the SICK WL12 for proximity detection, the Keyence IV Series for vision-based applications, the Fluke 62 MAX for temperature monitoring, the Sentek Drill & Drop for soil moisture measurement and the CropX Smart Sensor for analysing soil nutrients.
2. **Actuators for Industrial Automation and Robots:** Actuators are responsible for converting energy into motion, allowing robots to move, lift, or manipulate objects. They function as the muscles of an automation system, powering robotic arms, conveyor belts and industrial machinery. Electric actuators are widely used in industrial automation due to their precision control and energy efficiency. Hydraulic actuators create powerful movements and are ideal for heavy machinery, construction equipment and large-scale automation. Pneumatic actuators deliver quick response times and are advantageous in lightweight applications like robotic arms in agriculture and food processing. Example models include Festo DSNU (pneumatic), Parker H-Series (hydraulic), SMC LEY Series (electric) and SMC Pneumatic Actuators (for fruit picking applications).
3. **Controllers for Industrial Automation and Robots:** Controllers act as the brain behind automation and robotics systems by processing sensor information, carrying out logic-based tasks and transmitting commands to actuators for precise operation. Programmable Logic Controllers (PLCs) are a popular type of controller, designed for real-time industrial automation, engaging in sequential logic and managing inputs and outputs in harsh environments. They are frequently used in manufacturing, assembly lines and process automation. Other specialised controllers include CNC controllers, which precisely operate machining tools like lathes and mills; motion controllers, which regulate servo and stepper motors for high-precision movement in robotics and automation; robot controllers, which manage multi-axis robotic arms and handle kinematics, path planning and execution; and Proportional-Integral-Derivative (PID) controllers, which regulate process variables like pressure, temperature and speed for real-time stability. In agriculture, machine controllers automate GPS-guided tractors and

autonomous sprayers, while greenhouse automation controllers optimise irrigation, temperature and humidity for improved crop yield. Common models include Siemens SIMATIC S7-1500 (PLC), Fanuc R-30iB Plus (robot controller), Topcon X35 (precision farming controller) and Trimble TMX-2050 (agriculture automation system), each tailored for specific automation applications.

4. **End Effectors:** End effectors, called End of Arm Tooling (EOAT), are attachments that enable robots to interact with objects. They are the robotic equivalent of hands, performing tasks such as gripping, welding, cutting, painting and packaging. Grippers may be mechanical, suction-based, or magnetic, facilitating the pick and place of objects in manufacturing, logistics and food processing. Welding torches are prevalent for precision welding tasks in the automotive and aerospace sectors. Cutting tools play an important role in precision machining and automated harvesting. In agriculture, soft-touch grippers enable robots to harvest delicate fruits like strawberries and tomatoes without causing damage. Proper calibration of EOAT is essential to ensure that grippers, welding torches and suction cups exert optimal force, positioning and accuracy for their specific tasks. Notable models include the Schunk PGN-Plus (gripper), Fronius TPS/i (welding torch) and Octinion Rubion (soft-touch gripper for strawberry picking).
5. **Motion Control Systems:** Motion control systems are essential for automation and robotics, enabling precise movement, positioning and synchronisation of mechanical components in applications such as robotic arms, CNC machines and conveyor systems. These systems consist of controllers (PLCs, motion controllers, CNC controllers) that process commands, along with drives and motors (servo motors, stepper motors) that generate movement and feedback devices (encoders, sensors) that ensure accuracy through real-time adjustments. Motion control can be open-loop, executing predefined movements without feedback, or closed-loop, continuously adjusting based on sensor data for greater precision. These systems are widely used in industrial automation, CNC machining, automated transport and even medical robotics. Examples include Siemens SINAMICS Drives and Yaskawa Servo Motors.

DID YOU KNOW?

Servo motors provide precise position control with feedback, making them ideal for high-accuracy applications. In contrast, stepper motors move in fixed increments without feedback, offering simpler, cost-effective motion control.

6. **Light Detection and Ranging (LiDAR):** LiDAR is a remote sensing technology that uses laser pulses to measure distances and create high-resolution 3D maps of environments. It works by emitting laser beams that reflect off surfaces and return to the sensor; the time taken for the light to return is used to calculate precise distances. LiDAR systems consist of a laser source, scanner and detector, often paired with GPS for accurate positioning. It is widely employed in autonomous vehicles, robotics, geospatial mapping and industrial automation. It enables real-time object detection, navigation and spatial awareness, essential for advanced perception systems in smart automation and robotics. Top models of LiDAR include Velodyne Lidar Alpha Prime and Ouster OS1-128, among others.

7. **Vision Systems:** A vision system is a sophisticated image-processing technology utilised in industrial automation, robotics and quality control to enable machines to capture, analyse and interpret visual data for precise decision-making. It comprises high-resolution cameras, vision sensors, lighting and advanced image-processing software, often powered by artificial intelligence and machine learning to enhance object recognition and automation capabilities. These systems process images in real-time to identify object position, shape, defects, or barcodes, guiding robotic movements or ensuring product quality. Some leading vision system models include the Cognex In-Sight 9902L and Keyence CV-X Series.
8. **Human-Machine Interface (HMI):** HMIs offer a user-friendly interface that enables operators to monitor, configure and troubleshoot automation systems. They can be touchscreen panels featuring a graphical user interface (GUI) or control panels with physical buttons and indicators. In manufacturing and logistics, HMIs facilitate worker interaction with robots and the adjustment of operational parameters. In agriculture, HMIs are employed for remote monitoring of irrigation systems, tractors and automated greenhouses. Example models include Siemens SIMATIC HMI Comfort Panels, Allen-Bradley PanelView Plus 7 and Trimble GFX-750 (for smart farming applications).
9. **Supervisory Control and Data Acquisition (SCADA):** SCADA is a set of centralised applications to monitor, control and analyse real-time data from various equipment and processes. SCADA allows operators to oversee robotic performance, track sensor data and generate alerts for system failures, enhancing operational reliability. Examples of SCADA software include Ignition SCADA and Siemens WinCC.
10. **Manufacturing Execution Systems (MES):** MES plays a vital role in real-time production control, quality management, and data-driven operations. In robotic manufacturing environments, MES acts as a central hub, integrating with robots, PLCs, and automation controllers to capture, analyse, and optimise production data. Advanced MES platforms such as Opcenter from Siemens, DELMIA from Dassault Systèmes, and Rockwell's FactoryTalk, enable dynamic scheduling, robot task management, downtime tracking, and seamless traceability across automated production lines. By coordinating robotic work cells and ensuring compliance with Industry 4.0 standards, MES significantly enhances manufacturing agility, efficiency, product quality and traceability in production lines.
11. **AI-Driven Robotics:** AI-powered robotics transforms automation by enabling real-time decision-making, adaptability and precision. Unlike traditional robots, AI-driven robots can learn from their environment, adjust movements dynamically and optimise efficiency over time. In industrial automation, AI-powered robotic arms can automatically modify grip force according to object size and weight. In agriculture, AI-driven robots are employed for autonomous fruit picking, precision seeding and automated weeding, significantly lowering labour costs and enhancing productivity. Notable models include Boston Dynamics Spot (AI-driven robot), FarmBot Genesis (AI-powered farming robot) and Fanuc AI-powered robots.
12. **AI-Based Predictive Maintenance:** AI-driven predictive maintenance prevents equipment failures by analysing real-time data from sensors. By detecting anomalies in vibration, temperature and pressure levels, these systems can predict potential breakdowns and schedule maintenance before failures occur. In manufacturing, AI-based

monitoring systems track the condition of machinery, conveyors and robotic arms. In agriculture, predictive maintenance monitors tractors, irrigation pumps and greenhouse equipment, ensuring optimal performance. Example models include Siemens MindSphere (Predictive Maintenance Platform), PTC ThingWorx and John Deere JDLink (Smart tractor monitoring system).

13. **Cloud Computing and Industrial Internet of Things (IIoT):** Cloud computing and IIoT facilitate real-time data collection, remote monitoring and predictive analytics in automation. Cloud-based SCADA systems enable factory managers to control operations from any location. Notable examples include AWS IoT Core, Microsoft Azure IoT, Climate FieldView (for smart farming) and NetBeat by Netafim (an AI-powered irrigation system). We will discuss cloud computing in detail in Chapter 8.
14. **Edge Computing in Automation:** While cloud computing is essential, edge computing enables real-time data processing at the machine level without depending on the cloud, thus reducing latency. Solutions such as Siemens Industrial Edge and AWS Greengrass improve automation performance. We will discuss edge computing in detail in Chapter 8.
15. **Networking and Communication Protocols:** Automation depends on industrial communication protocols such as Ethernet/IP, Modbus, PROFIBUS and OPC-UA for seamless data exchange among sensors, controllers and cloud systems. These facilitate smooth coordination in smart factories and IIoT ecosystems.
16. **Cybersecurity for Industrial Automation:** With greater connectivity through IIoT and cloud computing, securing industrial automation systems from cyber threats is crucial. AI-driven security platforms and industrial firewalls (such as Cisco Cyber Vision and Palo Alto Networks for OT Security) protect automated systems against cyberattacks. Cybersecurity is discussed in detail in Chapter 13.
17. **Simulation and Digital Twins:** Before implementing in production, the use of robot simulation software and digital twin technology is necessary to test automation systems in a virtual environment. Siemens Tecnomatix, Dassault Systèmes DELMIA and ABB RobotStudio are commonly utilised for this purpose. Digital Twins and Simulation concepts are discussed in detail in Chapter 11.
18. **Robot Safety Standards:** Robot safety standards are essential to guarantee the safe design, operation and integration of robotic systems in industrial and collaborative settings. These standards help reduce risks, safeguard human workers and prevent accidents by outlining specific safety measures for robot movement, human interaction and emergency response protocols.

    Among the most widely recognised industrial robot safety standards, ISO 10218-1 and ISO 10218-2 establish critical safety requirements for robot manufacturers and system integrators. These standards outline protective measures such as emergency stop functions, speed limitations, restricted work zones and safety mechanisms like physical barriers, light curtains and safety interlocks. For collaborative robots (cobots), designed to work alongside humans, ISO/TS 15066 introduces power and force limiting guidelines, speed and separation monitoring, hand-guided operation and collision detection to prevent potential injuries. Regional standards include RIA R15.06 in North America, ANSI/RIA R15.08 for mobile robots and the European Machinery Directive 2006/42/EC, all ensuring adherence to global safety regulations. Modern robots utilise AI-driven monitoring, vision systems and safety-rated sensors to detect human presence, adjust speed and halt operations when necessary. Compliance with these standards is vital for workplace safety, risk reduction and effective human-robot collaboration in industrial automation.

19. **Ingress Protection (IP) Rating:** The IP Rating, defined by the International Electrotechnical Commission (IEC) Standard 60529, classifies the degree of protection that an electrical enclosure or mechanical casing offers against solid particles (like dust and debris) and liquid ingress (such as water and moisture). The IP code consists of two digits: the first digit (ranging from 0 to 6) indicates protection against solid particles, while the second digit (ranging from 0 to 9) signifies resistance to water penetration. This rating system allows for a more accurate classification than generic terms like 'waterproof' or 'dustproof', ensuring that equipment meets specific environmental protection standards for industrial, automotive and consumer applications. The higher the IP rating, the greater the protection against dust and water. This classification is widely used in industrial automation, robotics, outdoor electrical equipment and consumer electronics, ensuring devices are suitable for harsh environments. Table 5.1 presents the IP Rating Chart.

Table 5.1 IP Rating Chart

IP Rating	Solid Protection (First Digit)	Liquid Protection (Second Digit)
IP00	No protection	No protection
IP10	Protection from large objects (≥ 50 mm)	No protection
IP20	Protection from fingers (≥ 12.5 mm)	No protection
IP30	Protection from tools/wires (≥ 2.5 mm)	No protection
IP40	Protection from small tools/wires (≥ 1 mm)	No protection
IP50	Limited dust protection	No protection
IP60	Fully dust-tight	No protection
IP61	Fully dust-tight	Dripping water (vertically falling)
IP65	Fully dust-tight	Water jets from any direction
IP67	Fully dust-tight	Temporary immersion in water (up to 1 m)
IP68	Fully dust-tight	Continuous immersion in water (manufacturer-defined depth)
IP69	Fully dust-tight	High-pressure, high-temperature water jets

5.4 TYPES OF AUTOMATION AND ROBOTICS

So far, we have examined the evolution and essential building blocks of industrial automation and robotics. In this section, we will explore the types of automation and robotics commonly used in industries.

5.4.1 Types of Automation

There are four major types of automation, i.e., Fixed, Programmable, Flexible and Integrated Automation. Each serves different industrial needs, ranging from high-volume repetitive manufacturing to dynamic, real-time adaptable production. In the following sections, we will try to understand these automation types and their role in optimising efficiency, reducing costs and enhancing productivity while aligning with modern smart manufacturing trends for various industries.

Fixed Automation

Fixed or hard automation refers to a pre-programmed manufacturing system designed to perform a single repetitive task with minimal human intervention. Its primary objective is to create a fully integrated sequence of automated processes, ensuring high-speed, standardised production with maximum efficiency. Once configured, these systems are dedicated to a specific product design and cannot be easily reprogrammed, making them ideal for high-volume, repetitive manufacturing.

A notable example of fixed automation is its application in automobile manufacturing, particularly in body welding, chassis assembly and painting. In a body shop, pre-programmed robotic arms perform precise welding at predetermined locations, while fixed robotic stations carry out bolting, riveting and stamping in a structured sequence. These automated assembly lines ensure consistency and efficiency but cannot adapt quickly to different vehicle models without significant reconfiguration.

Fixed automation excels in mass production, where large quantities of standardised products are manufactured with speed, precision and repeatability. With regular maintenance, systems offer long-term reliability and can provide continuous operation with minimal downtime. Although the initial setup costs can be substantial, the long-term return on investment (ROI) is high, as they maximise production output while lowering per-unit costs.

However, the rigidity of fixed automation poses challenges for industries that require frequent product modifications or customisation. The inability to adapt to changing demands makes it less suitable for small-batch production or sectors with rapidly evolving product designs. The high capital investment in system design and installation can also create barriers, particularly for small and medium-sized manufacturers. Manufacturers risk underutilising production lines if market demand fluctuates, leading to financial inefficiencies. Businesses must carefully assess costs, production volume and long-term adaptability before incorporating fixed automation into their manufacturing strategy.

Programmable Automation

Programmable automation enables manufacturing systems to adjust operations for various product configurations through reprogramming and setup changes. This flexibility makes it ideal for batch production, where each new batch requires modifications like tool changes, fixture adjustments and updated machine settings before production resumes. While this adaptability supports different product designs, reprogramming and setup adjustments can be time-consuming, resulting in temporary downtime. Industries that benefit from programmable automation typically manufacture products in low to medium volumes, where customisation and adaptability outweigh the necessity for continuous, high-speed production.

In automotive manufacturing, CNC machines are vital in machining engine blocks, crankshafts and transmission components. These machines work based on pre-programmed instructions that can be adjusted for various engine models, allowing manufacturers to produce multiple variants without overhauling the entire system. Similarly, in bottling plants, automated machines package products of different sizes and types by updating program settings, enabling manufacturers to switch between packaging formats efficiently. These examples highlight how programmable automation fosters product diversity without necessitating entirely new equipment.

While this type of automation offers adaptability and cost-effectiveness for batch production, it also presents challenges. The necessity for frequent reprogramming leads to downtime, which reduces overall productivity compared to fixed automation. Additionally, the initial investment in programmable automation is relatively high due to its advanced control systems and flexibility. Businesses must weigh these factors when considering programmable automation for their manufacturing needs.

Flexible Automation

Flexible or soft automation represents an advanced programmable automation that reduces downtime and enables quick transitions between product variations with minimal manual intervention. In contrast to traditional programmable automation, which necessitates time-consuming reprogramming and setup adjustments, flexible automation efficiently streamlines these processes, ensuring continuous production. It is designed to effectively manage a limited range of product variations, making it ideal for industries that face diverse and changing production demands. By removing the need for batch processing, flexible automation supports the continuous manufacturing of different products without significant changeover delays.

A prime example of flexible automation can be seen in modern automotive manufacturing, where AI-driven robotic arms and smart sensors dynamically adjust to various vehicle models on the same production line. For instance, BMW's Smart Factory employs adaptive robotic systems that change their programming in real-time to assemble sedans, SUVs and electric vehicles without halting production. Vision systems identify the type of car frame and automatically switch tools, such as grippers for different door panels, ensuring smooth and efficient operations. This adaptability makes flexible automation a powerful solution for industries requiring frequent product customisation.

Although flexible automation improves efficiency and responsiveness to shifting market demands, its adoption presents challenges. The sophisticated machinery and intelligent control systems necessary for automated changeovers entail a high initial investment, which may pose a barrier for some manufacturers. Furthermore, the cost per unit remains higher than that of fixed automation, as flexible systems prioritise adaptability over mass production efficiency. Since it does not achieve the economies of scale typical of fixed automation, it is better suited for industries that value variety and customisation over large-scale uniform manufacturing.

Integrated Automation

An integrated automation system is a fully connected and self-regulating framework that automates manufacturing processes under centralised computer control. By seamlessly integrating machines, robots, sensors and software, this enables real-time data exchange, predictive maintenance and optimised production efficiency with minimal human intervention. As a key component of Industry 4.0, this system enhances operational agility and ensures greater precision, making it ideal for smart factories. By coordinating all aspects of production, integrated automation improves quality control, reduces downtime and streamlines workflow management.

Modern automotive plants, such as Tesla's Gigafactory, demonstrate the power of integrated automation in optimising vehicle manufacturing and supply chain management. PLCs and SCADA systems oversee robotic welding, painting and assembly by collecting real-time sensor data to ensure precision. AI-driven analytics enhance reliability by predicting potential equipment failures before they disrupt production. Additionally, flexible robotic assembly lines dynamically adjust to different car models without halting operations. At the same time, the MES continuously tracks each vehicle's progress, reducing defects and ensuring consistent quality.

Figure 5.3 depicts an automotive assembly plant with integrated automation. The plant has a mix of fixed and flexible automation powered by integration of hardware, software, data, HMIs across different functions like design, production, maintenance, inventory, logistics.

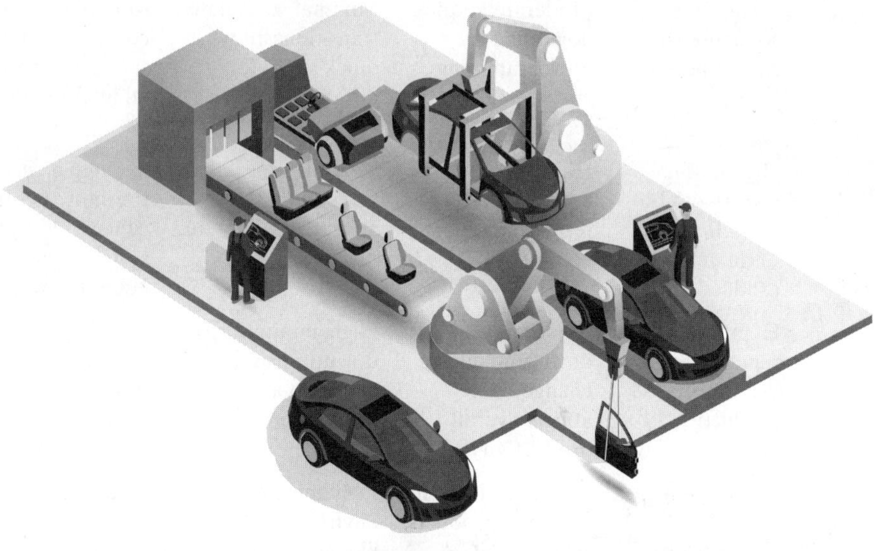

Credit: web_monster / Shutterstock

Figure 5.3 Integrated automation production line

While integrated automation greatly enhances production efficiency, it requires a substantial initial investment due to its complex system integration, advanced robotics and intelligent control platforms. Moreover, maintaining such sophisticated systems requires a highly skilled workforce capable of managing automation hardware and software integration. The necessity for continuous updates, system monitoring and specialised expertise can render long-term maintenance costly and resource-intensive. Nevertheless, businesses implementing integrated automation enjoy improved operational efficiency, greater flexibility and a more resilient manufacturing process.

Table 5.2 is the comparison matrix of the above types of automation. In this section, we have studied the various types of automation. Next, we will examine the different types of robots and their applications in industrial automation.

Table 5.2 Automation Type Comparison Matrix

Aspect	Fixed Automation	Programmable Automation	Flexible Automation	Integrated Automation
Flexibility	Very low; designed for one specific task or product.	Moderate; allows reprogramming for different batches but requires downtime.	High; It supports continuous production with quick and seamless product changeovers.	Very high; integrates and automates the production process, including discrete and continuous operations.
Production Volume	High; ideal for mass production of standardised products.	Low to medium; suitable for batch production.	Medium to high; supports variable production demands efficiently.	High; supports both batch and continuous production.
Setup Time	Long; once configured, cannot be easily adjusted.	Moderate; requires downtime for reprogramming and reconfiguration.	Minimal; automated transitions allow quick setup for different product styles.	Minimal; centralised control enables seamless operation of all integrated systems.
Cost	High initial investment but cost-effective for large-scale production.	Higher than fixed automation due to reprogrammable equipment.	Very high initial investment for advanced, adaptable machinery.	Very high cost for implementing centralised control systems and advanced equipment.
Efficiency	High for repetitive tasks; low for diverse production needs.	Moderate; downtime during changeovers affects overall productivity.	High; minimal downtime and continuous production ensure efficiency.	Very high; streamlines and optimises the production process with minimal human intervention.
Applications	Mass production of standardised products like automotive assembly lines and packaging plants.	Batch production of varied products, such as CNC machining, robotic welding and steel rolling.	Dynamic production environments like automotive assembly lines and circuit board manufacturing.	Smart factories, automated warehouses and pharmaceutical production and packaging.
Key Benefits	• High production speed and precision. • Cost-effective for high-volume tasks.	• Flexibility to produce different designs. • Equipment can be reused for multiple tasks or products.	• Continuous production of varied products with minimal downtime. • Eliminates batch processing; supports diverse schedules.	• Seamless integration of all systems for full automation. • Enables early detection of machine failures and efficient inventory management.

(*Continued*)

Table 5.2 *(Continued)*

Aspect	Fixed Automation	Programmable Automation	Flexible Automation	Integrated Automation
Key Shortcomings	• Lack of flexibility; inefficient for frequent design changes.	• Downtime during changeovers reduces productivity.	• High cost per unit due to lack of economies of scale and advanced technology.	• Extremely expensive to implement and manage.

5.4.2 Robot Types and Applications

There are many types of robots with a wide range of functionalities. In this section, we will discuss the types of robots frequently used in industrial setups like manufacturing, agriculture, supply chain and logistics.

Key Parameters of Robots

Before exploring the various types of robots, let's first examine the key parameters on which robots are typically categorised:

1. **Mobility:** This term refers to the ability of robots to move and navigate within their environment to perform tasks. Robots can be either stationary or fixed in place (ideal for welding, assembly, packaging, etc.) or mobile, meaning they can move freely along a predefined path or autonomously (ideal for material transport, warehouse logistics, agricultural fieldwork, etc.).
2. **Processing Speed:** Processing speed denotes how quickly a robot can complete a task, encompassing movement, decision-making and response time. It depends on factors such as actuator speed, sensor input processing and computational efficiency. High-speed robots are employed in fast-paced industries like food packaging and electronics assembly. In contrast, low-speed, high-precision robots prioritise accuracy over speed, ensuring quality and safety in fields like aerospace or surgery.
3. **Payload Size:** Payload size indicates the maximum weight a robot can lift, carry, or manipulate while maintaining stability and performance. This factor is crucial in selecting robots for industrial, logistics and agricultural applications. Heavy-duty robots manage large automotive parts, construction materials and heavy machinery, while light-payload robots are suited for delicate tasks like electronics assembly, medical applications and food handling.
4. **Modular Expandability:** Modular expandability describes a robot's capability to adapt and integrate additional tools (such as new grippers), sensors, interchangeable end-effectors, or software upgrades without necessitating a complete system replacement. This feature enhances flexibility, longevity and scalability in automation.
5. **Degree of Freedom (DOF):** DOF in a robot refers to the number of independent movements it can perform. Each DOF represents a joint or axis that allows motion in a

specific direction (e.g., up/down, left/right, rotation). Low DOF (3-4) is suitable for simple, linear movements like pick-and-place tasks, while high DOF (6+) permits complex, multi-directional movements for tasks like welding, painting and assembly.

6. **Work Envelope:** A robot's work envelope is the three-dimensional space within which it can move and operate. It is defined by the robot's reach, range of motion and DOF. A small work envelope is ideal for fast pick-and-place tasks in compact areas, while a large work envelope is used for welding, painting and material handling over a wide area.

With this understanding, let's explore the types of robots widely used in industries.

Cartesian Robots

Due to their simplicity and versatility, Cartesian robots are among the most widely used in industrial applications. They operate using a Cartesian coordinate system (X, Y, Z) for linear movements along three axes: forward/backward, up/down and side-to-side, which is why they are also referred to as 'linear robots'. Their cuboid-shaped work envelope is ideal for pick-and-place, material handling, packaging, storage, cutting and drilling tasks. When positioned on a raised platform above two parallel rails, these Cartesian robots are termed 'Gantry robots'. This configuration prevents the second axis from being cantilevered and enables gantries to achieve longer stroke lengths and, in many cases, larger payloads than standard Cartesian robots. Figure 5.4 depicts the Cartesian 'Gantry' robots.

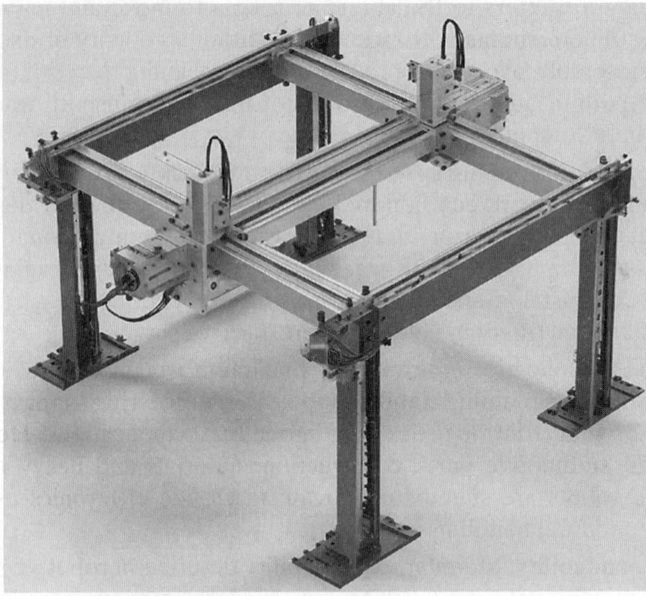

Credit: Images generated by AI

Figure 5.4 Cartesian 'Gantry' robots

Cartesian robots offer a simple and efficient control system, making them easy to program and operate. They deliver high accuracy and rigidity across all three linear axes, ensuring precision in industrial applications. These robots are also customisable and cost-effective, enabling manufacturers to adapt them to specific needs without incurring excessive costs. Additionally, gantry models can lift heavy objects, making them well-suited for material handling and assembly tasks in industries like automotive and aerospace.

Despite their advantages, Cartesian robots are confined to linear movements and lack rotational or multi-axis capabilities, limiting their flexibility in complex automation tasks. They also require significant floor or mounting space, especially for long stroke length applications. Moreover, their structure demands regular maintenance to prevent deflection or bending, which can arise from extended stroke lengths, impacting overall stability and performance.

Cartesian robots are widely utilised in assembly lines across various industries, including electronics and automotive, where precision and repeatability are critical criteria. They play a vital role in precision cutting, drilling and 3D printing, ensuring high accuracy in manufacturing processes. In warehouses and logistics, these robots enhance material handling by effectively transporting and organising goods.

SCARA Robots

Selective Compliance Assembly Robot Arm (SCARA) robots are advanced automation solutions that integrate X, Y, Z and rotary motion into a compact system, requiring only end-of-arm tooling for operation. Their cylindrical work envelope and two rotational joints provide enhanced flexibility, enabling faster and more complex movements than Cartesian robots, though with slightly lower precision.

A key feature of SCARA robots is their selective compliance, which allows for slight adjustments in horizontal motion while maintaining stability in vertical movements. This makes them ideal for component insertion, pick-and-place operations and circuit board assembly, where alignment flexibility is essential without compromising accuracy. Figure 5.5 depicts the SCARA robot.

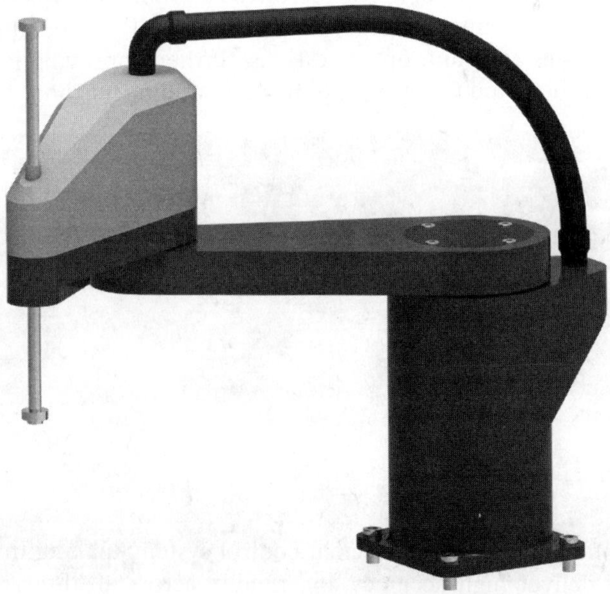

Credit: laufer / stock.adobe.com

Figure 5.5 SCARA robots

However, they have limitations in payload capacity and range of motion, making them unsuitable for lifting heavy objects or performing complex, multidirectional movements. Unlike articulated robotic arms, their DOF is limited (4), restricting them to simpler, short-range applications.

Despite these limitations, SCARA robots are widely utilised in electronics manufacturing, material handling, biomedical applications and packaging. They play an essential role in circuit board assembly, soldering and lightweight pick-and-place tasks, ensuring efficiency in electronics production. In the biomedical sector, they precisely handle delicate components, while in packaging, they facilitate quick sorting, alignment and packing in fast-paced environments.

Articulated (6 Axis) Robots

Articulated robots, commonly known as 6-axis robots, are among the most versatile and sophisticated industrial robots, capable of performing a wide range of complex tasks. Their design mimics a human arm, featuring multiple joints that allow movement across various planes with 6 or more DOF. This flexibility enables them to rotate and manoeuvre in multiple directions, making them ideal for intricate tasks that other robot types may struggle to accomplish. Some models may have fewer or additional axes depending on specific industrial applications. Their ability to handle diverse operations makes them indispensable in automotive manufacturing, aerospace, logistics and precision machining, where adaptability and precision are essential.

Initially, articulated robots required strict separation from human workers due to safety concerns, as their robust and fast-moving arms posed potential hazards. However, advancements in sensor technology, computer vision and collision detection systems have allowed modern articulated robots to operate safely alongside human workers. These developments have led to their increasing adoption in collaborative environments, where they enhance production efficiency without compromising workplace safety. By integrating AI-driven automation, these robots can learn and optimise tasks, further improving industrial workflows. Figure 5.6 illustrates articulated robots in action while welding machine parts.

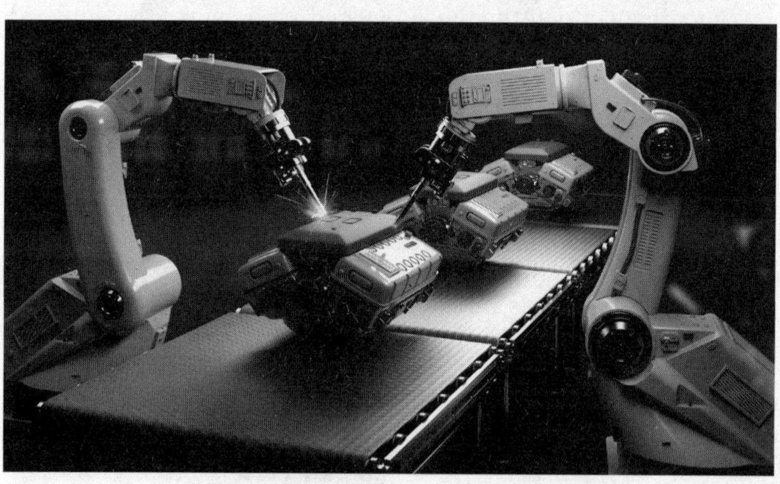

Credit: Phonlamai Photo / Shutterstock

Figure 5.6 Articulated robots in action

Articulated robots are highly valued for their multi-plane movement capabilities, precision and load-handling capacity. Their durability and strength enable them to lift and manipulate heavy objects, making them essential for demanding industrial applications. They can also be equipped with various EOATs, such as grippers, welding torches and cutting instruments. While these advantages make articulated robots a powerful automation solution, they also present challenges. Their high cost, complex programming and maintenance requirements can hinder smaller businesses. Additionally, their sophisticated motion control requires skilled operators to ensure efficient performance.

Despite these challenges, articulated robots are widely used in painting, welding, material handling, packaging and agriculture. In automotive manufacturing, they are widely used in paint shops. Also, they execute tasks like arc welding and spot welding in welding shops. In material handling, they assist in lifting, transferring and stacking objects. Their role in packaging and palletising automates sorting and arrangement in warehouses. Furthermore, in agriculture, articulated robots mounted on Autonomous Mobile Robots (AMRs) help in fruit picking, vegetable harvesting, pruning and weed removal, reducing reliance on manual labour and improving productivity. Their ability to operate across various industries makes them vital to modern automation.

Collaborative Robots (COBOTs)

Collaborative robots, or cobots, are designed to work safely alongside humans, eliminating the need for fenced enclosures. Operating at controlled speeds and forces, they enable seamless human-robot interaction in assembly, pick-and-place and material handling tasks with over 6 DOF. Unlike traditional industrial robots, cobots are highly adaptable and easy to program, allowing operators to teach them tasks through direct physical guidance instead of complex coding. Their flexibility and user-friendliness benefit large, small and medium-sized businesses, facilitating cost-effective automation with minimal training and infrastructure adjustments.

Cobots operate at varying levels of human-robot collaboration, from working independently in shared spaces to fully cooperating on joint tasks. They comply with relevant safety standards, as mentioned in Section 5.3. Cobots incorporate features such as automatic stopping upon human detection, hand-guided programming, adaptive speed control and force-limiting mechanisms. These safety measures enable cobots to operate efficiently in dynamic environments, enhancing productivity while maintaining worker safety. Figure 5.7 illustrates a cobot working on machine assembly alongside its human colleague.

Cobots provide numerous advantages, such as easy programming, cost-effective deployment and adaptability in shared workspaces. However, cobots have limitations in payload capacity and speed, rendering them less suitable for heavy-duty or high-throughput applications. While they excel in precision-driven tasks, they may not match the output efficiency of traditional industrial robots in fast-paced production environments.

Despite these limitations, cobots are extensively used in machine tending, where they assist in loading and unloading parts in CNC machines and other automated systems. They play an important role in pick-and-place operations, accurately sorting and positioning components. Cobots handle small and complex parts with high precision in assembly tasks, particularly in electronics manufacturing. Additionally, they are employed in arc welding for safe and controlled lightweight welding operations. Cobots equipped with vision systems also conduct

Credit: BigBlueStudio / Shutterstock

Figure 5.7 Cobot in action

quality inspections, identifying defects and ensuring consistent product standards in manufacturing lines.

Automated Guided Vehicles and Autonomous Mobile Robots

Automated Guided Vehicles (AGVs) are pre-programmed transport systems designed to follow fixed paths using magnetic strips, embedded wires, laser reflectors, or predefined markers. They are widely used in structured environments such as automotive assembly lines and large manufacturing plants, where material transport follows repetitive patterns. AGVs rely on basic sensors and collision detection systems to prevent accidents, but they lack the ability to reroute autonomously. If an obstacle blocks their path, they stop until manual intervention removes the obstruction. Their reliability makes them ideal for transporting pallets, raw materials and finished goods in settings where workflow layouts are consistent and predictable.

Autonomous Mobile Robots (AMRs) provide a more intelligent and flexible alternative to AGVs, functioning as self-navigating robotic systems equipped with LiDAR, computer vision, AI-driven mapping and advanced sensors. Unlike AGVs, AMRs do not rely on predefined paths; instead, they use Simultaneous Localisation and Mapping (SLAM) technology to create and update dynamic maps of their environment. This enables them to adapt to real-time changes, reroute around obstacles and optimise navigation based on traffic conditions and workload demands. AMRs are extensively used in e-commerce fulfilment centres, smart warehouses and hospitals, where real-time decision-making and adaptability are crucial. Figure 5.8 shows how an AMR and AGV behave when an obstacle is present on their path.

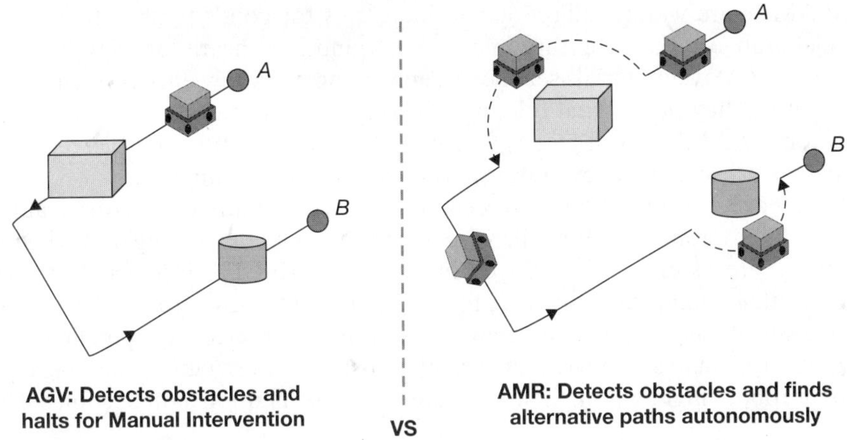

Figure 5.8 AGVs vs AMRs while facing obstacles

Both AGVs and AMRs offer significant advantages for industrial automation. Their ability to streamline material handling tasks, reduce human effort and enhance safety makes them valuable assets. While AGVs excel in predictable, high-volume workflows, AMRs provide greater flexibility by dynamically adjusting to changing environments. These systems are scalable, efficient and capable of operating in fleets, optimising logistics in warehouses and factories. However, challenges such as high initial investment costs, payload limitations and reliance on precise mapping must be considered when implementing these technologies. Additionally, AMRs need frequent software updates and maintenance to ensure optimal performance. Figure 5.9 shows an AMR working in warehouse setup for material handling operations.

Credit: MAX BLENDER 3D / Shutterstock

Figure 5.9 Autonomous mobile robots in action in warehouse

AGVs and AMRs are widely utilised across industries for goods transportation, order fulfilment, collaborative tasks, inventory management and healthcare logistics. In warehouses and factories, they move materials seamlessly, enhancing workflow efficiency. In e-commerce fulfilment centres, they play a vital role in delivering shelves or items to workers or robotic arms. Additionally, AMRs work alongside robotic arms for sorting and picking operations, assist in inventory tracking and even transport hospital medical supplies.

So far, we have learned about some commonly used robots in industrial automation. Besides these, many robots with similar types or functionalities (along with their benefits and limitations) are extensively employed in smart factories. For instance, Cylindrical and Delta Robots offer similar functionalities to Cartesian and SCARA robots. Likewise, drones utilised in logistics and agriculture and crawler robots used for oil and gas pipeline inspections are types of AMRs. Table 5.3 shows comparison snapshots for easy understanding of what we learned about robots, highlighting their key parameters as mentioned in Section 5.4.2.

Table 5.3 Comparison matrix of robots

Criteria	Cartesian	SCARA	Articulated	COBOT	AMRs
Design	Linear motion along X, Y, Z axes.	X, Y, Z + rotary motion with selective compliance.	Human-arm-like with 6 or more axes.	Designed to work alongside humans safely.	Wheeled robots with sensors
Mobility	Stationary	Stationary	Stationary	Stationary	Mobile
Processing Speed	Medium	High	Medium	Medium	Variable (Depends on Navigation)
Payload Size	High	Low-Medium	High	Low-Medium	Medium-High
Modular Expandability	Low	Low	High	High	High
DOF	3-4	4	6+	6+	Depends on Navigation
Work Envelope	Cuboidal (Rectangular/ Box-shaped)	Cylindrical	Spherical	Spherical or Semi-Spherical	No Fixed Work Envelope
Precision	High.	High but less than Cartesian.	Very high.	Moderate; precision varies by application.	High for navigation and task accuracy.
Applications	Cutting, drilling, 3D printing, material handling.	Assembly, soldering, electronics, packaging.	Welding, machine tending, assembly and material handling.	Assembly, pick-and-place, quality inspection.	Goods transportation, order fulfilment, inventory.

(*Continued*)

Table 5.2 (*Continued*)

Criteria	Cartesian	SCARA	Articulated	COBOT	AMRs
Advantages	Simple control, accurate and customisable.	Fast, cost-effective, easy integration.	High flexibility, powerful, adaptable tooling.	Safe, easy programming, cost-effective.	Flexible, scalable, autonomous navigation.
Disadvantages	No rotational motion, requires space, high maintenance.	Small payload, lower precision, fewer degrees of freedom.	High cost, complex programming, limited speed.	Limited speed/ strength, reduced productivity.	High initial cost, limited payload, mapping accuracy.
Prominent Models	IAI RCP6, Yamaha FLIP-X, Prusa i3 MK3S+.	FANUC SR-3iA, ABB IRB 910SC, Epson T3 SCARA.	ABB IRB 6700, FANUC M-710iC, Universal Robots UR10.	Universal Robots UR Series, ABB YuMi, KUKA LBR iiwa.	Amazon Robotics, MiR, Fetch Robotics Freight Series.

DID YOU KNOW?

Service robots are designed to assist humans in non-industrial settings, managing tasks in offices, homes, public spaces and factory maintenance. These autonomous or semi-autonomous machines are used in retail, healthcare, hospitality and domestic services, enhancing efficiency and convenience. Powered by AI, sensors and often integrated with AMRs, they can navigate environments, interact with people using Natural Language Processing (NLP) and carry out tasks with minimal human input. Examples include robotic vacuum cleaners, delivery robots, robotic waiters and elder care assistants, each tailored for specific functions. Unlike industrial robots, which follow structured, repetitive workflows, service robots function in dynamic, human-focused environments, requiring real-time decision-making and adaptability. Designed to work alongside people, they respond to unpredictable situations instead of strictly following pre-programmed sequences.

5.5 HOW TO ADOPT AUTOMATION AND ROBOTICS IN INDUSTRY 4.0? - STEP-BY-STEP APPROACH

In this section, we explore the systematic adoption of automation and robotics in various industrial setups within the framework of Industry 4.0. The objective is to understand the problem areas in industries and the factors we need to consider while approaching automation solutions for them. We will also examine how different types of automation and robotic technologies align with industrial needs, referencing the key building blocks of automation and robotics discussed in previous sections.

5.5.1 Step 1: Identify Problem Areas to Solve Through Automation and Robotics

Industries today face mounting challenges related to productivity, efficiency and sustainability. Labour shortages and rising operational costs make it difficult to maintain consistent output, while manual processes often lead to quality issues, defects and financial losses. The absence of real-time data and automation results in inefficiencies, unplanned downtime and production delays. Moreover, workplace hazards such as toxic fumes, hazardous chemicals, high temperatures and high energy consumption raise concerns about safety, compliance and environmental impact.

For instance, an automobile company can experience increase in defect rates and production disruptions due to manual assembly and unexpected equipment failures. By implementing automation, AI-driven quality control and predictive maintenance, the company can significantly enhance efficiency, reduce defects and minimise downtime. This illustrates how technology-driven solutions can support businesses in improving performance, adaptability and long-term sustainability.

5.5.2 Step 2: Define Objectives - What We Want to Achieve from Automation

Given these challenges, the next step is defining the key objectives the manufacturing unit seeks to achieve through automation. First, it is essential to identify business goals and Key Performance Indicators (KPIs), such as reducing labour costs, increasing productivity and improving product quality. The organisation should determine one or more goals based on its specific needs and establish measurable KPIs to gauge the success of automation by comparing targets with actual performance. Next, conducting a thorough assessment of current pain points in relation to these business goals is crucial, as not all issues can be resolved at once. Prioritisation is necessary to focus on the most critical problems, such as production delays, frequent equipment failures, or inefficiencies in human labour. Finally, defining a budget and expected return on investment (ROI) is vital for informed decision-making. This includes estimating costs, setting a budget, evaluating potential savings and analysing the long-term financial benefits that automation can provide to the organisation.

5.5.3 Step 3: Perform Feasibility Study

In this journey toward automation, it is important to determine which tasks should be automated, as not all processes in manufacturing can or should be automated. The primary candidates for automation include repetitive tasks such as assembly, welding, material handling and picking and placing materials, which demand high precision and consistency. Additionally, hazardous tasks like painting, handling toxic chemicals, working near furnaces, exposure to loud noise and high vibrations and lifting heavy objects must be automated to ensure worker safety. Moreover, areas prone to frequent failures, such as uncontrolled heat or pressure in boilers and wear and tear of turbine components, should also be prioritised for automation. Once these tasks are identified, the next step is to analyse the current workflow to understand how materials, machines and workers interact, identifying inefficiencies or bottlenecks that hinder productivity. Following this, an As-Is vs. To-Be mapping must be performed to evaluate the feasibility of automation by comparing the existing processes with the proposed automated solution. Simultaneously, the current infrastructure, i.e., machines, software and sensors, should be assessed to determine whether they can support automation or if upgrades are necessary. Another critical factor is determining the level of automation needed, whether to go for full automation or semi-automated solutions that include human intervention at critical points. For instance, baggage collection conveyors in airports often operate semi-automatically. They require manual activation and also run continuously regardless of luggage presence. In contrast, a fully automated conveyor system equipped with sensors, actuators and vision systems will activate only when luggage is detected, stop automatically in case of no luggage, malfunctions or spills. At this stage, a cost-benefit analysis is essential, to compare the initial investment against long-term savings in labour costs, energy efficiency and reduced waste. Lastly, the impact on the workforce must be assessed to determine if employees require training to operate or maintain automated systems. Planning upskilling and reskilling programs will ensure a seamless transition, allowing employees to adapt to new technologies while optimising productivity in an increasingly automated manufacturing environment.

5.5.4 Step 4: Choose the Right Automation and Robotics Technology

Selecting the right automation and robotics technology is essential for achieving an efficient and seamless transition to automated operations. Rather than adopting a one-size-fits-all approach, industries must evaluate their specific operational needs and implement a strategic combination of automation technologies. The key to this decision lies in understanding industry-specific tasks and aligning them with the most suitable robotic and automation solutions.

Industries such as automotive, electronics, warehousing and food processing typically utilise a mix of articulated robotic arms, AGVs, AMRs, conveyor systems, CNC machines and SCARA robots to automate the processes. For instance, articulated robots dominate automotive assembly lines for tasks such as welding, painting and assembling vehicles. Likewise, SCARA robots are employed in electronics manufacturing for precise chip placement and circuit board assembly, delivering high-speed and repeatable accuracy.

Industries like aerospace, chemical processing, metal foundries and heavy machinery manufacturing need advanced automation solutions tailored to high-risk environments. AI-driven robotics, vision systems, articulated robots, cobots and AGVs are crucial for handling hazardous tasks such as spray painting, chemical exposure and high-temperature operations near furnaces. For example, cobots assist with heavy lifting, reducing worker fatigue. Similarly, AI-based predictive maintenance solutions utilise smart sensors, actuators and AI-powered analytics to identify early signs of wear and tear, thereby minimising unexpected breakdowns.

Table 5.4 provides insight into the types of robots suitable for various industries performing specific tasks.

Table 5.4 Industry-wise task-wise suggested robot types.

Industry	Key Task	Best Robot Type
Automotive	Welding, painting, assembly	Articulated robots
Logistics and Warehousing	Order picking, material transport	AGVs, AMRs
Electronics	Circuit board assembly, chip placement	SCARA Robots
Pharmaceuticals	Lab automation, drug packaging	SCARA, Cobots
Aerospace	Precision welding, fuselage assembly	Articulated robots, Cartesian/ Gantry Robots

An essential automation component is material handling and transport, which is key in optimising manufacturing and logistics. Depending on industry-specific requirements, various automation technologies are utilised: Cartesian and Gantry Robots are used for precise, stationary lifting in CNC machine loading, heavy panel assembly and accurate stacking in industries such as automotive and aerospace. Operating within a fixed framework, they are highly reliable for structured environments but less suitable for dynamic spaces requiring mobility. AGVs and AMRs provide high flexibility in warehouses and manufacturing plants by autonomously transporting materials. These robots are particularly beneficial in logistics operations, as they adapt to real-time demand and reduce reliance on traditional conveyor-based systems. Conveyor Belts remain an industry standard for continuous, high-volume transport, especially in bottling plants, distribution centres and manufacturing lines. While efficient, they require dedicated floor space and lack the adaptability of AGVs and AMRs.

Table 5.5 shows suggested material handling technology options based on different scenarios.

Table 5.5 Material handling technology based on different industrial scenarios

Industrial Scenario	Suggested Technology
Heavy material transport over distances	AGVs, Forklift Robots, AMRs
Precise lifting and fixed placement	Gantry/Cartesian Robots
Dynamic warehousing and logistics	AGVs, AMRs, Cobots
Continuous, fixed-path transport	Conveyor Belts

Harsh manufacturing environments pose significant challenges to robotic systems. Hence, careful consideration of factors such as temperature, chemical exposure, dust and electrical hazards is needed. To ensure safe and reliable operation, it is essential to select robots with an appropriate IP rating (as explained in Section 5.3) that aligns with the specific conditions of the factory premises. The IP rating of robotic and automation systems varies based on their design, application and environmental exposure. Table 5.6 provides a general guideline for the same. Actual ratings may differ according to manufacturer specifications and industry requirements. Hence, it is crucial to evaluate each system's suitability for its intended use.

Table 5.6 IP Rating guidelines for different automation systems

Automation System	Typical IP Rating	Protection Level	Application Considerations
Cartesian Robots	IP54 - IP65	Dust-protected to water jet-resistant	Utilised in CNC machine loading, precision assembly and material handling. Higher IP-rated models are available for dusty or semi-wet environments.
SCARA Robots	IP40 - IP67	Basic dust protection to full water resistance	Common in electronics assembly and packaging. IP67-rated models are used in pharmaceutical, medical and food processing industries where washdown capability is required.
Articulated Robots	IP54 - IP67	Dust-resistant to water-resistant	Used in automotive, metalworking and general manufacturing. IP67-rated models can withstand coolants, paints, splashes and wet environments.
COBOTs	IP40 - IP67	Basic dust protection to full water resistance	Standard IP40-IP54 cobots are used indoors, clean environments, whereas IP67 cobots are specifically designed for food, pharmaceutical and hygienic applications.
AGVs	IP54 - IP65	Protection against dust and low-pressure water jets	Used in warehouses, logistics and factories. IP65-rated AGVs can handle semi-outdoor environments and resist water splashes.
AMRs	IP54 - IP67	Protection against dust and water sprays to full immersion resistance	Designed for transporting dynamic materials, higher IP-rated models are suitable for use in wet or industrial washdown areas.
Conveyor Systems	IP54 - IP66	Dust-protected to high-pressure water jet-resistant	IP54 conveyors are utilised in dry logistics and manufacturing, whereas IP66-rated conveyors can endure food processing and pharmaceutical washdowns.

A well-planned automation strategy ensures that manufacturing and logistics operations remain agile, adaptable and future-ready in response to evolving industry demands. By aligning the right automation and robotics technology with specific industry needs, businesses can maximise efficiency, reduce costs and enhance workplace safety.

5.5.5 Step 5: Integrate Hardware and Software

The seamless integration of hardware and software is crucial for the successful deployment of automation and robotics systems. Industrial automation software enables smooth coordination between various hardware components, ensuring a fully automated and optimised production process.

As explained in Section 5.3, PLCs manage robotic movements, conveyor operations and machine functions by continuously processing sensor inputs and triggering actuators accordingly. SCADA systems provide centralised monitoring and visualisation, allowing operators to effectively oversee real-time machine performance, adjust process parameters and respond to system alerts.

MES software tracks production data, optimises workflows and ensures quality control by integrating with PLCs and SCADA to analyse machine performance and material flow.

AI-driven predictive maintenance utilises sensor data to detect equipment anomalies, reducing unexpected downtime and enhancing system reliability.

AI-driven robotics enables adaptive automation, allowing robots to adjust to real-time dynamic conditions, improving efficiency and precision.

IIoT platforms enhance automation by connecting all devices to cloud-based systems, enabling remote monitoring, data analytics and predictive decision-making.

Secure industrial communication protocols such as Ethernet/IP, Modbus and OPC-UA ensure uninterrupted data exchange between these software platforms, creating an interconnected and intelligent automation ecosystem.

For critical systems, edge computing can be utilised for real-time data collection and analysis at the machine level, drastically reducing latency.

For example, a food packaging facility's robotic arm is integrated with a PLC-based automation system to handle product sorting and packaging. A SCADA system continuously monitors robot movements and temperature conditions, while an MES tracks production rates and packaging errors. AI-driven predictive maintenance detects motor wear and tear early, preventing unexpected failures. This combination of hardware, software and intelligent technologies results in a fully optimised, efficient and Industry 4.0-compliant automation system.

5.5.6 Step 6: Prepare Infrastructure and Safety Setup

Implementing automation and robotics in a manufacturing environment requires a well-planned infrastructure and a robust safety framework to ensure seamless integration, operational efficiency and worker safety. This involves establishing a reliable physical and digital infrastructure, designing optimised robotic workspaces and adhering to industrial safety standards to create a secure production environment.

A stable power and electrical system is fundamental, as industrial robots require high-voltage power supplies, with grounding and surge protection to prevent failures. Uninterruptible power supplies (UPS) are often installed to maintain continuous operation.

In addition, network and communication infrastructure is needed to enable real-time data exchange between robots, controllers, PLCs, MES and SCADA systems.

Factory layout and robotic work cell design are critical to ensuring efficiency, accessibility and safety. Workspaces must be configured to accommodate robots and other automated machines, allowing for optimal workflow and easy maintenance access. Robotic work cells should be equipped with barrier guards, presence-sensing devices, emergency stops, control interfaces and audio-visual alerts, all working together to minimise risks and maintain a safe working environment.

Barrier guards provide physical protection and come in various types, such as fixed guards, which prevent crushing or trapping hazards and interlocked guards, which halt robot operations when opened.

Presence-sensing devices create invisible safety zones around robots, ensuring that workers remain at a safe distance. Light curtains are a widely used safety feature that creates an optical barrier around hazardous areas using infrared beams. If an object or person interrupts the light field, the system immediately stops the robot, preventing accidental contact. Pressure-sensitive safety mats detect human presence and prevent hazardous interactions. Wearable safety devices integrated into personal protective equipment (PPE) track workers' proximity to dangerous machinery and trigger alerts when safety limits are exceeded.

Emergency Stop buttons, also called E Stop buttons, serve as critical fail-safes, allowing workers to immediately halt robotic operations in case of emergencies. These stops remain engaged even after power disruptions, preventing unintended restarts. HMIs function as centralised control systems, featuring status indicators, emergency shutdown controls and user-friendly displays that operators use to safely restart and monitor robotic systems.

Audio-visual warnings are essential components of robotics safety. Warning lights, signage and high-contrast floor markings inform workers about potential hazards and safe zones. Digital displays and real-time monitoring screens allow operators to track robot status and diagnose issues efficiently. Audible alarms ensure rapid response during emergency situations. For maximum effectiveness, these warning systems must be clearly visible, standardised across the facility and integrated with other safety mechanisms to provide layered protection. Figure 5.10 shows a typical robot cell setup with safety features.

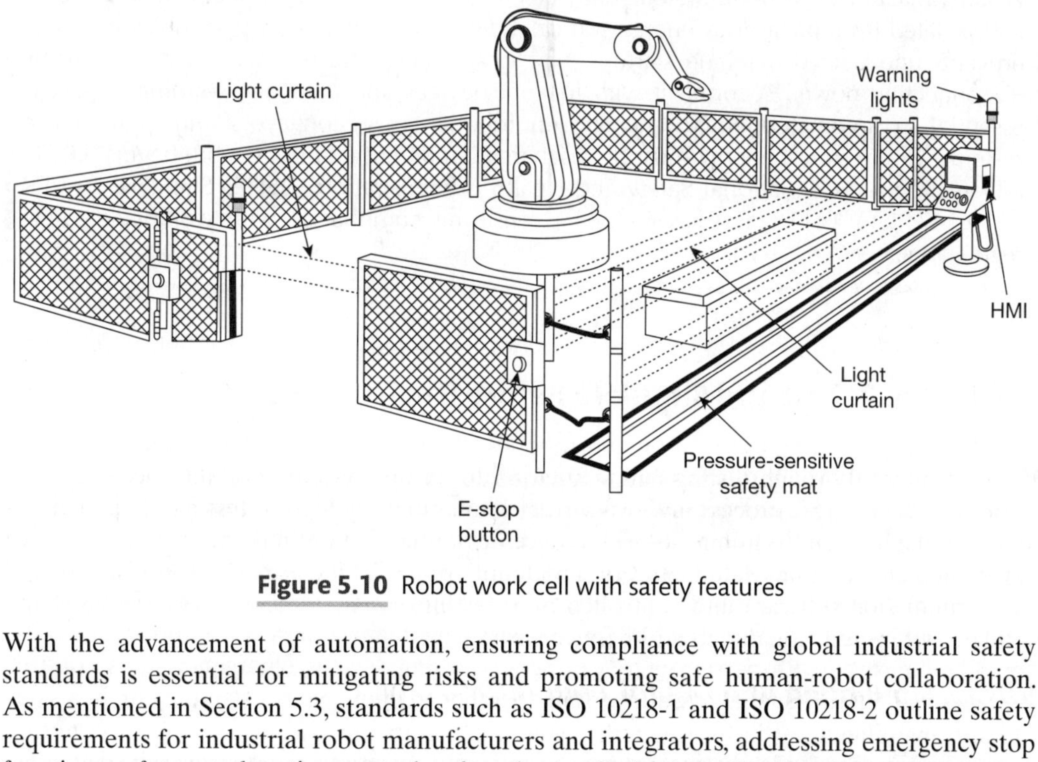

Figure 5.10 Robot work cell with safety features

With the advancement of automation, ensuring compliance with global industrial safety standards is essential for mitigating risks and promoting safe human-robot collaboration. As mentioned in Section 5.3, standards such as ISO 10218-1 and ISO 10218-2 outline safety requirements for industrial robot manufacturers and integrators, addressing emergency stop functions, safety-rated motion control and workspace limitations. Concurrently, ISO/TS 15066 establishes safety guidelines for collaborative robots (cobots), which include force and speed limits, collision detection and human interaction safety protocols. Functional safety assessments, such as Hazard Analysis and Risk Assessment (HARA), assist in identifying potential hazards, failure modes and safety controls before automation deployment.

Proper training and safety protocols are also crucial for workers operating automated systems. Employees must be educated on emergency stop and restart procedures, manual overrides and robot maintenance. Adherence to Lockout-Tagout (LOTO) procedures guarantees that electrical and mechanical systems are safely de-energised before maintenance, preventing unexpected machine startups.

DID YOU KNOW?

Hazard Analysis and Risk Assessment (HARA) is a structured process to identify, evaluate and mitigate potential industrial automation and robotics risks. It systematically analyses hazards associated with machinery, robotics and automated systems, assesses their likelihood and severity and implements safety measures to reduce or eliminate those risks. This process ensures compliance with safety standards such as ISO 12100 and ISO 10218, enhancing workplace safety and minimising accidents.

Lockout-Tagout (LOTO) is a critical safety procedure in industrial automation that ensures machinery, robotic systems and electrical equipment are effectively shut down and isolated from hazardous energy sources before maintenance or system setup. LOTO prevents unexpected machine startups, electrical shocks and mechanical movement by disconnecting power, securing it with lockout devices and attaching warning tags. It is essential in automated environments with robotic arms, conveyors and high-voltage systems, where stored energy must be discharged and verified before work begins. LOTO adheres to the Occupational Safety and Health Administration (OSHA) Standard 29 CFR 1910.147, ISO 14118 and ANSI Z244.1, ensuring compliance with workplace safety regulations.

5.5.7 Step 7: Test and Simulate Before Deployment

Before fully integrating automation and robotics into manufacturing operations, it is essential to conduct thorough testing and simulation to ensure system reliability, accuracy, efficiency and safety. This process involves virtual simulations, real-world testing, stress analysis and pilot deployment to minimise risks, reduce downtime and optimise system performance before implementation. Manufacturers can identify potential issues early by utilising digital twins, simulation software and controlled pilot testing, enhance automation processes and validate system functionality in a risk-free environment.

Virtual Simulation and Digital Twin Technology

Modern manufacturing industries utilise simulation software and digital twins to develop virtual models of robotic systems, production lines and automation workflows before deploying physical equipment. Robot simulation platforms assist engineers in simulating robotic motion, collision detection and human interactions. They also help predict mechanical wear, failures and inefficiencies using AI analytics and optimise robot path planning and production synchronisation before real-world execution.

Real-World Testing in a Controlled Environment

After completing virtual simulations, real-world testing is performed in a controlled production environment to evaluate stability, accuracy and system responsiveness. This phase includes prototype testing, calibration, performance validation and integration testing to guarantee optimal automation functionality.

Prototype testing begins with low-volume production trials to refine gripper alignment, motion parameters and material handling precision. Sensors, actuators and machine vision systems are calibrated to ensure accurate object detection, obstacle avoidance and quality control. Performance validation encompasses cycle time analysis, robot speed assessment and load testing to assess the system's capability to manage varying payloads without performance degradation.

To ensure adherence to ISO 10218-1 and ISO/TS 15066 safety standards, collision and safety testing are carried out, verifying emergency stop functionality and speed adjustments based on human proximity for secure human-robot collaboration. Integration testing connects the automation system with PLCs, SCADA and MES platforms to confirm real-time data exchange and process synchronisation, ensuring seamless communication between robots, sensors and enterprise-level systems.

Stress Testing and Disruptive Testing

Before final deployment, stress testing and disruptive testing assess the system's robustness under extreme conditions. The automation system undergoes high-speed and high-load testing to evaluate durability and performance limits. Simulated power failures and recovery tests measure the system's ability to resume operations without manual intervention during unexpected disruptions. Additionally, simulated sensor malfunctions and communication breakdowns examine how well the system detects and responds to faults or errors, ensuring resilience against unforeseen failures. This phase is critical for identifying weak points in automation workflows and ensuring that the system can operate reliably under real-world challenges.

Pilot Deployment and Operator Training

After successful testing, the automation system is gradually deployed on a limited scale before full-scale production. This pilot deployment enables engineers to monitor real-time performance, collect production data and make final adjustments before expansion. It ensures that all modifications are implemented efficiently, preventing disruptions during full-scale implementation.

Equally important is operator training and skill development. Workers and engineers should participate in structured training programs that cover robot operation, safety protocols, troubleshooting and system maintenance. This ensures personnel can manage automation workflows, respond to alerts and perform manual overrides when necessary. A well-trained workforce is crucial for maximising system efficiency, reducing downtime and ensuring safe collaboration between humans and robots.

5.5.8 Step 8: Go for Full-Scale Deployment and Monitoring

After rigorous testing and pilot implementation, the full-scale deployment of automation and robotics marks the final stage of integration into the manufacturing process. The deployment is generally carried out in phases to minimise operational disruptions, ensuring any unforeseen challenges that can be addressed without affecting overall production. A structured rollout strategy involves gradually increasing automation levels, closely monitoring system interactions and making real-time adjustments for a seamless transition. Once deployed, automation systems and robots must be continuously monitored to ensure they operate at peak performance. Table 5.7 displays some major Key Performance Indicators (KPIs) for tracking the performance of automation implementation.

Table 5.7 Automation and robotics KPIs

Category	KPI Name	Description
Operational Efficiency	Cycle Time	Time taken by a robot to complete one operation or task.
	Throughput Rate	Number of units processed per hour or shift.
	Utilisation Rate	Percentage of time a robot is actively working vs. idle.
Reliability and Maintenance	Mean Time to Repair (MTTR)	Average time required to restore a robot after failure.
	Downtime Percentage	Total time a robot is unavailable due to maintenance or failures.
	Predictive Maintenance Accuracy	Effectiveness of AI and IoT-based predictive maintenance.
Quality and Accuracy	Defect Rate	Percentage of defective units produced by an automated system.
	Rework Rate	Number of products requiring correction due to robotic errors.
	Pick-and-Place Accuracy	Precision of robotic arms in material handling tasks.
Safety and Compliance	Incident Rate	Number of safety-related incidents involving automation
	Emergency Stop Activations	Frequency of emergency shutdowns due to hazards.

IIoT sensors are vital in tracking key performance indicators and delivering real-time insights into robot movement, workload balancing and process optimisation. By integrating AI-driven analytics, manufacturers can utilise predictive maintenance to detect potential faults before they result in system failures. This approach minimises unplanned downtime, maximises resource utilisation and improves overall equipment effectiveness.

Thus far, we have explored the evolution and essential components of automation and robotics. We now also comprehend the various types of automation and robots and have examined the step-by-step process for implementing automation and robotics in an industrial setting. With all these insights established, let's delve into a real-life case study on implementing automation and robotics in a global retail giant.

5.6 CASE STUDY OF REAL-LIFE IMPLEMENTATION OF AUTOMATION AND ROBOTICS IN AMAZON WAREHOUSE[1]

5.6.1 Introduction

In 1994, Amazon was founded, initially focusing on selling books online. Over the years, the company has transformed into a global leader in e-commerce, driven by a relentless emphasis on innovation and efficiency. A key factor in its success has been its strategic investments in automation technologies, which have revolutionised fulfilment and logistics operations. This case study examines how the company's adoption of robotics, cloud computing and employee upskilling programs has led to remarkable growth in revenue, workforce and operational efficiency.

5.6.2 History and Early Investments

Pre-Automation Time

The company started in the mid-1990s, operating from a small office and rapidly expanding into many product categories. To meet growing demand, the company created large-scale distribution centres known as 'Fulfilment Centres', where orders are picked, packed and shipped. By 2006, the company launched cloud computing services, providing storage and computational power to other businesses and introduced a third-party seller program to enhance product variety and sales.

Automation Initiatives

In 2012, the company acquired Kiva Systems, which marked a significant turning point in its automation strategy. This acquisition established a foundation for advanced robotics in its Fulfilment Centres. Further investments in technologies such as AMRs, robotic arms and packaging systems further streamlined operations.

5.6.3 Robotics and Automation Systems

Kiva Systems and their AMRs

Kiva Systems, acquired by Amazon in 2012 and rebranded as Amazon Robotics, pioneered the development of AMRs to revolutionise warehouse automation. Initially introduced by Kiva, these drive units autonomously navigate Amazon's fulfilment centres, transporting entire shelves (pods) to human workers, eliminating the need for them to walk long distances to retrieve items. This goods-to-person system significantly reduces order processing time, enhances picking accuracy and optimises warehouse storage density.

---

1 Laber, J., Thamma, R. and Kirby, E. D. (2020). The impact of warehouse automation in Amazon's success. *Int. J. Innov. Sci. Eng. Technol.*, 7, 63–70.

Over time, Amazon Robotics introduced advanced AMR models such as Hercules, Pegasus and Xanthus, each designed for specialised warehouse functions. Hercules, a stronger and more compact AMR, improved payload capacity while maintaining high-speed navigation. Pegasus incorporated automated sorting capabilities, reducing package handling time in outbound logistics. Xanthus, Amazon's most modular and cost-efficient AMR, features a low-profile design and interchangeable top modules, enabling it to adapt to different warehouse workflows.

To ensure seamless human-robot collaboration, Amazon introduced safety innovations like the Robotic Tech Vest, a wearable system that allows AMRs to detect human presence and adjust their speed or path accordingly. These enhanced AI-driven obstacle avoidance and real-time safety protocols ensure that workers and robots coexist safely in dynamic warehouse environments.

Canvas Technology and their AMRs

In 2019, Amazon acquired Canvas Technology, a company specialising in computer vision-driven autonomous navigation for industrial environments. Canvas developed advanced AMRs equipped with 3D mapping, real-time object detection and AI-based path planning, allowing them to navigate warehouses independently without predefined paths or external guidance systems. Figure 5.11 depicts articulated robot arms placing packages on AMRs in the warehouse, which will then navigate autonomously to the desired locations in the fulfilment centres.

Credit: maxvanit / 123RF

Figure 5.11 Articulated robots and AMRs in action in a warehouse

5.6.4 Packaging Systems

Amazon has partnered with external firms like CMC Srl and SmartPac to automate packaging. CMC CartonWrap machines can pack up to 700 boxes per hour, significantly reducing the need for manual labour. SmartPac systems create patented envelopes. This automated packaging solution is designed to produce custom-fit, tamper-proof and eco-friendly shipping envelopes. The system efficiently adapts to the size and shape of each product, ensuring secure and compact packaging while minimising excess material usage.

DID YOU KNOW?

An automated carton packaging system customises boxes in real-time using fanfold-fed corrugated cardboard, ensuring each package is perfectly sized for its order. The system automatically adjusts box dimensions through product recognition or database integration, optimising flexibility and efficiency.

5.6.5 Usage of Soft Robotics

Amazon has explored grippers from Soft Robotics, a company specialising in adaptive, pneumatically actuated robotic grippers designed for delicate picking and placing tasks. These are used as EOAT fixed with articulated robots. Unlike traditional rigid grippers that rely on mechanical fingers or suction cups, Soft Robotics' technology uses flexible, air-filled actuators that conform to the shape of objects, enabling secure yet gentle handling of items with varying sizes, shapes and textures.

These grippers leverage soft material engineering, AI-driven control algorithms and real-time force modulation, making them ideal for e-commerce fulfilment, food handling and fragile item packaging. Amazon's interest in these bio-inspired robotic hands aligns with its push for advanced warehouse automation, where robots must efficiently sort, pick and pack a diverse range of products, from lightweight apparel to oddly shaped consumer goods.

By integrating computer vision and machine learning, Soft Robotics' grippers can dynamically adjust grip strength and optimise object handling strategies in fast-paced fulfilment centres. Figure 5.12 shows the grippers attached to robotic arms handling the picking and placing tasks in the warehouse.

5.6.6 Employee Development and Training

The company introduced several initiatives focused on education and skill development to address concerns about workforce displacement caused by automation. In 2012, it launched the Career Choice Program, which covers 95% of tuition costs for employees pursuing education in high-demand fields such as engineering and healthcare, empowering them with opportunities beyond their current roles.

In 2019, the company further solidified its commitment with Upskilling 2025, a $700 million investment to train 100,000 employees in software engineering, machine learning and IT support. This initiative includes programs like Amazon Technical Academy, which provides training for software engineering roles; Machine Learning University, offering six-week courses on AI and automation; and AWS Training and Certification, which equips employees with cloud expertise through structured classroom programs. These initiatives reflect the company's dedication to ensuring that automation enhances workforce potential rather than replacing it.

Figure 5.12 Robot arms with grippers doing pick and place jobs in warehouse

5.6.7 Usage of Cloud Computing and Robotics Software

Amazon Web Services (AWS)

Launched initially to assist third-party sellers, AWS transformed into a comprehensive cloud platform that provides computing power, storage, IoT and analytics services. By switching from Oracle databases to AWS solutions, the company decreased costs by 60% and administrative overhead by 70%.

AWS RoboMaker

This service assists developers in creating, testing and deploying robotic applications with the open-source Robot Operating System (ROS). Customers, including NASA and Stanley Black & Decker have utilised RoboMaker for innovative projects such as simulating rover designs and training autonomous construction vehicles.

5.6.8 Investments and Future Trends

In 2021, the company established the Robotics Innovation Hub, a 350,000-square-foot facility dedicated to advancing robotic technologies and fostering new opportunities in manufacturing and automation. This investment underscores the company's commitment to developing cutting-edge robotics solutions while creating high-tech jobs to support industry growth.

The company is actively investing in autonomous logistics solutions to tackle challenges in last-mile delivery. Notable innovations include Scout, a six-wheeled autonomous vehicle designed for efficient and contactless package delivery. Furthermore, the company's $1.2 billion acquisition of Zoox, a leader in self-driving vehicle technology, demonstrates its ambition to revolutionise autonomous transportation, enhancing the speed, reliability and efficiency of its logistics network.

Outcomes and Key Metrics

Since 2012, the company's investment in automation has led to remarkable workforce expansion, creating 700,000 new jobs across various roles. This shows that automation and job creation can coexist. At the same time, automation has fuelled significant revenue growth, with a 359% increase since its adoption, reinforcing its impact on business scalability and market competitiveness.

Beyond financial gains, automation has transformed operational efficiency by reducing the need for manual packaging labour, enhancing workplace safety in fulfilment centres and enabling faster delivery timelines. These improvements demonstrate how automation has optimised processes while maintaining a strong commitment to both employee well-being and customer satisfaction.

5.6.9 Key Takeaways

This case study illustrates how automation and robotics have played a crucial role in boosting efficiency, scalability and innovation in Amazon's fulfilment and logistics operations. The introduction of AMRs, automated packaging systems, AI-driven soft grippers and cloud-based robotics software has dramatically improved order processing and material handling. Investments in automation-friendly infrastructure, employee upskilling programs and AI-powered predictive systems showcase how automation can work alongside workforce development while promoting business growth and operational excellence. This case is an example of how a strategic, technology-driven approach to automation can reshape supply chain efficiency, reduce costs and cultivate a future-ready workforce.

5.7 CONCLUSION

In conclusion, industrial automation and robotics have transformed manufacturing, logistics and supply chain operations, enabling unprecedented levels of efficiency, precision and scalability. The integration of robotic arms, AMRs, AI-driven automation and cloud-based control systems has optimised productivity, operational flexibility and cost efficiency. As industries advance toward Industry 4.0 and beyond, emerging technologies like AI, IIoT, digital twins, edge computing and cloud robotics are set to redefine real-time decision-making, predictive maintenance and autonomous production.

Cloud and networked robotics are playing an increasingly vital role in the future of automation by enabling robots to leverage cloud computing for real-time data processing, AI model updates and remote coordination. Instead of being constrained by onboard processing power, cloud-connected robots can access high-performance computing resources, allowing for faster learning, dynamic task optimisation and seamless fleet management in smart factories and warehouses. The rise of 5G and edge computing further enhances these capabilities, ensuring low-latency communication and real-time synchronisation between distributed robotic systems.

The future of robotics will also be shaped by advancements in cobots, AI-powered vision systems, soft robotics and swarm robotics, where intelligent machines will perform complex tasks and adapt and interact more naturally with human workers. Meanwhile, fully autonomous supply chains and smart factories will become increasingly prevalent, leveraging cloud-driven robotic networks to optimise logistics, inventory management and production scheduling with minimal human intervention.

According to Fortune Business Insights, the industrial robotics market is expected to surge from $15.6 billion in 2023 to over $40 billion by 2030, driven by the increasing demand for automated manufacturing, logistics and AI-enhanced robotics. The AI in robotics market is also poised for rapid growth, with a CAGR of over 25%, as industries invest in self-learning, cloud-integrated robotic systems. With the evolution of autonomous mobile robots (AMRs), networked robotics and cloud-based automation, industries will transition to hyper-efficient, data-driven and highly adaptable operations.

Despite significant advancements, industrial automation and robotics face several challenges that hinder widespread adoption. High implementation costs, complex system integration and the demand for skilled labour make automation inaccessible for many companies. Legacy infrastructure often requires costly upgrades, while interoperability issues between different automation platforms slow down deployment. Cybersecurity risks, including data breaches and system vulnerabilities, threaten cloud-connected robotics.

Externally, regulatory hurdles, ethical concerns and economic uncertainties further complicate the growth of automation. Policymakers must balance technological innovation with labour market stability to prevent large-scale job displacement. Supply chain disruptions, semiconductor shortages and geopolitical tensions present additional barriers to manufacturing and deployment. Furthermore, industries like healthcare and transportation must address safety and liability concerns, ensuring that autonomous systems comply with rigorous international standards before large-scale implementation.

As businesses continue their digital transformation journeys, adopting intelligent, connected and cloud-powered automation solutions will be key to maintaining competitiveness in the global economy. Companies that embrace these next-generation robotics technologies will enhance operational agility and efficiency and unlock new opportunities for innovation, scalability and long-term resilience in the next era of industrial automation. However, overcoming these challenges will be crucial to unlocking the full potential of industrial automation, paving the way for a more intelligent, interconnected and sustainable industrial ecosystem.

SUMMARY

- Automation refers to the use of mechanical, electronic and computer systems to control industrial processes with minimal human intervention. It enhances efficiency, accuracy and consistency while reducing human errors and labour dependency.
- Robotics is a branch of automation that focuses on designing, developing and operating intelligent machines capable of performing tasks autonomously or semi-autonomously. Robots integrate sensors, controllers, actuators and end effectors to interact with their environment and execute precise actions. Robotics is considered a subset of automation.
- For a system to be classified as a robot, it must possess three key elements: sensing, processing and actuation (physical movement). Chatbots, AI, RPA, voice assistants, ATMs and smart appliances utilise automation, AI and intelligent decision-making. Still, they

lack mechanical movement and the ability to engage in autonomous physical interactions with their environment. Hence, they are not considered as robots.

- Evolution of Automation and Robotics (Key Milestones)
  - **First Industrial Revolution (18th Century):** Introduction of mechanised manufacturing powered by steam and water.
  - **Second Industrial Revolution (19th-20th Century):** Use of electricity, assembly lines and mass production techniques.
  - **Third Industrial Revolution (20th Century):** Adopting computers, PLCs and industrial robots to automate manufacturing.
  - **Industry 4.0 (21st Century):** Integrating AI, IIoT, cloud computing and cyber-physical systems for smart factories.
- Key Building Blocks of Industrial Automation and Robotics
  - **Sensors:** Gather real-time data to detect objects, temperature, force, motion and environmental conditions.
  - **Actuators:** Convert energy into motion, allowing robots to move, lift, or manipulate objects.
  - **Controllers:** Act as the brain of automation systems, processing sensor data and executing logic-based commands.
  - **End Effectors (EOAT):** Robotic hands or tools like grippers, welding torches and suction cups for object interaction.
  - **Motion Control Systems:** Enable precise movement and positioning of robotic components.
  - **LiDAR:** Uses laser pulses to measure distances, helping robots navigate autonomously.
  - **Vision Systems:** High-resolution cameras and AI-driven image processing for object detection, quality inspection and barcode reading.
  - **Human-Machine Interface (HMI):** Provides interactive touchscreens and control panels for monitoring automation systems.
  - **Supervisory Control and Data Acquisition (SCADA):** Centralised system for real-time monitoring and control of industrial processes.
  - **Manufacturing Execution Systems (MES):** Tracks and controls manufacturing operations for improved efficiency.
  - **AI-Driven Robotics:** Robots learn and optimise tasks dynamically using AI-based decision-making.
  - **AI-Based Predictive Maintenance:** Detects machine failures in advance using real-time sensor data.
  - **Cloud Computing and IIoT:** Enables remote monitoring and data-driven automation through cloud connectivity.
  - **Edge Computing:** Processes data locally at the machine level to reduce latency.
  - **Networking and Communication Protocols:** Ensure seamless data exchange between automation components.

- **Cybersecurity for Industrial Automation:** Protects automated systems from cyber threats.
- **Simulation and Digital Twins:** Create virtual models of robotic systems to test and optimise performance before deployment.
- **Robot Safety Standards:** Ensure safe robot operation and compliance with global safety regulations.
- **Ingress Protection (IP) Ratings:** Defines robot resistance to dust, water and environmental hazards.

- Types of Automation (Definition and Usage)
  - **Fixed Automation:** High-speed production systems with minimal flexibility.
  - **Programmable Automation:** It is designed for batch production and requires reprogramming and downtime for each new product.
  - **Flexible Automation:** Real-time adaptability to product changes, eliminating downtime.
  - **Integrated Automation:** Fully autonomous, interconnected systems with minimal human intervention.
- Types of Robots (Definition and Applications)
  - **Cartesian Robots:** Move in linear X, Y and Z axes and are ideal for pick-and-place, material handling, packaging, storage, cutting and drilling tasks.
  - **SCARA Robots:** Provide fast, precise horizontal movement; ideal for component insertion, pick-and-place operations and circuit board assembly.
  - **Articulated Robots:** Feature multiple rotary joints for highly flexible movement. They are used in welding, painting and assembly.
  - **Collaborative Robots (Cobots):** Work safely alongside humans in manufacturing, healthcare and logistics.
  - **Autonomous Mobile Robots (AMRs):** These navigate independently using AI and LiDAR and are used in warehousing and smart logistics.
  - **Automated Guided Vehicles (AGVs):** Follow predefined paths for material transport in manufacturing and logistics.
- Step-by-Step Process to Adopt Automation
  1. **Identify Industry Challenges:** Analyse inefficiencies and assess where automation can add value.
  2. **Conduct Feasibility Analysis:** Evaluate the costs, ROI and technical feasibility of automation.
  3. **Select the Right Automation Technology:** Based on industry-specific needs and functionalities supported by these technologies, choose between stationary robots like Cartesian, Scara and articulated robots, movable robots like AGVs and AMRs and smart technologies like AI, IIoT, vision systems, etc.
  4. **Integrate Hardware and Software:** Implement PLCs, SCADA, MES, AI-driven analytics and motion control systems.
  5. **Set Up Infrastructure and Safety:** Ensure power stability, network integration and robotic safety measures like LiDAR, presence sensing devices like light curtains

and pressure-sensitive mats, emergency stop buttons, barrier guards and audiovisual warnings.

6. **Simulation and Testing:** Use digital twins and real-world trials to optimise performance before deployment. Technologies Used are AI-based simulation software, digital twin software and stress testing tools.
7. **Deploy and Monitor Performance:** Implement KPIs and monitor performance through real-time dashboards and AI-driven analytics.

EXERCISE

Multiple Choice Questions (1 Mark)

1. What is the primary goal of industrial automation?
   (a) Increase manual labour
   (b) Improve efficiency, productivity and safety
   (c) Eliminate all human workers
   (d) Reduce electricity consumption

2. Which of the following is not considered an industrial robot?
   (a) SCARA Robot
   (b) Cartesian Robot
   (c) Smart Washing Machine
   (d) Articulated Robot

3. What distinguishes automation from robotics?
   (a) Automation requires human interaction, while robotics does not
   (b) Robotics focuses on machines with motion, while automation covers broader process control
   (c) Automation and robotics mean the same thing
   (d) Robotics only applies to manufacturing industries

4. What is the primary purpose of using digital twins in automation testing?
   (a) To replace human workers in the production line
   (b) To create virtual models for simulating automation workflows before deployment
   (c) To manufacture physical robots for testing
   (d) To train operators in a real-world factory environment

5. Which industrial robot is known for its cuboidal work envelope?
   (a) Cartesian Robot
   (b) SCARA Robot
   (c) Articulated Robot
   (d) Delta Robot

6. Which system oversees and controls large-scale industrial automation processes?
   (a) PLC
   (b) SCADA
   (c) CNC
   (d) AI

7. A factory is deploying robots that work alongside human workers to perform repetitive tasks such as lifting and assembling parts. The robots need to detect human presence and adjust their movements for safety. Which robot type is most suitable?
   (a) Cartesian Robots
   (b) SCARA Robots
   (c) Collaborative Robots (Cobots)
   (d) Articulated Robots

8. An automotive company needs to manufacture multiple car models on the same production line without downtime for reconfiguration. What type of automation should they adopt?
   (a) Fixed Automation
   (b) Programmable Automation
   (c) Flexible Automation
   (d) Integrated Automation

9. In a food processing automation system, which type of end effector is best suited for handling delicate fruits like strawberries?
   (a) Magnetic Gripper
   (b) Mechanical Claw Gripper
   (c) Suction Cup Gripper
   (d) Soft-Touch Gripper

10. A high-precision robotic arm requires continuous position feedback to ensure operation accuracy. Which type of motion control system should be implemented?
   (a) Open-Loop Motion Control
   (b) Closed-Loop Motion Control
   (c) Fixed Automation Control
   (d) Manual Motion Control

11. How does LiDAR technology enhance automation and robotics?
   (a) By improving robotic decision-making based on object detection and spatial mapping
   (b) By generating thermal data for industrial heating applications
   (c) By replacing actuators for motion control
   (d) By enhancing battery efficiency in mobile robots

12. A manufacturing company wants to automate defect detection in a production line using visual inspection. Which of the following should they use?

(a) CNC Controller
(b) LiDAR Sensor
(c) Vision System
(d) Pneumatic Actuator

13. Which of the following is a primary function of an HMI (Human-Machine Interface) in automation systems?
(a) Controlling robotic speed and acceleration
(b) Allowing operators to monitor and configure automation processes
(c) Detecting and analysing objects using vision sensors
(d) Providing power to robotic actuators

14. Which of the following is not a part of stress testing in automation deployment?
(a) High-speed and high-load testing
(b) Simulated power failures and recovery tests
(c) Collision and safety verification
(d) Simulated sensor malfunctions and communication breakdowns

15. Which type of robot is commonly used in welding, material handling and assembly tasks due to its multi-axis flexibility?
(a) Cartesian Robot
(b) SCARA Robot
(c) Articulated Robot
(d) AGV

16. A furniture manufacturing company produces custom-designed wooden cabinets. When switching between different cabinet designs, they need an automation system that allows batch production with some level of reprogramming with downtime. What type of automation should they adopt?
(a) Fixed Automation
(b) Programmable Automation
(c) Flexible Automation
(d) Integrated Automation

17. A car assembly plant needs robots that work alongside human workers to assemble vehicle interiors safely. These robots should have built-in safety mechanisms to prevent accidents. Which type of robot should they implement?
(a) SCARA Robots
(b) Collaborative Robots (Cobots)
(c) Articulated Robots
(d) Autonomous Mobile Robots (AMRs)

18. A global retail company wants to automate order picking and sorting in its distribution centre. The system should be able to retrieve products and deliver them to human pickers efficiently. Which solutions should they use?

(a) AMRs
(b) Traditional Forklifts
(c) Vision Systems for Object retrieving and Sorting operations
(d) Both A and C

19. An electronics manufacturer produces multiple smartphone models on the same production line. They need automation to switch between different phone models with minimal downtime quickly. Which of the following solutions would best meet their needs?
(a) Fixed Automation
(b) Programmable Automation
(c) Flexible Automation
(d) Integrated Automation

20. Which of the following KPIs measures the percentage of time a robot is actively working versus idle?
(a) Mean Time to Repair (MTTR)
(b) Utilisation Rate
(c) Pick-and-Place Accuracy
(d) Incident Rate

Short Answer Type Questions (2 Marks)

1. Why are robots considered a subset of industrial automation?
2. How does LiDAR technology improve the autonomy of industrial robots?
3. How does Flexible Automation differ from Fixed Automation?
4. How do Cobots differ from traditional industrial robots?
5. What role does Ingress Protection (IP) Rating play in industrial automation?
6. What are key differences between AGVs and AMRs?
7. How does a Manufacturing Execution System (MES) enhance industrial automation?
8. What is the significance of IIoT sensors in full-scale automation deployment?
9. Define two key performance indicators (KPIs) used in automation monitoring and their importance.
10. How is cloud computing transforming industrial robotics?

Long Answer Type Questions (5 Marks)

1. Define Industrial Automation and explain how it optimises the production process.
2. Define Robots and Robotics. How are robots a key part of industrial automation?

3. Discuss the role of LiDAR and Vision Systems in robotics and automation. How do they improve system performance?
4. Industrial automation relies on various building blocks. Explain how sensors, actuators and controllers work together in an automated system.
5. What are the Advantages and Limitations of Articulated Robots?
6. How can an organisation choose the right automation and robotics technology?
7. What Infrastructure and safety setup should be followed for implementing automation and robotics?
8. Explain the step-by-step process of simulation and testing of automation and robotics before live deployment.
9. How will emerging technologies redefine the future of industrial automation and robotics?
10. What economic and technological barriers affect the large-scale implementation of automation?

CHAPTER 6

Additive Manufacturing/ 3D Printing

OBJECTIVES

One of the important factors in connected manufacturing space is to make the manufacturing process distributive and as much automated as possible, end to end. Shopfloors for casting/ moulding activities were heavily dependent on large-scale manufacturing units and centralised processes, which could not implement *manufacturing near distribution units*. The advent of 3D printing paved the way for this distributive approach. Similarly, 3D printing highly influenced mass customisation and sustainable manufacturing practices.

By the end of this chapter, you should be able to:

- Understand the guiding principles of 3D printing
- Recognise different technologies of 3D printing and the types of software used in this, and
- Articulate the applications of 3D printing in the industrial vertical markets.

6.1 INTRODUCTION

In the previous chapter, we explored how automation and robotics have enhanced the precision, efficiency, and scalability of manufacturing processes. Building on that discussion, this chapter will delve into additional emerging trends in production systems – specifically, the growing emphasis on distributed manufacturing and mass customisation. We will examine how 3D printing has played a pivotal role in unlocking new possibilities in these areas.

Throughout human history, technology has consistently demonstrated its significant influence. Truly disruptive technologies have changed the course of human civilisation by opening up new avenues and possibilities. It is commonly believed that 3D printing, or additive manufacturing (AM), has enormous potential to become one of them. The term '3D printing' encompasses a wide variety of processes and techniques that produce parts and products in an 'additive' fashion. Traditional manufacturing processes such as turning, milling, drilling, grinding, cutting, and boring are 'subtractive' in nature because the material is removed from the base object to produce the final product. In contrast, 3D printing techniques generally operate on the principle of adding materials layer by layer to form an object.

The role of 3D printing is crucial for the speedy adoption of Industry 4.0 practices. As we discussed in earlier chapters, one of the key principles of Industry 4.0 is distributed manufacturing, in which the parts or products are produced as close to the end market as possible. This can only be achieved if the production is simplified and distributed through 3D printing techniques.

6.2 WHAT IS 3D PRINTING?

Three-dimensional printing or 3D printing is a method of creating a three-dimensional physical object layer by layer from a digitally created design. This is also known as *Additive Manufacturing*. The materials in many thin layers are typically deposited, joined, or solidified using computer-controlled processes. This is fundamentally different from any other existing traditional manufacturing techniques where the product is obtained either by removing the materials or is cast through moulds. As a result, when additive processes are adopted, the production processes are more efficient in terms of less wastage, more automation and a higher possibility of customisation.

Though followed as the basic processes for a long time, traditional manufacturing has many limitations.

- Manufacturing has evolved from manual labour-intensive processes to automated processes, but the fundamental methods like machining, casting, forming, and moulding remained the same. More automation added new complex processes, but fundamentally it is about subtracting material from a large block or casting or moulding to create the end product or a tool, either to create the end product or to produce a tool for the casting or moulding process. This can result in wastage of up to 90% of the block of material.
- Also, due to the traditional design and production processes, complex parts often require expensive tooling or assembly of a large number of parts, seriously limiting the capability of such production processes.

On the other hand, 3D printing creates the objects directly, adding the material layer by layer, which simplifies the whole production process.

- The process is toolless and through innovative designs, it is possible to create numerous parts and products directly as per need without incurring excessive costs or high lead time.
- Complex parts and intricate geometries can be directly created, avoiding the assembly requirements, which also proves to be cost-effective.
- 3D printing processes are both energy and material-efficient. The processes can utilise up to 90% of product material and the lighter and stronger design of the product keeps it environmentally efficient throughout the lifecycle.

- Till a few decades back, access to 3D printing capability was limited to large multinational companies as the cost and scale of owning and operating a 3D printer were prohibitive. But the advent of smaller desktop versions of the printers which can be acquired for less than $1000 has brought this technology within the reach of a wider range of users. This not only improved the adoption of the technology manifold but the possibilities and applications also have seen exponential growth.

There still exist a few limitations in the 3D printing method: (i) it is not yet cost-effective for large-scale mass production, (ii) the maximum part dimension is restricted by the size of the 3D printer used, (iii) some 3D printing processes remain relatively slow compared to traditional manufacturing methods., (iv) while material options are expanding through ongoing innovation, the range of compatible materials is still limited than conventional manufacturing.

6.3 HISTORY OF 3D PRINTING

6.3.1 Early Innovations

In recent times, 3D printing has become a much-discussed topic. But actually, the technology was first introduced in the late 1980s. During that time, the technology was conceived as a *rapid prototyping (RP)* technique, which can create prototypes faster and cheaper before the products are developed in the industry. The first 3D printing technology was stereolithography (SLA) which was patented by Charles Hull in 1986. He also co-founded a company to produce commercial SLA machines. We will discuss this technology later in Section 6.5.

DID YOU KNOW?

It is interesting to note that the very first patent application for RP technology was filed by Dr. Kodama, in Japan, in May 1980. He was the first to describe a layer-by-layer approach to manufacturing using a photosensitive resin polymerised by UV light. But the full patent specifications were not submitted subsequently within the one-year deadline and thus the application got cancelled!!

Research on other 3D printing technologies was also going ahead during the same time, and in 1989 Carl Deckard from the University of Texas received a patent for the Selective Laser Sintering (SLS) process. In 1992, Scott Crump received a patent for Fused Deposition Modelling (FDM). FDM is one of the most commonly used technologies today because of the open-source *RepRap model* that made this technology accessible to many entry-level machines.

DID YOU KNOW?

The RepRap is a self-replicating manufacturing machine. It is a desktop 3D printer capable of printing plastic objects. Interestingly, since many parts of the RepRap machine are made of plastic, it can print those parts itself. This allows RepRap to self-replicate by producing a kit of its own components, which can then be assembled into another machine. Thus, if someone owns a RepRap machine, they can not only print various useful objects but also create another RepRap machine. In 2017, RepRap's creator, Dr Adrian Bowyer, was awarded the 3D Printing Industry's 'Outstanding Contribution to 3D Printing' award.

6.3.2 Prototyping to Production

1990 to early 2000 was the time when there was a high focus on inventing new technologies of 3D printing. However, the focus was still on creating prototyping applications for industrial usage. Parallel research works were going on to substitute tooling, casting, and direct manufacturing applications with these new technologies. New terminologies like rapid tooling (RT), rapid casting, and rapid manufacturing (RM) were becoming more common at that time. A single terminology, *Additive Manufacturing (AM),* was coined as an umbrella term for all these technologies.

6.3.3 Democratisation of the Technology

The year 2005 was a significant milestone in the realm of 3D printing. Before this, the technology was primarily utilised by large-scale industries to produce intricate, high-value components. However, the prohibitive cost and substantial size of 3D printers made them inaccessible to office or personal users. This changed when Dr Bowyer from the University of Bath in England unveiled the RepRap concept in 2005. This open-source, self-replicating 3D printer could manufacture some of its own parts, drastically reducing costs. It opened up a new avenue for manufacturing complex products without the need for expensive and large industrial printers. The desktop version of the printers was called 'concept modellers' and the use of 3D printers for concept development and functional prototyping became accessible to a larger audience.

Since 2010, the visibility and applications of 3D printing have had considerable momentum. New and improved technologies kept on emerging as well, and a wide variety of materials had been introduced. Some of the key news that made a mark during this period are – the first 3D printed prototype of a car 'Urbee' was presented in 2010, the first prosthetic jaw was printed and successfully implanted in 2012 opening the domain of medical component printing, first family moved to a 3D printed house in 2018, etc. Additive Manufacturing has become an accepted form of sustainable manufacturing process and has been used widely in automotive, medical and architectural fields.

As 3D printing has become a key component of the 4th Industrial Revolution, its impact on the industrial sector and its potential to serve future consumers is undeniable. It still has many untapped potentials that will surely be unfolding soon. See Figure 6.1 for the chronology of advancement in 3D printing technology.

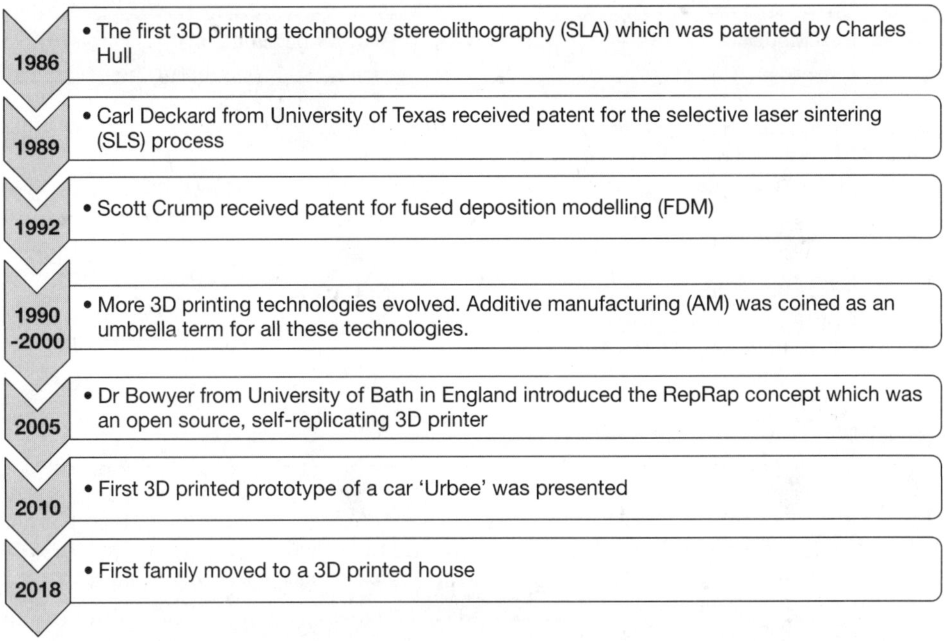

Figure 6.1 History of 3D printing

6.4 HOW 3D PRINTING WORKS?

The main concept of 3D printing is based on creating a 3D shape by incrementally adding layers. The movements and actions of the printers are controlled by a language termed *G-code*. This contains the commands to move the parts of the printer, and thus it is the way for humans to instruct the machine on what to do. The *slicer software* divides the 3D model of the target object into layers, and the related G-codes are produced and transmitted to the 3D printer, which can construct the object layer by layer. These concepts will be explained in detail in the subsequent sections. Please refer to Figure 6.2; the main stages of the 3D printing process are (i) creating the 3D model of the target object in the digital environment, (ii) converting the model file into a file format compatible with a 3D printer, such as STL, (iii) using 3D printer slicer software to slice the model into layers and producing the G-codes, (iv) transmitting the G-codes to the 3D printer and (v) printing of the object.

6.4.1 3D Modelling

The first step of the 3D printing process is to create a digital design of the object that is to be printed. It is very important to take into consideration all the parameters of the printer and the material while creating the 3D model so that the printing process can achieve the intended accuracy, finish, and complexity. Digital 3D models can be created with different computer-aided design (CAD) software or through 3D scanning systems. We will discuss more about the 3D Modelling software in Section 5.6.

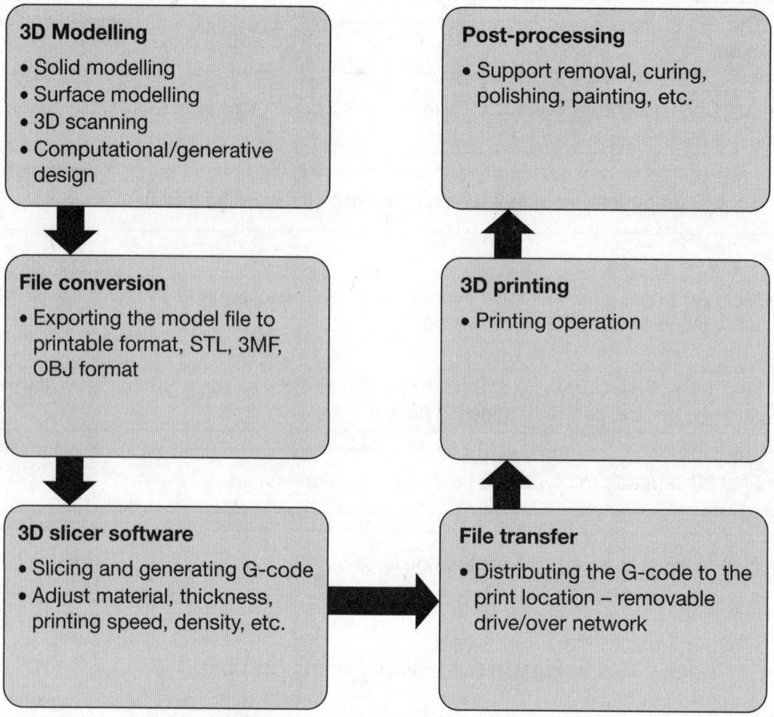

Figure 6.2 Stages of the 3D printing process

6.4.2 File Conversion

3D slicers can recognise only a few file formats like STL, obj, and 3MF. So, the output of the 3D Modelling software should be converted into these intended file formats, so that the slicer software can work with those. Most of the 3D modelling software can export the 3D models to an STL file. Once the file conversion is complete, the geometry of the object can't be changed, but the size and orientation of the model can be changed.

6.4.3 Slicing

During 3D printing, objects are printed in thin slices or layers. So, a slicer software is used to transform a 3D model into layers and generate G-codes which can be used by the 3D printer. Several parameters like the position of the object on the 3D printer table, thickness of each layer, printing material, temperature, and printing speed can be adjusted using the software. This software can also simulate and monitor the construction process. Some of the well-known open-source slicer software are Ultimaker Cura, CraftWare, and Z-Suite.

6.4.4 File Transfer

The G-codes thus created are transferred to the 3D printer through removable drives or over the network.

6.4.5 Printing

Based on the G-codes transferred in the previous stage, the 3D printer parameters are set up and the printing process is started. The time required for printing an object depends on parameters like the technology of the printer, geometry, size of the model, resolution, amount of support structure required, etc. It is possible to change the amount of support structure required by changing the orientation of the object on the printing tray, and that in turn can reduce the post-processing time.

6.4.6 Post-processing

After the printing is finished, the printed part goes through post-processing steps like cleaning the *support structures*, curing, *sintering*, painting and polishing to finalise the product. Support structures play a critical role in 3D printing for creating intricate geometries and offering design freedom. Complex geometry parts like overhangs, holes, and bridges are susceptible to collapse during printing and thus need support structures. As these structures are additional parts, it is mandatory to remove them after printing is complete. Some common support removal techniques include dissolving in chemicals, breaking away, smoothing the surface with sandpaper, water-jetting, and washing off powdered supports. In most of the polymer-based and powder-based printing technologies, sintering is an essential step. The printed product is heated at a high temperature to fuse the particles of the material and the resulting product will be harder, stronger, and more durable.

6.5 3D PRINTING TECHNOLOGIES

Over the years, 3D printing technologies have undergone significant advancements. What began with polymer-based materials has now expanded to include cutting-edge metal printing. Alongside this evolution, the precision and durability of printed products have steadily improved, driven by continuous technological innovation. The major types of 3D printing technologies can be broadly classified into five categories, as outlined below.

6.5.1 Vat Photopolymerisation

The principle of this technology is to use conventional or UV light to harden a polymer resin, which is kept in a container, to create the required shapes and objects.

Stereolithography (SLA)

SLA is the first commercialised 3D printing technology, that became popular for its ability to produce exceptionally high-quality prints. The process is based on precise curing of liquid photopolymer resin using an ultraviolet laser, guided by digital design data (typically an .stl file). The resin is contained in a vat equipped with a movable build platform. A laser beam scans the surface of the resin in the X-Y plane, selectively hardening the material layer by layer according to the design's specifications. After each layer is cured, the platform descends incrementally along the Z-axis, according to the thickness of the layer. This allows a fresh layer of resin to coat the surface before the laser traces the next cross-section. This cycle repeats until the entire object is formed, with each thin layer bonding seamlessly to the one beneath it. Once printing is complete, the platform rises out of the vat, and the finished part is removed and cleaned of excess resin. SLA's precision and layer-by-layer approach enable the fabrication of intricate geometries with outstanding accuracy, making it ideal for applications demanding fine details and smooth surface finishes. See Figure 6.3.

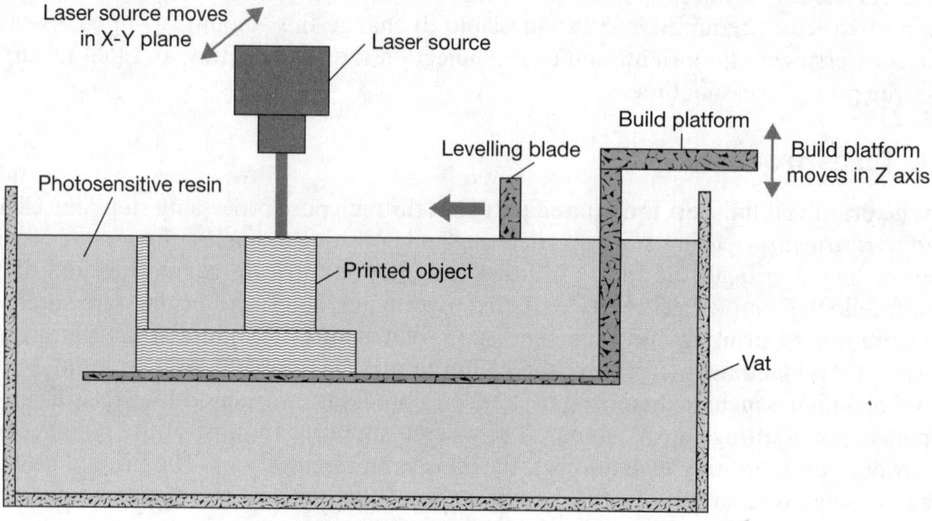

Figure 6.3 Stereolithography

Due to the nature of the SLA process, support structures are required for overhanging or undercut parts. These supports need to be removed and cleaned at the post-processing stage. As the main material is resin, post-processing also involves curing where the part is subjected to powerful light in an oven-like machine in order to harden the resin fully.

Materials Supported

- Epoxy resins with good electrical insulation and high heat resistance are the primary materials used.
- Other thermoset liquids like ceramic-like white materials, true silicone materials, and polycarbonate-like materials are also used.

Advantages

- Produces smooth surfaces with minimal stepping. The level of detail that can be printed is also very high. Thus, it is highly suitable for the production of objects with detailed geometries.
- Objects are isotropic, which means that they have the same strength in all directions and are not split or separated along the layers.
- A wide variety of resins, including coloured resins, are available to make different types of objects.
- Desktop and commercial models are affordable, and consumables are also not very expensive.

Disadvantages

- A limiting factor is the post-processing steps required to remove the support materials.
- The materials can become more brittle over time and thus the stability of the product can be an issue with this type of printing.

Applications

- As the materials used are mostly resins, this is most suited for creating high-end prototypes and moulds for industrial parts.
- This has proven to be suitable for the production of 3D polymer parts in the biomedical device sector and also for the production of 3D ceramic parts with desirable mechanical properties.

Digital Light Processing (DLP)

DLP or digital light processing is a similar 3D printing process that works with photopolymers. The major difference from SLA is the light source and the way the projection is controlled. Instead of the laser beam, a more conventional light source, such as an arc lamp, with a liquid crystal display (LCD) panel or a deformable mirror device (DMD), which acts as a projector, is used in DLP. This projector applies the light on the entire surface of the resin pool in a single pass, unlike a laser beam scanning the layer in SLA. See Figure 6.4. This makes the process faster. The accuracy and print resolution depend on the resolution of the projector as the projected image of each layer consists of pixels formed at the edges.

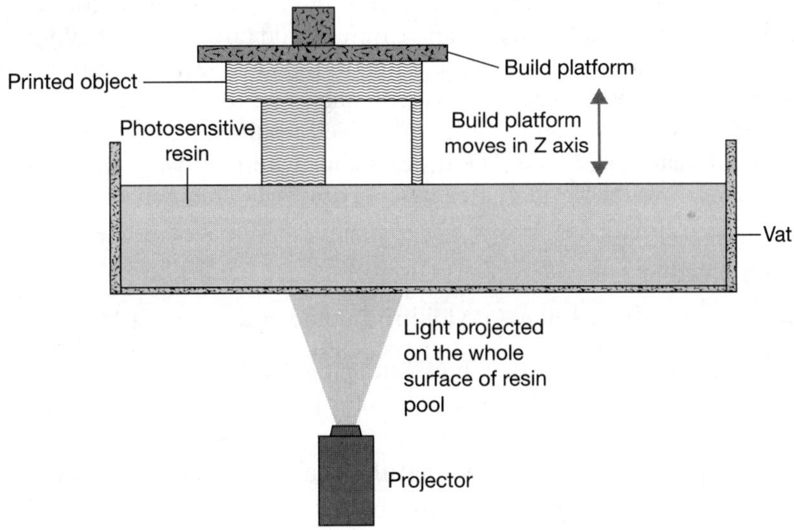

Figure 6.4 Digital light processing

Materials Supported

- All the photopolymers that are supported by SLA can also be used for DLP.

Advantages

- DLP can produce highly accurate parts with high-resolution.
- Because only a shallow vat of resin is required for this process, it generally results in less wastage and lower running costs. This is one of the advantages of DLP over SL.

Disadvantages

- The DLP process also requires support structures and post-curing similar to the SLA process and thus incurs similar additional costs for the steps.

6.5.2 Material Extrusion - Fused Deposition Modelling (FDM)

3D printing that works on the principle of extrusion of thermoplastic material is the most common and recognisable 3D printing process. This is one of the oldest and most common technologies used in 3D printing. Fused Deposition Modelling (FDM) or Fused Filament Fabrication (FFF) are the popular names of this process. The technique works very similarly to a computer-controlled glue gun. Plastic filaments are melted and deposited through a heated extruder, as thin 2D layers onto a build platform, one on top of the other to build a 3D object as per the design fed by the computer. Once deposited, each layer hardens and bonds to the previous layer. As a result of this layering principle, the layer traces can be seen on the outer surface of the printed object, which is known as the stair effect. The surface roughness can be reduced through hand sanding during the post-processing stage.

As the name 'fused filament fabrication' suggests, these 3D printers generally use filaments made from thermoplastic materials. The thermoplastic filament is inserted into the nozzle assembly, where heating elements raise it to the required temperature. Once molten, the material is forced through the nozzle and carefully laid down on the build plate to create a layer of the object. The nozzle moves according to the G-codes and forms the object layer by layer. The extruded parts can build parts with angles up to 45 degrees. But if the angle is more than the layers may not bond to each other well resulting in failed parts. In such cases, a support structure material is used. The process ends with removing the support structures by dissolving them in water through a chemical bath or in commercial wash stations. (*See* Figure 6.5)

Materials Supported

- Mainly plastic materials like Acrylonitrile butadiene styrene (ABS), Polylactic acid (PLA), Nylon, Chlorinated polyethylene (CPE), Polycarbonate (PC), Thermoplastic polyurethane (TPU)

Advantages

- Both desktop and industrial-grade machines are available at a reasonable cost. The consumable cost is also low, which makes FDM the most affordable 3D printer technology for manufacturing prototypes and functional parts.
- Multicolour printing can be done by multiple-nozzle FDM printers.
- Printed parts are robust and durable. Thus, functional parts can also be obtained by this technology.
- Dissolvable materials can be used for printing the support structures, so that it is easy to remove the support structures.

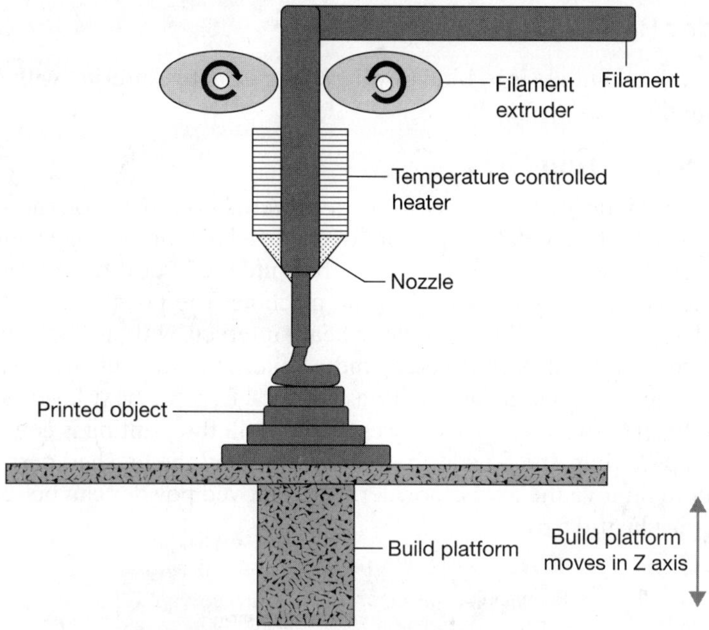

Figure 6.5 Fused deposition modelling

- FDM/FFF has become a preferred 3D printing technology due to its operational simplicity, environmental sustainability and economical operation. Also, the produced parts demonstrate robust mechanical properties, thermal stability, and chemical resistance, which makes the technology popular.

Disadvantages

- Due to the way one layer is spread on top of another, the resolution and dimensional accuracy are lower than some of the other technologies.
- The parts are anisotropic, which means that the part strength varies in different directions. Unless the printing direction is planned properly, the parts can split along the layers.
- It has a lower printing speed compared to SLA,
- The FDM/FFF processes require support structures for any objects with overhanging geometries. In complicated designs, removing the support structures can be difficult or can damage the end product.
- As the surface resolution is not optimal, significant post-processing may be required if a high surface finish is needed in the final product.

Applications

Due to the simplicity, affordability, and smaller footprint of the printers, they are mostly the first choice for small users like educational institutions, hobbyists and prototype builders, as well as for industrial use.

6.5.3 Powder Bed Fusion

The principle of this technique is to bind powdered material by sintering with a laser beam to create the required shapes.

Selective Laser Sintering (SLS)

The Selective Laser Sintering process uses a high-powered laser beam on the surface of a bed of powdered material, which fuses the powder to create a layer of the object. First powder bed of tightly compacted powdered material is prepared, and the laser is traced across the surface in the X-Y plane as per the 3D data fed to the machine. The powdered material sinters or fuses the particles forming a solid as the laser beam interacts with it. The powder bed drops incrementally after one layer is completed, and a roller is used over the surface of the bed to smoothen the powder. Then in the next pass of the laser, the next layer is formed which gets fused with the previous layer. This process repeats till the printing is complete. After the finished object is picked up, the powder bed is removed and the finished part is cleaned with a brush or air jet to remove the excess powder. The removed powder can be reused for future printing cycles. (*See* Figure 6.6)

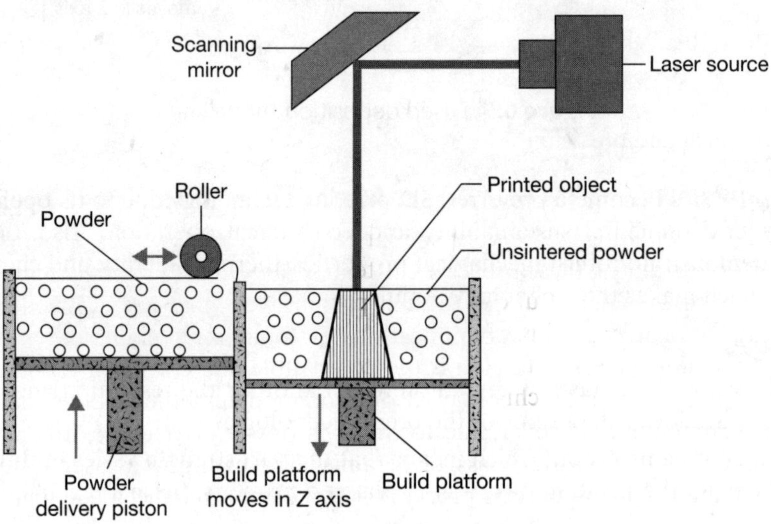

Figure 6.6 Selective laser sintering

Typically, the products are created in a single colour – white, grey, or black. The printed parts have a slightly rough surface finish, are porous, and absorb water.

Materials Supported

- A wide range of materials, such as, metal, nylon, glass powder and ceramic can be used in SLS machines. Metal objects can be polished and ceramic objects can be fired in a kiln and glazed to create the final product.

Advantages

- The powder bed itself serves as a stable support structure for overhanging parts and undercuts and thus complex designs are achievable with this process.

- Also, there is no need to cut out or remove the support structures as a post-processing step.
- The finished products are porous and absorb water. So, the finished products are often dyed with different colours, offering a variety of choices for the end product.
- Though the initial printed output has a slightly rough surface finish, with proper surface treatment measures, the finish quality can improve.
- One of the most economical options in 3D printing
- The parts are nearly isotropic, which means that the parts have the same strength in all directions.

Disadvantages

- The machines are expensive. Desktop versions of the machines are not available and thus the initial setup cost is high.
- There is considerable cooling time required because laser sintering needs high temperatures.

Applications

- Laser sintering is capable of processing both plastic and metal materials. However, sintering metal necessitates the use of a more powerful laser and elevated temperatures during the process. The parts manufactured using this method are considerably stronger than those produced with SL or DLP, although they typically lack the same level of surface finish and precision.

Selective Laser Melting (SLM)/ Direct Metal Laser Sintering (DMLS)

SLM and DMLS technologies are very similar to SLS technology, but a complete melting process is performed during these processes instead of sintering. Aviation and medical sectors use these technologies widely. But due to their prohibitively high costs, SLM and DMLS are not suitable for fabricating consumer applications. Titanium is a popular material that can be used in these technologies. As titanium is bio-compatible and corrosion resistant, it is ideal for biomedical implants. The technology also accommodates other metals including stainless steel and aluminium. However, due to the high power laser usage, it cannot use SLS-compatible materials like plastics, glass, or ceramics. The high-power laser systems demand precise thermal management, making process control complex and challenging.

6.5.4 Jet Technologies

Binder Jetting

In this technique, a liquid binding material is selectively sprayed on the surface of a powder bed of the part material, which gets fused to form a layer. Once a layer is finished, the powder bed is incrementally lowered and a roller smoothens the powder over the surface of the bed. The jet head then traces over the surface again, jetting the binder and thus the subsequent layers are formed and fused with the previous layer, creating the final object in the process.

It is possible to create coloured objects by adding a full-colour palette to the binder. The parts created out of this process are not as strong as the sintering process and can be porous and brittle. So, those need post-processing like sintering or infiltrating to improve durability. Figure 6.7 represents the working principle.

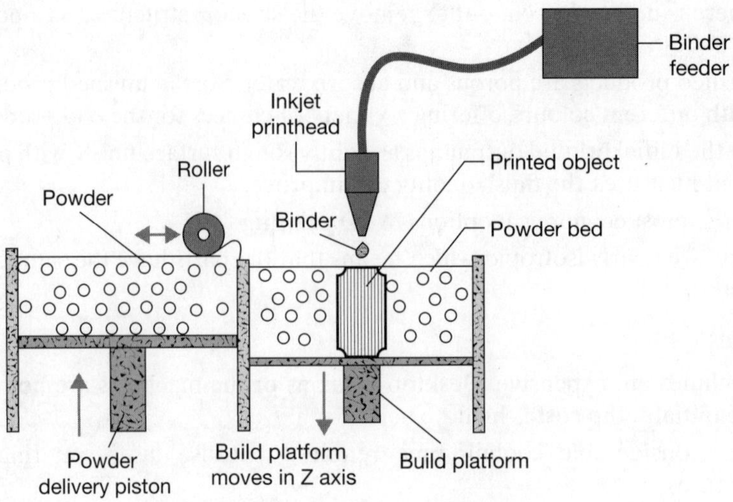

Figure 6.7 Binder jetting

Materials Supported

- Metals and ceramics are the most commonly used materials in binder jetting applications. Other powdered materials, like sand, can also be used. Many polymers like ABS or PLA are suitable materials as well for use in binder jetting operations.
- A few metal alloys of titanium, stainless steel, and copper are also used regularly as their mechanical properties allow for printing strong yet light parts.

Advantages

- One advantage, which is similar to the SLS process, comes from the use of a powder bed. There is no need for support as the powder bed itself provides this functionality.
- The range of materials that can be used is high.
- A distinctive advantage of the process is the ability to print colour objects directly.
- Another important advantage is that this process happens at room temperature. So, the chance of distortion of parts due to thermal effects is negated.
- By using multiple jet nozzles, it is possible to create multiple parts at once, and thus this is a very fast and economical additive manufacturing technique.

Disadvantages

- The post-processing requirement to improve the mechanical properties and durability of printed parts adds some overhead and cost to the process. Moreover, high-temperature used in post-processing steps like sintering can damage the surface finish or deform the unsupported parts that get softened due to heat.

Applications

- Full-colour prototypes, low-cost metal parts, and production of large sand-casting cores and moulds.
- Due to low cost and fast production speed, it is also used for creating replacement parts, as well as in the food and jewellery industries.

Material Jetting

The technique where the actual build material is deposited through jet heads in small layers to form the final object is called material jetting. The process is very similar to an inkjet printer. In inkjet printers, the printing head moves in the X-Y axis, depositing the ink to create 2D images. If we think of moving the head also in the Z axis (up and down), depositing a material, then it will replicate the principle of material jetting. Photopolymers in liquid or molten state are selectively jetted through single or multiple jet heads in tiny droplets which get spread on a build platform. A UV light source is located near the print head, which cures the deposited material, solidifying the structure and creating a layer. The build platform is lowered incrementally and the process is repeated to build the final object layer by layer. With the use of multiple jet heads, it is possible to simultaneously jet the support materials or multiple different materials of different properties, adding versatility to the parts. This is a very precise 3D printing method and can produce accurate parts with a very smooth finish. (*See* Figure 6.8).

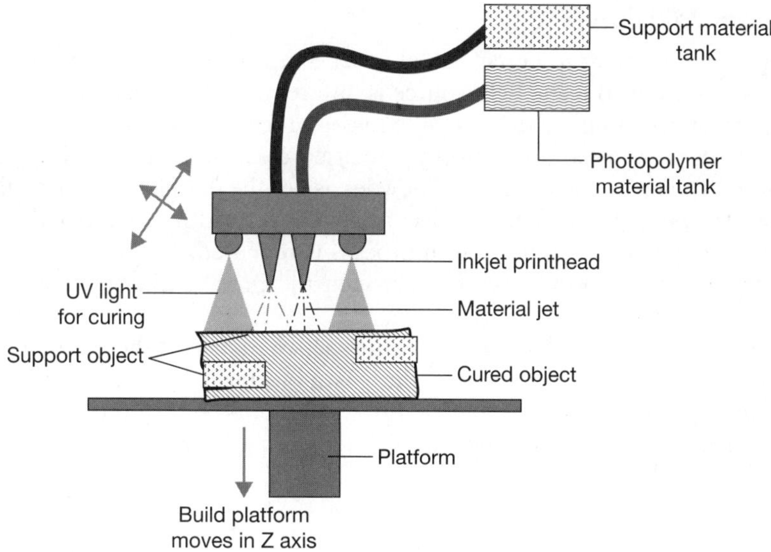

Figure 6.8 Material jetting

Materials Supported

Different types of resins are the main materials used in material jetting. These include base and composite resins, transparent and coloured opaque resins and their mixtures. As supporting structures are a must for this technique, dissolvable materials are commonly used for this purpose.

Advantages

- This is one of the techniques that can produce high-resolution objects
- As the material used is liquid, it can create a smooth surface finish
- It has a unique capability to combine mixed materials during printing
- Resin combinations can form different textured and coloured products

Disadvantages

- Photopolymers used in this technique may tend to degrade in mechanical properties while exposed to over-curing.
- These printers, as well as the build materials, are expensive, compared to other technologies.

Applications

- This has widespread applications in design, prototyping, demonstration models and training or educational purposes.
- Dental and biomedical objects, anatomical models, fashion industry (for glasses, shoes, and more) have extensive applications of this technique.

6.5.5 Metal Printing

Electron Beam Melting (EBM)

This technique is very similar to Direct Metal Laser Sintering as both are capable of forming parts from metal powder. But the light source is different. As the name suggests, an electron beam inside a vacuum chamber carries out the fusion process. Metal powders or filaments are fused under a high-pressure atmosphere by a focused electron beam providing high energy and temperature. The beam fuses a layer of powder as per the design fed, and once the current layer is formed, the build platform lowers according to layer width. The top of the layer is then covered with metal powder and the fusion process is repeated. A new layer is thus created which gets fused with the prior layer and through the process, the intended 3D object is obtained (refer to Figure 6.9).

This process is widely used for working with reactive metals as the vacuum environment helps prevent collisions of electrons with the gas molecules and thus provides a positive effect on the processing of these metals. Energy consumption is also reduced significantly in this technique. This technique can produce fully dense parts or medical-grade products with metals.

Materials Supported

- Commonly used metal powders are titanium and chromium-cobalt alloys. As the processes involve carrying electrical charges, non-conductive materials like ceramic or polymers can't be used as the material.

Advantages

- High-strength parts can be easily produced by EBM technology. Sometimes, the residual stress of the material is reduced by preheating the material and the print bed. Also, the porosity of the product is adjusted by adjusting the beam parameters.
- EBM possesses the ability to fabricate fully dense components in a range of metal alloys, up to a medical-grade standard. This capability has led to its notable success in various production applications within the medical industry, especially in the creation of implants.
- The process is very fast and requires less support than other techniques.
- Unused powder can be removed and stored for reuse, which reduces wastage.

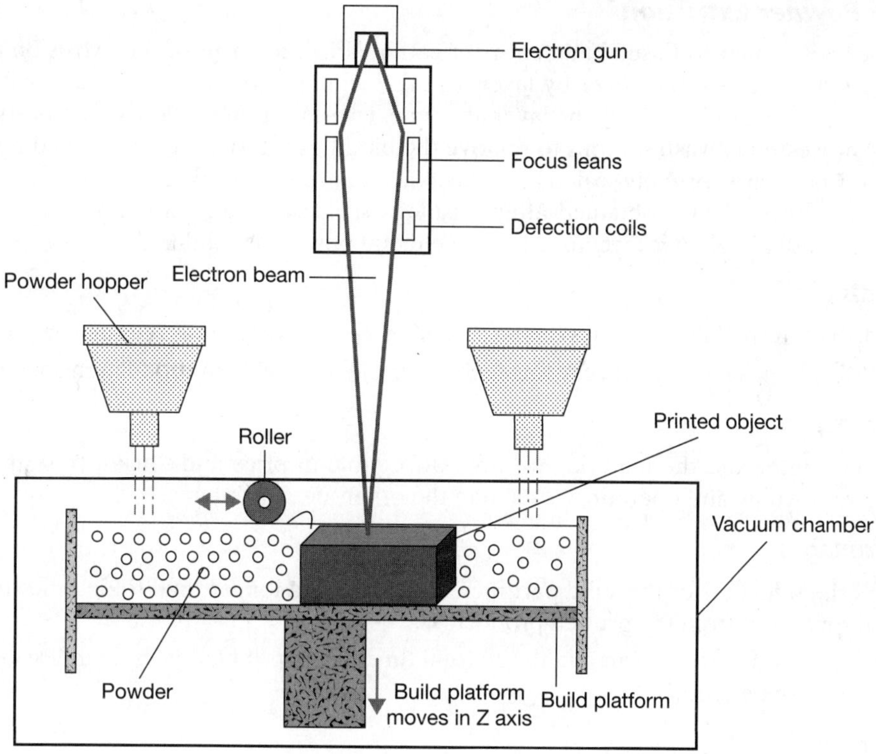

Figure 6.9 Electron beam melting

Disadvantages

- Printers and materials are expensive
- The choice of materials is limited

Applications

- As this technique can produce fully dense and medical-grade metal products, this is widely used in production applications and the medical industry, like implants.
- Aerospace and automotive industries are also exploring the options to use EBM technology for manufacturing fulfilment.

DID YOU KNOW?

The use of gold and silver in 3D printing has increased significantly over the last few years. These are very strong materials and are used in powdered form. The largest users are in the jewellery sector. Titanium is another strong material that is being directly 3D printed with sintering, melting, or EBM technology. The use is in industrial applications and tooling.

Bound Powder Extrusion

The process is similar to Fused Filament Fabrication (FFF) in terms of the extrusion process that prints the final product layer by layer. In this technique, the metal is bound with waxy polymers and the part is printed one layer at a time. These parts are called 'green' parts, which are post-processed in wash stations to remove the binding material using a de-binding liquid. The 'green' parts are normally scaled up to keep allowance for shrinkage during the sintering process. The 'brown' part, obtained after washing, is placed in a high-temperature furnace where the residual polymer is removed and the metal powder is solidified.

Materials Used

- Most of the metals and alloys can be used in this process as long as they can be powdered.
- With a bulk sintering process, this technique can achieve high production speed.

Advantages

- The printer and the material are both affordable in price and thus have significantly lower owning and operating costs than the other metal printers.

Disadvantages

- As the thickness of the solid part increases, it takes a longer time to de-bind the polymer material from the printed product.
- The sintering process can add to the total time of the production cycle unless bulk sintering can be used.

Applications

- This technique is very accessible and thus is used for a wide variety of manufacturing processes.

Casting

This technique is also called direct investment casting or lost wax casting. In the first step, the original 3D part is printed with castable resins in an SLA printer or with wax in a material jetting printer. Plaster is then poured over the 3D print and is hardened. After that, the plaster is heated to melt the wax or the resin out of the mould. Molten metal is poured into the plaster cast which solidifies on cooling. The paster is removed and the part is cleaned through polishing.

Materials Supported

- Precious and semi-precious metals like gold, silver, platinum, bronze, copper and brass are the commonly used materials.

Applications

- This process is widely used for creating prototypes and also in the jewellery industry.

As discussed above, the key factors in selecting a 3D printing technology include the range of materials it can process, along with the specific advantages and limitations that influence its suitability for various applications. In Table 6.1, a comparison of the 3D printing technologies is discussed.

Table 6.1 Comparison of 3D printing technologies

Technology	Process Description	Materials Supported	Advantages	Disadvantages	Applications
SLA	UV laser hardens photopolymer resin layer-by-layer.	Epoxy resins, ceramic-like materials, silicone, polycarbonate-like resins.	High detail, smooth surfaces, isotropic parts, affordable consumables.	Post-processing (support removal), brittle materials over time.	Prototypes, biomedical devices, ceramic parts.
DLP	Projects light (e.g., arc lamp) to cure entire resin layers at once.	Same as SLA.	Faster than SLA, high resolution, less resin waste.	Requires supports and post-curing like SLA.	Similar to SLA but with faster production.
FDM	Extrudes melted thermoplastic filament layer-by-layer.	ABS, PLA, Nylon, TPU, PC, CPE.	Low cost, durable parts, multi-colour printing, dissolvable supports.	Low resolution, anisotropic, slow, visible layer lines.	Prototyping, education, functional parts.
SLS	Laser sinters powdered material (nylon, metal, ceramic) layer-by-layer.	Nylon, metals, ceramics, glass.	No supports needed, complex geometries, reusable powder.	Rough surface finish, expensive machines, cooling time.	Industrial parts, porous/ water-absorbing products.
Binder Jetting	Binder liquid fuses powder layers (metal, ceramic, sand).	Metals (stainless steel, titanium), ceramics, sand, polymers (ABS, PLA).	Colour printing, no supports, room-temperature process, fast.	Brittle parts require post-processing (sintering).	Full-colour prototypes, jewellery, sand-casting molds.
Material Jetting	Jets photopolymer droplets, UV-cured layer-by-layer.	Composite resins, transparent/ coloured resins, dissolvable supports.	High resolution, smooth finish, multi-material capability.	Expensive, material degradation over time.	Dental models, prototypes, fashion industry.
EBM	Electron beam melts metal powder in a vacuum.	Titanium, chromium-cobalt alloys (conductive metals only).	Dense medical-grade parts, reduced residual stress, fast.	Limited materials, high cost.	Medical implants, aerospace components.

(Continued)

Table 6.1 (*Continued*)

Technology	Process Description	Materials Supported	Advantages	Disadvantages	Applications
Bound Powder Extrusion	Extrudes metal-polymer composite, debinded and sintered post-print.	Most powdered metals/alloys.	Low cost, scalable production.	Long debinding/ sintering time for thick parts.	Affordable metal parts manufacturing.
Casting	3D-printed wax/ resin mold is cast in plaster, then filled with molten metal.	Gold, silver, platinum, bronze, copper, brass.	High precision for jewellery.	Multi-step process (mold creation, casting).	Jewellery, artistic metalwork.

6.6 OVERVIEW OF 3D MODELLING SOFTWARE

Computer Aided Design (CAD) software with 3D capability and 3D modelers are the main types of software used for designing 3D printing models. Originally CAD software was used for architectural, industrial, and manufacturing designs that needed precision. Later 3D capabilities were added to it and have become a good choice to design 3D products for printing. Similarly, 3D Modelling software was primarily used for visualisation in film industries and video games, but the design generated can very well be used for 3D printing. 3D Modelling software can be broadly grouped into 4 segments based on the working principle:

1. **Solid Modelling:** AutoCAD, SOLIDWORKS, Inventor, Fusion 360 and TinkerCAD are a few 3D software that have good solid Modelling tools. A feature of Solid Modelling is that the designs always have volume. So, the designs are always water-tight, manifold, and thus ready for 3D printing. For creating the 3D model, this software uses mix and match techniques, like (i) combining primitive 3D objects like boxes, cylinder, pyramids, sphere, etc., (ii) extruding 2D sketches using the Modelling software, like Fusion 360, to create irregular 3D models, (iii) revolving 2D shape around an axis, or sweeping a shape along a path, (iv) overlapping different 2D and 3D shapes and using Boolean functions like union, subtraction, intersection to create new 3D models.

   Two types of design approaches are prevalent in solid modelling, which are as follows:

   - *Direct Modelling Approach* – the design is created or modified freely with the geometry and dimension of the model. This is a simple design approach and is considered when not many variations are expected during the printing process and in the final product dimensions.
   - *Parametric Modelling Approach* – the design is created by defining the relationships between the objects of the model. Variables and constraints can be added to the design to make it flexible and can be modified dynamically based on printing needs.

It requires more time to set up this type of model and thus is considered when the design is intended to be reused with variable dimensions.

As we discussed above, when we start the Modelling process with 2D sketches or 3D primitives, solid Modelling is an easy and fast technique that can be adopted. The resulting designs are watertight and thus require very little repair before the printing process. The parts created from these designs can achieve high precision.

2. **Surface and Mesh Modelling:** The geometrical shapes that are difficult to achieve with solid Modelling alone can use this technique. Surface Modelling software uses splines between points to create a smooth surface finish. So, the organic shapes or the shapes that require high accuracy can be designed through this software. Mesh Modelling on the other hand uses triangles or polygons that are connected to create the surface. So, the surfaces are not smooth and before printing the number of polygons on the surface is increased to make the mesh look smoother. A few popular software in this category are Maya, Rhino, 3ds Max, Cinema 4D, Blender, Sketchup, etc. Few CAD software like AutoCAD, SOLIDWORKS, Inventor, Fusion 360, etc., also have this capability.

   3D Sculpting is a specialised type of mesh Modelling where the on-screen experience is more like working with clay. Some of the software that provides this capability are ZBrush, Mudbox, Maya, Meshmixer, Blender, 3Dcoat, etc. This technique is used in industries like automotive design, jewellery design and while creating organic shapes.

3. **Computational Design:** This is a generative or algorithmic design technique where visual programming languages are used to create 3D content. The software has code blocks along with shapes and instructions that can be used to create a visual 3D object. Rhino Grasshopper, TinkerCAD codeblocks, nTopology, Dynamo and Mariomette are some of the software used in this type of design. This is a good choice when the design needs easy customisation.

4. **3D Scanning:** It is difficult to create solid models or computational models for real-life objects like sculptures, persons, or body parts like dental structures. 3D scanning processes come in handy in those cases. A few such processes are described below.

   - *Photogrammetry* – the process starts with capturing a series of 2D pictures of a 3D object through a normal camera, mobile, or tablet, from different angles, slowly moving in circles. The photogrammetry software then processes the photo set using triangulation methods to extrapolate a 3D mesh from 2D pictures. As no special equipment is required, this is a very easy way to scan 3D objects into a 3D mesh or a UV map with colour.
   - *Laser Triangulation* – this is a similar technique of photogrammetry, where a rectilinear laser beam is projected on the surface of a 3D object. A camera analyses the deformation of the laser beam and then the software triangulates and extrapolates to create a 3D mesh. This technique can create only 3D mesh and doesn't capture any colour information.
   - *Structured Light* – light with patterns or dots is projected on a 3D object and a camera captures the deformation of the patterns on the geometry of the object. The software then analyses the images to recreate a 3D mesh of the object. This can record a 3D mesh along with the colour.

DID YOU KNOW?

When we deal with 3D space, we use three coordinates - X, Y and Z. But while dealing with textures, it is sufficient to use two coordinates, referred to as U and V. Thus when a 3D model is created from a 2D texture, the process is called UV mapping.

Though 3D scanning is a useful process, most of the time the mesh created by the scanning software needs post-processing before it can be 3D printed. So, normally 3D scanning is used as the first step of creating the design and then complemented with other Modelling techniques to create the final printable design.

6.7 BENEFITS OF 3D PRINTING

We discussed in the previous sections how 3D printing has made a paradigm shift from traditional manufacturing (or prototyping) methods, whether at an industrial, local, or personal level. As a result, it brings a host of benefits that traditional manufacturing methods simply cannot offer.

6.7.1 Distributed Manufacturing

Local manufacturing model or distributed manufacturing model is a concept where the products may be centrally designed, but are produced on demand in the places where those are needed. That not only reduces the huge transportation costs but also removes the necessity to maintain high inventories. The principles of 3D printing are perfectly aligned with this model. Once the STL files are finalised, they can be easily transmitted over the network to printing facilities across the globe based on the customer's need.

An interesting real-life use of local manufacturing can be found in the US Army. As the army troop is frequently on the move, procuring replacement parts for their machinery has always been challenging. They solved the problem with the use of 3D printers. High-precision desktop 3D printers are carried with the troop and the replacement parts are printed as and when required to resolve their supply chain woes.

6.7.2 Mass Customisation

The term mass customisation has evolved based on the flexibility that 3D printing can offer during product design and production. With innovative design steps, it is even possible to produce parts and products of different natures, without adding an overhead on cost and complexity. So, personalisation of products based on customised needs is easily possible with 3D printing techniques.

A good example of the usage of mass customisation can be found in the jewellery industry. Through traditional moulding and casting methodologies, it used to be very difficult to create customised products as the creation of the mould itself is a time-consuming and expensive affair. 3D printing can quickly create wax patterns of the customised jewellery, which can be covered with mould materials like plaster and then the wax is melted to create customised moulds quickly. Molten precious metal is then poured into the mould and customised pieces of jewellery can be produced much faster.

6.7.3 Complexity

Traditional manufacturing had to depend on a few basic processes, like machining and casting. Complex products often need to depend on the assembling of parts, which possess a constraint to the level of complexity that can be achieved in the design. On the other hand, as 3D printing works on additive manufacturing principles, complex components for industrial applications can be directly printed, which are lighter and stronger than their traditionally produced counterparts.

6.7.4 Tool-less Production

Most of the labour and cost are expended in creating tools for production in the traditional manufacturing process. In contrast, 3D printing processes are additive in nature and thus do not need any tooling, saving the costs, lead time and labour needed for it. Similarly, the elimination of assembly requirements by intelligently designing and directly printing intricate parts saves huge expenses on complicated tooling processes.

6.7.5 Sustainable / Environmentally Friendly

3D printing has already proven to be more sustainable and environmentally friendly than the traditional production processes. Local printing options can reduce carbon footprint drastically. These processes are more efficient in energy usage and material usage as the wastage of material is reduced many folds compared to the traditional subtractive methods. Other characteristics, such as the reduction in space required for factory setup and the possibility of using biodegradable materials and renewable resources, contribute to their sustainability and environmental friendliness.

6.8 APPLICATIONS OF 3D PRINTING

6.8.1 Modern Manufacturing

In the previous section, we discussed the impact of 3D printing in modern manufacturing, like local manufacturing, mass customisation, easier and faster production of complex parts and environmental friendliness. Along with these, the social, economic, environmental and security aspects of the manufacturing process have also seen positive trends in its use. Simplification of the supply chain, the possibility of smaller production batch sizes, which reduces inventory size, cutting down on shipping charges of spare parts across the globe – all these save huge capital as well as operating expenditure for the manufacturers. Innovative and new ways of designing parts and products are already being explored by millions of small manufacturers, thanks to the affordability of 3D printing units and a completely different *additive* principle over traditional technologies that opened up huge options.

A real life case: A 70-year-old, 1600-employee-strong transmission and gear assembly unit in the USA, Ohio had been facing a continuous challenge of shut down of assembly lines due to frequent breakage of a few carriage components that carry materials to different sub-units of the plant through rails. These components cost around $180 each to produce using traditional manufacturing methods and more importantly, they cause considerable downtime due to the time required for production. In 2020, the management of that company decided to buy a desktop FDM printer that could create these components faster and much cheaper. They saved over $100,000 in 2020 alone by printing the parts in-house.

6.8.2 Medical and Dental

Personalisation and customisation requirements are particularly high in the medical and dental sectors. For example, prosthetic implants must be highly personalised, similar to dental restorations. As a result, the medical and dental industries adopted 3D printing early in its development. The ability of 3D printers to easily utilise biomedical materials has also played a crucial role in making them a natural choice for medical applications. Common uses include the production of hip and knee implants, hearing aids, dental crowns, and orthotic insoles for shoes. Advanced technologies are being developed to 3D print complex body implants such as skin, bone, tissue and organs like the liver. These advancements reduce surgery time, enhance surgical planning, and enable the fabrication of patient-specific implants and surgical devices tailored to the patient's anatomical structure. Due to its high market potential, the commercialisation of 3D printing in this sector remains a key focus.

DID YOU KNOW?

3D printers are used for printing food also!! Biotech company Modern Meadow produces steak chips and genuine, lab-grown leather in its 10,000-square-foot bio-fabrication facility in Brooklyn, New York. So, it does not need any slaughter to produce meat products. Also, compared to conventional farming methods, the production facility uses just 1 per cent of the land, 4 per cent of water and 55 per cent of energy. Is this the future then?

6.8.3 Aerospace and Aviation

Aerospace and aviation are other sectors that have adopted 3D printing from its very early days. It started with creating prototypes for aircraft engines and other parts and later started producing parts that were already being used in the aircraft. Boeing, Airbus, GE, Morris Technologies, Rolls-Royce and BAE Systems are some of the key players in this sector who are already using critical and industrial-grade 3D printing systems for creating jet engine prototypes and end-user parts.

A real life case: Recently, NASA has developed a prototype of a 3D-printed rocket that is 20-30% more efficient than the traditionally built rockets. The additive printing technology fabricated the rocket part and the necessary tools quicker and cheaper. Moreover, it enabled

NASA to use novel materials that could get 30% performance improvements of the rockets, which means huge savings on rocket fuel and drastic cost reduction. They could use a unique alloy material that is more heat resistant and stronger than other materials that were used by the traditional tooling process. This has enabled NASA to try out a new propulsion technology in the combustion chamber, which makes the rocket more efficient.

6.8.4 Automotive

The automotive industry has traditionally been an implementer of automation. Thus, it embraced 3D printing technologies very quickly, starting with rapid prototyping and then progressing to create end parts of the car. Car components manufactured with FDM and FFF technologies are lighter but stronger and thus provide performance improvement and lower fuel consumption due to a reduction in body weight. Volkswagen, BMW, and Ford are the pioneers in the automotive industry who use 3D printing technologies in manufacturing end-user parts. The industry also tested 3D printing a fully functional car in 2014 and the future looks very bright with the performance and cost-effectiveness angles.

Another area in the automotive sector that has seen high potential for the use of 3D printing is the supply of spare parts for sales functions. Printing spare parts locally and on-demand reduces the cost of holding and shipping huge inventories.

6.8.5 Jewellery

3D printing is disrupting the $280 billion jewellery industry significantly. This industry has traditionally been driven by handcrafting and wax-casting technologies. Personalisation and customisation requirements are very high in this sector. The most time and effort-consuming part of the traditional process was mould making and due to the manual nature of the process, it was difficult to preserve and replicate the designs easily. With 3D printing workflow, CAD software is used to design precision patterns that are printed with high-resolution 3D printers. The printed part is then used as the cast in a mould. After the pattern is melted out, the mould is ready to use in the traditional casting process. The use of digital technology helps in preserving and replicating the pattern, which improves the production speed and precision drastically.

6.8.6 Fashion

Fashion designers worldwide are fascinated with the numerous innovative options opened by 3D scanning and 3D printing technologies. Using 3D printing, components of the whole clothing, like embellishments, accessories, buttons, as well as small end-user accessories, are quickly and simply produced. Creating prototypes of accessories before producing the actual pieces is also a common process in the fashion industry. 3D printing is used widely for this. It can reduce production time and waste significantly. Many global brands are using 3D printing for the creation of accessories. Shoemaker Camper uses it for designing its end products, Adidas uses it for manufacturing midsole shoes, and fashion label ZER fabricates designer products and accessories using this technology. Overall, the use of a variety of materials, eco-friendly aspects, faster design and customisation possibilities have made 3D printing a great choice for the fashion industry.

So, the opportunities for 3D printing are boundless. With the use of robotics and digital twins, new possibilities for 3D printing are continually emerging. For example, cartesian robots due to their rigid construction and ability to move along 3 dimensions are well suited for assisting in 3D printing of large surface areas. Similarly, the designs for the 3D printing get refined with the use of digital twin technology. The prototypes created and fine-tuned through digital twins result in superior quality printing, as the design team gets opportunities to iteratively optimise model parameters beforehand.

6.9 CONCLUSION

The future of 3D printing is set to undergo a remarkable transformation, fuelled by rapid digitalisation and the integration of next-generation technologies. As smart factories, Industry 4.0 frameworks, machine learning and robotics become increasingly prevalent, they are accelerating the adoption of online 3D printing, especially in simulation-driven applications where precision and adaptability are essential.

Advancements in material science are further expanding the horizons of 3D printing. The continuous development of high-performance materials with improved strength, flexibility, and resilience is unlocking new applications across sectors such as automotive, aerospace, and healthcare. These innovations are enabling 3D printing to fulfil complex requirements that were once considered unattainable.

Generative AI is emerging as a game-changer in this domain, transforming how designs are conceptualised and executed. By enhancing design workflows and boosting production efficiency, generative AI shortens the design cycle by up to 25% and supports mass customisation to cater to individual customer demands. It also accelerates prototype development and improves the precision of outcome predictions, thereby minimising production risks.

These AI-driven algorithms facilitate the creation of highly optimised product designs by considering key parameters like structural strength, material usage, and weight. Their predictive capabilities contribute to reducing potential failures and refining the overall reliability of additive manufacturing processes.

With global market projections indicating a surge from USD 23.41 billion in 2025 to USD 101.74 billion by 2032, 3D printing is clearly evolving from a niche technology into a critical pillar of modern manufacturing. As new technologies continue to reshape this space, 3D printing will increasingly serve as a catalyst for smarter, faster, and more personalised production across industries.

SUMMARY

- Three-dimensional printing or 3D printing is a method of creating a 3-dimensional physical object layer by layer from a digitally created design. This is also known as Additive Manufacturing.
- Pros
  - The process is tool-less
  - Complex parts and intricate geometries can be directly created

  - Processes are both energy and material-efficient
  - Technology is affordable for end users
- Steps of 3D printing
  (i) Creating model
  (ii) converting the model file into an STL file
  (iii) slice the model into layers and produce the G-codes
  (iv) transmit the G-codes to the 3D printer
  (v) print the object.
- Major technologies:
  - Vat Photopolymerisation, such as Stereolithography (SLA) and Digital Light Processing (DLP).
  - Material Extrusion, such as Fused Deposition Modelling (FDM) or Fused Filament Fabrication (FFF).
  - Powder Bed Fusion, such as Selective Laser Sintering (SLS), Selective Laser Melting (SLM), and Direct Metal Laser Sintering (DMLS).
  - Jet Technologies, such as Binder Jetting and Material Jetting.
  - Metal Printing, e.g., Electron Beam Melting (EBM), Bound Powder Extrusion, Casting, etc.
  - 3D modelling software can be broadly categorised into four segments based on their working principles: Solid Modelling, Surface and Mesh Modelling, Computational Design and 3D Scanning.

EXERCISE

Multiple Choice Questions (1 Mark)

1. In which way does 3D printing differ from traditional manufacturing?
   (a) It uses a subtractive approach
   (b) It involves less labour-intensive processes
   (c) It requires creating moulds for production
   (d) It uses only metals for production

2. Which of the following is a key benefit of 3D printing in medical applications?
   (a) Mass production of identical parts
   (b) Customisation of medical devices for individual patients
   (c) Use of only plastic materials
   (d) High cost of production

3. What is one of the advantages of distributed manufacturing?
   (a) High transportation costs
   (b) Need for large inventories

(c) Reduced carbon footprint
(d) Centralised production in one location

4. What is one of the key reasons 3D printing is considered more sustainable than traditional manufacturing?
   (a) It uses a larger amount of material for each part
   (b) It relies on traditional factories for production
   (c) It minimises material waste and energy consumption
   (d) It requires more space and resources for production
5. Which of the following 3D printing technologies is ideal for creating small-scale metal parts in industries such as aerospace and automotive?
   (a) Selective Laser Sintering (SLS)
   (b) Direct Metal Laser Sintering (DMLS)
   (c) Fused Deposition Modelling (FDM)
   (d) Stereolithography (SLA)
6. Which of the following materials is typically used in Direct Metal Laser Sintering (DMLS)?
   (a) Thermoplastic materials
   (b) Liquid resin
   (c) Metal powders
   (d) Paper
7. What is the advantage of Selective Laser Sintering (SLS) over FDM in 3D printing?
   (a) SLS does not require support structures.
   (b) SLS produces parts with a better surface finish.
   (c) SLS uses thermoplastic filaments.
   (d) SLS can print in metal and polymer materials.
8. Which 3D printing technology uses thermoplastic filaments that are heated and extruded layer by layer?
   (a) Stereolithography (SLA)
   (b) Fused Deposition Modelling (FDM)
   (c) Selective Laser Sintering (SLS)
   (d) Direct Metal Laser Sintering (DMLS)
9. What is a major benefit of Electron Beam Melting (EBM) compared to traditional manufacturing techniques?
   (a) It uses cheaper materials.
   (b) It produces parts faster than conventional machining.
   (c) It uses a laser to create parts from metal powder.
   (d) It can create complex geometries while reducing material waste.

10. What type of material is typically used in Stereolithography (SLA) 3D printing?
    (a) Powdered metal
    (b) Photopolymer resin
    (c) Thermoplastic filament
    (d) Wax

11. Which 3D printing technology is often used for creating functional parts and prototypes from metal?
    (a) Fused Deposition Modelling (FDM)
    (b) Stereolithography (SLA)
    (c) Selective Laser Sintering (SLS)
    (d) Direct Metal Laser Sintering (DMLS)

12. A hospital wants to create custom implants for patients based on their unique anatomy. The traditional methods are slow and cost-prohibitive. What is the best solution using 3D printing?
    (a) Use 3D printing for mass production of standard implants
    (b) Use 3D printing to create personalised implants quickly and affordably
    (c) Use 3D printing for creating temporary implants only
    (d) Use 3D printing for prototyping before using traditional methods

13. A manufacturing company is struggling with high transportation costs and long shipping times for parts. What can they do to address this issue using 3D printing?
    (a) Use 3D printing to produce all parts locally and on-demand
    (b) Use 3D printing to create only prototypes
    (c) Use 3D printing for packaging parts only
    (d) Use 3D printing to create larger inventory for shipping

14. A jewellery company wants to offer custom designs to customers but finds traditional moulding and casting too expensive. How can they use 3D printing to improve production?
    (a) Use 3D printing to create custom moulds directly
    (b) Use 3D printing to create wax patterns for casting
    (c) Use 3D printing for only the assembly of jewellery
    (d) Use 3D printing to create metal parts directly

15. A small parts supplier in the automotive industry faces challenges with high demand for spare parts. How can 3D printing help them manage this demand?
    (a) Produce spare parts in bulk to maintain inventory
    (b) Print spare parts locally and on-demand as needed
    (c) Focus on mass production without using 3D printing
    (d) Increase lead time for spare parts production

16. A designer in the fashion industry wants to minimise waste during production and speed up the prototyping process. What benefit of 3D printing can address these issues?
    (a) It reduces waste by using only the exact amount of material needed
    (b) It allows designers to mass-produce designs
    (c) It requires less energy for production
    (d) It eliminates the need for prototyping altogether
17. A company in the aerospace industry needs to create new parts for rockets with improved performance. How can 3D printing contribute to this goal?
    (a) By using traditional methods to create prototypes
    (b) By using 3D printing to produce rocket parts more efficiently and with new materials
    (c) By only creating basic, low-performance prototypes
    (d) By reducing the quality of materials used in the production process
18. A car manufacturer wants to create lightweight components to improve fuel efficiency. What is the advantage of using 3D printing in this case?
    (a) It creates heavier components that improve strength
    (b) It produces lightweight yet strong parts that reduce fuel consumption
    (c) It limits the materials available for component production
    (d) It increases the manufacturing time for parts
19. A company is trying to streamline its supply chain by reducing inventory costs and transportation fees for parts. How can 3D printing be used to solve this problem?
    (a) By increasing the size of the parts inventory
    (b) By printing parts on-demand locally to avoid transportation and storage costs
    (c) By printing more spare parts than necessary for future demand
    (d) By moving manufacturing processes overseas for lower costs
20. A biotech company is producing lab-grown leather and wants to improve its efficiency. How can 3D printing help in this case?
    (a) By reducing the material usage and speeding up the production process
    (b) By allowing them to produce leather at a slower rate
    (c) By increasing the waste material generated during production
    (d) By focusing on manual production methods only

Short Answer Type Questions (2 Marks)

1. Why is 3D printing considered an ideal solution for producing complex parts?
2. What does the term 'distributed manufacturing' mean in the context of 3D printing?
3. What is one significant benefit of 3D printing for rapid prototyping?
4. What is a key advantage of 3D printing in medical applications?
5. What is Fused Deposition Modelling (FDM) in 3D printing?

6. How does Stereolithography (SLA) work in 3D printing?
7. What are the key differences between Selective Laser Sintering (SLS) and Fused Deposition Modelling (FDM)?
8. What is Direct Metal Laser Sintering (DMLS), and in which industries is it commonly used?
9. Why is Electron Beam Melting (EBM) preferred for aerospace applications?
10. How does 3D printing benefit the fashion industry?

Long Answer Type Questions (5 Marks)

1. What is the significance of additive manufacturing, and how does it differ from traditional manufacturing methods?
2. What are some key materials used in 3D printing, and how do they affect the properties of the final product?
3. What are the technological challenges and limitations of 3D printing in large-scale manufacturing?
4. How does 3D printing impact the supply chain and spare parts logistics?
5. Explain how Stereolithography (SLA) works and the advantages it offers over other 3D printing technologies.
6. How does MultiJet Fusion (MJF) work, and what makes it different from SLS?
7. How does FDM technology impact the affordability of 3D printing for small businesses?
8. What is the significance of using Metal 3D printing technologies, such as DMLS and EBM, in the medical field?
9. What is Fused Deposition Modelling (FDM) in 3D printing, and what are its key applications?
10. How does FDM technology impact the affordability of 3D printing for small businesses?

Augmented Reality, Virtual Reality and Mixed Reality

CHAPTER 7

OBJECTIVES

In the introductory chapters, we briefly discussed Augmented Reality (AR), Virtual Reality (VR), and Mixed Reality (MR) as crucial pillars of Industry 4.0. In recent years, these technologies have evolved significantly, and innovative use cases have been found that have made shop floors smarter, safer, and more productive. Other than in manufacturing and smart factories, this group of *reality* technologies demonstrated immense potential across many other spheres such as marketing, e-commerce, healthcare, gaming, education, and many more. However, there still exists generic confusion about how AR, VR, and MR work and the differences between these, which limits their widespread adoption. So, in this chapter, we will cover the fundamental building blocks of these technologies and discuss their use in the Industry 4.0 context.

By the end of this chapter, you should be able to:

- Elaborate on the working principles and key components of AR/VR/MR
- Articulate the important use cases where they are already being used successfully
- Understand how large organisations reaped tremendous benefits by integrating these with their overall Industry 4.0 tenets

7.1 INTRODUCTION

During a thrilling ICC World Cup 2023 cricket match between India and Australia, the umpire suddenly called for the Decision Review System (DRS). We all were glued to the TV screen watching the projected path of the ball superimposed on the actual field to determine if it was a leg before wicket (LBW) dismissal. The seamless integration of the computer-generated ball trajectory, captured by Hawk Eye, with the real-world images of the field and the batsman, is a demonstration of the transformative power of Augmented Reality (AR) in enhancing our sports viewing experience. There are several other examples where AR is being used to enhance the realities around us. The popular mobile game 'Pokémon Go' is one such example. The players can view the Pokémons in real-world surroundings when they look through their mobile phone screens. Similarly, a Virtual Reality (VR) headset is capable of creating an immersive simulation where, with the combination of hardware and software, it replaces the real-world view with a computer-generated environment that you can effectively interact with. We can see a few such examples in Figure 7.1.

Credit: Lance Bellers / Shutterstock; Gorodenkoff. Shutterstock

Figure 7.1 Augmented reality and virtual reality

AR/VR technologies have emerged as potent tools that can offer distinctive customer experiences, heighten user engagement, enhance workforce efficiency, and amplify business productivity. These technologies are making significant impacts across various sectors, including corporations, educational institutions, and social enterprises. With the kind of innovation happening in these fields, AR/VR will soon alter the way humans learn, make decisions, and interact with their physical surroundings. Not only individuals, but also businesses will optimise how they design and manufacture products, manage their supply chains, cater to their customers, train their employees, and overall, how they compete in the marketplace.

In the following section, we will discuss the types of realities and their differences.

7.2 FUNDAMENTALS OF ALL THE REALITIES

Augmented Reality, Virtual Reality, and Mixed Reality are all part of the *extended reality* technologies. They either enhance or replace a real-life environment with a simulated one. Before we dive deep into the discussions on key components of these technologies, we will understand what features each of these technologies offers and how they are different from each other.

In order to improve our experience of the real world, **AR** is a technical tool that first recognises the characteristics of our physical surroundings before superimposing digital items like text, photos, sounds, and videos on top of them. For instance, augmented reality software, such as ARKit for iOS devices and ARCore for Android, can recognise movements, flat surfaces, and the location of the device in relation to its surroundings. By using that information, AR devices (often used in conjunction with a smartphone's camera) enhance the environment by adding digital features. Essentially, AR uses digital visual 3D components, sound, and other sensory stimuli to create an augmented version of the real physical environment. This technology has opened up new possibilities for interaction and engagement with our surroundings.

Many of you must have experienced the *Night Sky* app, a mobile-based augmented reality application. It allows users to immerse themselves in the splendour of the celestial sphere, recognising various celestial bodies such as planets, stars, and constellations. As you point your device towards the heavens, the app identifies the location, and the contour of the sky and overlays different celestial bodies and constellations on the screen as should be seen from that place (*see* Figure 7.2).

Credit: Andrey_Popov / Shutterstock

Figure 7.2 Augmented reality in use

DID YOU KNOW?

You might have noticed many teens and young adults are wandering around your area, absorbed in their mobile devices searching for something! They may be busy playing the augmented reality game – Pokémon Go. The 2016 sensation, Pokémon Go, enabled players to capture their beloved Pokémon by peering through their smartphones at the actual world, augmented with overlaid images. The game achieved massive success, boasting as many as 65 million users during its height of fame.

Virtual Reality (VR) offers an immersive experience that replaces real-world environments and objects with a simulated counterpart. Essentially, VR is a software-created artificial environment that represents a 3D space. It is presented to the user so convincingly that the user suspends disbelief and accepts it as a real environment. This technology can be experienced using VR headsets (like Oculus), glasses (like Google Glass) and goggles. The primary senses used to experience VR are sight and sound. Advanced VR devices enable interaction with the virtual environment through real-time sensors and AI-powered software, further enhancing the immersive experience.

VR technology has transformed tennis training, providing a comprehensive experience. It offers a realistic simulation of a tennis game, which helps improve a player's performance on the field and sharpen his skills (*see* Figure 7.3).

Credit: Anna Sadovskaia / Shutterstock

Figure 7.3 Boy with a VR glass learning how to play tennis from a virtual trainer

Mixed Reality (MR), a subset of augmented reality, facilitates the blending of real and virtual worlds to create new environments and visualisations where real objects and digitally simulated elements can coexist and interact. This is typically achieved using specialised goggles or glasses that allow you to perceive virtual objects as if they were in our immediate environment. Our movements in the physical world are tracked and replicated in a digital reality, which provides interaction capability with the virtual world.

A few examples of mixed reality applications are – HoloLens, Vuzix, Meta, Magic Leap, ODG, Epson Smart and Leap Motion. The Microsoft HoloLens features a transparent display that enables users to view their real surroundings even while the headset is on. It integrates digital elements, like holograms, into the actual environment, making them appear as though they exist in real space. This allows users to interact with the digital content while maintaining a presence in the physical world.

In Figure 7.4, we can see that the engineer is viewing as well as interacting with the digital twin while being present on the shop floor. We will study digital twin concepts and its relation with AR later in this chapter and also in Chapter 11.

Credit: ekkasit919 / 123RF

Figure 7.4 Engineer using mixed reality device for education and training

The distinguishing feature of MR compared to other reality-modifying technologies is its ability to enable natural interaction with digital content as if it were part of your physical surroundings. This makes MR a potent tool for applications like gaming and education, where a high level of immersion is beneficial. In contrast to MR, AR doesn't necessitate any specialised hardware beyond a smartphone or tablet, making it accessible to a broad user base and facilitating the easy sharing of AR content. Table 7.1 shows the comparison of these three reality technologies (also refer to Figure 7.5).

Table 7.1 Comparison of AR, VR and MR

Aspect	VR	AR	MR
Core Concept	Immersive digital world	Digital content is overlaid on the real world.	Blends the virtual and real worlds seamlessly
Content type	Entirely virtual, with no real-world elements	Overlaid onto real-world surroundings	Integrated and interactive with the real world
Interaction with Real world	The user is isolated from the real world.	The user experiences an enhanced real world with digital elements. Users interact mainly with the real world augmented with digital content.	Virtual objects are integrated into the real world. Users interact with both virtual and real-world elements.
Device Type	Head Mounted Display (HMD)	Smartphones, tablets, smart glasses, such as	Headsets or smart glasses
Examples	Oculus Rift, HTC Vive, PlayStation VR	Pokémon GO, Snapchat filters, Google Maps AR, Microsoft HoloLens	Microsoft HoloLens, Magic Leap, HoloLens 2
Primary Use Cases	Gaming, simulations, virtual experiences	Navigation, education, retail, and mobile apps	Training, design, gaming, education, remote collaboration
Examples in Industry	Virtual training, gaming, architecture simulations	Navigation aids, remote assistance, advertising	Design, remote collaboration, medical visualisation

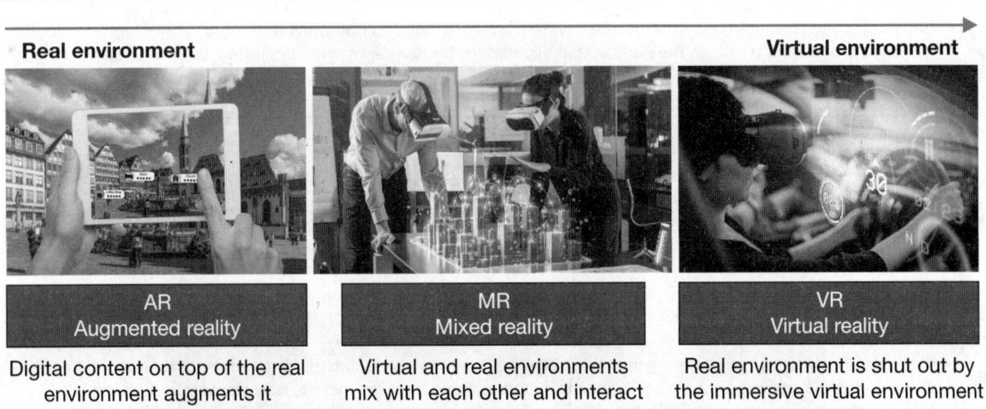

Credit: Prasit Rodphan / 123RF; Gorodenkoff / Shutterstock; Have a nice day Photo / Shutterstock

Figure 7.5 A comparison of AR, MR and VR

7.2.1 A Brief History of AR/VR

Due to the significant advancements in their technologies, AR/VR has started getting a lot of attention from the industries as well as personal users. But the technology was there for quite some time. Let's see over time how these technologies evolved to attain the maturity and usability of their present-day form.

1957
- A significant stride towards augmented reality was made by a cinematographer named Morton Heilig. He created the *Sensorama*, a device that offered an immersive experience by combining visuals, sounds, vibrations, and smells. Although it wasn't controlled by a computer, it was the first attempt to enhance an experience with additional data.

1960
- Heilig invented the *Telesphere Mask*, which was patented in the same year. This invention marked the first instance of a head-mounted display (HMD). However, it was designed for a non-interactive film medium and lacked motion tracking. The headset offered a wide field of view with stereoscopic 3D and stereo sound.

1961
- The first forerunner to today's Head-Mounted Display (HMD) was developed by two engineers from Philco Corporation, Comeau & Bryan. They created the *Headsight,* which featured a video screen for each eye and a magnetic motion tracking system connected to a closed-circuit camera. The *Headsight* was not designed for virtual reality, but to provide immersive remote viewing of hazardous situations for military use. The user's head movements would control a remote camera, allowing them to naturally observe their surroundings. Despite being a crucial step in the evolution of VR HMDs, the *Headsight* lacked computer integration and image generation.

1968
- Ivan Sutherland and his student Bob Sproull developed the first VR/AR HMD, known as the *Sword of Damocles*, which was connected to a computer rather than a camera. This device was large, intimidating, and too heavy for comfortable use, requiring suspension from the ceiling and the user to be strapped in. The computer-generated graphics were basic wireframe rooms and objects.

1975
- Myron Krueger, an American computer artist, developed the first 'virtual reality' interface, *Videoplace*, which allowed users to manipulate and interact with virtual objects in real time.

1987
- Jaron Lanier, the founder of the Visual Programming Lab (VPL), coined/ popularised the term 'virtual reality'.

1990
- Thomas P Caudell of Boeing introduced the phrase 'augmented reality'.

1992
- The first fully functional AR system *Virtual Fixtures*, was developed by Louis Rosenberg at the USAF Armstrong's Research Lab. This highly complex robotic system was designed to compensate for the lack of high-speed 3D graphics processing power in the early 90s. It allowed for the overlay of sensory information on a workspace to enhance human productivity.

2012
- Palmer Lucky began prototyping the *Oculus headset*, sparking the modern VR revolution.

2014
- Facebook's acquisition of Oculus and Sony's announcement of their VR project paved the way for large enterprises to foray into the AR/VR space.

2015
- Microsoft announced support for augmented reality and their *HoloLens headset*.

2019 – Present day
- There has been rapid development in the AR/VR field, with high potential for commercial use. Standalone VRs, which do not require a computer or mobile device to function, have become increasingly popular and affordable. Major technology companies like Meta, Google, Sony, Samsung, and Apple have invested in AR/VR/MR projects, leading to rapid advancements in device capabilities.

It can be safely assumed that the best of AR/VR technologies is yet to come. The enterprise AR/VR market is witnessing an expansion in the number and scale of companies it serves. This includes public corporations such as Microsoft and PTC, major consulting firms like Deloitte, Accenture, and McKinsey, as well as startups like REFLEKT, Ubimax, Scope AR, Atheer and Upskill, which have their own platforms for AR/VR development and delivery.

7.2.2 How Does Augmented Reality Work?

With the omnipresent smartphones, most of us knowingly or unknowingly use augmented reality apps regularly. We have discussed a few such simple applications of augmented reality in previous sections. To understand the probable uses of augmented reality, we should first understand the inherent technology behind it.

AR is initiated by a device, like a smartphone, tablet, or smart glasses, that is equipped with a camera and loaded with AR software. When the user aims the device at an object, the software recognises it through computer vision technology that scrutinises the video feed. The device then fetches data about the object from the cloud. The AR data is then displayed as a 3-D experience layered over the object, forming a fusion of actual and digital visuals for the user. As the user's perspective shifts, the AR display's dimensions and orientation automatically adapt to the evolving context, introducing new graphic or text data into view while other data recedes.

A 3-D digital replica, or the object's 'digital twin', stored in the cloud, acts as the bridge between the intelligent object and the AR. This model is either created through computer-aided design during product development or by digitising physical objects. The digital twin collects data from the product, business systems, and external sources to reflect the product's present state. It allows the AR software to accurately position and scale the most recent data on the object. Consequently, AR can offer a view of real-time data streaming from products and enable users to control them via touchscreen, voice, or gesture. For example, a user could tap a start button on the digital graphic overlay within an AR experience or simply say 'start' to send a command to a product via the cloud. An operator using an AR headset to interact with an industrial robot might see superimposed data about the robot's performance and gain access to its controls. Refer to Figure 7.6 which explains this flow.

Moreover, due to the capability to dynamically generate digital images, AR devices can offer varied user experiences to different users. For instance, in industrial environments, users in different roles, such as a machine operator and a maintenance technician, can observe the same object but be presented with different AR experiences customised to their needs.

7.2.3 Types of Augmented Realities

Based on how an AR device detects objects to create augmentation, it can be divided into two categories: Trigger-based augmentation and View-based augmentation.

Trigger-based Augmentation

The augmentation is activated by detecting triggers, such as AR markers, symbols, icons, GPS locations, and so on. When pointed at these markers, the AR device identifies the augmented object, processes the 3D image and then projects it on the user device. The specific nuances of different triggers are as follows:

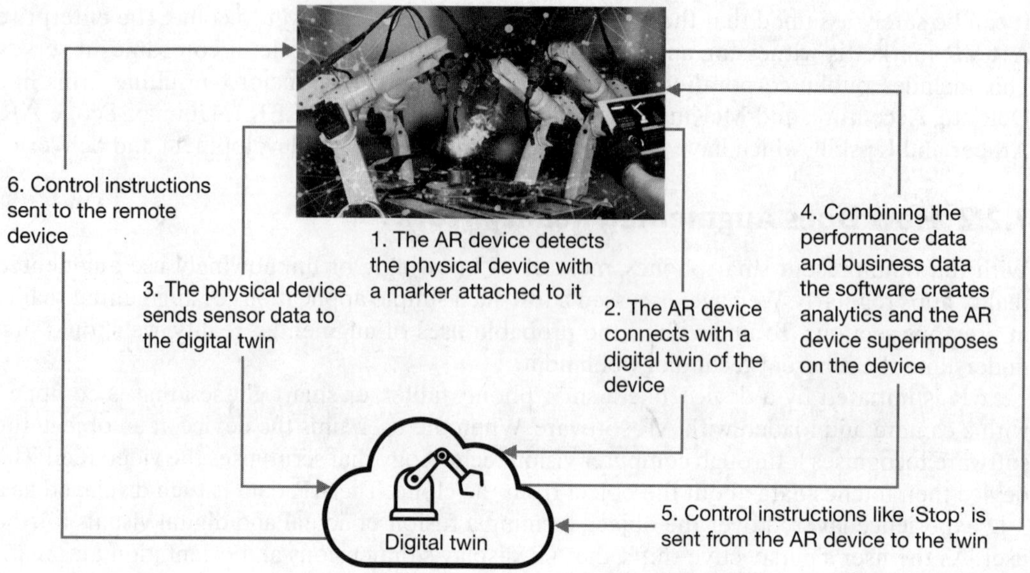

Credit: pitinan / 123RF

Figure 7.6 An augmented reality use case in the industry

- **Marker-based Trigger:** Marker-based augmentation operates by detecting and identifying AR markers. These AR markers are printed on paper and feature unique patterns. They somewhat resemble bar codes and allow the AR application to generate digitally amplified 360-degree visuals on the AR device. Vuforia, Wikitude and EasyAR are a few examples of marker-based AR.
- **Location-based Trigger:** In this augmentation technique, the AR application detects the device's real-time location, such as GPS-based and merges it with dynamic data retrieved from cloud servers or the app's backend. This is also called 'unregistered' AR experience because the pertinent location information is displayed without being tied to a specific object. Most AR-enabled maps and navigation systems, or vehicle parking aids, operate on the principle of location-based augmentation. In Figure 7.7, we can see that the mobile app is using the location trigger to pull up information about the surroundings.
- **Dynamic Trigger:** Dynamic triggers refer to the real-time changes or movements in the environment that trigger the initiation of the augmentation. This is the most responsive form of augmented reality as the motion tracking sensors in the AR device can detect the change in the images of the real world and superimpose them with the digital content providing an interactive and immersive experience.

  A French cosmetic company has made great use of dynamic trigger-based augmented reality. In their shops, they use AR mirror apps having on-screen pickers for various makeup products. Users can try those products and view them on the screen, how they would look in real life. Here, the digitally generated makeup items are dynamically superimposed on the face of the users and adjusted by tracking the movement and shift

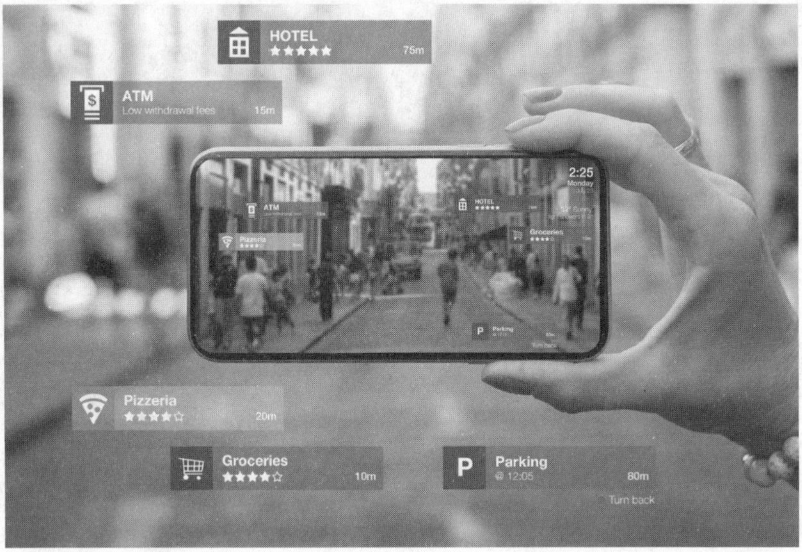

Credit: Kaspars Grinvalds / Shutterstock

Figure 7.7 Location-based augmented reality

in positions. This innovative idea not only saved the wastage of makeup products during trials but also saved the effort of the users in applying and removing the makeup.

View-based Augmentation

In AR, view-based augmentation is a process where the AR application identifies changeable surfaces and then links this dynamic view to its backend system to align reference points. Based on the already stored information, it subsequently overlays pertinent data onto the user's screen.

This form of augmentation enhances the AR experience by making it more engaging and immersive, as it adjusts to the user's shifting perspective of their surroundings. It proves especially beneficial in applications where the user's viewpoint is continually evolving, such as in navigation or exploration apps.

Superimposition-based Augmentation

This is the most sophisticated form of augmentation technology. The AR device uses optical sensors and computer vision to recognise a 3D object and match the same with its preloaded library. After detecting the object, it relays the information about the object that is stored in its database on the screen. Normally, this technology works on static objects.

For example, large automakers are now coming up with AR-based owner's manuals. The users need to point their AR device at a part of the car and the relevant information about those components along with the maintenance instructions are displayed on the screen. This eliminates the need for voluminous physical copies of the owner's manuals and is beneficial both for the automakers and the users.

In Figure 7.8, we can see that the performance information of the machine is overlaid on itself and the worker could track the parameters easily.

Credit: ekkasit919 / 123RF

Figure 7.8 Superimposition-based augmented reality

Generic Digital Augmentation

A generic digital augmentation is a creative form of AR in which any 3D object can be used to create an immersive experience. The digital object can be imposed on the actual space, providing an augmented surrounding.

An e-commerce-based home décor company has made very good use of generic augmentation techniques. The buyers can visualise how the home décor items will look in their rooms by superimposing the item onto the camera view of the rooms, as we can see in Figure 7.9.

Credit: Zapp2Photo / Shutterstock

Figure 7.9 Generic digital augmentation: simulates furniture and colour in a retail store

7.2.4 How Does Virtual Reality Work?

Virtual reality (VR) establishes a comprehensive system that immerses users in a simulated environment, effectively replacing the natural world with a digital one. This immersive experience is achieved through a blend of hardware and software that replaces our usual view with a virtual one.

The functionality of VR devices has seen significant advancements. Where early VR headsets provided a rudimentary virtual experience focused primarily on visual alterations, modern innovations have revolutionised VR with responsive software and enhanced hardware.

Contemporary VR systems encompass more than just the headset, screen, and motion controllers. They incorporate sensors that detect and track the user's movements, eye direction, and hand gestures in real-time, fostering a more profound synergy between the user's actions and the AI-driven software that powers the VR environment. VR now swiftly adapts to our actions, unleashing a new level of 'spatial computing' where interaction with computer systems feels intuitive and natural.

For an immersive and convincing experience, VR technologies utilise a mix of infrared LEDs, motion sensors, cameras and displays. These components enable the headset to collect pertinent data and convey it visually to the user. Additional elements like haptic feedback devices and spatial audio are sometimes employed to enhance the sense of immersion. These technologies will be described in details later in this chapter when we talk about the key technologies behind VR.

Rather than merely overlaying new visuals onto our field of vision, authentic VR crafts an interactive simulated space. It employs AI, sensors and algorithms to align the virtual realm with the physical world, tricking the mind into believing it has entered an entirely different setting.

7.2.5 Types of Virtual Realities

As we discussed earlier, Virtual Reality devices (like wearable headsets) work on the principle of immersing the users in a digital experience by using a combination of sensors, Artificial Intelligence, software and hardware. However, the level of immersion differs based on the application. The VR devices can be categorised into four main types based on the immersion level and their field of use.

Non-immersive VR

This is the earliest form of VR. Though the name suggests that this type of VR is not immersive (contrary to the definition of VR which mentions that all VR technologies are immersive), what it actually means is that the level of immersion is very fundamental. There is no scope for using spatial computing features, such as sensory experience, gesture controls, etc., but the user can move certain characters in the environment through controllers. Thus, the user is not immersed in that environment.

For example, a driving simulator has controls like a steering wheel and a pedal, using which a user can interact with the environment similar to a real car, but the user won't use any headset and won't feel an immersive experience during simulation.

Semi-Immersive VR

This VR technology is very similar to 'mixed reality'. The digital components are projected on top of real objects or environments and users can have a partial sense of immersion.

For example, the interactive exhibits in a museum, and flight simulators where the semi-immersive heads-up displays are used to see digital content projected on windshields are semi-immersive VRs.

Fully Immersive VR

This is the most advanced form of VR. The real world gets replaced by a virtual environment with images, sounds, and physical sensations, creating an illusion by simulating as many senses as possible. Normally the Head Mounted Displays (HMDs) use AI and sensor technologies and respond to the user's actions to generate a sense that the user is 'moving' within the virtual environment. It also takes the help of haptic accessories to trick the brain that the user is touching and interacting with the objects in that virtual space.

Meta Quest 2 and Quest 3, Oculus Rift, and HTC Vive are examples of fully immersive headsets. Users can get a 360-degree simulated experience using these devices.

7.3 KEY TECHNOLOGIES BEHIND AR AND MR

Till now we have discussed the features of different types of realities and their categories. It is now time to deep dive into the components and technologies that enable them to create or alter the real environment. The technologies mentioned below do not work as standalone technologies, but they interact significantly and exchange data with each other for overlaying the digital content on the physical environment in the required dimension and location.

7.3.1 SLAM (Simultaneous Location and Mapping)

SLAM empowers AR devices to perceive the world in three dimensions, identify real-world objects, and seamlessly project virtual content onto AR displays. It serves as the foundation for creating immersive and interactive AR environments.

Localisation

When you launch an AR app using platforms like Google ARCore, Apple ARKit, or Microsoft Mixed Reality, the system begins with limited knowledge of the environment. The device processes data from various sources, primarily the camera. To enhance accuracy, it combines information from sensors like the accelerometer and gyroscope. The primary objective is to localise the device's 3D position within the real world.

Mapping

SLAM aims to construct a map of the environment. This map includes spatial relationships between key points (recognisable features or markers). Real-world scenarios introduce uncertainties. Imperfect knowledge leads to probability distributions for each position. Algorithms such as the Extended Kalman Filter, Maximum a Posteriori (MAP) estimation, or Bundle

Adjustment (BA) handle these uncertainties. As new sensor updates arrive, the entire map dynamically updates, maintaining consistency.

Aligning Measurements

SLAM ensures that new measurements align with existing knowledge. Each sensor measurement contains inaccuracies. Whether derived from camera images or frame-to-frame movement estimation (using accelerometers or odometry), these measurements require precise alignment. Accurate alignment is crucial for reliable and seamless AR experiences.

7.3.2 Depth Tracking

Depth tracking mechanisms or Depth APIs play a crucial role in enhancing the realism of the AR experience by providing depth information to the digitally generated visuals. It helps the device's camera to understand the distance, shape, and size of real-life objects within the environment. The depth map thus created helps the device to place the digital content in a way that the user gets a more immersive experience. An example of the use of depth tracking is occlusion. This ensures that the virtual objects are placed in the right place concerning the real-world object. If a virtual object is to be placed behind a tree, then the depth information helps the AR device render the object in the right place and create a more convincing experience.

7.3.3 Image Processing and Projection

After SLAM and depth tracking algorithms have accurately mapped the environment and determined the depth information, the AR system is ready to create a seamless blend of virtual and real-world elements. Image processing and projection is the next step to create the augmented scene. When the user points their device (whether it's a dedicated AR headset like Microsoft HoloLens or any other AR-enabled device) at a scene, the AR app captures the real-world image through the device's lens. The collected image is then processed in the backend by the AR application. This processing involves aligning the virtual content with the real-world scene based on depth information. The AR program ensures that the virtual objects are rendered in the right dimensions and at the correct location within the user's view.

Let's take an example of a location-based AR. When the user aims their device at the surroundings, the AR app fetches location details (such as GPS coordinates) for that scene. It also retrieves additional information like types of buildings, road names, nearby restaurants, etc., from the cloud database. Using SLAM and depth tracking data, the app precisely places this digital information onto the physical objects as viewed through the screen. The user sees relevant details overlaid on the real-world scene, creating an informative and engaging experience. We have seen in Figure 7.7 how augmented reality is presented on top of a real environment.

7.3.4 3D Modelling

3D modelling is the key technology behind both AR and VR, that enhances their immersive experience. The 3D assets created by this technology are overlaid on real-life scenes to create augmented views. 3D assets can also offer features like interactivity through screens, realistic digital experience, customisation, and personalisation. Some of the powerful 3D modelling tools are Blender, CAD software, and Photogrammetry, each of these offering a different set of features and flexibilities for creating 3D assets. An example of 3D modelling is the use of AR to visualise the architecture of a building, its interiors, and its landscapes.

7.3.5 Manipulation and Interaction with Objects

AR environments provide exciting ways to engage with virtual objects in the real world, such as scaling, rotating and positioning the objects. Some of the methods used by AR for manipulating virtual objects are touch interaction on smartphone screens, virtual buttons on smart glasses, smartwatch touchscreen controls, and gesture controls. It is possible to use multiple manipulation methods to improve the interactive experience. For example, selecting an object by tapping on the screen and then manipulating it (like rotating or changing position) directly with hand gestures offers precise and interactive adjustments.

7.4 KEY TECHNOLOGIES BEHIND VR

A combination of hardware and software provides a seamless virtual reality experience. The list and capability of the technologies supporting VR are growing continuously. But we will discuss the key components that are the building blocks of VR. The key components of VR technology are discussed below.

7.4.1 VR Headsets (HMD)

Virtual Reality (VR) headsets, also known as Head Mounted Devices (HMDs), play a crucial role in many VR experiences. VR headsets are wearable devices that wrap around the user's eyes. They block out the external world and immerse you in a virtual simulation. By replacing your natural field of view with computer-generated visuals, they transport you to a different reality. The main parts of the headsets are screens, lenses, cameras, motion sensors and infrared LEDs.

Mainly two types of VR Headsets are there. Wired headsets, like HTC Vive and Oculus Rift, need to be connected to a PC or console and these offer high-quality experiences. Standalone headsets, like the Oculus Quest operate independently, without external connections. In Figure 7.10 we can see different types of VR headset models.

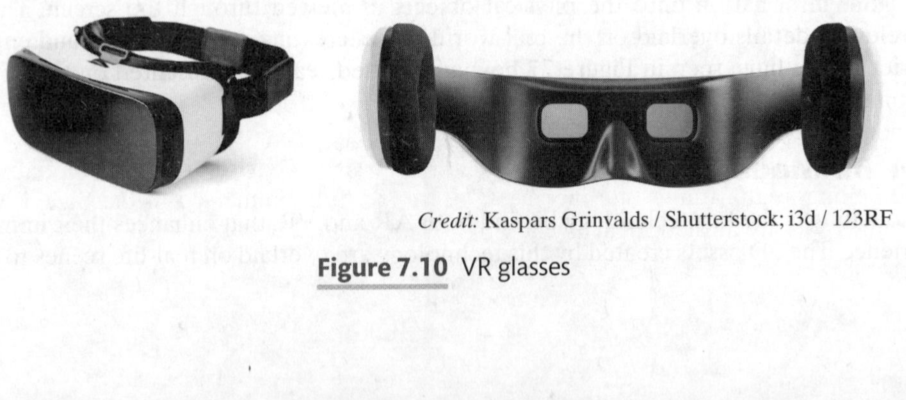

Credit: Kaspars Grinvalds / Shutterstock; i3d / 123RF

Figure 7.10 VR glasses

7.4.2 Lenses and Screens

Within VR headsets, the lenses are positioned between the screen (usually an LED display) and the user's eyes. Their primary purpose is to distort the images in a way that makes them appear three-dimensional (3D). When you wear a VR headset, it passes two slightly different images, one for each eye, through these lenses. The slight disparity between the images creates the illusion of depth, mimicking how our eyes perceive the real world.

Also, inside the headset, there are infrared cameras that monitor the user's movements. As the user turns her head or shifts position, the headset adjusts the light and content on the screen accordingly. This dynamic adaptation ensures that the virtual environment remains synchronised with real-world movements.

Some advanced headsets incorporate eye-tracking technology. By tracking the eye movements, the headset can optimise rendering quality where we are looking. This technique, known as foveated rendering, allocates more processing power to the central area of our vision (the 'fovea'), enhancing realism while conserving resources elsewhere.

7.4.3 Latency, Field of View and Frame Rate

VR headsets align virtual information with the real world and create near-real-life experiences by controlling Latency, Field of View (FOV) and Frame Rate.

Latency

Latency refers to the delay between an action (such as moving our head) and the corresponding change in the virtual environment. When we turn our heads or move, the VR headset must update the visuals instantly to maintain immersion. Minimal latency ensures that what we see aligns seamlessly with our movements. High latency can lead to discomfort, motion sickness, and a less convincing experience.

Field of View (FOV)

The Field of View represents the extent of our vision within the VR headset. In the real world, humans perceive their surroundings in a wide arc around their heads, approximately 200 to 220 degrees. Our left and right eyes overlap, allowing us to see in three dimensions. VR headsets strive to replicate this natural FOV to convince our minds that we're in a different world. The wider the FOV, the more immersive the experience feels.

Frame Rate

Frame rate refers to the number of individual frames or images displayed per second in a VR headset. It is measured in frames per second (FPS). Smoother and more realistic images are indicated by a greater frame rate. VR developers strive for high frame rates to give users a smooth and natural experience. Motion sickness can result from low frame rates that disrupt immersion. For head-mounted displays, the minimum suggested frame rates are typically 90 FPS, while for standalone systems, they are 60 FPS. The virtual environment gets more lifelike and compelling as the frame rate increases. Users feel fully immersed in the digital world when frame rates are at their ideal levels.

7.4.4 Spatial Audio

The contribution of spatial audio is crucial when it comes to persuading our thoughts that we are actually in another universe. It analyses sound that is placed in three dimensions to replicate how our ears perceive sound in the actual world. The natural parameters, like the surroundings, distance, and direction of sound sources, are considered so that an accurate spatial audio effect can improve immersion in virtual reality. Thus, our visual signals and the realistic audio placement work together to provide a seamless experience.

Modern VR systems use intelligent algorithms to dynamically adjust audio based on your head movements. As you turn or tilt your head, the system recalibrates sound sources to match your perspective. This synchronisation ensures that audio aligns seamlessly with visual cues. For example, YouTube, Spotify and Google's VR audio system have embraced this '3D' audio for years.

7.4.5 Head, Movement and Position Tracking

The magic of VR lies in its ability to adapt to our changing position and actions. We can change our position, interact and also communicate within a space using VR. The head and position-tracking technologies utilise sensors, gyroscopes, and AI and influence what we see as we explore. The goal is to create a life-sized, 3D virtual environment without the usual constraints of TV or computer screens.

The basic models of the headsets initially used three degrees of freedom (3DoF) that allowed us to look left, right, up, and down. Today's most advanced headsets support six degrees of freedom (6DoF). With 6DoF support, we can look around in a 360-degree setup, just as we would in real life. Varjo XR-4 is a recent model of VR headset that supports 6DoF.

7.4.6 Virtual Reality Controllers

While some pioneers in the VR field are exploring spatial computing technologies like sensors and trackers to reduce the reliance on controllers, many VR headsets still come equipped with these devices. These hardware elements enable users to interact and perform actions within a virtual setting. Consider the controllers used with devices like the Meta Quest 2 or 3. These controllers serve as a direct line of communication, allowing us to convey our intentions to the VR software.

Companies are persistently advancing their tracking capabilities for body movements, leading to the development of Haptic interfaces and devices. Haptic interfaces are tools that facilitate manual interaction with virtual environments or remotely operated systems. They are typically used for tasks that would normally require hands in the physical world, such as exploring and manipulating objects. Generally, these interfaces receive motor action commands from the user and present suitable tactile feedback. Simple haptic interfaces include computer keyboards, mice, and trackballs. More complex haptic interfaces available in the market include gloves and exoskeletons that monitor hand positions, and joysticks. These can provide force feedback to the user.

7.4.7 Virtual Reality Software

The components of VR systems like latency, position tracking, frame rate, and 3D environments are integrated by virtual reality software. Like other types of software, VR software is highly adaptable and can be tailored to meet a variety of requirements. Nowadays, 'virtual reality as a service' solutions are also available enabling businesses to develop user experiences specific to their industry. Collaborative virtual reality spaces are also gaining popularity now, like Microsoft Teams' Immersive spaces, training simulations and more.

Most VR software encompasses all the necessary data and code to generate a 3D virtual world and avatars. They also ensure that information can be effectively streamed via a headset. Some of the latest VR software in the market can also integrate with existing tools, such as collaboration platforms, communication systems, and content management systems. Examples of virtual reality applications include Mobile VR for iOS and Android, as well as Samsung Gear VR.

AR/VR is revolutionising various fields, from manufacturing to entertainment and education, with their potential expanding every day. Let's delve into the real-life applications of AR/VR and explore how these technologies are transforming and benefiting industries.

7.5 AR/VR IN SMART FACTORIES

The smart factories started to intensively leverage AR/VR technologies in plant operations and workforce training during the COVID-19 pandemic. At that point in time, the main aim was to ensure the safety of the workforce, but it also improved efficiency, and thus was imbibed as part of the main process later. In a manufacturing setting, these technologies provide immersive experiences by enriching the physical space with data, digital elements and digital interactions. In the below use cases, we will see how industry leaders are exploring the potential of these technologies to improve worker safety, accelerate product launches, decrease training expenditures, and drive greater productivity.

7.5.1 Warehouse Management

By leveraging AR-enabled devices such as headsets or glasses, warehouse workers gain the ability to visualise and interact with software-generated overlays superimposed on their physical environments. These digital overlays, displayed through AR glasses or headsets, provide essential item details, quantities and optimal routes, guiding workers precisely to the location of specific items, as we can see in Figure 7.11. This streamlined navigation significantly enhances efficiency. Additionally, AR-powered devices can scan item-level RFID tags and barcodes, presenting information like expiration dates and storage details. By automating these repetitive manual tasks, the risk of errors is reduced. Warehouse managers benefit from real-time, accurate information, enabling informed decisions related to demand forecasts, storage optimisation and stock replenishment.

For example, a leading global logistics company integrated AR into its inventory management using smart glasses. Graphics displayed on these smart glasses guided workers through the warehouse, minimising errors and expediting the picking process. Crucially, the smart glasses recorded essential tracking data without requiring employees to pause and manually input information into a computer. The adoption of AR resulted in a remarkable 15 per cent efficiency increase during the picking process, underscoring the added value it brings to logistics operations.

Credit: pitinan / 123RF

Figure 7.11 AR in warehouse management and inventory tracking

7.5.2 Design and Build

Virtual technologies offer significant enhancements in cost, time, and quality during the design and build phases, particularly in product engineering. By leveraging AR/VR, product designers and engineers can explore options that were previously cost- or time-prohibitive. One of the key advantages of extended realities in product engineering lies in their ability to visualise products in a highly realistic and interactive manner. Designers and engineers can create virtual prototypes and assess them within a real-world context, facilitating better evaluation of aesthetics, functionality, and ergonomics. The adoption of AR/VR eliminates the need for physical prototypes, leading to substantial reductions in time and costs associated with traditional design iterations. With virtual prototypes, engineers can virtually test various configurations, materials, and dimensions, enabling faster and more precise decision-making (*see* Figure 7.12).

For example, leading automobile companies are increasingly embracing VR as an industry standard. A prominent global car manufacturer has established a virtual reality laboratory, empowering engineers to enhance the quality of early concepts by directly designing new car models. With VR, the manufacturing team can now create car designs before producing physical prototypes. By testing models in specific contexts that simulate real-world scenarios, VR accelerates time-to-market and lowers production expenses.

We will also discuss in detail in Chapter 11 how AR/VR and digital twin collaborate to make design, build and maintenance tasks streamlined.

Credit: Gorodenkoff / Shutterstock

Figure 7.12 VR in automobile design

7.5.3 Maintenance and Assembly Training

The integration of augmented reality AR/VR technologies has revolutionised assembly and operations training in the manufacturing industry. An AR system overlays interactive digital simulations onto real-world manufacturing machinery. Thus, the trainees can now engage with virtual equipment, practising operations and troubleshooting in a risk-free environment. Assembly line workers benefit from real-time AR-guided instructions projected onto their workspace, as shown in Figure 7.13. These step-by-step visual cues assist with assembly and maintenance procedures, reducing the learning curve and minimising errors. Whether through AR glasses or handheld devices, workers can see virtual elements, such as diagrams, text, or animations, that guide them through each critical assembly process step.

Similarly, using VR systems, the workers can practice complex assembly sequences risk-free, detect and resolve fit/alignment issues virtually, simulate user interactions to improve component's ergonomics and functionality and also can analyse wear patterns to extend product lifespans. Thus, incorporating VR during the assembly stage leads to products that are not only more durable but also perform better for customers. This proactive approach minimises the need for customer service and repairs down the line. Moreover, VR in the development phase allows you to assess a project's potential success before committing excessive time and resources to it.

A prominent aircraft manufacturer has harnessed AR technology to enhance efficiency and precision in the intricate process of aircraft wiring assembly. Technicians wear AR glasses that provide wire routing instructions, eliminating the reliance on printed manuals and manual inspections. The AR system overlays wire paths onto the actual aircraft structure, ensuring a clearer visualisation of precise installations. As a result, assembly time is reduced by 25%, and

Figure 7.13 AR is being used in equipment training

errors are significantly minimised. These improvements lead to substantial cost savings due to reduced rework and enhanced overall efficiency.

7.5.4 Factory Floor Planning

Determining an effective factory floor plan needs intricate design, testing and trials of optimal placement of tools, equipment and personnel. Efficient floor planning is critical for mass-production manufacturing. Inefficient design of the production line may not only cause delays, but may also cause unexpected shutdown of the line, incurring substantial costs. But it is not possible to experiment with various floor plans physically, due to factors like prohibitive cost and complex infrastructure and manpower requirements to support such trials. In these scenarios, virtual technologies can offer streamlined solutions. Different production flows simulating how workers and robots execute tasks can be meticulously designed and tested without implementing changes in the physical layout. A virtual floor plan can be completed in a shorter duration. It ensures seamless and efficient operations in the newly configured production line and enables the expedited introduction of new products to the assembly line.

7.5.5 Improved Worker Safety

We have discussed that AR/VR technologies can overlay information on physical assets, which allows workers to gain crucial knowledge about those assets. For example, an AR system superimposes digital instructions onto a real-world view, illustrating the correct safety procedures. Similarly, VR creates a controlled environment that simulates high-risk situations, offering immersive training experiences. Manufacturers can document best practices, safety processes and disseminate them across facilities using AR/VR. These help in enhancing worker safety by increasing their awareness about the potential hazards without facing actual danger. It can also help in streamlining the training process, improving decision-making, and reducing risks, all of which contribute to improved productivity.

For instance, a well-known car manufacturer has employed VR/AR to instruct workers on safety processes to reduce accidents, and it has led to a significant decrease in production line injuries by 70%.

7.5.6 How a German Automaker Leverages AR/VR in Production

Let's explore a practical example of how a prominent German car manufacturer utilises AR/VR:

- AR application and digital twin are used at the vehicle conceptualisation and prototyping stage. A true-to-scale holographic 3D model, which is the digital twin of a new car model, is created with the intricate design of each section. Then the real geometries are overlaid on it using AR goggles. With this, the design team can quickly analyse whether the production assembly line workers will be able to properly fit the components. The key benefits achieved are the reduction in the period between the design of individual parts to the production planning and the reduction in integration time for new vehicles, resulting in overall time and cost savings.
- With a similar approach, the company also reduces dependency on trainers who work with individual workers in the engine assembly units. AR goggles are used in training sessions, where the participants are guided through each process step with visualisations. Multiple people can undergo AR training simultaneously.
- VR is utilised by this automaker extensively to plan the plant floor, assembly operations, logistics and construction. A digitised 3D plant structure is created, which is leveraged while setting up new workstations or new assembly lines. This approach provides flexibility in experimenting with multiple layouts and operations planning without physically investing in hardware.

We can observe that AR/ VR technologies enable an advanced form of automation that integrates digital technology with the physical assets of manufacturing plants. The key to the successful use of these technologies is how well the real-world data is combined with interactive elements and computer-generated graphics to create a fully immersive 360-degree experience.

7.6 OTHER APPLICATION AREAS OF AR / VR / MR

As we can see in Figure 7.14, AR/VR has a significant presence in almost all the major industries. Although AR and VR have witnessed the highest traction in the industrial and manufacturing industry, the gaming industry is also an ardent adopter of AR/VR followed by other sectors such as automotive, healthcare and education.

7.6.1 Gaming

As depicted in the chart in Figure 7.14, the gaming industry commands a very high market share in AR/VR revenue, and this trend is on the rise. The VR gaming market size is projected to reach USD 45.20 billion by 2027, up from USD 5.12 billion in 2019, with a remarkable

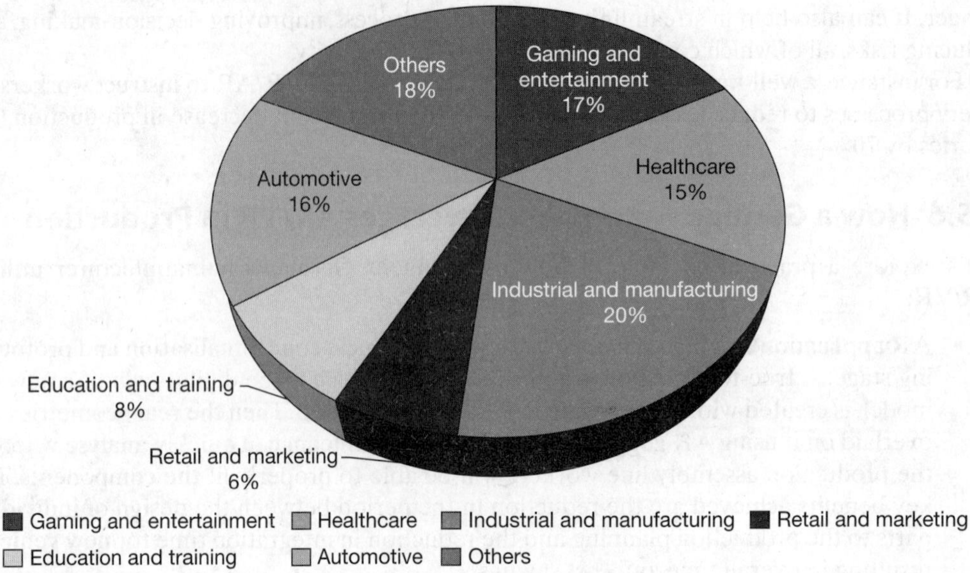

Figure 7.14 AR/VR market split by industry

compound annual growth rate (CAGR) of 31.8% during the forecast period (source: Fortune Business Insights report).

Looking ahead, the convergence of AR and VR games, coupled with the availability of cloud infrastructure, leaves no doubt that AR/VR represents the future of gaming. Let's delve into the specifics of VR and AR gaming:

- **VR Gaming:** Immersive Experiences is the key feature of VR games. Equipped with VR gear, users can explore high-quality, computer-simulated 3D environments. They can freely move within these virtual worlds, interact with objects, and engage in various experiences. VR gaming relies on specialised headsets and powerful hardware to deliver lifelike sensations.

  A few notable VR games are – 'Beat Saber' by Meta has achieved an impressive revenue of 255 million since its release in 2019. 'Half-Life: Alyx' by Valve Software has sold over 2 million units since its 2020 launch.
- **AR Gaming:** Multidimensional interaction capability makes AR games very popular. AR gaming creates a space where players can seamlessly interact with 3D graphics within real-world environments. AR games integrate digital content into the physical world, allowing players to engage with virtual objects and characters.

  For instance, games like Pokémon GO use AR to place virtual creatures in real-world locations, encouraging exploration and interaction.

7.6.2 Healthcare

Advancements in AR and VR have transformed healthcare, enhancing patient care and medical encounters. As technology progresses, we anticipate even more impactful applications in the coming years.

- **Augmented Surgery:** In surgical settings, the integration of VR/AR provides better visualisation and preparedness for procedures. Surgeons can precisely identify structures within the surgical field, explore alternative approaches and streamline workflow. Benefits include shorter procedure times, improved outcomes and reduced complications.
- **Pain Management:** VR is explored as a pain management solution, especially for chronic pain. AR can alleviate phantom limb pain experienced by amputees. By projecting a digital limb onto patients, their brains perceive the amputated limb as still present. This technique, combined with electrodes, helps patients relax and ease pain.
- **Assisting Healthcare Staff:** AR superimposes digital information on a patient's body. For instance, AccuVein uses AR to assist nurses in locating veins during IV installations. This improves success rates and simplifies the process for both nurses and patients.
- **Medical Training:** AR/VR aids in surgical education. It can replace traditional cadavers and animal models for practising procedures. Simulating live surgeries prepares providers better and expands access to medical education, especially in areas with limited resources.
- **Equipment Maintenance and Compliance:** AR guides maintenance personnel in servicing medical equipment, ensuring adherence to regulatory standards.

These applications demonstrate how AR and VR are shaping the future of healthcare, benefiting patients, practitioners and the industry as a whole. As we can see in Figure 7.15, the global AR/VR market in healthcare is projected to grow by 3.5 times by 2028.

Figure 7.15 The global AR/VR market in healthcare is projected to grow by 3.5 times by 2028

7.6.3 Education and Training

IIn addition to their applications in manufacturing and healthcare training, AR and VR are increasingly being harnessed in the broader field of education and training. These technologies create immersive and interactive learning experiences, transforming how knowledge and skills are acquired, especially by creating dynamic, hands-on experiences that empower learners and enhance their confidence.

- **Standardised Training Environments:** AR and VR provide a consistent, reproducible environment for repeated and optimised training. Whether it's practising surgical techniques or emergency response procedures, trainees can immerse themselves in lifelike scenarios.
- **Mission-Critical Industries:** In sectors like aviation, real estate, military, and oil & gas, AR and VR simulate emergency situations and dangerous events. Thus, the trainees

can practice responding to such critical circumstances and hone their skills while being within a safe and controlled setting.

- **Interactive Product Demonstrations:** AR and VR allow for engaging product demonstrations. Trainees can explore new products virtually, gaining insights and understanding features interactively.
- **Gamified Learning:** By incorporating game elements, AR and VR make training more engaging. Trainees participate in virtual simulations and challenges, test their knowledge and skills and can learn from mistakes without real-world consequences.

7.6.4 Retail and Marketing

The retail and marketing industries have been early adopters of new technologies, and in recent years, those have undergone rapid transformation due to advancements in fields like web 3.0, AI, extended reality (XR) and the metaverse. These innovations have significantly enhanced the shopping experience for buyers. Let's delve into some prominent use cases of extended reality in the retail and marketing domains:

- **Home In-store Experience:** This group of innovative applications allows customers to virtually try out products before making a purchase. Augmented Reality (AR) solutions enable online shoppers to understand what they're paying for and how items will work for them.

  For instance, eyewear manufacturer Lenskart offers an option for customers to try out their range of spectacles virtually. Using the app and their mobile phone camera, customers can see how the glasses look on their faces from the comfort of their homes. This exemplifies the use of augmented reality in retail.
- **Home Decor and Furniture Fitting:** Choosing the right furniture can be daunting, especially when time is limited. Augmented Reality comes to the rescue by allowing customers to explore ideal furniture options online or in-store and enable hassle-free selection.
- Take IKEA, the Swedish furniture giant, as an example. Their 'Try before you buy' applications let buyers visualise 3D furniture images superimposed in different locations within their homes. This immersive experience helps customers make informed decisions.
- **Virtual In-store Experience:** Enhancing Fitting Rooms. Retailers leverage 3D technology to create virtual fitting rooms. These spaces allow customers to try on clothing and accessories virtually.

  SizeStream, a maker of fitting room technology, provides a mobile device in fitting booths that captures customers' body measurements and allow them to visualise how different garments would fit.
- **Virtual Car Showroom:** This is a concept where online visitors can explore detailed representations of a car's exteriors and interiors. They have the freedom to walk around the virtual vehicles, examining them from every angle. Additionally, they can open doors, sit in the seats, start up the engines and interact with features like the in-car entertainment system and sunroof. This immersive experience allows potential

customers to get a feel for the cars from the comfort of their homes. Participating in a VR test drive is as straightforward as using a VR headset and downloading the relevant applications.

Mercedes Benz has already implemented this technology successfully, which resulted in a boost in sales and enhanced customer satisfaction.

Extended reality not only enhances customer experiences but also reduces returns by providing clearer expectations through digital twins. It's an exciting frontier for retail and marketing fields.

7.6.5 Services

Streamlining access to technical and expert information on-site significantly enhances the productivity of field service professionals. AR and VR devices facilitate easy access to crucial information in an immersive format, complete with 3D illustrations, video-driven instructions, and real-time feedback. When connected to an in-house technician, these devices empower field technicians to work more efficiently.

For instance, ThyssenKrupp, a leading elevator manufacturer, leverages Microsoft HoloLens technology in its service operations. With HoloLens, service technicians can visualise and diagnose elevator issues before physically working on them. Additionally, they enjoy remote, unrestricted access to technical expertise right at the job site, enhancing safety and efficiency. By addressing service requests proactively and utilising hands-free holographic guidance during on-site visits, ThyssenKrupp has successfully reduced the average duration of service calls by up to fourfold.

In Figure 7.16, we can see a summary of applications of AR/VR in various domains.

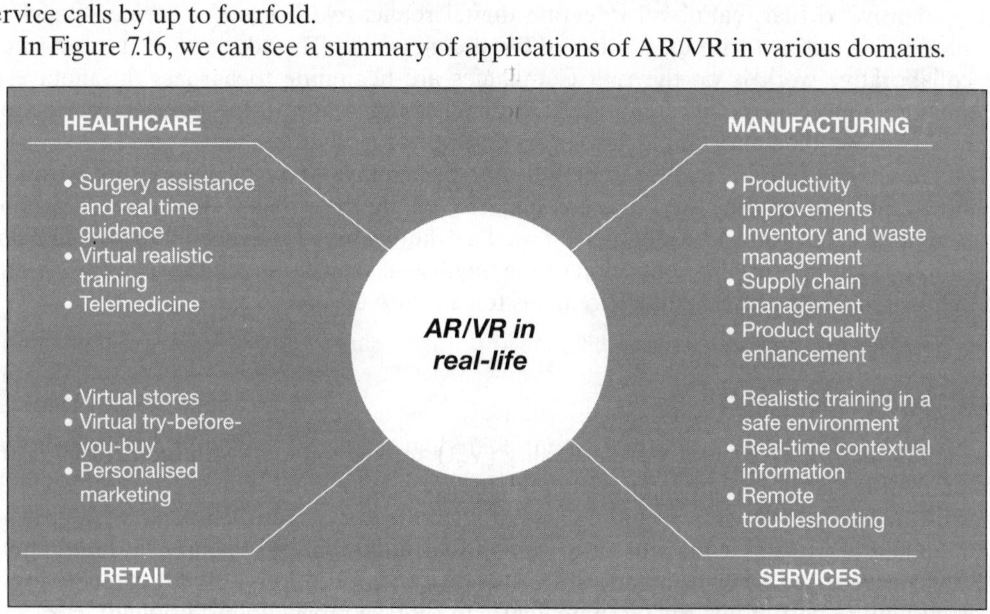

Figure 7.16 Applications of AR/VR in different domains

7.7 CONCEPT OF METAVERSE

The metaverse is a virtual shared space where people from different locations can interact, collaborate, and share experiences in a fully immersive manner. It builds upon traditional virtual reality, allowing users to explore and engage in a 3D environment. By combining AR and VR technologies, the metaverse creates a powerful virtual environment where users can seamlessly transition between the real world and the virtual world. In the metaverse, users not only experience immersion but also interact and collaborate. For instance, in a typical video meeting, users can see and talk to each other, but in a metaverse meeting, they can virtually collaborate on tasks and socialise. This evolving concept has the potential to revolutionise communication, collaboration, gaming, online shopping, and more.

The term 'metaverse' was coined by author Neal Stephenson in his 1992 sci-fi novel 'Snow Crash', which depicted a dystopian future world built within an alternative 3D-connected reality. Tech giants like Meta (formerly Facebook), Microsoft, Google, Nvidia, and Qualcomm envision a network of real-time 3D worlds that multiple people can enter simultaneously. In the metaverse, you can work, learn, chat, relax, attend virtual concerts, and engage in various activities. It aims to simulate the real world, and technologies such as AI, AR, and VR play a significant role. Augmented reality allows virtual objects to be integrated into the physical environment, while VR provides a fully immersive 3D virtual experience. Although wearing a VR headset isn't mandatory in the metaverse, experts believe that VR technology will become integral to this new environment.

VR and digital twinning are fundamental components of the nascent industrial metaverse. This expansive virtual realm will integrate digital replicas with broader elements such as machinery, production facilities, goods, and logistics networks. The adoption of the metaverse for collaborative work is on the rise. Companies are beginning to harness the metaverse to enhance remote work by creating three-dimensional spaces for employee interaction. Figure 7.17 shows the important technologies that work together in a metaverse.

McKinsey has optimistically forecasted that the metaverse market could be worth $5 trillion by 2030. It is anticipated that e-commerce will be the primary driver, while gaming, entertainment, education, and metaverse-based marketing are also expected to be significant contributors. The potential acceleration of the metaverse's development through the synergy of Augmented Reality (AR) and VR remains to be observed.

7.8 CONCLUSION

Augmented Reality (AR) and Virtual Reality (VR) are beginning to gain momentum in the manufacturing sector. It is anticipated that the industry will increasingly adopt these technologies for numerous purposes, including assembly and factory layout. The market is growing exponentially as we can see in Figure 7.18, from the current 60 billion to nearly 250 billion USD by 2030. This remarkable growth underscores the diverse applications of AR/VR technologies across various sectors. While the ultimate goal is to achieve broad, mainstream acceptance of AR/VR, this objective has not yet been realised.

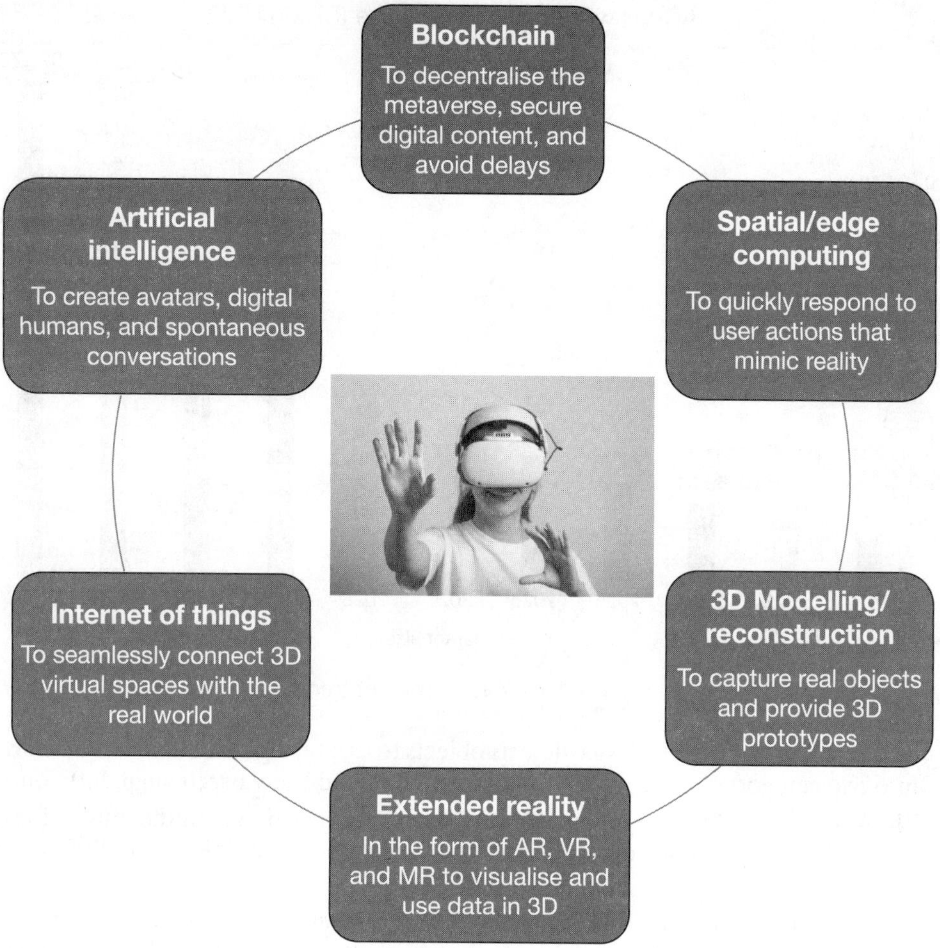

Credit: ArmadilloPhotograp / Shutterstock

Figure 7.17 The prominent technologies behind the metaverse

SUMMARY

- Augmented Reality, Virtual Reality, and Mixed Reality are extended reality technologies that either enhance or replace a real-life environment with a simulated one.
- AR is a technological tool that enriches our perception of the physical world by overlaying digital elements such as images, text, sounds, and videos.
- VR offers an immersive experience that replaces real-world environments and objects with a simulated counterpart.
- MR, a subset of augmented reality, facilitates the blending of real and virtual worlds to create new environments and visualisations where real objects and digitally simulated elements can coexist and interact.

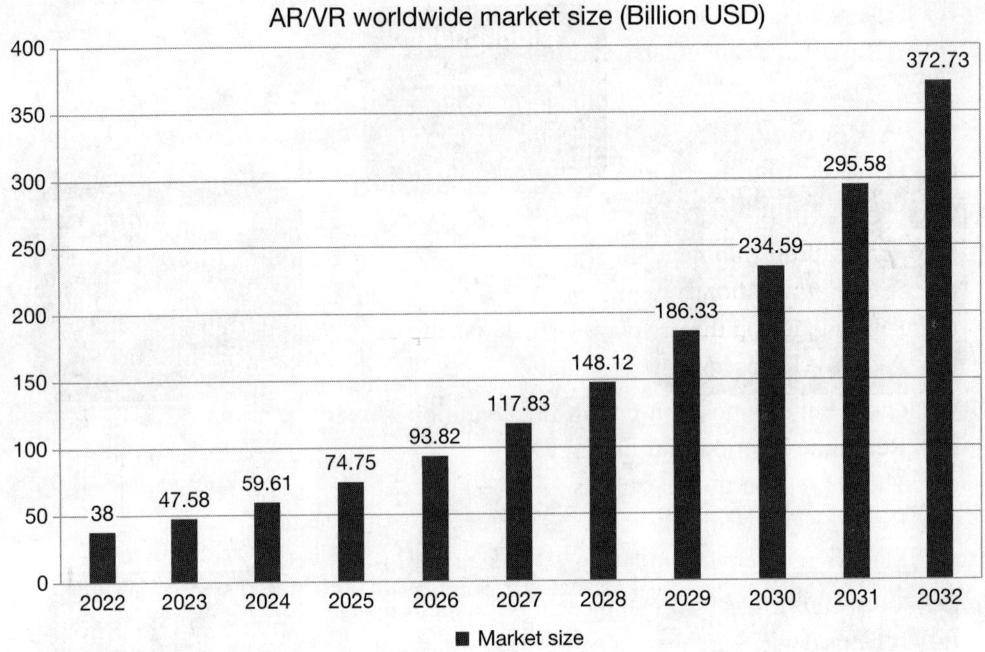

Figure 7.18 AR/VR market trend

- Based on the way an AR device detects objects to create augmentation, it can be divided into two categories – Trigger-based augmentation and View-based augmentation.
- The VR devices can be categorised into four main types based on the immersion level and their field of use – Non-immersive VR, Semi-immersive VR and Fully Immersive VR.
- The key technologies behind AR are – SLAM (simultaneous location and mapping), Depth tracking, Image processing and projection, 3D Modelling, Manipulation and Interaction.
- The key technologies behind VR are – VR headsets (HMD), Lenses and Screens, Latency, Field of View and Frame Rate, Spatial Audio, Head, Movement, and Position Tracking, Virtual Reality Controllers and Virtual Reality Software.
- The metaverse is a virtual shared space where people from different locations can interact, collaborate and share experiences in a fully immersive manner.

EXERCISE

Multiple Choice Questions (1 Mark)

1. Which of the following is an example of Marker-based Trigger AR?

(a) GPS-based AR for navigation

(b) A makeup app that superimposes virtual makeup

(c) A printed QR code used to display digital visuals
(d) AR used in car manuals

2. Which AR technique uses real-time location data to provide relevant information?
(a) Marker-based Trigger
(b) Dynamic Trigger
(c) Location-based Trigger
(d) View-based Trigger

3. What is an example of Dynamic Trigger AR?
(a) A museum exhibit with digital content projected on physical objects
(b) A GPS navigation system
(c) A cosmetic app that displays virtual makeup on a user's face
(d) A car's AR-based owner manual

4. What does Superimposition-based Augmentation primarily work on?
(a) Real-time location of users
(b) The 3D recognition of objects
(c) Recognising AR markers and GPS data
(d) Mapping out environments in 2D

5. How does Depth Tracking enhance an AR experience?
(a) It helps devices create maps of the environment
(b) It adjusts the virtual objects in relation to real-world objects
(c) It controls the position of virtual objects on the screen
(d) It processes sensor data from a GPS

6. What is the key purpose of SLAM (Simultaneous Localisation and Mapping) in AR?
(a) To map out 2D objects
(b) To create a real-time digital model of the user's surroundings
(c) To track gestures and movements
(d) To provide depth tracking

7. What role does VR play in worker safety?
(a) Provides physical protection equipment
(b) Simulates high-risk situations for training
(c) Increases accident rates
(d) Reduces the number of workers on-site

8. What is the advantage of using AR for engine assembly training?
(a) Reduces the need for digital tools
(b) Allows multiple trainees to be trained simultaneously
(c) Increases dependency on trainers
(d) Does not provide step-by-step instructions

9. Which of the following is a key feature of VR gaming?
(a) Passive experience
(b) Lifelike immersive experiences
(c) No hardware required
(d) Limited interactivity

10. What is one application of VR in healthcare?
    (a) Assisting in vein location during IV installations
    (b) Managing chronic pain
    (c) Conducting remote consultations
    (d) Scheduling appointments
11. What advantage does AR/VR provide in retail?
    (a) Increases product returns
    (b) Allows customers to view products in real-world settings
    (c) Eliminates the need for customer feedback
    (d) Reduces the need for marketing
12. What is the primary benefit of using AR/VR in service operations?
    (a) Reduced service durations and improved efficiency
    (b) Increased employee turnover
    (c) Higher costs of service calls
    (d) Reduces the need for customer service representatives
13. What is the metaverse?
    (a) A type of virtual game
    (b) A fully immersive, shared virtual space for socialising and working
    (c) A new virtual reality headset
    (d) A mobile app for virtual shopping
14. Scenario: A medical school has integrated virtual reality (VR) into its surgical training program. Students now practice performing surgeries in a virtual environment that mimics real-life operations, without any risk to patients. The VR system provides immediate feedback and allows for repetition until students are proficient.
    Question: What is the most significant benefit of using VR for surgical training in this scenario?
    (a) It reduces the need for experienced surgeons in the operating room.
    (b) It allows students to perform surgeries on real patients.
    (c) It provides a risk-free environment for practising complex procedures.
    (d) It eliminates the need for any in-person training.
15. Scenario: A major automobile manufacturer uses augmented reality (AR) glasses on the production line. Workers wear these glasses, which overlay step-by-step instructions directly onto their view of the vehicle parts, showing them where to place components, check measurements and complete the assembly process.
    Question: How does the use of AR glasses in the automotive assembly line benefit the workers?
    (a) It reduces the number of components needed for each vehicle.
    (b) It provides real-time guidance and reduces errors in assembly.
    (c) It eliminates the need for human workers.
    (d) It speeds up the delivery of finished cars to customers.

16. Scenario: A high-end fashion retailer has introduced a virtual reality (VR) platform that lets customers try on clothing items virtually in their homes. Using a VR headset, customers can see how different clothes fit and look without ever having to go to a physical store.

    Question: What is the key advantage for customers using the VR platform for shopping?

    (a) They do not need to worry about sizes as they will automatically fit.

    (b) They can virtually try on clothes from the comfort of their own home.

    (c) They can purchase items without seeing them in person.

    (d) They are guaranteed a perfect fit every time.

17. Scenario: A customer support centre has integrated virtual reality (VR) training modules for new employees. The VR training simulates real customer interactions, helping trainees practice their responses to various customer inquiries and issues in a controlled environment.

    Question: How does VR improve the customer support training process?

    (a) It allows trainees to handle real customer calls from the start.

    (b) It simulates real-world interactions, helping trainees build experience without risk.

    (c) It eliminates the need for any in-person training sessions.

    (d) It guarantees that every trainee will provide perfect customer service.

18. Scenario: An educational institution has introduced a VR program that allows students to explore historical landmarks and scientific concepts virtually. Students can 'visit' ancient civilisations, study astronomy, or walk through human anatomy – all in a fully immersive VR environment.

    Question: What is the primary educational advantage of using VR in this scenario?

    (a) It makes studying historical subjects easier.

    (b) It provides immersive, hands-on learning experiences without leaving the classroom.

    (c) It eliminates the need for textbooks.

    (d) It speeds up the curriculum and reduces the number of lessons.

19. Scenario: A technician working in the field for a utility company uses AR glasses to get real-time information about the equipment they are servicing. The glasses overlay schematics and instructions onto the technician's view of the equipment, helping them perform the repair more efficiently.

    Question: What is the key advantage of using AR glasses for field service work?

    (a) The technician can diagnose problems more quickly using detailed overlays.

    (b) The technician can work without any tools or equipment.

    (c) The technician can perform repairs without any previous knowledge of the equipment.

    (d) The technician no longer needs to interact with customers.

20. Which of the following is a key technology used for creating 3D assets in AR and VR?

    (a) GPS sensors

    (b) 3D Modelling

    (c) Gesture Controls

    (d) Motion Sensors

Short Answer Type Questions (2 Marks)

1. How does Augmented Reality (AR) combine real and digital visuals for the user?
2. Explain the difference between Trigger-based Augmentation and View-based Augmentation in AR.
3. How does SLAM (Simultaneous Localisation and Mapping) contribute to the functioning of AR?
4. What distinguishes Fully Immersive VR from other types of Virtual Reality technologies?
5. What is the benefit of using VR in architectural design?
6. How does augmented reality (AR) enhance the efficiency and accuracy of assembly lines in manufacturing industries?
7. What is the role of lenses in VR headsets?
8. What is foveated rendering in VR?
9. How does spatial audio enhance immersion in VR?
10. What are the two main types of VR headsets?

Long Answer Type Questions (5 Marks)

1. Explain the process of how Augmented Reality (AR) works, from detection to display.
2. Describe the different types of Augmented Reality (AR) based on how objects are detected and augmented.
3. What are the main differences between Augmented Reality (AR) and Virtual Reality (VR) in terms of user experience and technology?
4. Discuss the different types of Virtual Reality (VR) technologies and their use cases.
5. Explain the role of AR/VR in improving the learning and development of employees in the corporate sector.
6. How do AR and VR contribute to creating more efficient supply chain management?
7. How does AR/VR technology improve design and prototyping processes in the automotive industry?
8. What are the challenges faced by industries in adopting AR/VR technologies and how can they be overcome?
9. How can AR/VR technologies optimise factory floor planning and manufacturing operations?
10. What is the role of SLAM (Simultaneous Localisation and Mapping) in Augmented Reality and how does it enable interactive experiences?

CHAPTER 8

Cloud Computing

OBJECTIVES

This chapter aims to delve into the realm of cloud computing. In this era of the 4th Industrial Revolution, the impact of the cloud is humongous. Cloud computing powers Industry 4.0 by delivering a scalable, flexible and cost-effective infrastructure that seamlessly integrates advanced technologies such as IoT, AI and big data analytics, fostering automation, real-time insights and digital transformation in industrial operations.

By the end of this chapter, you should be able to:

- Understand the evolution of cloud computing and various deployment and service models of cloud implementation.
- Comprehend the critical components of cloud computing, including Virtualisation, Containerisation, Cloud Orchestration and Management, Cloud Native Applications and Edge and Fog Computing.
- Drive implementation of various layers of cloud computing for manufacturing systems.
- Understand the security and privacy aspects that should be considered during cloud implementation.

8.1 INTRODUCTION

Cloud computing has fundamentally changed how businesses and individuals access and manage computing resources. This involves providing computing services over the Internet (referred to as 'the cloud'). These services include servers, networking, storage, software, databases, analytics and intelligence. The goal is to facilitate rapid innovation, provide flexible resources and achieve economies of scale. Unlike traditional on-premises infrastructure, users generally pay for the specific cloud services they use, which enables cost reduction, improved operational efficiency and the ability to scale under changing business needs. The introduction of cloud computing represents a significant transformation in the IT landscape, empowering organisations to navigate and thrive in an era where digital agility is of prime importance.

8.2 WHAT IS CLOUD COMPUTING

8.2.1 Definition and Features

The National Institute of Standards and Technology (NIST) furnishes a widely accepted academic definition of cloud computing: '*Cloud computing is a model for enabling ubiquitous, convenient, on-demand network access to a shared pool of configurable computing resources (e.g., networks, servers, storage, applications and services) that can be rapidly provisioned and released with minimal management effort or service provider interaction*'.

Cloud computing encompasses the following key features:

- **On-Demand Self-Service:** Users can independently allocate and oversee computing resources, such as server time and network storage, as required, without necessitating human intervention from service providers. Thus, agility is improved as the users can rapidly re-provision infrastructure resources with minimum expense.
- **Broad Network Access:** Cloud services are accessible and easily reachable via the network. They can be accessed through standard mechanisms using diverse devices, including laptops, smartphones, tablets and workstations. This flexibility in accessibility ensures that the cloud is always at your fingertips.
- **Resource Pooling and Multi-tenancy:** Computing resources are usually consolidated into clusters to cater to multiple users (tenants), with diverse physical and virtual resources dynamically assigned and reassigned in response to demand. Even though cloud customers utilise shared resources, they remain oblivious to one another and their data is strictly isolated. This ensures the highest level of data privacy and security.
- **Rapid Elasticity:** Computing resources can be provisioned to expand and released to contract quickly and flexibly, based on demand fluctuation.
- **Measured Service:** Cloud services can optimise utilisation of resources by measuring and adjusting usage based on service type like compute, storage, bandwidth consumption and active users.
- **Self-Service Catalogue:** Cloud often provides pre-configured services catalog. Users, even without deep technical knowledge, can quickly deploy and manage these services.
- **Automation and Orchestration:** Cloud environments support the automation and orchestration of tasks and workflows, enabling efficient resource management and streamlined operations.
- **Scalability:** Cloud computing supports scalability and flexibility to handle varying workloads and changing business needs. This helps organisations to grow and innovate without any constraints of physical infrastructure.
- **Availability and Reliability:** Cloud providers offer redundant infrastructure and data replications across diverse geographical locations to ensure high availability and reliability. This feature drastically minimises the risk of downtime and data loss in case of any system failure.
- **Security and Compliance:** Cloud providers implement robust security measures, such as encryption, access controls and compliance certifications, to protect data and ensure regulatory compliance.

Figure 8.1 below shows the key features of cloud computing.

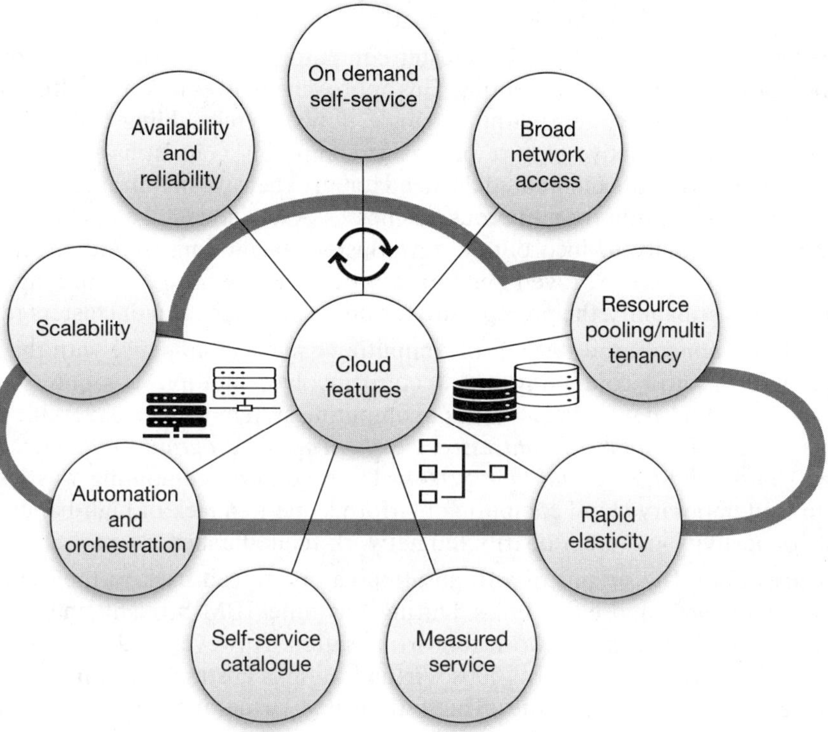

Figure 8.1 Features of cloud computing

8.2.2 Evolution of Cloud Computing

Cloud computing revolves around leasing computing services, a notion that originated in the 1950s. The current state of cloud computing is the result of the significant contributions of the following key technologies:

1. **Distributed Systems:** Multiple independent systems are integrated to form distributed systems. From user perspective, it offers a unified, single view. The primary objective is optimisation of resource sharing, concurrency, scalability, constant availability, and fault tolerance. Initially, physical proximity of the independent systems was key to enable distributed systems. However, advanced computing concepts like mainframe, cluster, and grid computing, worked towards addressing the geographical limitations.
2. **Mainframe Computing:** Mainframe computing technology came during 1951. It introduced powerful machines with massive memory and super processing capabilities. These systems can handle billions of calculations and transactions on real-time. This capability makes them ideal for mission-critical applications, that demand high reliability, security, and scalability. Modern mainframes, for example, IBM Z® is capable

of processing over a trillion secure web transactions daily while adhering to stringent privacy regulations. Despite high capability, the forbidding cost of mainframe led to the development of much affordable alternative like cluster computing.

3. **Cluster Computing:** In the 1980s, cluster computing emerged as a more cost-effective substitute for mainframe computing. This approach involves linking multiple computers to create a unified system within a single network, resembling distributed systems. However, the distinctive feature of cluster computing is its utilisation of commodity hardware. This feature eliminated the need for an expensive mainframe server and made it an appealing option for numerous businesses. A key advantage of cluster computing lies in its flexibility to incorporate new clusters, facilitating smoother scalability for organisations. While it proved ideal for entities needing robust computing power and on-demand scalability, the prerequisite for physical co-location of clusters remained.
4. **Grid Computing:** In the 1990s, grid computing gained prominence with the rise of the Internet. The grid computing architecture involves multiple computers connecting through the Internet to create a mega-computing system with considerable computing capabilities. *This approach addressed the challenge of geographical constraints associated with establishing cloud-like systems*. However, grid computing was less efficient than contemporary cloud computing platforms due to a lack of high-bandwidth internet connectivity and other distributed network-related issues.
5. **Virtualisation:** The origins of virtualisation can be traced back to the mainframe era of the late 1960s and early 1970s. During this time, IBM brought time-sharing solutions. This laid the groundwork for virtualisation. However, widespread adoption of the concept occurred in early 2000s. Virtualisation is abstraction of physical resources like servers into virtual entities. These are called virtual machines. This technological breakthrough significantly reduced the cost of computing. Organisations and individuals now can access computing resources without owning the underlying hardware. Virtualisation has become a foundation of cloud computing.
6. **Utility Computing:** Utility computing is a flexible service model that provides computing resources on-demand basis with a pay-per-use pricing structure. Utility computing allows scaling up or scaling down of resources dynamically to meet fluctuating demands. Unlike traditional fixed-rate plans, in this case, users pay only for the specific resources they consume. This was inspired by the pricing model used for traditional utilities like electricity and water.
7. **Web 2.0:** Web 2.0 is dynamic, interactive and more responsive 2nd generation World Wide Web. It supports collaboration and information sharing through various online channels, like social media, web-based communities, and blogs. The web 2.0 provides the primary channel through which cloud computing delivers services to users, facilitating a connected and engaging experience. This concept was popularised around 2004. Notable examples of Web 2.0 applications include Facebook, Google Maps, Twitter, and others.
8. **Service Orientation:** Service-oriented architecture (SOA) is a design approach that focuses on building applications using reusable services. These services are self-contained units that can be combined to create complex systems. Key concepts in

SOA are Quality of Service (QoS) and Software as a Service (SaaS). QoS guarantees that services meet specific performance and reliability standards. SaaS delivers software applications over the internet on subscription basis.

The evolution of cloud computing is shown in Figure 8.2 below.

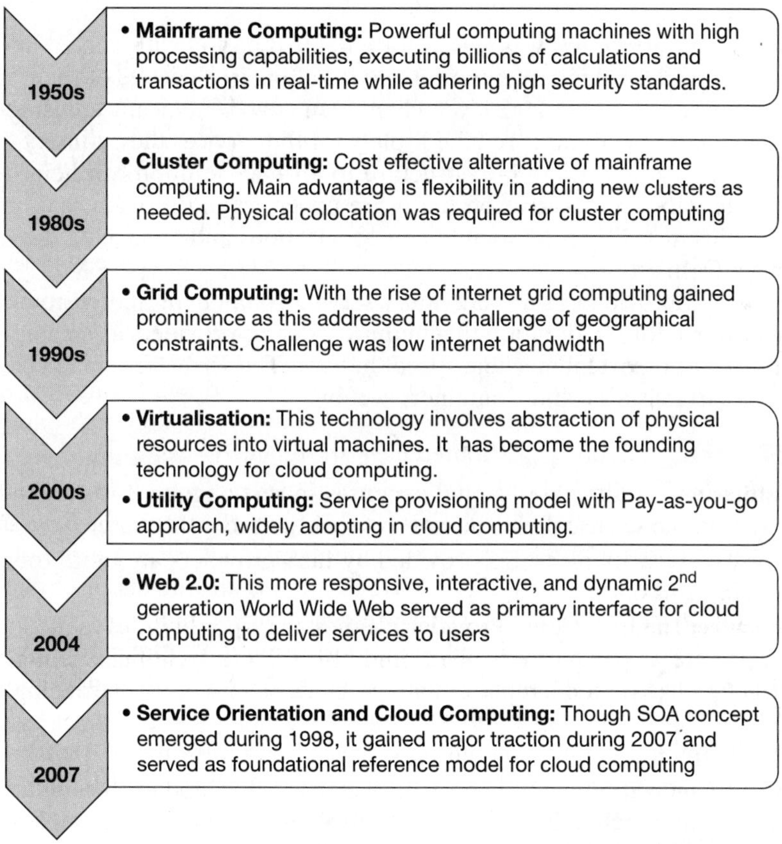

Figure 8.2 Evolution of cloud computing

8.3 CLOUD SERVICE MODELS

Cloud service models define what kind of resources or capabilities are being provided over the Internet and how much control the user has over those resources. The varying degrees of control, security and scalability are designed to meet the diverse needs of users and businesses. They empower users and organisations with high flexibility and control to tailor the computing solutions as per their needs. In this section, we will discuss three types of cloud service models i.e., Infrastructure as a Service (IaaS), Platform as a Service (PaaS), and Software as a Service (SaaS). Each of them provides distinct levels of abstraction and control over the underlying infrastructure and applications.

8.3.1 Infrastructure as a Service (IaaS)

Infrastructure as a Service (IaaS) is also known as Hardware as a Service (HaaS). It offers essential resources like computing, storage, security measures, and networking, to support the development and deployment of applications and services on an on-demand basis with pay-as-you-go model.

In the past, IT infrastructure was leased for a defined period with a fixed hardware configuration. Clients paid for this configuration and duration, irrespective of their actual usage. However, with the advent of the IaaS layer, clients can now dynamically adjust their configuration to meet evolving needs and are billed solely for the services they utilise.

Transitioning an organisation's infrastructure to an IaaS solution can help decrease the upkeep of on-site data centres, save on hardware expenses and provide real-time business insights. It also aids in swiftly provisioning new applications and enhancing the reliability of the underlying infrastructure.

Each resource is provided as an individual service component and customers only pay for a specific resource for the duration they require it. Some providers also charge customers based on the amount of virtual machine space they use. The virtual machine concept will be discussed in the Virtualisation and Containers section.

Advantages

- **Cost Efficiency:** It eliminates capital expenses, lowers ongoing costs and adopts a per-user payment model, usually on an hourly, weekly, or monthly basis.
- **Security:** The security measures provided by IaaS providers are more robust than the in-house infrastructure.
- **Maintenance:** The IaaS Cloud Provider handles all responsibilities related to managing the data centre as per business needs and also the life cycle management (LCM) of infrastructure like periodic replacements, upgrades and security hardening.

Disadvantages

- **Restricted Control:** Since IaaS providers manage and update the underlying infrastructure, it can potentially limit users' control and their ability to implement specific customisations.
- **Security Challenges:** In IaaS model, users are responsible for safeguarding their data and applications. This requires strict monitoring and proactive measures from users.

Examples of IaaS are Google Cloud Platform, Amazon EC2, Microsoft Azure, Digital Ocean, OpenStack, Rackspace, VMware, etc.

8.3.2 Platform as a Service (PaaS)

Platform as a Service (PaaS) is a cloud service model that provides developers with a platform and environment to build and deploy applications and services on top of it. These services are hosted in the cloud and accessed through web browsers.

In a PaaS model, the provider manages hardware and software infrastructure hosting, eliminating the need for users to set up and maintain in-house hardware or software for developing and running applications. This separation allows application development and deployment to occur independently of specific hardware requirements.

While users do not have control over the underlying cloud infrastructure, including networks, servers, operating systems, middleware and storage, they maintain control over their deployed applications and, in some cases, the configuration settings of the hosting environment. For example, when planning an annual event, you could either build a venue from scratch or rent one; however, the event itself remains the same regardless of the venue choice

Advantages

- **User-friendly:** It offers easily accessible infrastructure and IT services through web browsers, enabling usage from any location.
- **Cost-effective:** Pricing is based on actual usage, avoiding the costs of maintaining on-premises hardware and software.
- **Lifecycle Management:** Designed to facilitate the entire web application lifecycle, including build, test, deploy, manage and update.
- **Efficiency:** PaaS model simplifies complex processes by enabling higher-level programming. This improves the overall effectiveness of application development.

Disadvantages

- **Reduced Control Over Infrastructure:** PaaS environments gives users lesser control over infrastructure compared to IaaS. Though the overall platform management becomes simpler, it restricts user customisation options.
- **Dependence on Provider:** PaaS users rely on the cloud vendor to maintain platform availability, scalability and reliability. This reliance can expose users to risks associated with service interruptions, security breach or other provider-related issues.
- **Limited Flexibility:** PaaS solutions may not be suitable for certain workloads or applications, limiting their value for specific organisations.

Some major PaaS solutions are Salesforce, IBM Smart Cloud, Amazon Web Services (AWS), Windows Azure and Google App Engine.

8.3.3 Software as a Service (SaaS)

Software as a Service (SaaS) model provides software applications over the internet. Instead of installing software on individual devices, users access these applications directly through a web browser. This eliminates the need for complex software and hardware management, reducing costs and simplifying IT operations.

SaaS offers a flexible, pay-as-you-go model, allowing users to access software applications directly from a SaaS service provider.

Advantages

- **Cost Efficiency:** SaaS is cost efficient as users pay only for the software services, minimising unnecessary infrastructure and platform related expenses.
- **Time Savings:** Users can directly run most SaaS applications from their web browsers. This eliminates the need for downloading and installing software, thereby minimising the time and knowledge needed to handle potential deployment issues.
- **Accessibility:** SaaS model ensures accessibility of application data from any location, providing flexibility and convenience to the users.

- **Automatic Updates:** Customers can rely on SaaS providers to handle updates automatically instead of purchasing new software versions. This ensures that they always get the latest features and improvements.
- **Scalability:** SaaS allows users to get services and features on-demand, providing scalability to accommodate changing needs.

Disadvantages

- **Restricted Customisation:** SaaS solutions often provide less customisation flexibility compared to on-premises software. Users may be limited by the SaaS provider's solution, reducing their ability to tailor the software to specific needs.
- **Dependence on Internet Connectivity:** A reliable Internet connection is essential for the effective operation of SaaS solutions, which are generally cloud-based. This can create challenges for users in areas with limited connectivity or those needing to access the software offline.
- **Security Concerns:** Although SaaS providers are accountable for data security on their servers, there remains a risk of data breaches and other security issues.
- **Limited Data Control:** SaaS providers may have access user data. This raises concerns for organisations needing strict control to protect their data.

Figure 8.3 shows a snapshot of Cloud Service Models, their coverage for business and popular instances in the market.

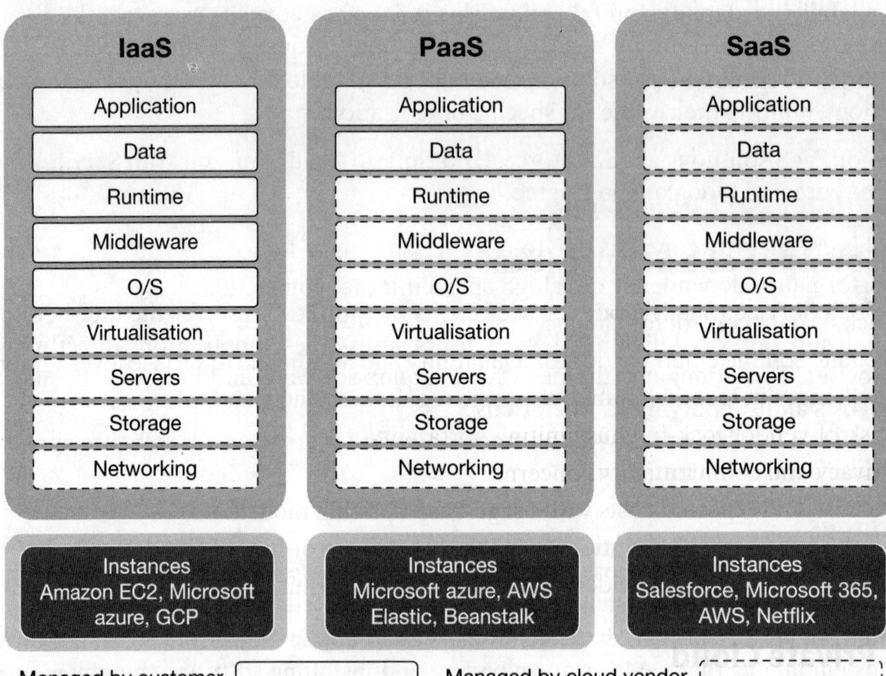

Figure 8.3 Cloud service models

8.4 CLOUD DEPLOYMENT MODELS

Cloud deployment models help us to categorise cloud environments according to their usage, size, accessibility, and purpose. These models are also classified based on their physical location. In this section, we will take a closer look at the different types of cloud deployment models and their alignment with the organisations' requirements.

8.4.1 Public Cloud

The public cloud refers to computing services delivered by third-party cloud service providers (CSPs) through the public Internet. This makes these services accessible to anyone interested in using or acquiring them. These services can be free or offered on a pay-as-you-go basis, allowing customers to pay only for the specific CPU cycles, storage, or bandwidth they utilise. Public cloud services ensure reliability, redundancy and robust security measures, including encryption and access controls, to protect user data.

Some examples of public clouds are Windows Azure Services Platform, Elastic Compute Cloud (EC2) from Amazon and IBM Smart Cloud Enterprise.

Advantages of the Public Cloud

- Lower ownership costs. The cloud service provider handles maintenance.
- Easy integration and rapid deployment of services and applications.
- Location independence.
- High scalability.
- Accessibility to the public. The cloud provider provides a variety of services and resources.
- Built-in redundancy for enhanced reliability.

Disadvantages of Public Cloud

- Inherent security concerns due to resource sharing.
- Performance dependency on a high-speed internet connection.
- Lack of data control for clients.
- Reliance on the cloud vendor for availability and adherence to service level agreements.
- Challenges in meeting industry-specific regulations and standards.
- Risk of vendor lock-in, thus limiting portability.
- Privacy and confidentiality concerns.
- Potential unexpected costs with usage-based pricing models.
- Limited customisation options compared to private or hybrid clouds.
- Dependence on the provider's support for issue resolution.

8.4.2 Private Cloud

Private cloud computing offers comparable benefits to the public cloud, such as scalability and self-service, but operates on a proprietary architecture. Also referred to as an internal or corporate cloud, the private cloud is a single-tenant infrastructure where dedicated resources like network, computing and storage are tailored to meet the needs of a single organisation.

Private clouds offer a solution when public cloud services fall short of an organisation's unique requirements. For instance, if a public cloud cannot guarantee the necessary service availability or uptime, or if hosting critical workloads in a shared environment poses unacceptable risks, a private cloud can be a viable option. Additionally, concerns regarding security or regulatory compliance in a multi-tenant environment may drive organisations towards a private cloud, enabling them to harness the advantages of cloud computing while retaining full control and ownership of their infrastructure.

According to Global Market Research, the Private Cloud services market is projected to have a Compound Annual Growth Rate (CAGR) of 29.7%, achieving a staggering $528 billion from 2022 to 2029.

Advantages of Private Cloud

There are different types of private clouds, each providing distinct benefits to businesses:

1. **On-premises Private Cloud:** This refers to creating a private, virtualised environment within the organisation's own data centre. This offers maximum control and security over data and infrastructure.
2. **Hosted Private Cloud:** This type offers dedicated resources through a third-party provider, ensuring that servers remain exclusive to each organisation and independent from other customers' activities.
3. **Virtual Private Cloud (VPC):** VPC is hosted in an isolated environment within a public cloud. It allows organisations to run their workloads using private resources while servers are shared with other customers.

Disadvantages of Private Cloud

A private cloud setup offers substantial benefits related to data security, regulatory compliance, and high availability for mission-critical systems. But it substantially increases cost, both Capital Expenditure (CAPEX) and Operating Expenditure (OPEX), along with complexity of LCM operations.

8.4.3 Hybrid Cloud

A hybrid cloud model is a combination of both private and public cloud, allowing organisations to manage workloads effectively across both environments. To achieve successful hybrid cloud deployment, it is essential to ensure high compatibility between the software and services used in both the public and private clouds.

Advantages of Hybrid Cloud

Hybrid cloud allows organisations to utilise both private and public cloud solutions, considering business needs and the best of both worlds, i.e., control and scalability. With a flexible, policy-driven approach, businesses can store sensitive data in a secure private cloud while keeping not sensitive or publicly shareable data in cost-effective and high-performance public cloud. This setup also ensures regulatory compliance by placing the appropriate information in the right locations. For example, credit card information can be securely stored in a private cloud, while less sensitive data can be maintained in the public cloud.

Disadvantages of Hybrid Cloud

The biggest challenge for hybrid clouds is the complexity of managing different types of clouds as a single environment with integration, compliance and compatibility.

8.4.4 Community Cloud

The Community Cloud is a variation of the private cloud deployment model. It is shared among multiple organisations with common interests and requirements, such as industry type, mission, performance benchmarks and compliance with regulations and security. This model functions as a type of multi-tenant architecture that focuses on sharing a private pool of cloud resources, which are isolated from external access.

This shared approach helps minimise CAPEX for establishing the community cloud, as these costs are distributed among the participating organisations. Operations can take place in-house or may involve a third party on-site. Businesses utilise the infrastructure provided by a CSP for software and development tools tailored to meet community-specific needs. Additionally, each organisation maintains its own private cloud space designed to fulfil the shared requirements for security, privacy and compliance within the community.

Government agencies and industries in finance, healthcare and legal sectors find community clouds appealing due to strict regulatory compliance requirements.

Community Clouds facilitate collaborative project management by offering a shared platform for development, software application sharing and addressing common challenges within the community. *They balance the customisation of a private cloud and the cost-sharing benefits of a public cloud, making them suitable for organisations with shared interests and compliance requirements.*

Let's quickly compare the public, private (on-premise, hosted), hybrid and community cloud models in Table 8.1 below.

Table 8.1 Comparison among Cloud Deployment Models

	Public Cloud	On-Premise Private Cloud	Hosted Private Cloud	Hybrid Cloud	Community Cloud
Hardware Deployment and Management	CSPs	Customer	CSPs	Jointly by CSPs and Customer	CSPs
Sharing Hardware	Shared	Dedicated	Dedicated	A mix of shared and dedicated	Dedicated to a business community
Scalability	Very High	Limited	High	Very High	Limited
Cost	Low	High majorly	Often low	Often low	Often low
Predictable Cost	No	Yes	Yes	No	Yes
Utility Billing	Yes	No	Depends on CSPs	Partial	Depends on CSPs

(Continued)

Table 8.1 (*Continued*)

	Public Cloud	On-Premise Private Cloud	Hosted Private Cloud	Hybrid Cloud	Community Cloud
Flexibility	Yes	Limited	Limited	Yes	Limited
Customisation	No	Yes	Depends on vendor	Partial	Yes
Security and Compliance	Depends on CSPs	Highly secure	Secure	Secure	Secure
Instant Provisioning	Yes	Yes	Yes	Yes	Yes

8.5 KEY COMPONENTS OF CLOUD ARCHITECTURE

Cloud architecture refers to the structure and design of the various elements that make up a cloud computing environment. It consists of different layers and components that work together to provide instant access to computing resources via the Internet. While a detailed explanation of generic cloud computing architecture and its components is beyond the scope of this chapter, we will focus on the key components that enable the successful integration of Internet of Things (IoT) solutions within the cloud ecosystem. This integration is crucial for driving digital transformation across various industrial sectors and realising the vision of a connected world under Industry 4.0. Additionally, we will discuss the concepts of Edge Computing and Fog Computing and their roles in establishing effective edge connectivity.

8.5.1 Virtualisation

Virtualisation is a technological approach facilitating the creation of virtual counterparts for servers, storage, networks and diverse physical machines. Through virtual software, it replicates the functionalities of tangible hardware, enabling the simultaneous operation of multiple virtual machines on a single physical machine. Essentially, it involves partitioning a solitary physical server into multiple logical servers. Once divided, each logical server behaves like a physical server, independently running an operating system (OS) and applications. This virtual environment is referred to as a Virtual Machine (VM). The instance of OS running on the VM is termed as Guest OS, while the OS running directly on the physical hardware is called Host OS. Figure 8.4 below depicts a simplified understanding of Virtualisation.

Although the concept emerged in the 1960s, it did not gain substantial traction until the 1990s. The advent of the Internet revolution in that era necessitated a vast array of servers hosting diverse applications and large databases for Internet-based solutions. These extensive collections of servers referred to as 'server farms', consumed significant space and power, posing challenges related to their life cycle management (LCM). Moreover, there was substantial wastage as most servers remained underutilised. Introducing virtualisation proved to be a logical solution to consolidate these applications on shared servers. Additional

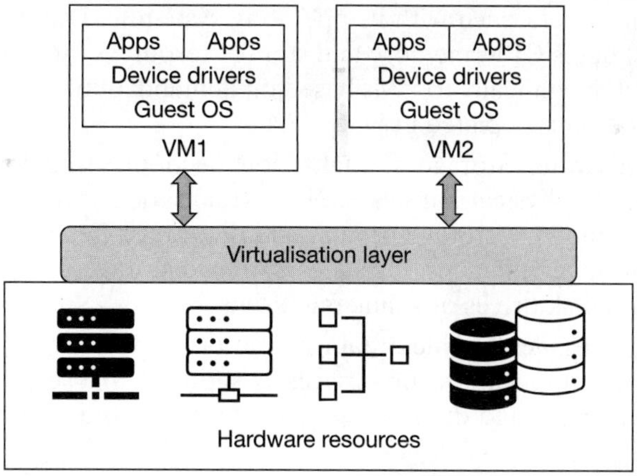

Figure 8.4 Virtualisation

requirements for robust isolation, high availability, traffic-based load balancing and resiliency also emerged. Virtualisation technology rapidly evolved to meet these demands, becoming integral to optimising resource utilisation and addressing the challenges associated with the growing demands of internet-based solutions.

Virtualisation Benefits

Virtualisation is the core of any data centre (DC). With fewer servers doing the same job and hosting multiple applications concurrently, the number of servers in a DC is drastically reduced. This has enormous benefits for businesses, as depicted in Figure 8.5.

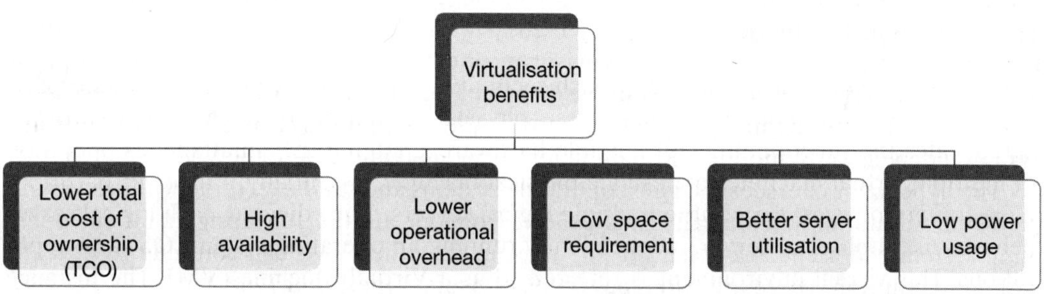

Figure 8.5 Benefits of virtualisation

Types of Virtualisations

In this section, we will discuss a few significant areas of virtualisation. Apart from server virtualisation, this technology is implemented in other areas like network and storage.

1. **Server Virtualisation:** Server virtualisation is a technology where virtual representations of physical servers are created, allowing multiple virtual servers to run on one physical machine. This setup increases efficiency by allowing a single server to host

multiple virtual servers, each with its operating system and applications. Server virtualisation consolidates CPU, memory and storage resources into a centralised pool that can be allocated dynamically to VMs based on demand, optimising the resource usage. Server virtualisation is depicted in Figure 8.4.

2. **Storage Virtualisation:** Storage virtualisation decouples the logical representation of storage from its physical hardware. The virtualisation layer is positioned between applications and physical storage devices and provides a flexible and efficient method for managing storage resources. It creates a unified storage pool from diverse storage devices, like hard disk drives or solid-state drives.

   This technology enables seamless data migration between different storage devices without disrupting applications or services. It supports load balancing and allows the creation of data copies and duplication across various storage devices for backup and disaster recovery.

3. **Network Virtualisation:** Network virtualisation splits a single physical network into multiple networks. This is achieved by abstracting and decoupling network resources from their underlying hardware. It creates a virtualised layer, allowing multiple virtual networks to exist on a shared physical network infrastructure. Isolated, logically segmented networks are designed to operate independently of the physical network configuration. Examples include Virtual LAN (VLAN), Virtual Extensible LAN (VXLAN) and IP Layer 3 Virtual Private Network (L3VPN). Network virtualisation enables simultaneous services like Broadband Internet, Voice over IP (VoIP) and video streaming. It often aligns with Software-Defined Networking (SDN), enhancing network agility and programmability.

Hypervisors and Virtual Machines

Hypervisors, also known as Virtual Machine Managers (VMMs), create and manage VMs. The term was coined by IBM in the early 1970s. This technology serves as the virtualisation layer, as illustrated in Figure 8.4. Hypervisor technology makes the virtualisation of physical servers possible. It acts as a supervisor or enabler to create VMs, allocate necessary resources (compute, memory, storage, etc.) to the VMs so that they act as a virtual hardware system independently, modify VM parameters as needed and delete the VMs.

The hypervisor oversees the guest operating systems and their execution within a virtual platform. Additionally, it logically isolates VMs from one another, ensuring that if one operating system experiences a crash for any reason, the other VMs can continue to operate without disruption. There are two types of hypervisors: Type 1 and Type 2.

1. **Type 1 Hypervisor:** A Type 1 Hypervisor is also called a Bare Metal hypervisor. This type of hypervisor runs directly on the hardware without any host OS. In this case, the hypervisor conducts the host OS roles, i.e., administers the physical devices. However, the reduced communication overhead compensates for the higher complexity of a Type 1 hypervisor and, thereby, higher development time and cost.

2. **Type 2 Hypervisor:** A type 2 Hypervisor is a hosted hypervisor. It runs as an application on the host OS with virtualisation capabilities. The host OS provides the interface to communicate with the underlying hardware devices, such as compute, memory, storage, etc.

   The Type 1 and Type 2 Hypervisors are depicted in Figure 8.6.

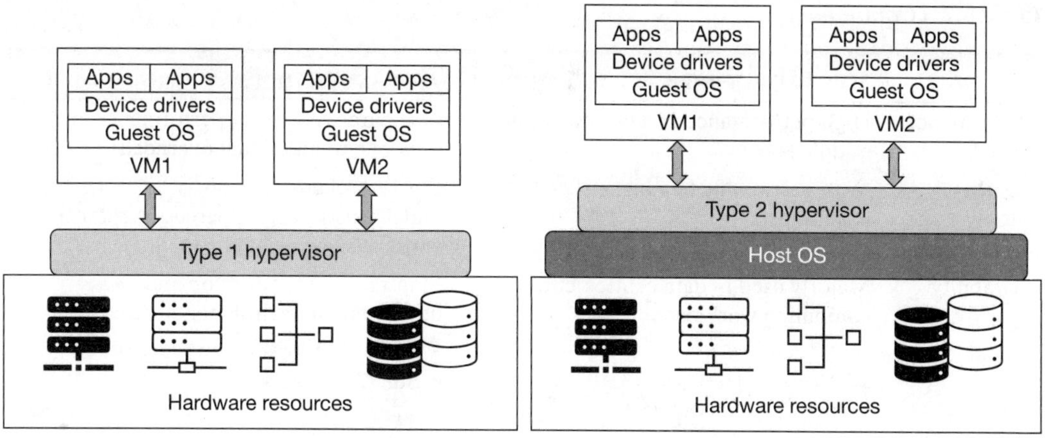

Figure 8.6 Type 1 and Type 2 hypervisors

Despite the similarities in the common objective of running and managing VMs, there are significant differences between Type 1 and Type 2 Hypervisors.

Resource Allocation

Type 1 hypervisors can directly access machine resources and employ custom allocation strategies for their VMs. On the other hand, Type 2 hypervisors rely on negotiation with the operating system for resource allocation, resulting in a slower and less efficient process.

Managing Type 2 Hypervisors is much simpler than Type 1 Hypervisors.

- **System Performance:** Type 1 hypervisors provide improved performance for VMs because they operate directly on the hardware, without the need to share resources with an underlying OS. In contrast, Type 2 hypervisors rely on the host operating system to allocate resources, which can result in reduced performance for the VMs.
- **Isolation:** Isolation in Type 1 hypervisors is at a higher level as each virtual environment is more securely separated without a shared layer, in contrast to Type 2 hypervisors. Hence, Type 2 is less secure than Type 1. This security advantage in Type 1 requires regular updating and patching of virtual machine OSs to keep the system safe.

Table 8.2 is the summary of differences between Type 1 and Type 2 Hypervisor.

Table 8.2 Differences between Type 1 and Type 2 Hypervisor

Categories	Type 1 Hypervisor	Type 2 Hypervisor
Way of Working	Runs directly on the hardware without dependency on the host OS. Hence, its other names are Bare Metal Hypervisor or Embedded Hypervisor.	Runs on standard OS. It is also called Hosted Hypervisor.
Resource Allocation	Has direct access to machine resources and can employ custom allocation strategies for their VMs	Relies on negotiation with the operating system for resource allocation

(*Continued*)

Table 8.2 (*Continued*)

Categories	Type 1 Hypervisor	Type 2 Hypervisor
Performance/ Speed	High performance and fast due to no middle layer	Comparatively lower performance and slow due to additional overhead
Security	More secure due to a higher level of isolation	It is less secure than any issue in the underlying OS has repercussions for the entire system, including the hypervisor.
Suitability	Majorly used in data centres, enterprise computing workloads	Majorly used for desktop and development environments
Examples	• VMWare ESXi • Microsoft Hyper-V • Citrix XenServer	• VMWare Workstation Player • Sun's VirtualBox

Type 1 hypervisors are pivotal in the Internet of Things (IoT) ecosystem, ensuring the smooth operation of diverse operating systems on IoT devices. They provide advantages in isolation, security and resource optimisation, making them suitable for automotive, healthcare, industrial automation and smart home device applications. The rising prevalence of IoT devices in various sectors has generated heightened demand for Type 1 hypervisors, enabling the effective operation of complex software stacks on IoT devices with limited resources.

Type 1 hypervisors enable the incorporation of real-time operating systems (RTOSs) alongside general-purpose OS, allowing the management of real-time constraints and deterministic behaviour. RTOS and Traditional OS differ primarily in their ability to handle time-sensitive tasks and provide predictable response times. The significant differences are listed in Table 8.3.

Table 8.3 Comparison between RTOS and Traditional OS

Categories	RTOS	Traditional OS
Time Constraints	They are specifically designed to adhere to strict timing constraints and deadlines. It prioritises tasks based on urgency, ensuring that critical tasks are executed within specified time limits.	Prioritisation is based on fairness and efficiency and does not ensure timely task execution.
Scheduling	Utilises deterministic scheduling for priority-based task execution	Relies on general-purpose scheduling algorithms with less emphasis on strict deadlines
Predictability	Offers predictable and deterministic behaviour crucial for real-time applications	It may not guarantee response time. Behaviour varies based on system load and task prioritisation.
Kernel Efficiency	It has a smaller, streamlined kernel for efficiency in time-critical tasks.	Features a larger, more complex kernel for versatility in handling diverse applications
Overhead and Complexity	Minimises overhead for reduced latency and efficient real-time task management	It might have additional features and overhead suitable for general-purpose computing, potentially introducing delays in time-sensitive applications.

(*Continued*)

Table 8.3 (*Continued*)

Categories	RTOS	Traditional OS
Use Case	Suited for applications with strict timing requirements, such as embedded systems, robotics and safety-critical devices	Well-suited for general-purpose computing where real-time constraints are less critical, like desktop computers and servers
Products	• FreeRTOS (Amazon) • Keil RTX (ARM) • Neutrino (Blackberry) • Zephyr (Linux Foundation)	• MS-Windows • Linux • Ubuntu • Solaris

Containerisations

In the preceding section, we explored Hypervisors and VMs. VMs offer a highly isolated and self-contained environment with highly efficient virtualisation implementation. However, VMs incur performance and resource costs due to the overhead required for emulating virtual hardware. In scenarios where the extensive isolation provided by VMs is unnecessary, a lighter isolation level may be needed to mitigate some of the performance degradation associated with hypervisor-based virtualisation. This can be accomplished through another virtualisation technique known as Container-based virtualisation.

A container is a lightweight software unit encompassing code, system libraries, settings and elements necessary for operation in a computing environment. They can be deployed across various environments, from personal laptops to private data centres and public clouds. Containerisation facilitates swift and efficient software deployment, accelerating both horizontal and vertical integration of applications across numerous IoT platforms at an unprecedented scale.

The degree of virtualisation achieved with containers differs from that of VMs. Unlike VMs, containerisation does not generate a new virtual machine within the host operating system by emulating the presence of actual hardware. Instead, containers isolate the utilisation of system resources and impose restrictions on their usage. This distinction with VMs is why containers are commonly labelled virtual environments. The Architecture of VMs vs containers is shown in Figure 8.7.

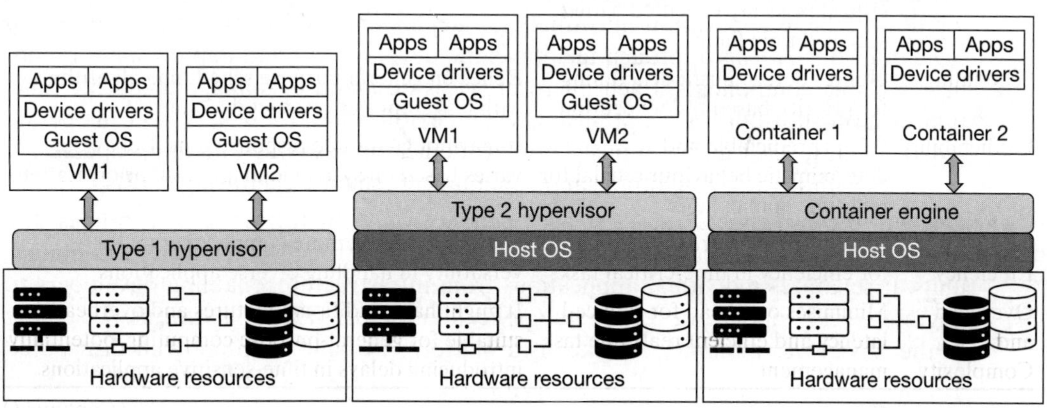

Figure 8.7 Virtual machines vs containers

Table 8.4 below shows the significant differences between VMs and containers.

Table 8.4 Differences between VMs and Containers

Categories	VMs	Containers
Definition	Replicates the functionality of an operating system, with the hypervisor virtualising the underlying hardware, files and configurations	It is a software unit that includes code, system libraries, settings and essential elements required for operation within a computing environment. Unlike virtual machines, containers operate on host servers where the operating system is virtualised, encompassing only the dependencies and the application.
Composition	It comprises virtual computing, storage, memory, networking, libraries and configuration files.	Contains only applications and their files and libraries
Size and Loading time	Sizes are in multiple GBs and it takes a few minutes to load	Sizes are in MBs and it takes a few seconds to load
Security	It provides a high level of security and isolation. It is pretty unlikely that an application in a VM would impact other VMs	It does not provide pure isolation, as many components, including the kernels, are shared. Also, root access is required to share OS, which may lead to vulnerabilities. Applications in one container may impact the entire host or other containers.
Performance	Usually slower than container as HW emulation takes place through the hypervisor.	It is faster since there is no middle layer or hypervisor.
Products	• VMWare vSphere • Zen • Hyper-V	• Docker • Hyper-V Containers • AWS • Windows Server Containers

- **MicroServices:** Container technology provides a standardised and efficient way to package, deploy and manage microservices. Microservices architecture involves breaking down a large application into smaller, independent services that can operate and scale individually. This approach facilitates the development and deployment of modular, scalable components. Each microservice is designed to manage a specific function, ensuring flexibility and optimising resource utilisation. These services can operate independently, scale autonomously, possess dedicated resources, undergo updates or upgrades without affecting others and communicate seamlessly through APIs.
- **Docker Technology:** Docker is an open software platform. It streamlines the development and execution of software packages through standardised container units. It separates individual applications from infrastructure, enabling developers to migrate, deploy, update, or terminate applications across various platforms. According to the Docker Community, 'Docker is a platform to build, ship and run distributed

applications'. It serves as a toolbox with a unified API, offering simple commands and automation procedures for constructing highly reliable container applications.

With Docker and containerisation methods, developers can drastically reduce the time and effort required to move from coding to production deployment. Each Docker container runs a single application in isolation. It packages specific application code along with its dependencies for that application. This isolation improves security and enhances overall service availability by preventing system failures. If one container fails or an application crashes, the other containers continue to operate without disruption. This capability enables developers to quickly implement security patches and updates, minimising bugs or conflicts when integrating various IoT devices and applications.

With the above concepts in mind, let us explore the use of container technology in IoT devices:

- **Use of Container Technology with IoT Devices:** Many embedded devices are designed with fixed functionalities due to hardware constraints, making it challenging to support evolving workloads. Adopting a distributed cloud architecture can significantly enhance device performance and system integration capabilities. This approach enables the deployment of lightweight, scalable, cloud-native solutions on embedded IoT devices. It effectively overcomes physical constraints and reduces platform dependencies.

  Containerisation streamlines the development, testing and deployment of consistent applications in large-scale IoT systems. Containerisation utilises isolated environments to ensure portability and security.

  Containerisation is needed in IoT, especially when integrating cloud computing with embedded devices through microservices. Microservices utilise the limited resources of devices. Containers enable applications to operate as microservices across various IoT devices and platforms. Major cloud providers like Azure and AWS offer container image deployment models for edge devices.

8.5.2 Cloud Orchestration and Management

Cloud orchestration automates tasks to manage workloads on the cloud. This automation integrates processes into workflows, ensuring specific business functions are executed seamlessly. Cloud orchestration tools enforce policies, validating proper permissions for task execution or workload connections. Typical tasks include provisioning server workloads, managing storage capacity and deploying virtual machines or containers by orchestrating services, workloads and resources. The rise of containerised, microservices-based applications has fuelled the need to automate application deployment and management across the cloud. Cloud orchestration software addresses the complexity of dependencies across multiple clouds, providing policy-driven security and management capabilities. These tools enable the integration of different cloud models (public, private, or hybrid) and cloud delivery models (IaaS, PaaS, or SaaS), offering unified automation across diverse cloud environments.

Kubernetes

Kubernetes has emerged as the standard container orchestration framework for cloud-native deployments, aiding development teams in transitioning to microservices architectures and embracing DevOps practices. Amid digital transformation efforts, organisations leverage Kubernetes to modernise their IT and OT systems, integrating technologies like IoT platforms, big data analytics and machine learning. Kubernetes helps in addressing the complex challenges of building and deploying scalable, reliable and distributed cloud-native IoT solutions:

- **DevOps Enablement for IoT:** Kubernetes supports a unified deployment model, allowing DevOps teams to rapidly test and deploy new IoT services with zero-downtime rolling updates. This agility aligns with customer and market demands for quick feature deployment.
- **Scalability:** Kubernetes facilitates the automatic scaling of Pods across network clusters. This capability is crucial for handling large-scale device connections, data processing and real-time analytics like IoT Solutions.
- **High Availability:** Kubernetes provides the necessary tools for deploying highly available services, which are crucial for business-critical IoT solutions. Its architecture enables independent workload execution, allowing for restarts or recreations without impacting end-users.
- **Deployment to the IoT Edge:** As IoT services increasingly move to the edge for improved responsiveness, Kubernetes offers a common platform for deploying services at the edge.

DID YOU KNOW?

K8s, derived from 'Kubernetes', is a shorthand form with eight characters between K and S. Typically, when people mention Kubernetes or K8s, they refer to the original project designed by Google for high availability and scalability. In contrast, K3s is a lightweight Kubernetes distribution by Rancher Labs, certified by the Cloud Native Computing Foundation (CNCF) and optimised for smaller deployments, edge computing and IoT. Originating from frustration with slow Kubernetes cluster setup, Darren Shepherd of Rancher Labs streamlined the Kubernetes source code, resulting in K3s, a distribution under 50MB that runs all Kubernetes components in under 512MB of RAM, ideal for swift development and testing.

8.5.3 Cloud Native Applications

In this context, let us quickly understand what **Cloud Native application** is and how it differs from **Cloud Enabled application**. The term 'cloud-native' refers to applications specifically designed for cloud environments from the very beginning, incorporating technologies like microservices, container orchestrators and auto-scaling. Cloud-native development incorporates agile practices such as DevOps and Continuous Integration and Continuous Delivery (CICD) model to build, test, integrate and deploy highly scalable applications. 'Cloud-enabled' applications refer to traditional enterprise applications originally hosted in on-premises data centres and later modified to operate in the cloud. This adaptation involves altering specific software modules to facilitate the migration of the application to cloud servers, while preserving its original features. They lack the flexibility, resiliency and scalability inherent in cloud-native counterparts, as they retain their monolithic structure despite being moved to the cloud.

These cloud-native applications are built using microservices and deployed via container technology, which is lighter than traditional virtual machines. However, 'always-on' components and scaling challenges persist in microservice deployment. A new cloud-native model called Function-as-a-Service (FaaS) or Serverless computing has been introduced to address these challenges. Serverless computing offers several advantages over traditional server-centric setups in cloud computing. It provides improved scalability, flexibility and quicker deployment at a reduced cost for developers. With serverless architectures, developers don't need to purchase, provision, and manage backend servers as the service provider handles these responsibilities. This eliminates the need for investment in DevOps, reduces expenses and allows developers to focus on app development without being restricted by server capacity. Similar to the 'pay-as-you-go' model, developers are charged only for the resources they use. Code runs only when backend functions are needed, automatically scaling up as necessary. Provisioning is dynamic, precise and real-time, unlike traditional setups where developers must predict and purchase server capacity regardless of usage. For example, Coca-Cola, the leading soft drink company, has fully embraced serverless technology with its successful integration into vending machines, leading to substantial cost savings. When a beverage is bought, the payment gateway connects to the AWS API Gateway, initiating an AWS Lambda function to finalise the transaction. Given the need for vending machines to communicate with headquarters for inventory and marketing, the ability to pay per request rather than constantly operate at full capacity significantly reduced expenses.

Serverless computing offers benefits such as reduced implementation constraints and auto-scaling services. However, it also has limitations, including latency issues, delays when a function or service is invoked for the first time or after a period of inactivity (referred to as 'cold starts'), lack of infrastructure control, security concerns, testing complexity and potential vendor lock-in.

8.5.4 Edge and Fog Computing

The cloud is integral to IoT's growth and market viability, enabling the connectivity and data sharing of billions of previously isolated endpoint devices. These devices, ranging from sensors to cameras, gain value through their ability to communicate with each other in a common language. The challenge arises when Personal Area Network (PAN) and Wireless PAN (WPAN) devices lack IP compliance, requiring edge gateways to bridge the translation gap. Additionally, the proximity of sensors introduces real-time requirements, contrasting with the latency introduced by the WAN and cloud components. Edge or fog computing technologies are developed to address the latency issue, meet the real-time decision-making at the edge and handle the meaningful data stored in the cloud for deeper analysis.

Edge Computing

Edge computing brings data processing closer to its source, processing it at the network's edge rather than relying on distant cloud data centres. The International Organisation for Standardisation (ISO) defines edge computing as a distributed computing approach where processing and storage occur on networked devices near the data source. This approach reduces the amount of data sent to the cloud, minimising latency and enhancing overall system performance.

The exponential growth of connected devices, which is expected to exceed three times the global population by 2023, has contributed to the rise of the IoT. Of these connected devices, approximately half are Machine-to-Machine (M2M) connections, totalling around 14.7 billion. This increase in connected devices leads to a significant rise in data traffic across the network.

While cloud offloading has traditionally addressed the computational limitations of IoT devices, it comes with drawbacks such as increased latency, network overhead and security risks. The edge computing model has emerged, allowing computing tasks to be offloaded to nodes closer to the end devices. This improves the QOS by reducing latency and enhancing proximity to data sources and consumers.

Simply connecting a sensor to a computer device at a remote location is not edge computing. However, as soon as that sensor delivers data to a cloud service, they are working with edge computing devices that must act in concert with centralised cloud services.

Edge computing adds significant values in the following areas:

- **Latency Reduction:** Edge systems are positioned near end users and services. They reduce latency by minimising network hops and transmission. This is crucial for applications like cloud-based gaming, video streaming and devices or machines requiring real-time decision-making.
- **Bandwidth Management:** Edge computing addresses bandwidth limitations in specific environments by using filtering, caching and data compression techniques. This approach optimises available bandwidth, particularly in scenarios in which bandwidth is restricted or where data costs for cloud or data centres become prohibitive.
- **Computing Resiliency:** Resilient computing is crucial in scenarios with unreliable communication, such as transportation or logistics applications tracking a fleet. These systems use edge local caching technique to store data temporarily until communication is restored. Additionally, they may implement failover techniques to switch to different carriers in the event of a primary carrier loss, ensuring data integrity and continuity.
- **Privacy and Security:** In certain scenarios, securing or filtering specific data before transmitting it to the cloud or other edge locations is needed for security and privacy concerns. This is particularly critical for applications dealing with health data, images, or personal privacy, which are often regulated by government standards and involve significant computing resources at the edge.

With the above context, let us explore some of the use cases of Edge Computing in Table 8.5.

Table 8.5 Use Cases of Edge Computing

Use Case	Usage of Edge
Data Processing from IOT	Edge systems enable efficient real-time local processing of bulk data through filtering, anomaly removal, data compression, rule engines, etc., eliminating the cost of transferring the data to the cloud.
Device Management	Firmware upgrades and patching for IoT and Edge devices without human intervention.
Patient Monitoring System	Edge systems can securely and resiliently communicate with upstream healthcare provider systems while coexisting with patients.

DID YOU KNOW?

An instance of Edge Analytics for near-real-time intelligence of 5G radio network performance was deployed by a Telecom Service Provider (TSP) during a notable temple inauguration event in India. The TSP supported uninterrupted video surveillance of numerous camera feeds using edge analytics infrastructure deployed at 16 identified sites. The purpose of running an analytics solution at the network's edge was to ensure super-fast data acquisition and processing at the nearest point where the data is generated. Some of the essential components of the software stack that made such a swift and agile analytics solution are - K3s to orchestrate multiple Kubernetes pods on a standalone cluster (single server), Apache Kafka for streaming radio network signalling data, Apache Spark for running heavy-duty in-memory analytics and correlation of events from the signalling data and Apache Drill to analyse multi-structured and nested fault signatures without the overhead (data loading, schema creation and maintenance, transformations, etc.).

These edge servers conducted real-time analysis of the vast amount of video frame data packets produced by the cameras. The TSP communicated any abnormalities detected in the packets in real-time to the video surveillance team and 5G network performance operations teams for further action.

Fog Computing

Fog computing extends cloud computing to the network's edge, distributing resources and services across a network fabric. This contrasts with edge computing, which may or may not involve the cloud. This node between the cloud and IoT devices includes storage, computing and networking components. Fog computing's proximity to edge devices reduces latency, enhancing real-time applications like industrial IoT and autonomous vehicles performance.

The fog computing model provides a flexible architecture for distributed, latency-aware applications and services. It incorporates fog nodes, which can be either physical or virtual, positioned between smart end devices and centralised cloud services. Importantly, the fog layer is not essential for the cloud ecosystem to function in the Internet of Things (IoT) and similarly, centralised cloud services are not required for the fog layer to support the functionality of smart end devices. This flexibility enables various architectural options tailored to specific use case scenarios and the best approach for enhancing end-device performance.

The **fog node** is a central part of fog computing, which exists as either a physical (e.g., gateways, servers) or virtual component (e.g., virtual machines) closely connected to intelligent end devices. It offers computing resources, knows its location in a cluster and facilitates data management and communication between IoT devices at edge and fog computing service or centralised cloud system. Fog nodes can operate in a centralised or decentralised way, acting independently or forming clusters for scalability across different locations. Fog computing can be used in wired and wireless networks, though it is ideal for wireless IoT networks, as it can support distributed analytics for large-scale wireless sensor applications. Like the traditional cloud computing model, Fog computing supports implementation in different deployment models (public, private, hybrid, community) and service models (IaaS, PaaS, SaaS).

Healthcare institutions are leveraging the power of fog computing to revolutionise patient care and results. Wearable IoT devices with sensors continuously monitor vital signs like heart rate and blood pressure. The real-time processing and analysis of this data at the patient's location, facilitated by fog computing, enable prompt detection of anomalies and timely intervention as needed. This capability of fog computing in healthcare underscores its potential to make a significant impact in the industry.

The following are key attributes of the fog computing Model to support the above characteristics as per NIST:

- **Autonomy:** Fog nodes can make local decisions independently at the node level or within a cluster of nodes.
- **Heterogeneity:** Fog nodes come in various forms and can be deployed in diverse environments.
- **Hierarchical Clustering Support:** Fog nodes support hierarchical structures, with different layers offering distinct service functions while collaborating as a unified system.
- **Manageability:** Fog nodes are managed and orchestrated by sophisticated systems that can automate routine operations.
- **Programmability:** Fog nodes are inherently programmable at multiple levels, allowing various stakeholders like network operators, domain experts, equipment providers, or end users to make modifications through programming.

Figure 8.8 illustrates the edge, fog and cloud layers, their respective functions and the data flow between them. The number of devices decreases as we ascend to higher layers, whereas processing speed or response time increases as we descend to lower layers.

Some of the key differences between Fog and Edge computing are mentioned in Table 8.6.

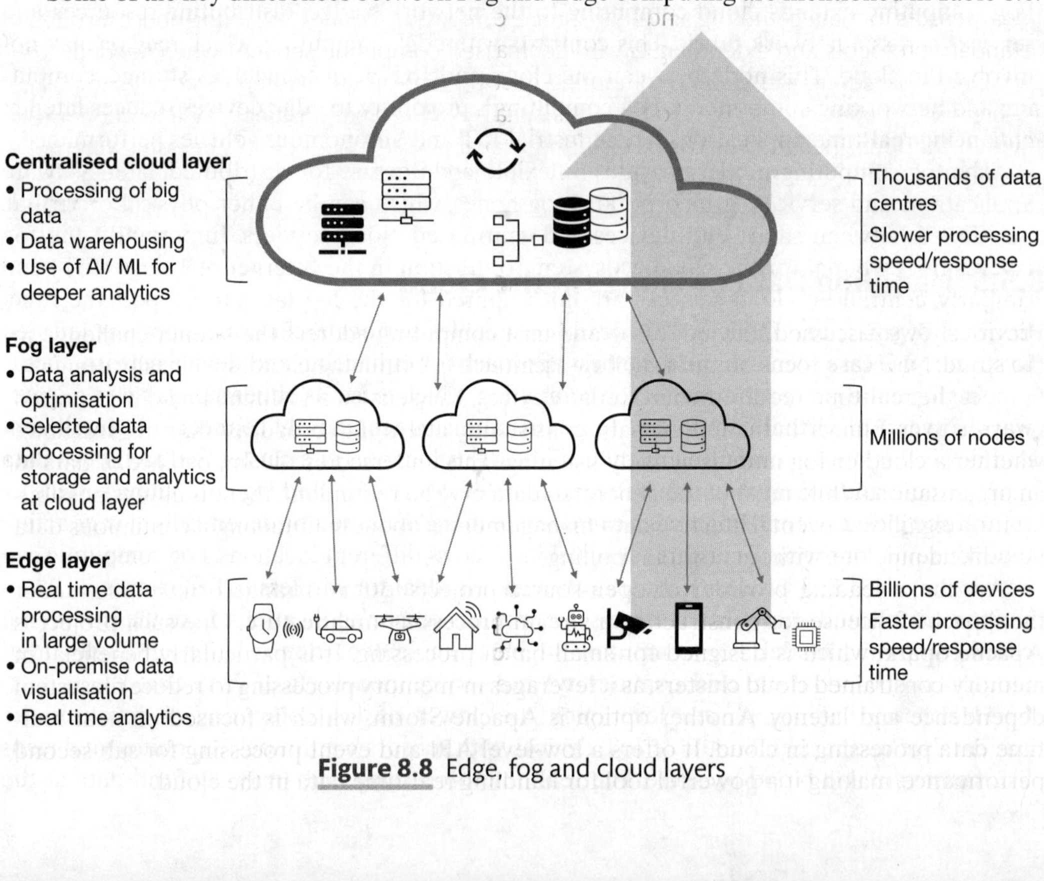

Figure 8.8 Edge, fog and cloud layers

Table 8.6 Difference between Fog and Edge Computing

Fog Computing	Edge Computing
Fog computing employs a multi-layer architecture, separating hardware and software functions for dynamic reconfigurations. It offers intelligent computing and transmission services.	Edge computing runs specific applications in a fixed logic location with direct transmission service and is more limited to a few peripheral devices.
Fog computing aids in extracting crucial information from the extensive data collected by devices, filtering it and then transmitting the filtered data to the cloud for storage.	Edge computing accelerates data processing by simultaneously handling information received from devices.
High bandwidth is essential as data from edge nodes is transmitted to the cloud.	Data only comes from edge nodes, so the bandwidth requirement is low.

DID YOU KNOW?

The evolution of fog computing led to new concepts like mist nodes, specialised nodes placed closer to peripheral devices for geographically distributed low-latency processing. They are also termed 'Cloudlets'. Mist computing is a lightweight form of fog computing located at the edge of the network fabric. It utilises microcomputers and microcontrollers near sensors for data gathering and near-source computing, connecting to fog nodes through standard protocols and extending toward centralised computing services when needed.

Dew Computing is a similar computing method that provides Internet services to end-users without continuous connectivity. This entails having small instances of Internet nodes within the local network of users, ensuring a seamless experience in accessing Internet services.

8.5.5 Handling Data of Things in the Cloud

Previously, we discussed how edge, fog and mist computing address the latency challenge in the cloud. Now, let's focus on the data flow architecture within the cloud, specifically designed to meet the real-time requirements of edge devices. The sensors and connected devices, often referred to as 'Things' in the IoT domain, consistently attempt to broadcast data, regardless of whether a cloud or fog node is actively listening. This underscores the importance of data in an organisation. While most of the generated data may be redundant, there is always possibility for a significant event. Effective data management is about maintaining a continuous data flow, like a pipeline, without constant polling.

Apache, a leading provider of open-source projects, offers a range of solutions under the Apache 2 license for constructing a stream processing architecture. One such project is Apache Spark, which is designed for small-batch processing. It is particularly beneficial in memory-constrained cloud clusters, as it leverages in-memory processing to reduce filesystem dependence and latency. Another option is Apache Storm, which is focused on near-real-time data processing in cloud. It offers a low-level API and event processing for sub-second performance, making it a powerful tool for handling real-time data in the cloud.

In large-scale IoT deployments with thousands of sensors, a cloud environment often employs a **data lake**, a vast storage facility containing raw data from diverse sources. Unlike traditional hierarchical file systems, data lakes utilise flat filesystems and organise data by adding metadata elements (tags) to each entry. Apache Hadoop is the data lake model which is widely adopted by cloud providers. Data lakes, store structured and unstructured data, assuming all data is valuable and should be retained permanently. *However, the data lake and data warehouse are not identical.*

A **data warehouse** stores processed data that can be used for analytics, operational reports and specific Business Intelligence (BI) use cases. Data Lake and Data Warehouse are complementary tools and industries might need both to support a wide range of use cases and reporting needs.

The actual value of an IoT system does not solely reside in individual sensor events or vast data archives. It is derived from the interpretation of data and the decisions drawn from it. While an extensive network of interconnected devices is beneficial, the essence of value lies in understanding the data, uncovering patterns and leveraging **data science and analytics**. The cloud plays a significant role in storing and analysing data. Data can be in many forms: Structured, Unstructured, or Semi-structured. This might require interpretation in real-time and, in the process, need to be correlated with other data sources or archived in the cloud for deeper analytics.

Thus, to build a Cloud Analytics Architecture for IoTs, we must plan how to handle random and continuous data streams and send the interpretations in real-time as much as possible.

'Chapter 9 - Big Data' and 'Chapter 10 - Analytics' discuss data lakes, data warehouses and data Analytics in detail.

8.5.6 Solution Considerations for IoT with Cloud

Let us explore some use cases across various industries embracing IoT and cloud. During the design phase, factors such as scale, bandwidth, real-time requirements and data types must be considered to determine the appropriate cloud architecture for IoT solutions. Table 8.7 shows some general examples and a deeper level of analysis and understanding would be required to prepare a robust solution in practice.

Table 8.7 Solution considerations for IoT with Cloud

Industry	Use cases	Cloud services	Usual Bandwidth Requirement	Real-time Needs
Manufacturing	• Factory Operations Management • Shop Floor Automation	• Dashboards • Data lakes • Bulk storage • Prediction	500 GB/day/factory part produced	< 1 sec
Oil and Gas	• Remote monitoring • Failure Prediction	• Dashboards • Data lakes • Bulk storage for historical rate • Prediction	1 to 2 TB/day/oil rig	< 1 sec

(Continued)

Industry	Use cases	Cloud services	Usual Bandwidth Requirement	Real-time Needs
Logistics and transport	• Tracking Geolocation • Sensing equipment function	• Dashboards • Logging • Storage • Prediction	• Vehicles: 4 TB/day/ vehicle (50 sensors) • Aircraft: 2.5 to 10 TB/day (6000 sensors)	< 1 sec
Healthcare	• Asset tracking • Patient tracking • Home health monitoring • Wireless health equipment	• Private Cloud • Edge / Fog Computing • Storage and archival	1 MB/day/sensor	Life critical < 1 sec Non-life critical = on each change
Retail	• Temperature-controlled supply chain monitoring • Functioning of POS Machines • Security Systems	• Public / Hybrid Cloud • Micro-segmentation of Network • Dashboards	• Security: 500 GB/ day/camera • General: 1–1000 MB/day/device	POS/ Credit transactions – 100 ms

8.6 CLOUD COMPUTING FOR INDUSTRY 4.0: IMPLEMENTATION IN MANUFACTURING SYSTEMS

Cloud computing is a key enabler for Industry 4.0, providing scalable, on-demand resources that facilitate the digital transformation of industries. It offers the flexibility, connectivity, and computational power necessary to manage the enormous data volumes and transactions from advanced smart factories, interconnected devices, and intelligent systems. Cloud technology enables real-time analytics, remote monitoring, and seamless collaboration throughout supply chains, leading to more agile, data-driven industrial processes. With this broader perspective, this section will look into how cloud computing is specifically integrated into manufacturing systems to facilitate factory automation and improve operational efficiency.

Factory automation can be segregated into different layers, each with specific requirements. We discussed CIM (Computer-Integrated Manufacturing) Model in detail in Section 4.4.2 of Chapter 4 - Industrial Internet of Things (IIoT). The central concept of CIM is to divide tasks within a manufacturing process into distinct hierarchies, reflecting the management levels of a company. The lowest manufacturing pyramid levels, close to the shop floor, manage components like sensors and actuators. Automation controllers and PLC are positioned above these levels, with increasing abstraction from the shop floor as one ascends the pyramid, culminating in Enterprise Resource Planning (ERP) at the top. Each level controls different values, with lower levels managing physical aspects such as temperature and pressure, while higher levels deal with business values. Timing parameters also vary across levels, with control loops at the lowest levels operating on millisecond cycles, contrasting with the highest levels with cycle times spanning days or weeks. With that background, let us explore how the cloud can influence these automation layers through flexible delivery models, massive storage and computing power and software services in a virtualised environment finally optimising the cost. Figure 8.9 shows the cloudification of different levels of CIM.

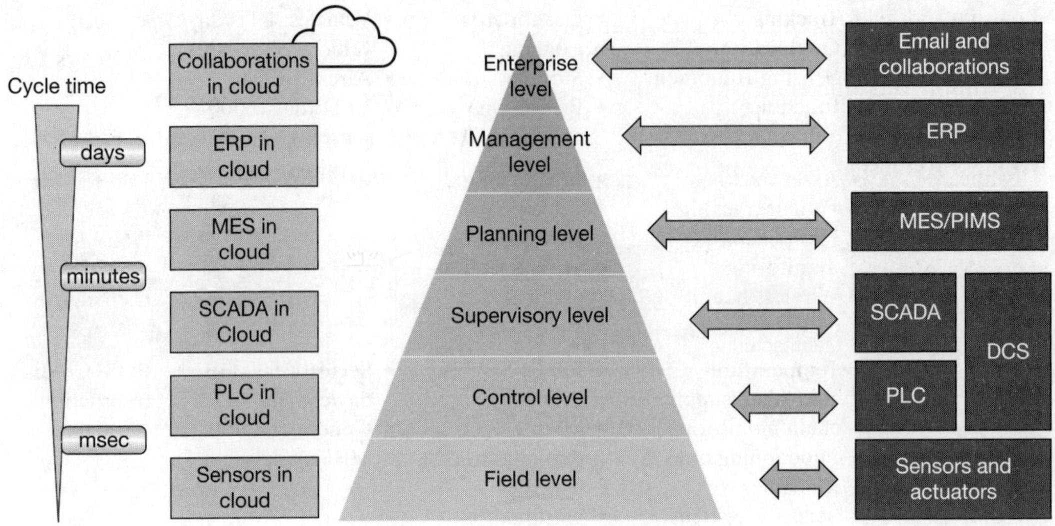

Figure 8.9 Cloudification of MES levels

It is important to understand the cloudification of each of these levels. The top layer, the Enterprise Level, is associated with email and other collaboration applications, which are well-known cloud products and do not need much elaboration. Therefore, we will start from the ERP level.

8.6.1 Management of Manufacturing through Cloud ERP

The manufacturing process involves the creation of finished products by assembling raw materials using machinery, human labour and other tools. The complexity of this process has significantly increased due to recent global events like wars, geopolitics, pandemics and natural calamities. Compounded by the trend of offshoring in the manufacturing industry, these issues have led to shortages of components and distribution challenges. Hence, manufacturers have a definite need for an ERP solution, especially a cloud-based ERP solution, to manage the complex ecosystem. The high cost of on-premises ERP implementation has led businesses to prefer cloud ERP solutions increasingly. Cloud ERP solution offers remarkable benefits for business:

- It is a comprehensive business management solution that streamlines operations, automates processes and facilitates real-time data analysis for informed decision-making.
- In manufacturing, a cloud ERP solution with a robust database specifically addresses production planning, scheduling and inventory management.
- Tailored for the industry's unique processes, it optimises resource management and forecasting accuracy.

- It integrates applications to automate processes and provide actionable business insights, incorporating features like AI, ML, Material Requirements Planning and Business Intelligence.
- Unlike legacy ERP Systems that rely on disconnected systems, cloud ERP software fosters collaboration, connecting businesses with customers, dealers and partners for improved efficiency.

IFS Cloud ERP, Oracle NetSuite ERP, Microsoft Dynamics 365 Business Central and SAP Business ByDesign are examples of Cloud ERP for Manufacturing.

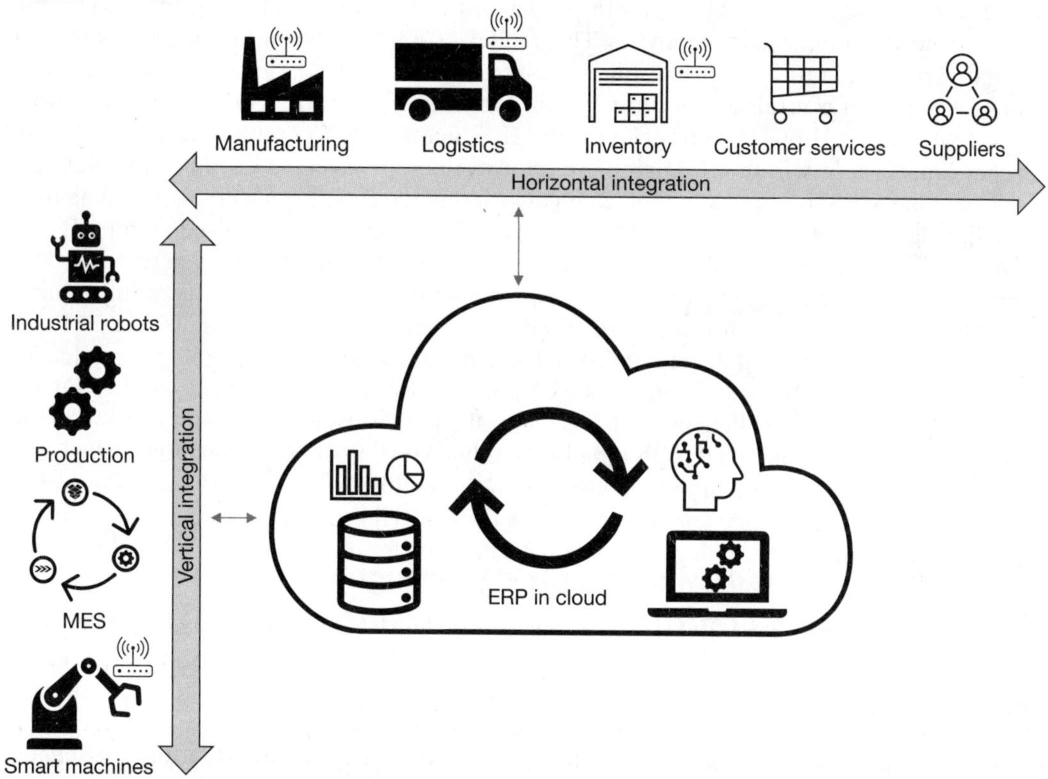

Figure 8.10 Horizontal and vertical integration of cloud ERP

The above Figure 8.10 depicts the horizontal and vertical integration with the Cloud ERP solution.

Horizontal integration establishes connections across the entire value chain, including internal logistics, production, distribution, sales, services and external partners and suppliers.

Vertical integration links various company levels with IT systems, integrating production, distribution, manufacturing and low-level PLC machine controllers and sensors for enhanced flexibility and performance in planning management.

ERP systems supporting these integration approaches empower autonomous production with intelligent robots, prioritising safety, flexibility, versatility and collaboration. This integration offers benefits in customer value creation, competitive advantage, increased company productivity and global economic growth.

8.6.2 Manufacturing Execution Systems (MES) in Cloud

The advent of cloud computing has facilitated the development of an integrated manufacturing Software-as-a-Service (SaaS) platform known as Cloud-MES, addressing various manufacturing needs. This platform offers insights into execution, quality, production, plant maintenance notifications and seamless integration with ERP systems. Integrating smart factories with Cloud MES enables smooth production execution through end-to-end visibility and advanced manufacturing analytics. This modern Cloud MES solution is inherently connected with machines through IoT, quality control processes, ERP and other cloud applications, serving as a centralised business system for efficient and collaborative manufacturing operations. MES 101 of 42Q, Tulip Next Gen MES, Siemens OpCenter powered by AWS and Proficy Smart Factory from GE Digital are some of the top players in Cloud MES solutions.

The cloud MES solution is a game-changer, offering the entire application, including functionality, data storage, processing and backup, on the cloud as a centralised system. It connects the shop floor to overall business systems, providing broader visibility across all plants, a crucial aspect of managing global and complex manufacturing. As manufacturing expands globally and integrates with customers, on-premises solutions become challenging. Mature manufacturers are turning to cloud MES for efficient, accessible, centralised production management across multiple plants. This shift is driven by lower investment, faster ROI and improved accessibility, making cloud MES increasingly popular in the manufacturing industry. By consolidating manufacturing data in the cloud from multiple plants across the globe, manufacturers gain visibility into individual plants and utilise *a unified dashboard for various plants, obtaining insights that enhance operational efficiencies across the entire manufacturing landscape*. Below, Figure 8.11 depicts a schematic diagram for on-premises MES vs cloud MES.

8.6.3 Supervisory Control and Data Acquisition (SCADA) in Cloud

SCADA (Supervisory Control and Data Acquisition) systems are computerised applications for remote and real-time control of industrial machinery and processes, aligning with the CIM's supervisory control concept. Initially, SCADA systems operated as standalone and proprietary setups, depending on a central unit and a backup mainframe for remote terminal sites. However, technological advancements have since led to the development of distributed network models, significantly improving both resiliency and safety. Later iterations focused on security, cost-effectiveness and standardisation, employing a networked systems approach. The latest generation embraces web and cloud-based technologies, enabling secure remote access and control from any internet-connected device. The key distinction lies in the location of software and data, with on-premises SCADA at the plant site and cloud-based SCADA relying on web-based applications managed by the SCADA provider.

Businesses interested in managing their own SCADA may find on-premises solutions appealing. However, the downside includes managing all hardware, architecture and software, leading to higher costs, risks and obsolescence if not done efficiently. Many industries opt for cloud-based SCADA, outsourcing hardware, software, security responsibilities and

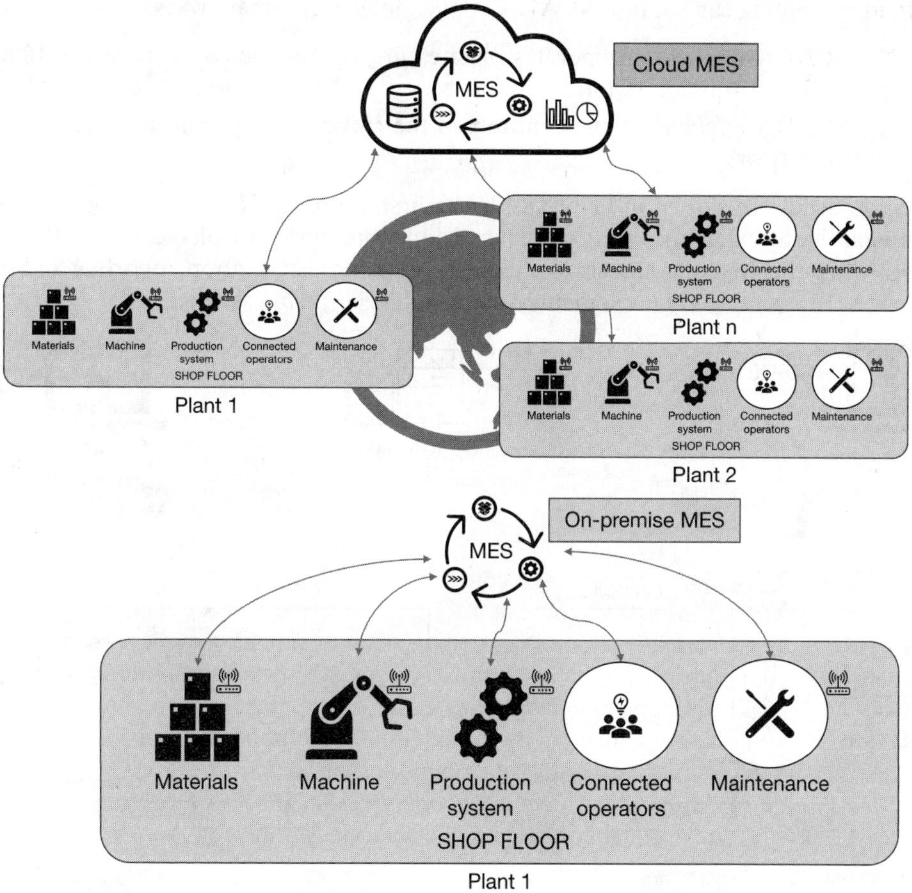

Figure 8.11 On-premises MES vs cloud MES

management functions to a trusted third-party service provider. This ensures continuous improvements and scalability at a lower overhead cost. On-premises SCADA is well-suited for businesses requiring proximity for system management, such as oil refineries and power plants. In contrast, cloud-based systems are ideal for operators managing dispersed assets over large distances, particularly in emerging industries like sustainable hydrogen production and solar power generation. Security can become a significant point of concern for SCADA systems. In an on-premises system, having a dedicated security team that stays updated on potential threats and viruses, with a resilient plan in place, is ideal. Larger companies often have competent security departments for this purpose. However, if in-house security expertise is lacking, alternatives like Emerson's Zedi SaaS SCADA may be preferable. The Zedi IIoT platform, designed 'born in the cloud', was developed entirely within a cloud system and operated on Microsoft Azure.

Cloud computing can support SCADA applications in two main ways:

- The SCADA application operates on-site, directly connected to the control network and sends information to the cloud for storage and distribution.
- The SCADA application runs entirely in the cloud and is connected remotely to the control network.

Both approaches are shown in Figure 8.12. The first approach is the more common one. In this model, the SCADA application's control functions remain isolated within the control network while the system connects to a cloud service for visualisation, reporting and remote user access. This is typically implemented using public cloud infrastructures.

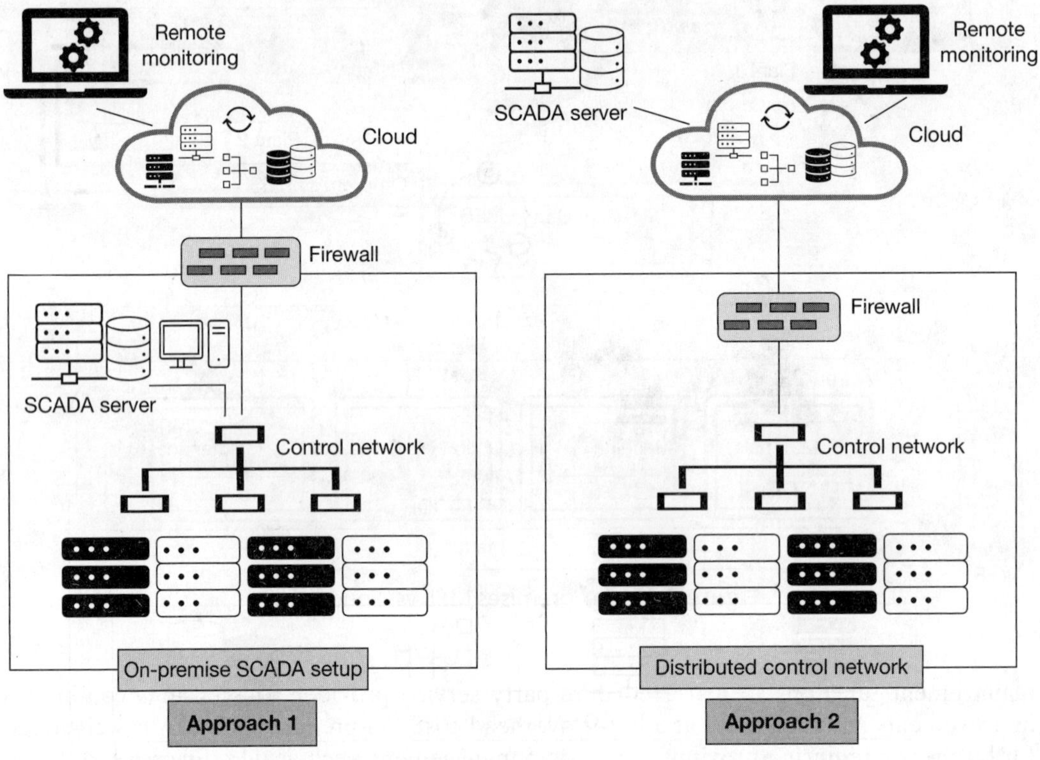

Figure 8.12 SCADA in Cloud - Solution Approach

The second approach is a standard implementation for distributed SCADA applications, mainly when a local deployment is impractical. In this case, controllers connect via Wide Area Network (WAN) links to a SCADA application that operates entirely in the cloud. These applications often utilise private or hybrid cloud architectures.

Applications like Utility (Electricity, Water and Gas) SCADA, Automatic Meter Reading, Building Management Systems, Renewable Energy (Solar, Wind, Small Hydro Plant) and Asset Management/Condition Monitoring systems generally use SCADA on the Cloud. FactoryTalk Optix from Rockwell Automation, Simatic WinCC from Siemens and AVEVA Edge are some of the well-known SCADA software on the cloud.

8.6.4 Programmable Logic Controllers (PLCs) in Cloud

This section will explore the Programmable Logic Controllers (PLCs) and their virtualisation as main enablers for Level 2 of CIM, i.e., Control Level. PLC is a programmable computing device primarily utilised for managing electromechanical processes, particularly in industrial setups. It monitors input device states, such as signals from a light switch and makes decisions regarding the next state of an output device, such as turning a light on or off. Due to their rapid response time and robustness, PLCs are frequently employed to monitor and report device status, diagnose faults in hardware devices like industrial machines and tools and execute device events.

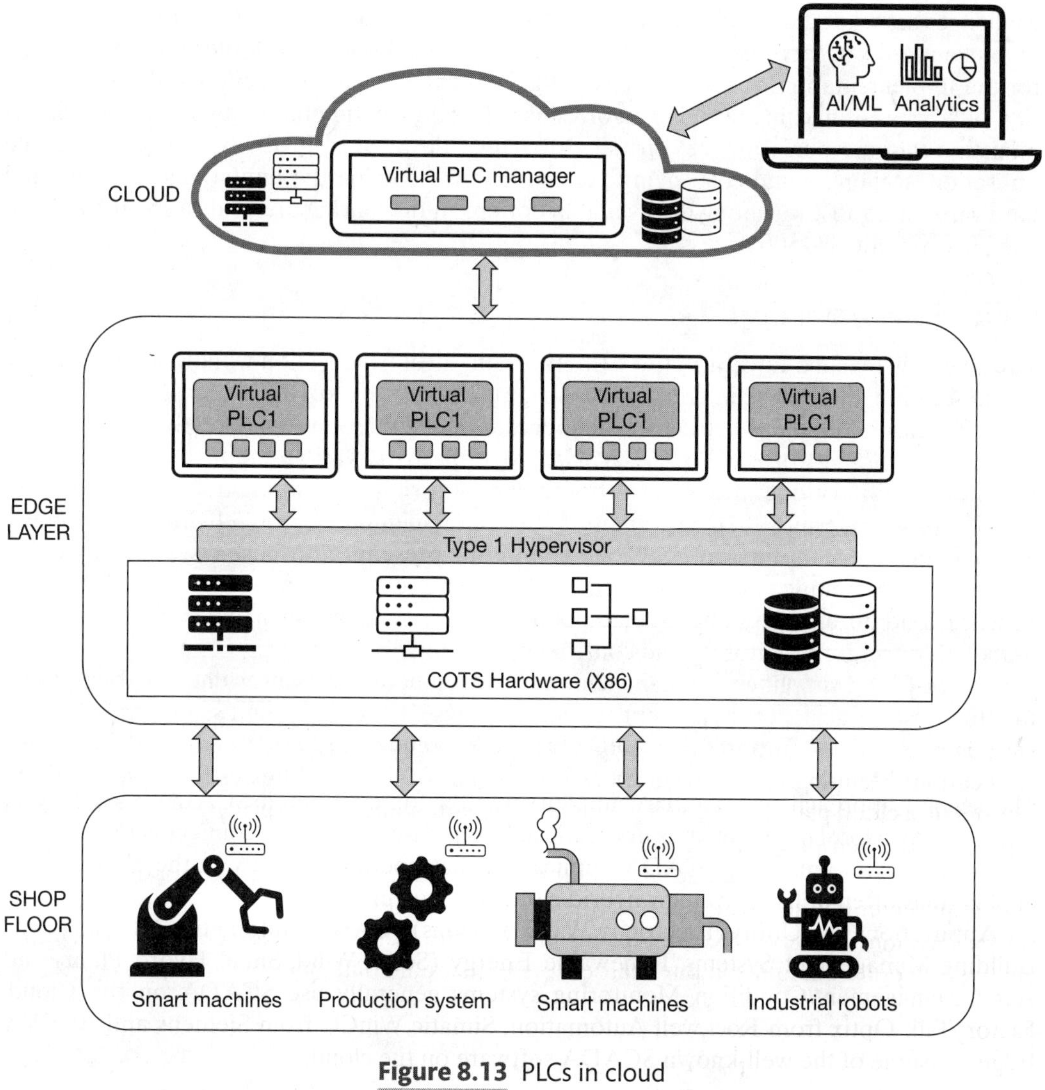

Figure 8.13 PLCs in cloud

However, current PLCs and the existing industry structure cannot meet future scaling needs. Integrated hardware and software-based PLCs struggle to support resource-intensive tasks like machine learning and artificial intelligence algorithms. Moreover, hardware-based PLCs result in individual maintenance and hardware costs, with each technical process requiring a dedicated PLC for each subsystem. Here comes the need for virtual PLCs (vPLCs) or Soft PLCs, which would solve the purpose while exploiting the benefits of cloud computing. vPLCs can function similarly to traditional PLCs in controlling industrial assets. A Commercial Off-The-Shelf (COTS) server like x86 situated at the edge can host multiple virtual PLCs for real-time control of assets. These virtual PLCs can be supervised and administered from a cloud instance. We can see a simple architecture of vPLCs in Figure 8.13.

The separation of hardware and software enhances scalability and addresses supply chain challenges associated with proprietary PLC hardware by leveraging the widespread availability of standard x86 server hardware. Integrating cloud technology with the shop floor allows for the implementation of software-based PLC operations, data collection and the application of advanced machine learning algorithms, all while meeting the requirements for deterministic real-time performance. Virtual PLCs provide greater flexibility, efficient resource utilisation, scalability and cost savings, effectively overcoming the limitations of traditional hardware-based PLCs. Among the available options in the market are Codesys Soft PLC and Siemens' Simatic S7-1500V.

8.6.5 Sensors in Cloud

We have now reached Level 1 of the CIM, the Field level. This section will discuss the Sensor Cloud solution and its role in efficiently managing the massive volume of sensor data.

The Sensor-Cloud infrastructure enhances cloud computing by effectively managing wireless sensor networks (WSNs). WSNs are utilised in various applications, including healthcare, military operations, infrastructure monitoring, environmental monitoring and manufacturing. Despite their usefulness, WSNs face memory, energy efficiency, scalability and data management limitations. Cloud computing presents a promising solution by offering scalable computing resources, storage and software services at a low cost. As a result, the Sensor-Cloud infrastructure has gained popularity, providing an open, flexible and reconfigurable platform for monitoring and control applications.

Sensor-Cloud simplifies the management of vast amounts of sensor data, enabling users to effortlessly collect, access, process, visualise, analyse, store, share and search this data for various applications. In traditional sensor networks, sensors are often dedicated to specific applications, resulting in underutilised resources. Sensor-Cloud addresses this by virtualising sensors on a cloud platform, allowing dynamic and automatic service provisioning. Figure 8.14 represents the Sensor-Cloud architecture.

Virtual sensors on the cloud are dynamic, facilitating automatic provisioning based on user requirements. Users can supervise these virtual sensors without concerns about physical sensor locations. However, before utilising virtual sensors, users should verify the availability and inspect potential faults in physical sensors to maintain data quality.

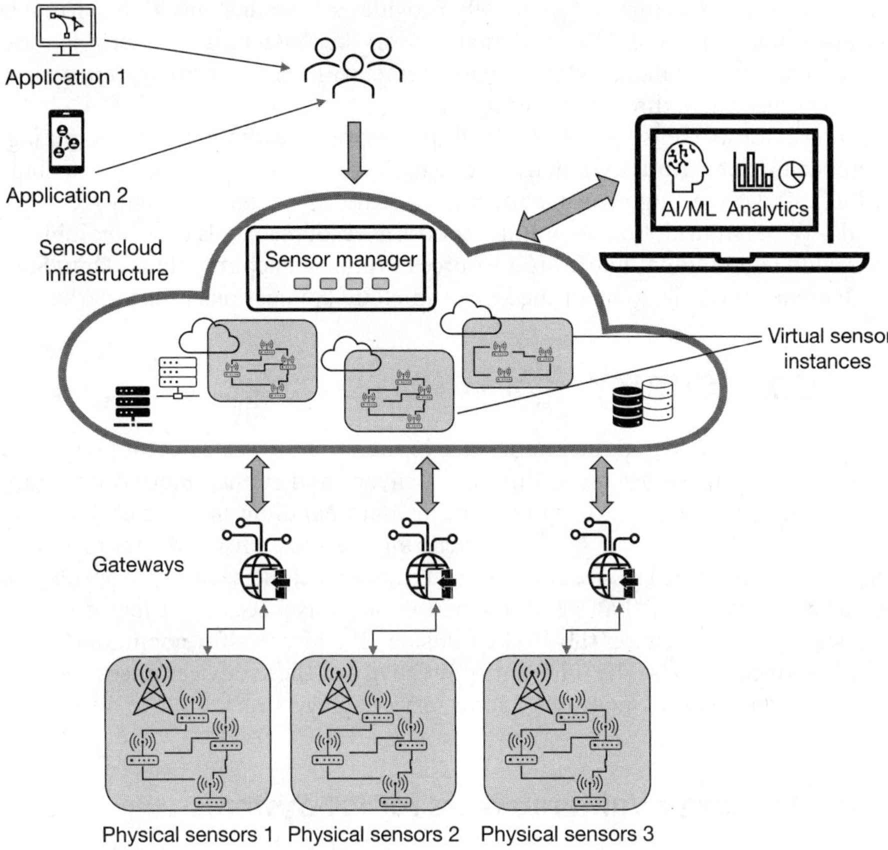

Figure 8.14 Sensor Cloud Architecture

DID YOU KNOW?

In 2010, James Kuffner from Google coined the term 'Cloud Robotics'. This concept refers to robotic and automation systems that depend on data or code delivered over a network for their operation. Instead of relying solely on a single, standalone system for sensing, computation and memory, Cloud Robotics utilises a more interconnected approach.

Within this framework, another concept known as 'Multiplicity' has emerged. Unlike the idea of a singular AI surpassing human intelligence, Multiplicity emphasises the collaborative efforts of diverse groups of humans working together with various machines to address complex challenges in areas such as search, transportation, healthcare, design and discovery. This approach harnesses the collective wisdom of crowds combined with the computational power of cloud technology.

Sensor-Cloud infrastructure automatically provides service instances, i.e., virtual sensors, to end-users when requested. These virtual sensors become part of users' IT resources, like disk storage, CPU and memory. Users can interact with these service instances and access the associated sensor data through a user interface.

Integrating extensive sensor data from multiple networks with the cloud computing model becomes appealing for various user analyses due to the provision of scalable processing power. Sensor-Cloud allows existing sensor networks to scale significantly, meaning organisations can expand services from cloud computing vendors as resource needs grow, avoiding substantial investments in additional hardware resources. Nimbits, Pachube, IDigi, ThingSpeak and Medidata Patient Cloud are some of the sensor-in-cloud applications in the market.

8.7 CLOUD SECURITY

In earlier sections, we discussed the importance of the Cloud in realising Industry 4.0's vision of a Connected World. We discussed different delivery and service models of cloud, cloud computing architecture, edge and fog computing, data handling in the cloud and different layers of cloud implementation for manufacturing systems. However, more connectivity means more vulnerabilities. This section will focus more on cloud security's approach towards enabling IoT systems rather than generic cloud security aspects. Smart factories and interconnected supply chains rely on the IIoTs, utilising IP addresses for communication within and beyond production lines. The inherent connectivity of these devices poses a security risk without proper cybersecurity measures. Cloud services play a crucial role in addressing these issues.

8.7.1 Cloud Security Vulnerabilities for IoT Systems

Numerous interconnections and services in a cloud and fog-enabled IoT system create potential vulnerabilities for malicious access. Introducing edge-based services from CSPs allows attackers to target device execution logic, causing malfunctions or operational denials. Including a fog layer adds further complexity and targets for potential attacks. The following Table 8.8 outlines some potential security threat areas CSPs generally try to address within their trust boundary when facilitating cloud-based IoT applications.

Table 8.8 Potential Security Threats

Application Layers	Potential Threats
Edge/Device Layer	1. Sensor disablement 2. Tampering/ Modifying sensor data (input and output) 3. Illegal access to API keys employed for connecting to cloud services
Fog Computing Layer	1. Unauthorised access to fog platforms 2. Tampering with policy management services 3. Snooping on IoT device communications with the fog platform, communications from fog to cloud, communications within fog platforms

(*Continued*)

Table 8.8 (*Continued*)

Application Layers	Potential Threats
Cloud Computing Layer	1. Gaining unauthorised write access to cloud-hosted virtual device configurations 2. Gaining unauthorised access to key and certificate provisioning systems 3. Gaining unauthorised read or write access to cloud databases 4. Gaining access to data shared with third parties 5. Unauthorised access to entitlements management services in the cloud 6. Denying device-to-cloud or cloud-to-device access 7. Hacking cloud administrator passwords, tokens, or SSH keys 8. Taking advantage of vulnerabilities in virtual machines and containers 9. Tampering with the device registration process at the cloud layer

However, the responsibility for the security hardening of devices is with the device vendors and is not part of CSPs. Organisations working as Prime Integrators (deploying, connecting and configuring E2E systems) would face the challenge of implementing a secured E2E solution from Edge to Cloud.

8.7.2 Cloud Security Services for IoT Applications

CSPs offer a plethora of security services to enable IoT Implementations. Cloud services are a unified force for distributed edge devices, offering analytics, intelligence, data storage and event processing for IoT systems. CSPs deliver tailored IoT services, simplifying the management and control of numerous devices. The interfaces supporting IoT at the edge must remain secure and the cloud enhances the utility of IoT edge devices by enabling data collection and transformation into actionable intelligence.

In addition to supporting device functionality, the cloud provides cybersecurity services for edge devices, covering identity provisioning, certificate management and anomaly monitoring in operations. CSPs are actively enhancing the security of edge-to-cloud connections.

Partnerships between cloud providers and hardware manufacturers, such as Microsoft Azure or Google's collaboration with Microchip, prioritise hardware-generated private key material, ensuring its confinement within the hardware for increased security. Also, IaaS services tend to maintain secure-by-default configurations for VMs, devices, containers and networks.

In this section, we will briefly go through such security services.

Onboarding of Devices

CSPs allow administrators to onboard devices individually or in bulk, transitioning them from untrusted to trusted. This transition process poses a security risk, potentially allowing malicious actors to introduce rogue devices. To address this, certain CSPs now offer preconfigured hardware options that embed cloud credentials directly into secure elements of the microprocessor. This approach, termed zero-touch provisioning by Microsoft, ensures rapid and secure registration of devices into the cloud. Microsoft is collaborating with chip developers like Microchip to implement this strategy.

Certificate Management

CSPs provide various options for provisioning keys and certificates on IoT devices, usually during onboarding. This process involves generating public/private key pairs and a Certificate Signing Request (CSR), which is then submitted to the CSP's Public Key Infrastructure (PKI) for signing. AWS IoT, for example, allows administrators to create their key pairs and manage the associated cryptographic materials regularly.

Policy Control

Under policy control, CSPs facilitate role definition (e.g., application, gateway, user roles), group creation and permission allocation, enhancing access control in cloud IoT implementations. Devices and services can also have roles. Administrators also have the flexibility to create custom roles to align with specific system requirements. CSPs enable administrators to grant permissions to devices and device groups. Azure IoT service, for instance, provides four defined permissions: RegistryRead, RegistryReadWrite, ServiceConnect and DeviceConnect, managed on a per-IoT Hub basis in the Azure cloud.

Real-time Configuration Management

Configuration management is crucial for IoT system security. CSPs provide features for real-time management of device configurations, even when devices are offline. AWS and Azure enable users to access and modify the state of edge devices, with the device configuration state stored in the cloud as a JSON document. For instance, if a smart light bulb (MQTT enabled) is offline, a command can be sent to the device configuration repository to change its colour and the change will take effect when the light bulb returns online. MQTT, or Message Queueing Telemetry Transport, is a lightweight messaging protocol for devices and networks characterised by high latency, unreliable connections and low bandwidth.

Gateway Security Management

The gateway serves as the bridge to the cloud and incorporates various security measures. It includes features like blacklisting and whitelisting. It enables the exclusion or authorisation of specific devices for communication. Cloud gateways help in policy configurations that dictate the application of encryption and authentication controls. For example, when connecting to the gateway, administrators can set authentication preferences for MQTT clients, i.e., devices communicating via MQTT over a specific network. Additionally, they can enforce limitations on the topics these clients can publish or subscribe to. Token-based authentication is a secure method that allows authentication without transmitting sensitive security parameters over the network. Platforms like Azure and Google Cloud IoT Core support token-based authentication.

Compliance and Security Monitoring

Ensuring security in a cloud environment necessitates robust compliance monitoring, incorporating policy configurations and utilising compliance analysis tools. Examples of compliance analysis include policy compliance, blacklist and whitelist adherence and connection security compliance.

Regarding security monitoring, IoT gateways must be configured to detect any suspicious behaviour exhibited by endpoints. Monitoring closely for repeated connection or authentication attempts, identifying malicious clients that connect without sending data and flagging abnormal connection terminations are crucial behaviours worth watching. Utilising log management features like AWS IoT suite's CloudWatch is a valuable tool for logging and maintaining an audit trail of process events, especially in tracking the inflow of messages from devices to the cloud. For AWS-based IoT deployments, leveraging Amazon CloudTrail is beneficial, as it supports account-level AWS API calls, enabling security analysis, analytics and compliance tracking.

In essence, safeguarding the security of cloud-based IoT systems demands a holistic and collaborative approach, prioritising proactive risk management, continuous improvement and collective vigilance. We can only realise the promise of IIoT-driven transformation while safeguarding the integrity, confidentiality and availability of critical industrial assets and data through concerted efforts to address vulnerabilities and fortify defences.

8.8 CONCLUSION

In conclusion, the future of cloud computing in the context of Industry 4.0 is expected to be transformative, offering many opportunities and advancements. It would be characterised by scalability, edge computing integration, ultra-low latency powered by 5G, advanced analytics and AI, security and compliance, collaboration and ecosystem integration and industry-specific solutions. Cloud platforms will continue to evolve and innovate to meet the growing needs of industrial enterprises in the digital era. This will integrate domain-specific capabilities, such as industrial automation, predictive maintenance and digital twins (to be discussed in detail in Chapter 11 - Digital Twins), to address the specific challenges and opportunities of manufacturing, logistics, energy and other industries. Cloud providers will continue to invest in advanced security technologies, such as encryption, identity and access management and threat detection, to safeguard industrial data and infrastructure. Additionally, cloud platforms will adhere to industry-specific compliance standards to address regulatory requirements in the manufacturing, healthcare and energy sectors. Cloud platforms will offer interoperability capabilities to enable seamless data exchange and integration across diverse industrial systems, suppliers and partners. As mentioned earlier, cloud robotics would be realised through the machine-to-cloud (M2C) communication principle, where robots would be given access to higher levels of computing, big data, lessons learned from past activities and AI/ML technologies. This interconnectedness fosters innovation, streamlines supply chains and drives new business models and partnerships.

SUMMARY

- **Definition:** Cloud computing offers on-demand access to a shared pool of network, server, storage, application and service resources. As defined by the National Institute of Standards and Technology (NIST), this model enables rapid provisioning and release of resources with minimal management effort.
- **Key Features of Cloud Computing:**
  - **On-demand Self-service:** Users can independently access computing resources as needed without needing to interact with service providers.

  - **Broad Network Access:** Cloud services can be accessed over the network using standard protocols, making them accessible from a variety of devices.
  - **Resource Pooling and Multi-tenancy:** Multiple consumers share pooled resources in a multi-tenant model, with physical and virtual resources dynamically allocated based on demand.
  - **Rapid Elasticity:** Capabilities can be scaled up or down automatically to meet fluctuating demand, providing elastic provisioning and release of resources.
  - **Measured Service:** Cloud systems automatically monitor and optimise resource usage through metering capability, offering transparency to both service providers and consumers.
  - **Self-Service Catalogue:** Users can browse and request services from a catalog, facilitating easy and immediate access to resources.
  - **Automation and Orchestration:** Automated management and orchestration of tasks streamline operations, enhancing efficiency.
  - **Scalability:** Cloud infrastructure allows for the adjustment of resources, enabling them to be increased or decreased in response to changing workload demands.
  - **Availability and Reliability:** High availability and reliability are achieved through redundancy and robust infrastructure.
  - **Security and Compliance:** Cloud providers implement stringent security measures and compliance protocols to protect data and ensure regulatory adherence.
- **Cloud Service Models:**
  - **Infrastructure as a Service (IaaS):** This model provides essential computing resources such as virtual machines, storage and networks on a pay-as-you-go basis. It offers the most control over IT resources.
  - **Platform as a Service (PaaS):** This platform empowers customers to develop, run and manage applications without the complexities of managing underlying infrastructure.
  - **Software as a Service (SaaS):** This service model delivers software applications over the internet on a subscription basis, eliminating the need for complex software and hardware management.
- **Cloud Deployment Models:**
  - **Public Cloud:** Services offered by third-party providers over the public Internet are available to anyone willing to pay. They are cost-effective and scalable.
  - **Private Cloud:** Exclusive cloud infrastructure operated solely for a single organisation, providing greater control over data and security.
  - **Hybrid Cloud:** This type of cloud combines public and private clouds, allowing data and applications to be shared between them for greater flexibility and optimisation of existing infrastructure.
  - **Community Cloud:** Shared infrastructure for several organisations with common concerns (e.g., security, compliance), managed and used collectively.

- **Components of Cloud Computing Architecture:**
  - **Virtualisation:** This technology enables the creation of virtual instances of physical resources, enhancing flexibility and efficiency. Key areas include server, storage and network virtualisation.
  - **Hypervisors and VMs:** Hypervisors manage virtual machines (VMs). Type 1 hypervisors run directly on hardware, while Type 2 runs on a host operating system.
  - **Containerisation:** This technique encapsulates applications and their dependencies into containers for consistent deployment across different environments, enhancing efficiency.
  - **Microservices:** This approach breaks down applications into smaller, independent services, allowing for more flexible development and scaling.
  - **Docker Technology:** A platform that uses containers to develop, ship and run applications, ensuring consistency across multiple environments.
- **Cloud Orchestration and Management:** Cloud orchestration involves automating workflows to manage cloud resources. Kubernetes is a leading framework for container orchestration, aiding in developing cloud-native applications and facilitating DevOps practices.
- **Advanced Cloud Concepts:**
  - **Cloud-Native Applications:** Applications specifically designed to leverage cloud environments, utilising technologies like microservices and auto-scaling.
  - **Serverless Computing (FaaS):** This allows developers to run code without managing servers and billing is based on actual usage.
  - **Edge Computing:** Processes data near the source, reducing latency and enhancing performance.
  - **Fog Computing:** Extends cloud computing closer to IoT devices, improving real-time processing and reducing latency.
- **Cloud in Industrial Applications:**
  - **Cloud ERP:** Enhances manufacturing processes by improving planning, scheduling, inventory management and resource optimisation.
  - **MES in Cloud:** Provides comprehensive manufacturing execution systems as SaaS, integrating with ERP for streamlined operations.
  - **SCADA in Cloud:** Moves traditional SCADA systems to the cloud, reducing costs and improving scalability.
  - **PLCs in Cloud:** Utilises virtual PLCs to meet scaling needs, reducing maintenance costs.
  - **Sensors in Cloud:** Virtualises sensors for better resource utilisation and real-time data processing.
- **Cloud Security:** Cloud security encompasses services provided by cloud service providers to secure IoT implementations, including identity provisioning, certificate management, policy control and real-time security monitoring.

EXERCISE

Multiple Choice Questions (1 Mark)

1. Which of the following is not a characteristic of cloud computing?
   (a) Scalability
   (b) High latency
   (c) Resource pooling
   (d) Pay-as-you-go model
2. Which delivery model provides infrastructure resources like storage and networking over the cloud?
   (a) SaaS
   (b) PaaS
   (c) IaaS
   (d) None of the above
3. What is the key difference between a Type 1 and Type 2 hypervisor?
   (a) Type 1 hypervisors run directly on hardware, while Type 2 hypervisors run on a host OS.
   (b) Type 2 hypervisors are more secure than Type 1 hypervisors.
   (c) Type 1 hypervisors are less performant than Type 2 hypervisors.
   (d) Type 2 hypervisors are more complex to manage than Type 1 hypervisors.
4. What is the primary goal of cloud orchestration?
   (a) To increase manual intervention
   (b) To reduce automation
   (c) To automate resource management
   (d) To decrease scalability
5. What is the primary goal of cloud-native applications?
   (a) To increase complexity
   (b) To reduce scalability
   (c) To improve application performance and scalability
   (d) To decrease security
6. What is the main function of edge computing in the cloud architecture?
   (a) Managing centralised cloud resources
   (b) Processing data closer to the source of generation
   (c) Storing data indefinitely
   (d) Replacing IoT devices

7. At which level of the CIM hierarchy does the cloud enable real-time control of assets?
   (a) Level 0: Physical Level
   (b) Level 1: Field Level
   (c) Level 2: Control Level
   (d) Level 4: Enterprise Level
8. What does the application layer in cloud CIM implementations focus on?
   (a) Physical device monitoring
   (b) Strategic business decisions
   (c) Real-time data analytics
   (d) Device configuration
9. What type of server can be used to host virtual PLCs?
   (a) Proprietary hardware
   (b) ARM servers
   (c) x86 COTS servers
   (d) ASIC devices
10. Which of the following is a key benefit of the Sensor-Cloud architecture?
    (a) Dedicated hardware for each sensor
    (b) Dynamic provisioning of virtual sensors
    (c) Elimination of data sharing
    (d) Physical sensors becoming obsolete
11. What is the primary security challenge in cloud-enabled IoT systems?
    (a) Lack of connectivity
    (b) Limited scalability
    (c) Vulnerability to malicious access
    (d) Inadequate storage
12. Which of the following is an example of a Sensor-in-Cloud application?
    (a) Kubernetes
    (b) ThingSpeak
    (c) TensorFlow
    (d) Docker
13. A factory manager wants to improve real-time monitoring and predictive maintenance using a scalable, cost-effective solution, as the plant is facing downtime due to limited storage and manual data handling.
    What would be the most suitable approach?
    (a) Using standalone systems for data storage
    (b) Implementing a cloud-based data processing platform

(c) Relying solely on manual inspections
(d) Expanding physical server infrastructure

14. A factory wants to minimise the risk of downtime due to server failures.
What feature of cloud computing ensures operational continuity?
(a) Redundancy
(b) Low-Cost Hosting
(c) High-Speed Network Access
(d) On-Demand Compute Power

15. An organisation needs a cost-effective solution for rapid prototyping and testing of industrial software.
What aspect of cloud computing should they leverage?
(a) SaaS for application use
(b) PaaS for development and testing
(c) IaaS for infrastructure management
(d) Middleware for device integration

16. A business wants to use pre-built tools for enterprise management without managing the underlying infrastructure.
Which cloud model is best suited?
(a) IaaS
(b) PaaS
(c) SaaS
(d) Middleware

17. A small manufacturer wants to adopt cloud services but is unsure of where to start due to cost considerations.
Which service model provides the easiest entry point?
(a) IaaS
(b) PaaS
(c) SaaS
(d) All of the above

18. A production unit needs enhanced protection for IoT devices against cyber threats.
What cloud feature is critical for ensuring IoT device security?
(a) Token-Based Authentication
(b) High Availability Zones
(c) Multi-Cloud Integration
(d) Edge Computing

19. A factory wants to test new production management software without disrupting ongoing operations.
What cloud feature can help achieve this?

(a) Virtualisation
(b) Scalability
(c) Interoperability
(d) Multi-Tenancy

20. A company integrates AI for predictive maintenance using cloud analytics. At which CIM level is this functionality implemented?
(a) Field Level
(b) Control Level
(c) Enterprise Level
(d) Process Level

Short Answer Type Questions (2 Marks)

1. Define the role of cloud computing in Industry 4.0.
2. What is the role of containerisation in cloud-native applications?
3. Explain the concept of Fog Computing in the context of Industry 4.0.
4. How does Edge Computing differ from traditional cloud computing?
5. What is the significance of serverless computing in cloud-native applications?
6. What are the primary security threats at the edge layer in cloud-enabled IoT systems?
7. Explain zero-touch provisioning in cloud security for IoT devices.
8. What are the advantages of integrating advanced machine learning algorithms with cloud computing for industrial automation?
9. List some real-world applications of the Sensor-Cloud architecture.
10. Why is compliance monitoring essential in cloud-based IoT systems?

Long Answer Type Questions (5 Marks)

1. What is the significance of cloud computing in the modern digital age?
2. Discuss the different types of cloud deployment models and their advantages and disadvantages.
3. Explain the concept of virtualisation and its role in cloud computing.
4. Discuss the integration of cloud computing at different levels of the CIM hierarchy.
5. An IoT company is building a cloud-native platform to manage smart devices across homes and industries. They require seamless scalability, high availability and rapid deployment of new services without downtime. Additionally, they plan to implement edge computing for faster responses.

Which orchestration tool would you recommend for their platform and why? How would Kubernetes or its lightweight version, K3s, help in addressing their needs, particularly in edge deployments and resource optimisation?

6. How can organisations ensure the security and compliance of their cloud-based applications and data?
7. What is the significance of edge computing in IoT?
8. Discuss the evolution of SCADA systems from standalone setups to cloud-based technologies. Highlight the benefits and challenges associated with adopting cloud-based SCADA for industries like renewable energy or utility management.
9. Explain the differences between cloud-native and cloud-enabled applications and describe how serverless computing addresses challenges in microservice deployment. Include examples and limitations of serverless computing.
10. Discuss the key differences between Type 1 and Type 2 hypervisors and compare virtual machines (VMs) and containers in terms of architecture and performance. Explain how containerisation enhances IoT device functionality and integration in dynamic environments.

Big Data

CHAPTER 9

OBJECTIVES

You might have heard the phrase 'Data is the new oil' (Clive Humby) indicating that in coming decades the data, and not oil, will be the driver of the economy. We are generating more data than ever, and this enormous amount of data is referred to simply as Big Data. Big Data is changing the nature of business and is a very vital component in Industry 4.0. In this chapter, we introduce the readers to the concept and terminologies of Big Data and the business motivation to use it.

By the end of this chapter, you will be able to:

- Understand the key characteristics of Big Data and their impact on data storage and processing.
- Differentiate between traditional RDBMS and modern Big Data storage solutions such as NoSQL databases, data lakes and cloud platforms.
- Explain the ETL and ELT processes and their significance in data integration and analytics.
- Describe the role of distributed computing and frameworks like Hadoop and Apache Spark in handling large datasets.
- Compare batch processing and stream processing, along with their respective use cases.
- Analyse how Big Data is applied across industries, including manufacturing, healthcare, finance and e-commerce.
- Identify challenges in implementing Big Data solutions, including data security, compliance and scalability.

9.1 INTRODUCTION

In the earlier chapters, we highlighted how data acts as the central thread in an Industry 4.0 ecosystem, enabling seamless integration across diverse systems. We also explored the growing need for cloud-based infrastructures to manage the vast volumes of data generated in such environments. In this section, we turn our focus to the storage and processing aspects. Big Data refers to the specialised field concerned with managing and analysing large-scale data sets, often originating from a variety of sources. Big Data solutions are required when traditional data storage and processing technologies are insufficient. It introduces newer techniques that leverage distributed storage and computing resources, as data continue to grow in size, speed, diversity and complexity. The data was measured by Terabytes in 2005, Petabytes in 2010, Exabytes in 2015 and Zettabytes in 2020 – which gives an idea of explosive data growths in the last two decades.

DID YOU KNOW?

Kilobytes are not 1,000 bytes like other metric measurements (gram, meter, or litre), but 1,024 (2^{10}) bytes. The same multiple continues for Mega, Giga, Tera, Peta, Exa, Zetta and Yottabytes, as well.

9.1.1 Evolution of Big Data

Though Big Data may appear as a new discipline, it can be seen as a natural evolution of the human need for record-keeping and calculations. The management and processing of datasets have been a long-standing problem; for example, societies worldwide from ancient times till today have collected census data and used statistics based on censuses for a variety of planning activities.

The definition of a Big Data problem is also evolving as software and hardware technologies advance. Three to four decades ago, one gigabyte of data would have been considered a Big Data challenge, necessitating the use of special-purpose computing resources. Nowadays, consumer-oriented devices like smartphones or gaming consoles can easily store, process and transmit tens or even hundreds of gigabytes of data.

The earliest known example of how prehistoric societies stored data is the Lebombo Bone, discovered in the Lebombo mountains located on the borders of Mozambique, Eswatini (earlier known as Swaziland) and South Africa. This bone tool, determined to belong to 40,000 BCE by carbon dating, is believed to be a tally stick and 29 notches suggested it may have been used as a lunar calendar. In Ishango bones (Republic of Congo), dating back to 18,000 BCE, Palaeolithic tribes believed to track their supplies and trade activities by marking them with notches using a sharp piece of quartz affixed at one end. Some scholars believe that the ancient Egyptians likely used various forms of data collection and analysis to plan and manage the construction of the pyramids.

In 3,800 BCE, the Babylonian Empire took the first known census, counting livestock and quantities of produce like milk and wool. The first population census was done in 2 CE during China's Han Dynasty, revealing 57.7 million people living in 12.4 million households. In 1703, Iceland recorded 50,366 people, which is considered the first modern census and was about as accurate as a census today.

Till late 19th century, it was a humongous task to process census data, due to lack of any process or tools. It took over seven years for 1880 US census to complete their final report. After that, first tabulating machine for data processing was invented by Herman Hollerith. This machine used punch cards for systematically processing data and it reduced the processing time significantly. With this machine, 1890 US census completed in just 18 months. Followed by Hollerith's path breaking invention, computing technology has dramatically evolved, and it enhanced the method of data storage and processing as well.

Earlier, the data was collected and stored for basic record keeping and transaction purposes mainly. Then came the relational database management system (RDBMS) in 1970s, introduced by Edgar F. Codd. This relational model organised data into tables (or relations) consisting of rows and columns, and provided flexible and efficient approach to manage data in structured way using Structured Query Language (SQL) for querying and manipulation. RDBMS became the backbone of enterprise data management by ensuring data integrity, supporting complex queries and enabling critical transactions for businesses like finance, human resources or inventory management.

During 1980s, the Data Warehousing technology was developed to address the challenging needs for centralised repository for business intelligence and analytics. The main advantage of data warehouses was that they could aggregate and store data from multiple sources, enabling organisations to perform complex queries and generate advanced analytical multi-dimensional reports. First-Generation Data Warehouses, like Teradata and IBM Netezza, were primarily built on RDBMS and provided high-performance analytics by using parallel processing functionalities and specialised hardware.

The explosion of the Internet in the 1990s marked a significant turning point in the evolution of Big Data. The proliferation of websites, online services and digital communication channels led to an exponential increase in data generation. Data began to come from diverse sources, including emails, social media posts, website logs and digital transactions. This era saw the emergence of unstructured data, which did not fit neatly into traditional relational databases. The need for new storage and processing solutions became apparent. This led to the emergence of NoSQL databases in the late 2000s. NoSQL databases were designed to handle large volumes of diverse data types, including key-value stores, document stores, column-family stores and graph databases. Key-value stores, such as Redis and DynamoDB, are optimised for simple get-and-put requests. Document stores like MongoDB and CouchDB are designed to store and retrieve semi-structured data in formats such as JavaScript Object Notation (JSON) or eXtensible Markup Language (XML). Column-family stores, including Apache Cassandra and HBase, support wide tables with millions of columns, making them suitable for distributed data storage. Graph databases like Neo4j and Amazon Neptune enable the management of data with complex relationships, which is particularly useful for social networks, recommendation systems and fraud detection.

The introduction of MapReduce and Hadoop in the mid-2000s revolutionised big data processing. MapReduce, a programming model introduced by Google, was designed to process large datasets in a distributed and parallel manner. MapReduce enabled efficient processing of massive datasets by distributing tasks across multiple nodes in a cluster, ensuring scalability and fault tolerance. Hadoop, an open-source framework developed by the Apache Software Foundation, implemented the MapReduce model and provided a scalable, distributed storage and processing platform. Hadoop democratised big data processing by enabling organisations to handle vast amounts of data using commodity hardware. It laid the foundation for the development of a rich ecosystem of big data tools and technologies, including:

- **Hive:** A data warehousing solution built on top of Hadoop, providing SQL-like query capabilities.
- **Pig:** A high-level scripting language for data processing on Hadoop.

- **HBase:** A distributed, scalable, NoSQL database built on HDFS, suitable for real-time read/write access to large datasets.
- **Zookeeper:** A coordination service for distributed applications, ensuring synchronisation and configuration management.

Data lakes emerged in the 2010s as a solution to handle the increasing variety and volume of data. Unlike data warehouses, which are optimised for structured data and predefined schemas, data lakes are designed to store raw, unprocessed data in its native format. A data lake allows organisations to ingest data from various sources, including logs, social media, IoT devices and traditional databases. The data is stored in a flat architecture, typically on distributed storage systems like Hadoop Distributed File System (HDFS) or cloud storage like Amazon S3. Prominent examples of data lakes include AWS Lake Formation, Azure Data Lake Storage, Google Cloud Storage, IBM Cloud Object Storage and Snowflake. These solutions provide advanced features such as data cataloguing, hierarchical namespaces, global edge-caching, flexible storage classes and integration with other cloud services. Figure 9.1 summarises this evolution of data storage and processing through human history.

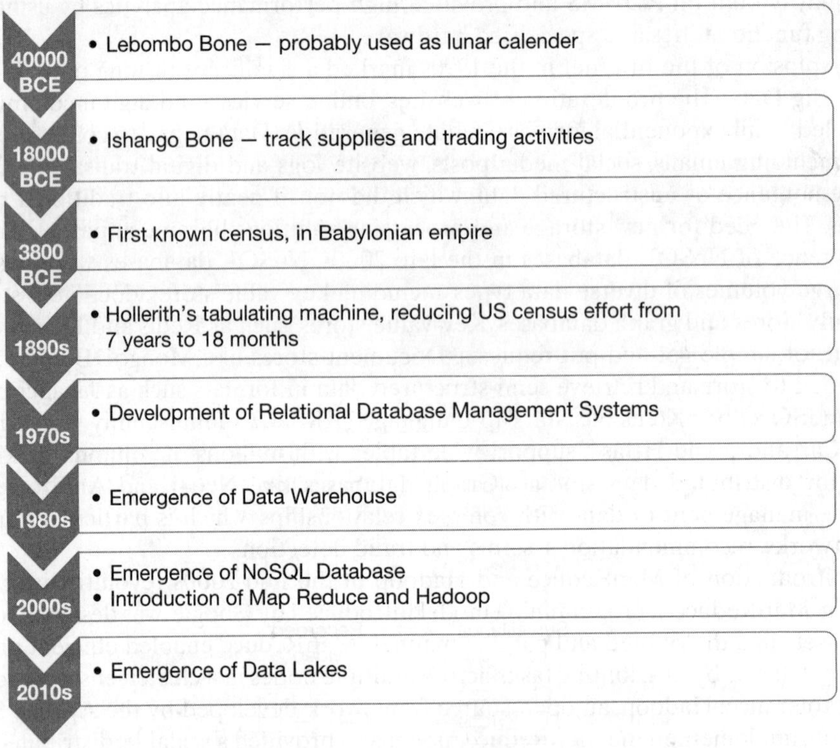

Figure 9.1 Evolution of data storage through human history

9.1.2 Different Types of Data

Data can originate from either human interactions or automated systems and its nature depends on how it is created and processed. Human-generated data comes from direct user interactions with digital platforms, applications and devices. Every time a person fills out an online form, posts an update on social media or takes a picture using a smartphone, they produce data (Figure 9.2). Please note that data can also be produced in non-digital ways, such as writing on paper. In such cases, while the data exists, it cannot be processed digitally until it is scanned, transcribed, or otherwise converted into a machine-readable format.

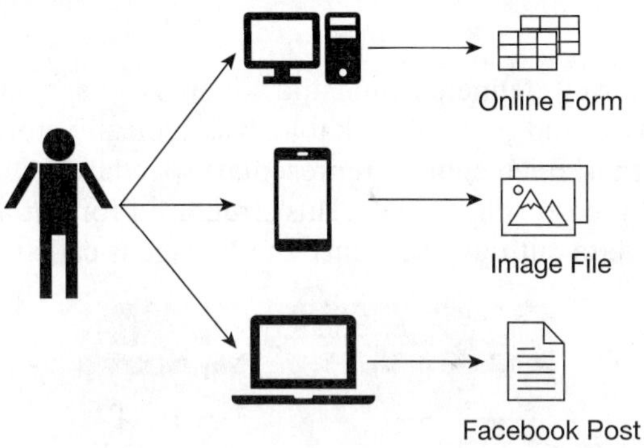

Figure 9.2 Examples of human generated data

In contrast, machine-generated data is produced automatically by software applications and hardware devices, often in response to real-world conditions (Figure 9.3). This data does not require direct human input but is instead created through automated monitoring, logging and computational processes. For example, when a customer purchases an item at a store, the point-of-sale (POS) system records the transaction and updates the inventory in real-time. Similarly, security cameras continuously capture and store video footage, which can be analysed to detect possible nefarious activities. Another example is an automated thermostat, which uses built-in sensors to track temperature and humidity, logging real-time environmental changes without any manual intervention.

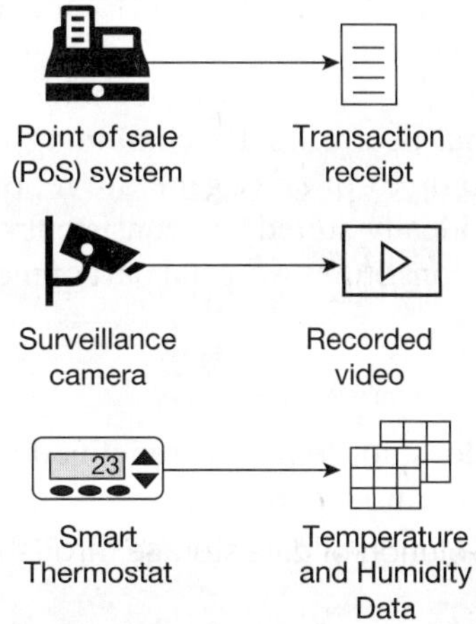

Figure 9.3 Examples of machine-generated data: transaction receipts from a PoS system, video from a surveillance camera, temperature and humidity data from a smart thermostat

These two categories, human-generated and machine-generated data, feed into modern Big Data systems in various formats. While human-generated data often includes text, images and videos, machine-generated data typically consists of numerical readings, event logs and sensor outputs. This data is categorised into three main types – structured, unstructured, and semi-structured, depending on its structure.

Structured Data

Structured data follows a well-defined format and adheres to a specific data model or schema, making it highly organised and easily searchable. It is typically stored in tabular form, consisting of rows and columns. Each column represents a specific attribute, and each row corresponds to an individual record (Figure 9.4). This structured format allows for clear relationships between different data entities, and hence the storage is called relational databases.

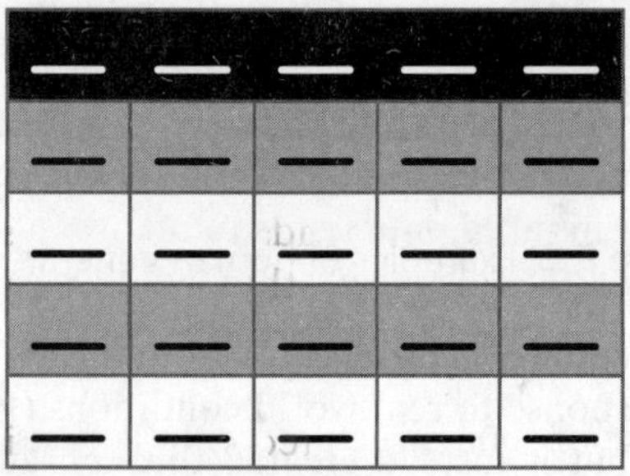

Figure 9.4 Structured data, stored in a table format

Enterprises and organisations generate structured data through various business applications and information systems. Enterprise Resource Planning (ERP) systems produce structured records detailing transactions, inventory levels, and employee data for finance, operations, and human resources business units, respectively. Similarly, Customer Relationship Management (CRM) systems store structured data related to customer interactions, including purchase history, support tickets and contact details.

The structured data is managed using SQL, which enables users to create, retrieve, modify, and delete data within relational databases. By combining these and other functions SQL allows businesses to perform complex queries, generate reports, and maintain data integrity. The structured form makes it ideally suited for consistency, accuracy, and fast search and retrieval requirements and thus constitutes a crucial part of modern information management systems.

Unstructured Data

In contrast, unstructured data does not follow a predefined format, making it challenging to store, process, and analyse. Unstructured data lacks a fixed model and hence cannot be stored into tables. It also requires specialised tools and techniques for processing and cannot be handled using SQL.

However, unstructured data is estimated to be 80% of the organisation's data, and hence it is critical to consider the same for decision-making despite the challenges. Additionally, it's growing at a rate far outpacing that of structured data, largely due to increased usage of digital communication, multimedia content, and sensor-generated information.

Unstructured data can be further categorised into two main types: textual and binary. Textual data includes documents, emails, social media posts, and reports, whereas binary data consists of non-text formats such as images, audio, and video files (Figure 9.5).

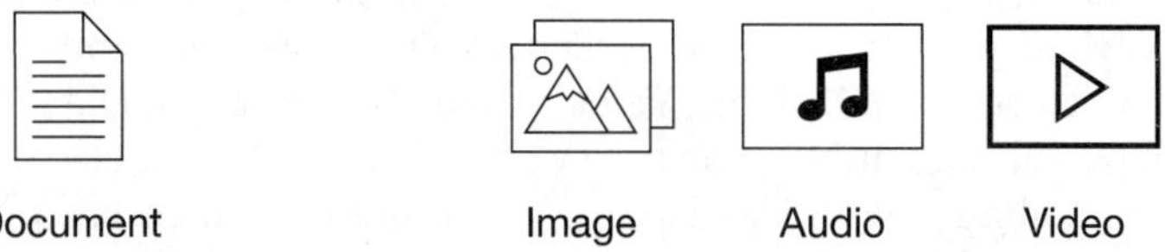

Figure 9.5 Examples of unstructured data: document file is textual data, whereas image, audio and video files are binary data

Semi-structured Data

Semi-structured data lies in between the structured and the unstructured data, in the sense that it possesses some structure but does not follow the fixed tabular format of relational databases. Semi-structured data tends to be organised more flexibly, following a hierarchical or graph-based form. This means that, while it has some level of organisation, it doesn't adhere to the strict row-column format and relationships found in relational databases.

Another key characteristic of semi-structured data is that it can include both structured and unstructured elements. For example, it might contain tags or markers to indicate data attributes, but the data itself can be flexible or vary in terms of content. Semi-structured data is often stored in text files, such as XML and JSON (Figure 9.6). These widely-used formats are designed to be readable by both humans and machines, and therefore extremely useful for exchanging data across disparate systems or applications. They are especially useful for serialising structured data for it to be transmitted over a network, for example, between a browser and a web server or between two software systems.

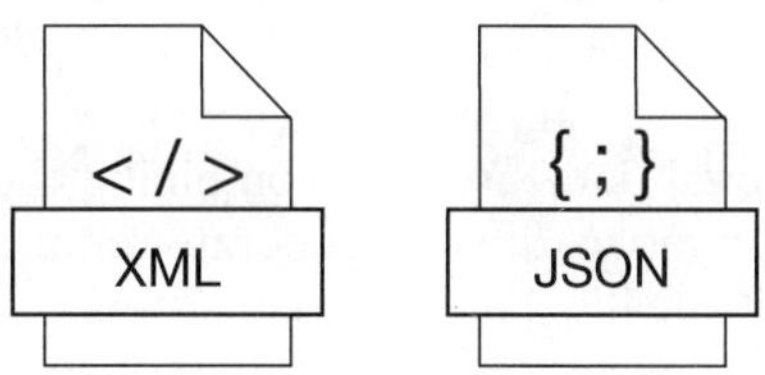

Figure 9.6 Examples of semi-structured data: XML and JSON files

Other semi-structured data types include Electronic Data Interchange (EDI) files, in which business documents are transmitted in a standardised electronic format, and Really Simple Syndication (RSS) feeds, in which news articles or blog posts are syndicated in a structured yet flexible format. Also, sensor data from devices like IIoT sensors often falls into this category, where the data is not strictly structured, but includes key information (e.g., temperature, pressure, location) in a format that can be easily read and processed.

9.1.3 Big Data Characteristics

When do you call your data Big Data? For that, it needs to satisfy certain defining characteristics that demand special attention in the solution design. These characteristics distinguish Big Data from traditional data and necessitate tailored approaches for handling, processing and storing it effectively. In 2001, Doug Lenny, while discussing the e-commerce surge in his article '3D Data Management: Controlling Data Volume, Velocity and Variety' introduced the first three characteristics of Volume, Velocity and Variety of Big Data. These got popular as 3Vs of Big Data. Subsequently, Veracity (4V) and Value (5V) are also added (Figure 9.7). Some experts kept adding new Vs (Variability, Visualisation, Vulnerability, Validity, Volatility and so on), even up to 17 Vs, but we will limit our discussion to the five most fundamental characteristics.

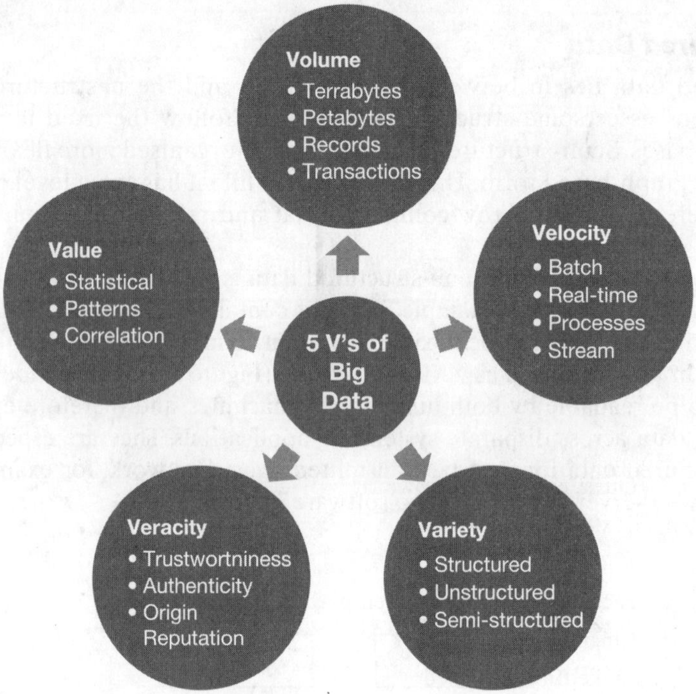

Figure 9.7 5Vs of Big Data

Volume

Big Data solutions must be capable of handling vast and continuously growing volumes of data. This high volume demands specialised storage and processing capabilities, along with enhanced data preparation, curation and management practices. Data sources that can generate such high-volume data, can be:

- **Social Media:** Facebook, YouTube, etc.
- **Sensors:** Smart meters, RFIDs, GPS sensors
- **Online Transactions:** Purchase data (Web, App, PoS), Banking transactions

As enterprises increasingly recognise data as a valuable asset, they become reluctant to discard it. The declining cost of data storage has further encouraged this mindset, leading to the accumulation of data both within individual enterprises and on a global scale. For perspective, the entire print collection of the US Library of Congress is estimated to be around 10 TB of uncompressed textual data, while the global volume of data is projected to reach 180 zettabytes by 2025, 18 trillion time of the former (Figure 9.8).

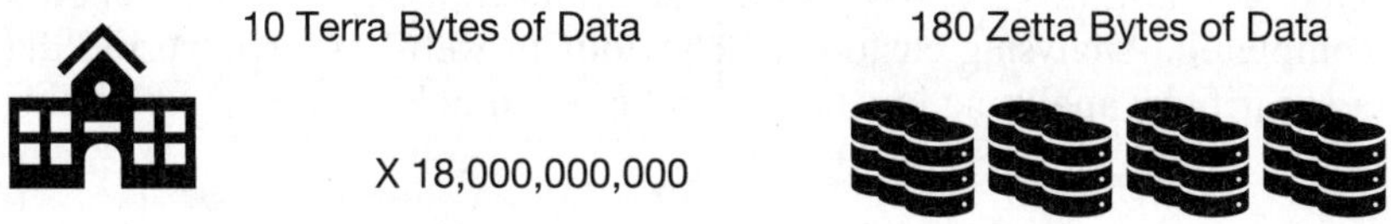

Figure 9.8 Perspective of the global volume of data.

Velocity

In Big Data solutions, the challenge extends beyond the sheer volume of data. It also includes the speed at which the data is generated. Table 9.1 gives a glimpse of data produced on average every minute as per 2024 DOMO Report. A Big Data solution designed to handle such data must be capable of processing the data as it arrives. Based on how the data is handled, it can be classified into two categories – batch processing and stream processing.

Table 9.1 Average data production per minute in some popular platforms

Activity	Data Produced
TikTok videos uploaded	16,000
Netflix hours streamed	362,962
Google searches	5.9 M
Questions answered by Siri	1,041,666
Text Messages sent	18.8 M
YouTube videos watched	3,472,222
Facebook + Instagram reels played	138.9 M
Emails sent	251.1 M
LinkedIn job applied	9,000
Meeting minutes recorded in Teams	229 M
Snapchat snaps	3.3 M
Zoom downloads	288
DoorDash order placed	$126,763
Slack messages sent	1.04 M
Fortnite hours live-streamed	1,563

1. **Batch Processing:** Under batch processing, the data is collected for an amount of time and then sent for processing together. One set of such data is called the batch and typically the time interval of the processing is constant, for example, a daily batch job

is run to archive the log files. Batch processing works well when the data generation requirement itself is not continuous or you do not need immediate analysis of the data. Payroll, utility billing, or order shipment are examples of batch processing which can be handled in monthly, quarterly and daily batches, respectively.

2. **Stream Processing:** Stream processing is key if you need to analyse the data in real-time. For tasks like fraud detection, the transaction data is analysed and anomalies are detected on real-time basis so that the fraudulent transactions are stopped before they are completed. Analysing customer behaviour in websites, or intrusion detection for cyber security by analysing log data are other examples of stream processing.

Variety

Data variety refers to the multiple types and formats of data that Big Data solutions must be equipped to handle, as data is often generated from diverse sources with different structures and representations. As mentioned in the previous subsection, data can be classified into three primary types: structured, unstructured and semi-structured (Figure 9.9). A Big Data solution often needs to manage a combination of these different types of data, which adds complexity to the system. Data variety also encompasses the various formats in which the data can be represented. Even when the data type remains the same, different systems or sources may store the data in different formats. For example, dates can be stored as dd/mm/yyyy in one system and yyyy/mm/dd in another; or a customer's name might be represented in one database as "First Last" and in another as "Last, First". These variations can also occur in formatting numeric values (e.g., currency with or without symbols, comma separators used or not), or using different units of measurement (e.g., Celsius vs Fahrenheit in measuring temperature). The presence of such variety poses significant challenges for data integration, as combining data from multiple sources requires converting or transforming data into a consistent format that can be used together. Additionally, processing and storing this diverse data is complicated, as different types of data might require different storage mechanisms and different processing methods.

Figure 9.9 Variety of Big Data

Veracity

Veracity refers to the accuracy, reliability and quality of data within Big Data environments. The incoming data may contain invalid, incomplete, or inconsistent information. Therefore, the incoming data needs to be processed to detect such erroneous data and subsequently correct or eliminate, to ensure it accurately reflects reality.

Consider the example of a thermometer at a remote location that records and transmits ambient temperature readings to a weather station. If the thermometer malfunctions, it may register an incorrect temperature, such as 80 degrees Celsius, which is clearly invalid data.

Alternatively, network issues could prevent some temperature readings from reaching the weather station, resulting in missing data. In both cases, the data quality is compromised.

With respect to veracity, 'signal' represents reliable and valid data that contributes to meaningful information. At the same time 'noise' refers to erroneous or missing data that cannot be transformed into useful information. A dataset with a high signal-to-noise ratio has greater veracity than data with a lower ratio. The veracity of data typically depends on its source and method of collection. Data gathered from controlled environments, such as online customer registrations, typically contains less noise and is more accurate. In contrast, data collected from uncontrolled sources, like customer reviews or social media posts, may have more inconsistencies and require more effort to clean and verify.

Value

Value denotes the significance and applicability of data for a business, highlighting the advantages it can offer in supporting the organisation's goals. Please note that the value of data is not only determined by its inherent qualities but also by how it aligns with the business's needs and objectives.

From our earlier discussion on veracity, we can infer that data with higher accuracy and reliability (higher veracity) will generally have greater value. This is because trustworthy data provides a more solid foundation for decision-making, analysis and strategic planning. If data contains errors, inconsistencies, or omissions, its usefulness diminishes, making it harder for organisations to derive meaningful insights or take appropriate actions.

Another crucial factor contributing to the value of data is its timeliness. The relevance of data is often tied to when it is available. For example, in the financial markets, a stock quote delayed by just 20 minutes may be nearly worthless, as the market can shift significantly during that time. In contrast, real-time data, such as sensor data from an industrial machine, can have substantial value as it allows for immediate analysis and prompt decision-making.

The value of data also influences how it should be treated within an organisation. High-value data typically justifies more investment in its storage, processing and protection. This could mean more frequent updates, more sophisticated processing tools and more stringent security measures. On the other hand, data that provides limited or no value may be subject to greater scrutiny regarding whether it should be stored, processed, or discarded. Organisations must evaluate which data is truly valuable and worth investing resources into, versus data that can be archived or deleted to reduce storage costs and improve operational efficiency.

9.2 BIG DATA STORAGE SOLUTIONS

Big data storage solutions are designed to handle the massive volume, variety, and velocity of data generated in today's digital world. In this section, we will discuss how modern solution augmenting the traditional storage systems to enable organisations to store, manage, and retrieve large datasets efficiently, supporting advanced analytics and real-time decision-making.

9.2.1 Relational Databases

A RDBMS is a type of database management system (DBMS) that organises data into structured tables (also known as relations) and uses SQL to manage and query that data. RDBMSs are widely adopted in multiple applications, from simple desktop software to large-scale enterprise systems.

Basic Concepts

1. **Tables:** An RDBMS stores data in tables that are made up of rows and columns (Figure 9.10). Each table represents an entity, like "Students", and each row within a table is one unique record, like an individual student. Columns within a table are the attributes of the entity. For example, a "Students" table can have columns named 'StudentID', 'Name', 'DateOfBirth', and 'Major'. Each row in the table contains the values for those attributes. In Figure 9.10 below, each row in the "Students" table is for one particular student's ID, name, date of birth, and major.

StudentID	Name	DateOfBirth	Major
101289	Ritu	11.07.2006	Economics
101290	Rahul	20.12.2005	Mathematics
101291	Srinivas	18.03.2006	Physics

Figure 9.10 Example of a RDBMS table

2. **Keys:**
   - **Primary Key:** A primary key is a unique identifier for each record in a table. For example, 'StudentID' is typically a unique ID given to the students, and is used as the primary key for a "Students" table, as multiple students may have the same 'Name' or 'DateOfBirth'.
   - **Foreign Key:** A foreign key is a column or a group of columns in one table that links to the primary key of another table, establishing a relationship between the two tables. For instance, if you have an 'Enrolments' table to track which students are enrolled in which courses, a foreign key might link 'StudentID' in 'Enrolments' to 'StudentID' in the 'Students' table.

c. **Relationships:**
   - **One-to-One:** Each record in one table is linked to exactly one record in another table. For example, each student might have one unique Dining Hall card.
   - **One-to-Many:** One record in a table is linked to multiple records in another table. For example, one professor might teach many courses.
   - **Many-to-Many:** Multiple records in one table are related to multiple records in another table. For example, students can enrol in many courses and each course can have many students enrolled.

SQL (Structured Query Language)

SQL is the language used to interact with an RDBMS. Here are some of the basic SQL operations:

- **SELECT:** Used to query and retrieve data from the database. For example, "SELECT Name FROM Students WHERE Major = 'Economics';" retrieves the names of all students majoring in Economics.

- **INSERT:** Used to add new records to a table. For example, "INSERT INTO Students (StudentID, Name, DateOfBirth, Major) VALUES (101292, 'Venkat', '05.10.2005', 'Mathematics');" adds a new student to the "Students" table.
- **UPDATE:** Used to modify existing records. For example, "UPDATE Students SET Major = 'Physics' WHERE StudentID = 101292;" changes Venkat's major to Physics from Mathematics.
- **DELETE:** Used to remove records from a table. For example, "DELETE FROM Students WHERE StudentID = 101292;" deletes the student with the ID 101292 from the "Students" table.

Popular RDBMS Examples

- **MySQL:** An open-source RDBMS widely used for web applications.
- **PostgreSQL:** An open-source RDBMS known for its advanced features and extensibility.
- **Oracle Database:** A powerful, commercial RDBMS used by many large enterprises.
- **Microsoft SQL Server:** A widely used commercial RDBMS with deep integration into Microsoft's ecosystem.

ACID Properties of RDBMS

ACID is an acronym that stands for *Atomicity, Consistency, Isolation, and Durability*. These properties ensure that database transactions in RDBMS are processed in a way that maintains data integrity, even in the face of power failures, crashes, or other system issues.

1. **Atomicity:** This property guarantees that a transaction is processed as an indivisible unit, ensuring that either all operations within the transaction are executed successfully or none at all. If an error occurs during the transaction, the database is rolled back to its state before the transaction begins. This guarantees that partial updates (i.e., where only some of the operations in a transaction are executed) do not occur.
2. **Consistency:** Consistency ensures that a transaction will bring the database from one valid state to another, adhering to predefined rules or constraints, such as data types, referential integrity and business rules. It guarantees that after a transaction, the database will always be in a consistent state, regardless of whether the transaction was successful or failed.
3. **Isolation:** This property ensures that concurrently executed transactions do not affect each other in a way that could lead to data inconsistencies. Each transaction is processed as if it is the sole transaction running in the system and its operations are isolated from others. This is important for preventing issues like dirty reads, where one transaction reads data that is in the process of being modified by another transaction. The level of isolation can vary, from allowing transactions to see changes made by others to ensure complete isolation until the transaction is finished.
4. **Durability:** Durability guarantees that once a transaction is committed, its results are permanently written to the database, even in the event of system failures like crashes or power loss. The changes made by a committed transaction are stored in such a way that they will survive any hardware failure, ensuring the reliability of the system.

Together, these ACID properties ensure that relational databases provide strong transaction management and data integrity, making them well-suited for applications where consistency and reliability are critical, such as banking systems, e-commerce platforms and other financial applications. However, ACID's strict requirements can limit scalability and performance, which is why NoSQL systems (refer Section 9.2.2), with their BASE model, are often preferred for large-scale, distributed systems where speed and flexibility are more important than immediate consistency.

However, RDBMS face significant limitations in a big data environment due to their reliance on a rigid schema, which requires data to adhere to a predefined structure. This makes it inefficient to store and manage semi-structured or unstructured data, which is common in big data scenarios. Furthermore, the system checks for conformity with the structure while inserting or updating the data in RDBMS, resulting in delays. This latency becomes a bottleneck working with high-velocity data streams that require rapid data ingestion and processing. This lack of flexibility and the inherent latency issues make traditional RDBMS unsuitable as the primary storage solution in big data applications, where scalability, speed, and the ability to handle unstructured data are the most critical requirements in such applications.

9.2.2 NoSQL Databases

NoSQL databases do not rely on a fixed schema, which allows them to handle a variety of data types, including unstructured and semi-structured data. This flexibility is particularly useful for applications where data is rapidly changing or when dealing with big data. Based on their data storage mechanism, NoSQL databases can be broadly grouped into four types: key-value stores, document stores, column-family stores, or graph databases.

Key-value Stores

Key-value stores are the simplest form of NoSQL databases, where data is stored as a collection of key-value pairs (Figure 9.11). The key is a unique identifier for the value, which can be anything from a simple string to a complex object. Examples of key-value stores are Redis, Amazon DynamoDB, CouchBase and Riak.

Key	Value
101289	Ritu, 11.07.2006, Economics, Statistics
101290	Rahul, 20.12.2005, Mathematics
101291	Srinivas, 18.03.2006, Physics

Figure 9.11 An example of data stored as key-value pairs in key-value stores

Since key-value storage systems typically do not maintain indexes, write operations are very fast. However, as the value is stored as an aggregate, partial updates are not supported; any update must be done by either deleting the existing record and inserting a new one.

Document Stores

Document stores manage data in the form of documents, typically using formats like JSON or BSON. Each document is a self-contained unit of data that may contain nested structures, arrays and different types of data. Examples of document stores are MongoDB, CouchDB and Terrastore (Figure 9.12).

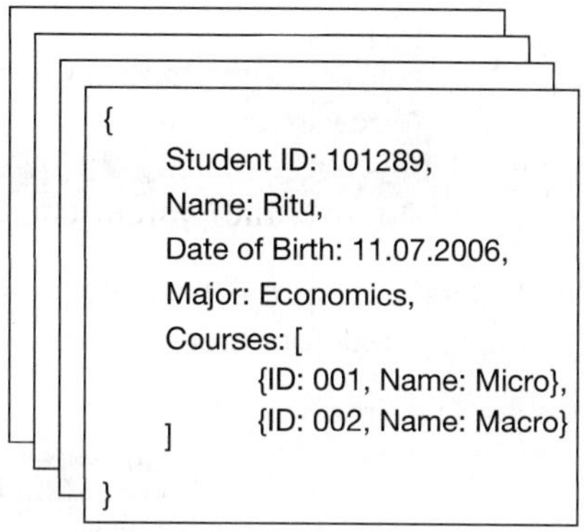

Figure 9.12 An example of document stores database

DID YOU KNOW?

BSON (Binary JSON) is a binary-encoded serialisation format that is similar to JSON (JavaScript Object Notation), but it is more efficient for storing and transmitting data. It is used primarily by MongoDB to store documents in a format that allows for efficient storage and retrieval of data. BSON supports a wider range of data types compared to JSON, such as Binary data (e.g., images, files), Dates (as opposed to representing dates as strings in JSON), Embedded documents and arrays, Floating-point numbers, ObjectId (unique identifiers).

BSON's format makes it easier for MongoDB to handle larger documents and objects by using a compact binary format, allowing for faster processing and reduced storage space compared to JSON, especially with complex or deeply nested structures.

Column-Family Stores

Column-family stores organise data into columns rather than rows. Each column family contains multiple columns that are related and each row in the database can have a different number of columns (Figure 9.13). Examples include Apache Cassandra, HBase and Amazon SimpleDB.

Graph Databases

Graph databases are designed to handle data with complex relationships by representing data as nodes (entities) and edges (relationships). In contrast to other types of NoSQL databases, which focus only on the structure of entities, graph databases also prioritise storing the connections between entities (Figure 9.14).

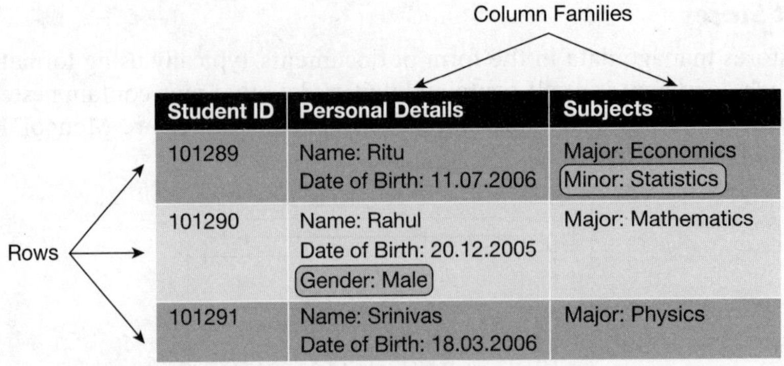

Figure 9.13 Column family database has the flexibility of having different columns (highlighted) for different rows

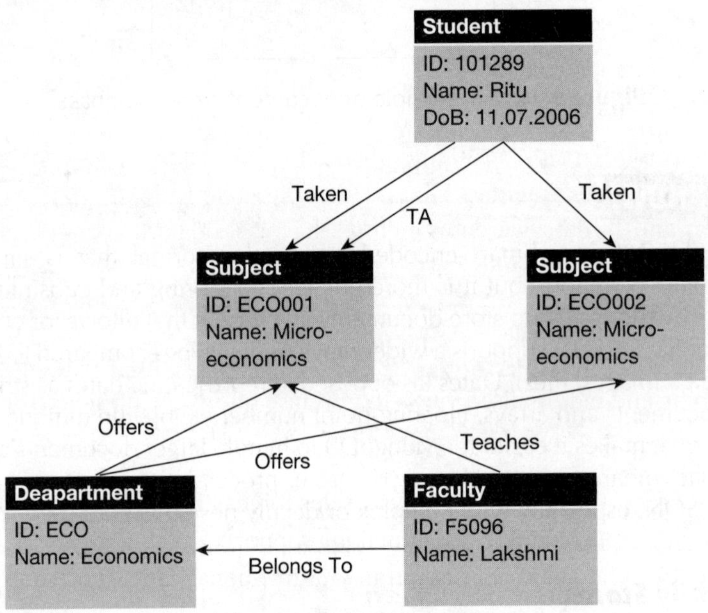

Figure 9.14 Graph Database stores the entities and the relationship between them

This model is particularly well-suited for applications like social networks, recommendation engines and fraud detection. Examples include Neo4j, InfiniteGraph and Amazon Neptune.

NoSQL storage systems are designed for high scalability, availability and fault tolerance, making them well-suited for handling large volumes of data across distributed environments. They are especially efficient in read-and-write operations, making them perfect for applications that require rapid processing of large volumes of data. However, unlike ACID-compliant relational databases, NoSQL systems do not offer the same level of guarantees for transactions and consistency. Instead, they follow the BASE (Basically Available, Soft state, Eventually consistent) model, which focuses on providing availability and eventual consistency rather than immediate consistency.

In practice, this means that NoSQL systems might be in an interim inconsistent state while they work toward eventual consistency. This 'soft' state makes NoSQL systems less suitable for scenarios that require strict transactional integrity, such as financial transactions or other use cases where real-time accuracy is critical. Therefore, while NoSQL excels in certain areas, it is not the best choice for large-scale transactional systems that rely on consistent, accurate data at all times.

9.2.3 Data Warehouse

A data warehouse is a centralised repository that stores large amounts of structured data from various sources. A data warehouse is optimised for read-heavy operations and is intended to support decision-making through historical analysis and reporting.

Key Characteristics

1. **Subject-Oriented:** Data warehouses are organised around key business domains or subjects (e.g., sales, finance, customer).
2. **Integrated Data:** Data from multiple sources (OLTP systems, external databases, spreadsheets) is integrated and stored in a consistent format.
3. **Time-Variant:** A data warehouse stores historical data, making it possible to track trends and changes over time.
4. **Non-Volatile:** Data in a warehouse is rarely updated or deleted. Once data is loaded, it is only queried and analysed, making it stable.
5. **Optimised for Query Performance:** Data warehouses use techniques like indexing, partitioning and materialised views to optimise query performance for large datasets.

Use Cases

1. Corporate reporting, financial analysis, sales forecasting and performance tracking.
2. Business intelligence and decision-making support.
3. Consolidating and analysing data from multiple departments and sources.

Example Systems

Amazon Redshift, Google BigQuery, Snowflake, Microsoft Azure Synapse Analytics, Teradata.

9.2.4 Data Marts

A data mart is a subset of a data warehouse, typically meant for a specific business function or department. It allows users to access relevant data without navigating the entire warehouse, providing a more streamlined and focused approach for departments like sales, marketing, or HR.

Key Characteristics

1. **Departmental Focus:** Each data mart is focused on a specific business area (e.g., sales, finance, customer service).
2. **Subset of Data Warehouse:** A data mart draws data from the main data warehouse, containing only the relevant data needed for specific business processes.
3. **Faster Access:** Since data marts are smaller and more focused, they often provide faster query performance than querying the entire data warehouse.
4. **Logical or Physical:** Data marts can be physically separate from the data warehouse or just a logical view on top of the main data warehouse.

Use Cases

1. Sales and marketing teams are analysing product performance and customer behaviour.
2. HR department is analysing employee performance and workforce analytics.
3. Financial teams tracking budgets, expenditures and revenue trends.

Example Systems

A data mart can be implemented as part of a data warehouse system like **Snowflake**, **Amazon Redshift**, or **Microsoft SQL Server**.

9.2.5 Data Lakes

A data lake is a large storage repository that holds vast amounts of raw data in its native format, including structured, semi-structured and unstructured data. Unlike data warehouses, which store cleaned and processed data in a structured format, data lakes are more flexible, allowing data to be stored before it is structured or processed.

Key Characteristics

1. **Schema-on-Read:** Data lakes allow for **schema-on-read** processing, where the structure of the data is applied only when it is read or analysed, making it more flexible than the predefined schema required by data warehouses.
2. **Handles All Data Types:** Data lakes store various types of data, including logs, images, videos, social media content, sensor data and more.
3. **Scalability:** Data lakes are built to scale massively, enabling organisations to store and manage petabytes or even exabytes of data.
4. **Cost-effective Storage:** They use low-cost storage for large datasets that may not need immediate processing but could be valuable for future analysis.
5. **Big Data and Machine Learning:** Data lakes are often used for data science and machine learning applications, where data is extracted, cleaned and analysed from a large pool of raw data.

Use Cases

1. Streamed data from IIoT sensors.
2. Unstructured social media or web data for future analysis, for example, any mention of the product in social media or review websites.
3. Audio files from customer centre service calls.

Example Systems

Amazon S3 (Simple Storage Service), Azure Data Lake Storage, Google Cloud Storage, HDFS.

9.3 BIG DATA MANAGEMENT TECHNIQUES

9.3.1 Online Transaction Processing (OLTP)

Online Transaction Processing (OLTP) systems are designed to manage and execute a large number of short, fast and real-time transactions. OLTP systems typically support day-to-day business activities, such as order processing, payments, reservations and inventory management, where the data must be updated frequently and responses need to be fast.

Key Characteristics

1. **High Transaction Volume:** OLTP systems handle a high number of transactions every second, each involving inserts, updates, or deletes.
2. **Real-Time Data Processing:** Transactions are processed instantly and are typically completed within milliseconds or seconds.
3. **Data Consistency:** OLTP systems ensure that all database transactions maintain the ACID properties (Atomicity, Consistency, Isolation, Durability) to guarantee the integrity of the data.
4. **Normalised Databases:** OLTP systems typically use normalised databases to avoid redundancy and maintain data integrity across tables.
5. **Concurrency:** OLTP systems need to manage simultaneous transactions without conflicts, ensuring the system can handle many users accessing the same data concurrently.

Use Cases

1. Banking and financial transactions (e.g., transferring money between accounts).
2. E-commerce platforms for order processing.
3. Retail stores for inventory tracking and point-of-sale transactions.
4. Customer relationship management (CRM) systems.

Example Systems

OLTP systems are often based on traditional relational databases like MySQL, PostgreSQL, Oracle, or SQL Server.

9.3.2 Online Analytical Processing (OLAP)

Online Analytical Processing (OLAP) systems are designed to support complex queries and analyses, often for business intelligence (BI) purposes. OLAP systems focus on querying and reporting data, performing complex calculations, aggregations and multidimensional analysis, rather than handling high volumes of transactions like OLTP.

Key Characteristics

1. **Multidimensional Data:** OLAP uses a multidimensional model, called OLAP Cube, to view data across multiple dimensions. This model allows users to view data from multiple perspectives simultaneously, for example, analysing sales by product over time across different regions. The user can drill down in any of these dimensions, namely product, time, and region, to view the sales numbers.
2. **Read-Intensive Workloads:** Unlike OLTP, OLAP systems focus on data retrieval rather than frequent updates or inserts, supporting complex queries and aggregations.
3. **Denormalised Data:** OLAP systems often use denormalised databases, with data organised in star or snowflake schemas, to optimise query performance.
4. **Historical Data Analysis:** OLAP systems are typically used for querying large datasets to analyse historical trends and patterns, supporting strategic decisions.
5. **Pre-aggregated Data:** OLAP systems may store pre-aggregated or summary data to speed up query responses for reports and dashboards.

Use Cases

1. Business reporting (e.g., sales reports, financial forecasts).
2. Data mining and market research.
3. Trend analysis and KPI (Key Performance Indicator) tracking.
4. Multidimensional analysis for budgeting and forecasting.

Example Systems

Microsoft SQL Server Analysis Services (SSAS), Oracle OLAP, IBM Cognos.

9.3.3 Extract Transform Load (ETL)

Extract, Transform, Load (ETL) is a process used to move data from multiple sources into a data warehouse or another centralised system. ETL is essential for integrating data from different formats and ensuring it is consistent, clean and ready for analysis.

ETL Process Stages

1. **Extract:** Data is extracted from various sources such as databases, files, APIs, or even web services. The sources can be in different formats (e.g., relational databases, XML, JSON, flat files).
   - **Challenges:** Handling different data formats, structures and sometimes incomplete or inconsistent data.

2. **Transform:** In this phase, the extracted data is cleaned, filtered and transformed into the correct format for analysis. Common transformations include:
   - Data cleansing (handling missing or erroneous data).
   - Filtering out irrelevant data.
   - Applying business rules.
   - Converting data types and aggregating data.
3. **Load:** The transformed data is loaded into the target system, typically a data warehouse. The load can be done in two ways:
   - **Full Load:** Loading all data into the system at once.
   - **Incremental Load:** Loading only the new or updated data to keep the warehouse updated without reloading everything.

Use Cases

1. Consolidating data from multiple operational systems (e.g., sales, marketing, HR) for unified reporting.
2. Data migration projects.
3. Data warehousing for business intelligence purposes.

Example Tools

Apache Nifi, Talend, Informatica, Microsoft SQL Server Integration Services (SSIS), Apache Airflow.

9.3.4 Extract Load Transform (ELT)

Extract, load, transform (ELT) is a modern approach to data integration that differs from ETL by loading raw data into a target system first and then performing transformations within that system. Refer to Table 9.2 for a comparison between ETL and ELT. ELT is particularly beneficial for big data architectures and cloud-based data warehouses.

Table 9.2 A quick comparison between ETL and ELT

Feature	ETL (Extract, Transform, Load)	ELT (Extract, Load, Transform)
Processing Order	Transform → Load	Load → Transform
Transformation Location	External ETL tools before storage	Inside the data warehouse or data lake
Data Handling	Structured, pre-processed	Raw, semi-structured and unstructured
Storage Type	Data warehouses	Data lakes, cloud warehouses
Performance	Slower due to pre-processing	Faster ingestion; transformation scales with cloud computing
Flexibility	Requires a predefined schema	Supports schema-on-read (flexible)
Best For	BI, compliance-heavy applications	Big data, machine learning, real-time analytics

ELT Process Stages

1. **Extract:** Similar to ETL, data is extracted from different sources, such as databases, files, APIs and streaming services.
2. **Load:** Instead of transforming data beforehand, ELT loads raw data directly into a centralised data repository, such as a data lake or cloud-based data warehouse (e.g., Snowflake, BigQuery, or Redshift).
   - **Advantages:** Faster ingestion since no transformation is needed before loading.
3. **Transform:** Data transformation occurs inside the target system using its processing power. This allows for:
   - On-demand transformations suited for different use cases.
   - Greater flexibility for analysts to perform transformations as needed.
   - Leverage of cloud-based computing power for scalability.

Use Cases

1. Cloud-based data warehousing, where processing power can be scaled dynamically.
2. Big data environments where structured and unstructured data are stored together.
3. Scenarios requiring flexible, schema-on-read approaches where transformations happen at the time of analysis.

Example Tools

- **Google BigQuery:** A serverless cloud data warehouse that allows direct loading of raw data and SQL-based transformations.
- **Amazon Redshift:** A cloud-based data warehouse that supports ELT by leveraging its internal processing power for transformations.
- **Snowflake:** A cloud data platform that allows data loading first and transformations on-demand using SQL.
- **Azure Synapse Analytics:** A Microsoft solution that integrates data storage and transformation within the cloud.
- **Databricks:** A platform that combines data lake and data warehouse capabilities, using Apache Spark for transformations.

9.4 BIG DATA PROCESSING

When handling Big Data, one needs to process large volumes of Data. Partitioning large datasets into smaller ones can speed up processing if one can execute them simultaneously. Think about an addition operation that adds a million numbers. If it is done sequentially, it will take a million iterations. However, if you can break it down into 100 chunks of 10,000 numbers and process them simultaneously, you will need 10,000 iterations to do those sums and then

100 iterations to add up the intermediate subtotals. So, the processing time can come down by almost 100 times in that case.

Two important approaches that enable big data processing at scale are *Parallel Data Processing* and *Distributed Data Processing*. These techniques facilitate the processing of large data by dividing it into small, tractable units and processing them concurrently.

9.4.1 Parallel Data Processing

Parallel data processing involves executing multiple tasks or operations simaltaneously, rather than sequentially. This approach is possible because most data tasks are independent in nature and can be executed in isolation. In this case, a single machine with multiple cores or processors can be used to handle various tasks simultaneously in those processors. This method is key to speeding up data analysis for big data by efficient use of computational resources.

Key Concepts in Parallel Data Processing

1. **Data Partitioning:** One of the core ideas of parallel processing involves dividing a dataset into smaller partitions and processing them independently with each processor or core working on a specific partition in parallel with others. This process will ensure the tasks are completed much faster than executing with one processor.
2. **Threading and Multicore Processors:** Modern central processing units (CPU) come with multiple cores, enabling systems to process multiple threads of execution simultaneously. Threading enables a single program to perform more than one task concurrently. Threading and multicore processors can significantly reduce the overall time to complete big data processing.
3. **MapReduce Framework:** One of the most common parallel data processing programming models is MapReduce (discussed in detail in Section 9.4.4). This framework is widely used in tasks like web page indexing for search engines, processing massive social network datasets, and analysis of log files.
4. **Load Balancing:** In a parallel processing environment, each processor must have an (mostly) equal amount of work to execute. Otherwise, some processors may complete their tasks early and lie idle, while other processors lag, resulting in increased overall processing time. The process of distributing this computing load to ensure each processor or core has an equal amount of work is called load balancing.

Benefits of Parallel Data Processing

- **Speed:** Parallel processing dramatically reduces the time needed for data operations by processing multiple data subsets concurrently.
- **Resource Efficiency:** It maximises the use of all available processing power within a machine.
- **Scalability:** Parallel processing can be scaled by adding more processors as datasets continue to grow.

However, parallel processing has limitations when working with datasets that are too large to fit within the memory or storage constraints of a single machine. This is where distributed data processing comes in.

9.4.2 Distributed Data Processing

While parallel processing involves processing it in a single machine, distributed data processing is an extension of the idea to the use of numerous machines, or *nodes*, in a network, for the processing of multiple chunks of a dataset. It's used when dealing with extremely large datasets beyond the capabilities of a single machine, in terms of either memory or processing power.

Key Concepts in Distributed Data Processing

1. **Cluster Computing:** Distributed processing of the data is generally performed by a collection of computers (or servers), also called a cluster, acting as a single system. Each machine in the cluster is given a portion of the data for processing. These computers communicate with each other to exchange intermediate results and synchronise their operations.
2. **Fault Tolerance:** Distributed systems are designed with fault tolerance mechanisms that allow other machines to take over the tasks of a failed node, so that the failure of one node does not lead to the failure of the entire system. This ensures the process continues without data loss or interruptions, improving system reliability.
3. **Distributed File Systems:** Distributed data processing relies on distributed file systems like *HDFS* or *Amazon S3* to handle large datasets, These systems store large amounts of data across multiple machines while maintaining redundancy, such that the data remains accessible even if some nodes fail.
4. **Apache Hadoop and Apache Spark:** Two of the most commonly used frameworks for distributed data processing are Apache Hadoop and Apache Spark:
   - **Hadoop:** Uses the MapReduce programming model to divide data into blocks, process them across different nodes and then combine the results.
   - **Spark:** An advanced distributed data processing engine that offers faster performance than Hadoop by using in-memory processing and supports a wide range of operations like streaming, machine learning and graph processing.
5. **Coordination and Synchronisation:** In distributed systems, coordination between nodes is crucial to avoid conflicts and ensure that the nodes work in harmony. Synchronisation techniques, such as distributed locking and consensus protocols (e.g., Zookeeper), ensure that the system runs smoothly even when nodes need to share data or resources.

Benefits of Distributed Data Processing

1. **Scalability:** Distributed systems can easily scale by adding more machines to handle larger datasets or higher processing demands.

2. **Fault Tolerance:** Distributed systems are designed to handle failures gracefully, ensuring high availability and reliability.
3. **Geographical Distribution:** Distributed systems allow data to be processed across different geographical locations, reducing latency for users in different regions.

Challenges of Distributed Data Processing

1. **Network Latency:** Communication between machines can slow down processing due to network delays.
2. **Complexity:** Managing a distributed system is more complex, with additional challenges in coordination, synchronisation and fault tolerance.
3. **Data Consistency:** Maintaining consistent data across multiple nodes can be difficult, especially in real-time applications.

9.4.3 Processing Workloads

Batch Processing

Batch processing involves processing large volumes of data that have been collected over a period of time. This technique is well-suited for jobs that do not require real-time results but need to process large datasets efficiently. In batch processing, data is collected, stored and processed in large chunks or 'batches'.

- **Key Characteristics:** Non-interactive, high-throughput, handles large volumes of historical data.
- **Use Case:** Data warehousing, processing logs and financial transaction processing.
- **Popular Framework:** Apache Hadoop is widely used for batch processing in big data environments.

Real-Time (Stream) Processing

In contrast to batch processing, real-time data processing (also known as stream processing) involves continuously processing data as it is generated or received. This technique is critical when timely data processing and near-real-time results are required, such as in fraud detection, sensor data analysis, or online recommendation systems.

- **Key Characteristics:** Low latency, continuous data flow, real-time insights.
- **Use Case:** Monitoring social media activity, processing IoT sensor data and financial market analysis.
- **Popular Frameworks:** Apache Kafka for managing real-time data streams.

9.4.4 Batch Processing Using MapReduce

Batch processing refers to the execution of a series of programs or jobs on a computer without manual intervention. This processing method is crucial for handling large-scale data operations, where immediate results are not necessary but efficient, bulk data handling is required. Batch processing is ideal for tasks such as transaction processing, data analysis and report generation. It operates by executing batches of jobs together, minimising idle time and making efficient use of computational resources.

MapReduce is a programming model developed by Google to support distributed computing on large data sets across clusters of computers. It simplifies the process of working with massive amounts of data by breaking down complex tasks into smaller sub-tasks that can be processed in parallel. The model divides the work into two main functions: the *Map* function and the *Reduce* function.

The Map Function

This function processes the input data and produces intermediate key-value pairs. Each node (computer) in the cluster receives a portion of the input data, performs a specific task on it and outputs the result in the form of key-value pairs. This phase can be thought of as the part where data is organised and prepared for combination.

Suppose we need to count the number of times each word appears in a large collection of books. In the Map phase, each computer processes a subset of books, creating a list of key-value pairs where the key is a word and the value is its count:

- **Computer 1:** {('data', 5), ('science', 3), ('learning', 4)}
- **Computer 2:** {('science', 7), ('machine', 2), ('data', 6)}
- **Computer 3:** {('learning', 5), ('data', 4), ('model', 3)}

The Reduce Function

Once the Map phase is complete, the Reduce function takes over. The Reduce function's role is to aggregate, filter, or summarise data. It receives the intermediate key-value pairs from all nodes and combines them according to the keys.

Using the word count example, the intermediate outputs from the Map phase are collected and grouped by key:

- **Input to Reduce:**
  - Key 'data': [5, 6, 4]
  - Key 'science': [3, 7]
  - Key 'learning': [4, 5]
  - Key 'machine': [2]
  - Key 'model': [3]

  The Reduce function aggregates these values:
- **Output:**
  - ('data', 15)
  - ('science', 10)
  - ('learning', 9)
  - ('machine', 2)
  - ('model', 3)

The power of MapReduce lies in its ability to distribute processing across many machines and efficiently handle the aggregation of results. This framework abstracts the complexity of parallel and distributed computing, allowing developers to focus on the logic for data processing without worrying about the intricacies of task scheduling, data distribution, or fault tolerance.

Step-by-Step Workflow of MapReduce

1. **Data Input:** The input data is usually stored in a distributed file system (e.g., HDFS in Hadoop).
2. **Splitting:** The input data is divided into splits, each containing a subset of the total data.
3. **Map Phase:** Each split is processed by a Mapper, which produces key-value pairs.
4. **Shuffling and Sorting:** The framework automatically groups and sorts the intermediate data by key.
5. **Reduce Phase:** Reducers process each group of key-value pairs to produce the final output.
6. **Output:** The results are written back to the distributed file system.

Figure 9.15 visually illustrates this step by step process of MapReduce algorithm.

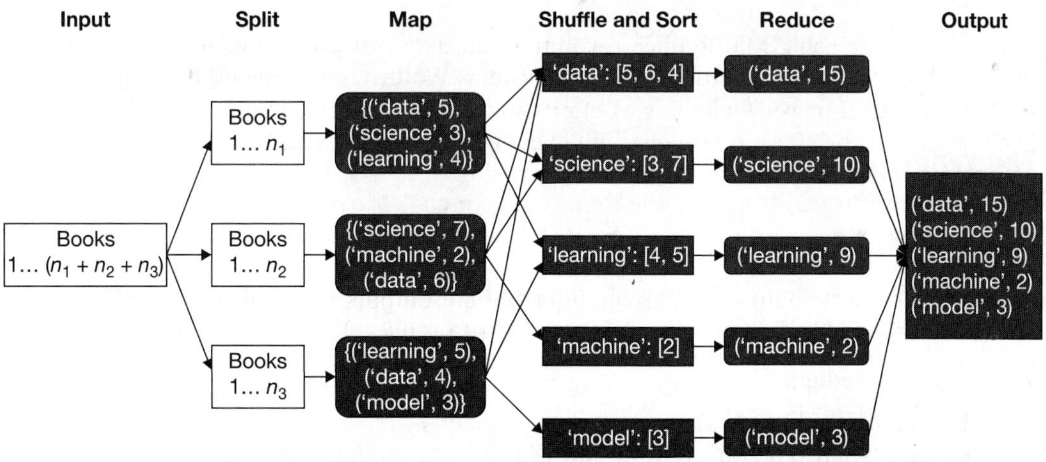

Figure 9.15 Illustration for MapReduce algorithm

MapReduce has been widely adopted in various fields that require the processing of vast data sets:

- **Web Indexing:** Search engines use MapReduce to crawl, index and rank billions of web pages.
- **Data Warehousing:** Companies use MapReduce to analyse large logs and records for insights into consumer behaviour.
- **Machine Learning:** The model is used for training algorithms on large-scale datasets, such as in recommendation systems.
- **Bioinformatics:** Researchers employ MapReduce for genome sequencing and analysis, where massive data sets need to be processed efficiently.

9.4.5 Advantages of Using MapReduce

1. **Scalability:** Easily handles petabytes of data across hundreds or thousands of machines.
2. **Fault Tolerance:** Automatically handles node failures by reassigning tasks.
3. **Simplicity:** Developers write simple Map and Reduce functions without managing underlying parallelism.
4. **Flexibility:** Can be adapted for a variety of data processing tasks, from simple counting to complex machine learning algorithms.

Challenges and Limitations

While MapReduce has significant benefits, it is not without its challenges:

- **Latency:** Batch processing can take considerable time for very large datasets.
- **Iterative Processing:** MapReduce is less efficient for algorithms that require multiple iterations, such as many machine learning tasks.
- **Complexity for Developers:** While simpler than managing distributed computing directly, writing effective MapReduce jobs requires understanding data partitioning and balancing.

Batch processing using MapReduce revolutionised data processing by enabling large-scale data handling and analysis across distributed systems. While it is no longer the only model used for big data (with newer technologies offering more in-memory computation), MapReduce laid the groundwork for modern distributed computing practices and remains a foundational model for understanding large-scale data processing.

9.4.6 Hadoop

Hadoop is an open-source framework designed to enable the distributed processing and storage of large datasets across clusters of computers, using simple programming models. It is developed under Apache Software Foundation and is built to scale from a single computer to thousands with each machine offering local computation and storage. This scalability makes it ideal for handling massive amounts of data efficiently.

Computationally, Hadoop follows the *MapReduce* model that segments the task into smaller sub-tasks to process in parallel across the cluster. The framework allows for the processing of large datasets by distributing the computing across multiple nodes as opposed to a single machine.

The *Hadoop Distributed File System (HDFS)* serves as Hadoop's storage layer. HDFS stores large amount of data by breaking it into smaller blocks and then storing the blocks across multiple nodes in the cluster. HDFS is also fault-tolerant, it maintains the redundancy of these blocks by keeping multiple copies to prevent data loss on hardware failure.

Another crucial component of the Hadoop ecosystem is the cluster resource management layer - *Yet Another Resource Negotiator (YARN)*. YARN allocates resources, such as CPU, memory and storage, to the distributed applications running on the Hadoop cluster. It ensures that resources are efficiently utilised and that tasks are assigned to appropriate nodes based on available capacity. It assigns tasks to appropriate nodes based on capacity, ensuring efficient utilisation of resources.

Finally, *Hadoop Common* is a set of common utilities and libraries that support the other components of the Hadoop ecosystem. These utilities provide core functionalities like file system access and configuration management, which are shared across Hadoop modules. Collectively, these components make Hadoop a powerful and flexible platform for handling big data.

How Hadoop Works

1. **Data Storage:** Data is split into smaller blocks (typically 128 MB or 256 MB) and distributed across multiple nodes in the cluster using HDFS. Each block is replicated to multiple nodes to ensure fault tolerance.
2. **Data Processing:** The processing of data is carried out using the MapReduce programming model:
    - **Map Phase:** The input data is processed in parallel by mappers, which generate intermediate key-value pairs.
    - **Shuffle and Sort:** The intermediate data is shuffled and sorted by key.
    - **Reduce Phase:** Reducers process the intermediate key-value pairs to produce the final result.
3. **Resource Management:** YARN oversees the allocation of cluster resources and schedules tasks to ensure balanced usage across nodes.

Advantages of Hadoop

- **Scalability:** Easily scales to handle petabytes of data by adding more nodes to the cluster.
- **Cost-effective:** Utilises commodity hardware, making it cheaper than traditional high-end solutions.
- **Fault Tolerance:** Data replication ensures that operations can continue even if nodes fail.
- **Flexibility:** Can store and process structured, semi-structured and unstructured data.

Limitations of Hadoop

- **Complexity:** Managing and programming in Hadoop requires knowledge of distributed systems and Java-based programming.
- **Latency:** MapReduce is disk-intensive and may have higher latency compared to in-memory processing frameworks like Apache Spark.
- **Iterative Processing:** Not ideal for tasks that require iterative computation, like machine learning algorithms, due to the multiple read/write operations to disk.

While Hadoop remains popular for batch processing and large-scale data storage, newer technologies like **Apache Spark** have become more prominent for use cases requiring faster, in-memory data processing. Hadoop can integrate with Spark and other tools in the big data ecosystem, like **Apache Hive** (SQL-like query engine), **Apache Pig** (data flow language) and **Apache HBase** (NoSQL database).

Hadoop laid the foundation for distributed data processing and big data analysis. Its core strengths, scalability, cost-effectiveness and fault tolerance, continue to be leveraged in various industries. While new technologies have emerged, Hadoop's modular architecture allows for integration and co-existence with other big data tools, making it a lasting component of the big data ecosystem.

9.5 BIG DATA IMPLEMENTATION APPROACH

The life cycle of a large data project is marked by a systematic process that enables effective data collection, storage, processing, and dissemination. While the ultimate goal of many projects is to produce insights or support decision-making, the success of any data project is inherently dependent on an understanding of business needs, competent data handling, and the development of a solid infrastructure with optimised processes. These essential stages (Figure 9.16) are discussed in depth in this section.

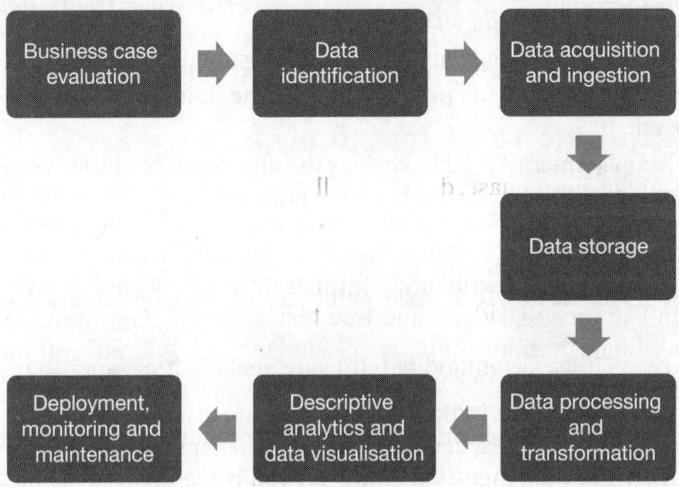

Figure 9.16 Big Data Implementation Approach

9.5.1 Business Case Evaluation

Each major data project starts with a clear statement about its intended purposes. Organisations need to start by determining what problems they seek to solve and how data can help alleviate them. This step requires collaboration with stakeholders, executives, engineers and users to understand what they expect and to define tangible measures by which they can measure success. A well-defined problem statement is critical, especially because big data-related solutions are likely to be inherently complex and require significant investment.

Project feasibility evaluation forms a core element of the planning phase. It is critical to determine whether data are available at this early point – does the organisation have access to the data required, or will they need to be sourced from outside? Furthermore, technical considerations, such as data storage capacity, computational power, and team expertise, require critical assessment. Regulatory requirements, especially related to sensitive data, i.e., personal, financial, or health data, are also a consideration for project teams.

Compliance with regulations, like General Data Protection Regulation (GDPR) and Health Insurance Portability and Accountability Act (HIPAA), is required to reduce both legal and ethical risk.

9.5.2 Data Identification

After defining the scope of the project, the next challenge is obtaining necessary data. Accessing a variety of data sources can greatly improve both depth and accuracy in the analysis, ultimately creating a more extensive understanding of phenomena being assessed. This works best for exploratory analysis where particulars about areas of inquiry are yet to be clearly outlined. A wider range of related data sets can provide useful information. These data sets can be from either internal or external sources pertaining to subjects under inquiry. Internal data sets are usually more structured in nature and are in line with predetermined specifications. On the other hand, external data sets can evolve from a variety of sources, including government data pools, third-party vendors, online documents, social media platforms, or even data from IoT sensor initiatives, forming a variety of structural forms.

9.5.3 Data Acquisition and Ingestion

During the Data Acquisition phase, data are collected from all data sources identified during the project's previous stage. Based upon project purposes, organisations might need data retrieval against heterogeneous systems. This includes data available in structured formats in a relational database, semi-structured formats like JSON and XML, and completely unstructured data like images, videos, and free text. External data sources can include data collected from third-party vendors or data collected through API integration, where data can be received in batch or real-time formats.

Following data collection, it is imperative to store data in a suitable environment within the organisation. The data ingestion scheme largely depends upon two big data characteristics – volume and velocity. The data ingestion strategies can be categorised as batch processing or streaming. In *batch ingestion*, data is collected for a time interval and then ingested in bulk. This method is suitable when instantaneous updates are not essential, such as performing historical analysis. *Streaming data ingestion*, conversely, is appropriate for applications needing real-time processing, e.g., tracking financial transactions for fraud detection purposes or tracking sentiments in the social network in real-time. Technologies like Apache Kafka, Apache Flink, and Spark Streaming are commonly used to handle continuous flow of data. These technologies enable data to be ingested and processed as it arrives, allowing prompt decision-making in dynamic environments.

9.5.4 Data Storage

Following data intake, it is essential to maintain information within a system marked by optimal performance, cost-effectiveness, and access. One proposed approach consists of the initial preservation of a precise replica of unprocessed data before proceeding further with any processing. Many organisations use data lakes to enable this purpose. A data lake can be described as a central repository holding unprocessed data in its original form, thus ensuring functional flexibility in handling disparate data types, including structured, semi-structured, and unstructured data. Renowned technologies like HDFS, Amazon S3, and Azure Data Lake Storage provide elastic storage capabilities for large volumes of data.

In rare cases where data fits a strictly formatted schema, use of a data warehouse would be sufficient. Google BigQuery, Amazon Redshift, and Snowflake offer organisations the ability to hold this form of structured data.

Additionally, metadata is added to data stored within a system during this stage. Metadata includes critical information, such as where data came from, when it was created or collected, how large it was, and how it was structured. Metadata use makes it easier for users to access and understand available data sets, improving data governance and discovery.

9.5.5 Data Processing and Transformation

Data collected and stored in the previous stage are rarely applicable at once. The next stage in data life cycle addresses data manipulation and transformation to improve its usability. In big data processing, most organisations use an ELT approach, contrasting with the ETL approach commonly applied to processing transactional data.

Data transformation consists of a collection of critical processes. Unlike traditional data, which typically has a predetermined structure and was previously validated, data brought together into a Big Data system often lacks structured organisation and can be unvalidated. Therefore, data cleansing and normalisation are critical processes with a view to resolving problems like incomplete values, duplications, or improper formats. Since Big Data systems pull data from multiple sources, having redundancy in data collected is a common occurrence. Employing this redundancy among related data can aid in validating data while helping to correct incomplete values, thus improving the data set's quality and thoroughness.

Data frequently requires normalisation or aggregation to match organisational requirements. In addition, data from different sources require schema reconciliation, a form of resolving structural and terminological discrepancies. For example, a name may be a single attribute in one database, yet in another it might be split into two different attributes – firstname and lastname.

Within extensive data processing, multi-node data can be managed by distributed computing platforms like Apache Spark and Hadoop MapReduce, effectively streamlining data processing. Stream processing technologies like Flink and Kafka Streams are implemented to specifically optimise real-time transformation of data, ensuring incoming data get structured and verified before use in future stages. Upon completion of transformation, the transformed data gets stored in a secondary data warehouse or data lake for future use.

9.5.6 Descriptive Analytics and Data Visualisation

After completing the procedure for loading the last data set, this phase is done to evaluate its accuracy and to spot any major discrepancies. At this phase, essential descriptive statistics are carried out, and appropriate graphical plots are created to allow for thorough validation.

9.5.7 Deployment, Monitoring and Maintenance

Enterprise data systems require careful planning to ensure scalability while at the same time ensuring superior performance. Cloud computing options, such as AWS Glue, Google

Dataflow and Azure Data Factory provide managed data pipeline processing services, relieving overheads associated with infrastructure management. On-premises deployments, on the other hand, mainly rely on Kubernetes or Hadoop clusters for efficient distribution of workloads. Once you start production, it is crucial to monitor continuously to ensure smooth operation. The company should carefully watch key metrics, such as data loading speeds, pipeline processing times, and storage capacity usage, to detect inefficiencies. Additionally, it is advisable to configure real-time notifications to alert engineers to malfunctions or anomalies.

9.6 USE OF BIG DATA ACROSS INDUSTRIES

In the previous subsections, we have covered what is Big Data and its storage and processing mechanisms. In the following subsections, we focus on how this technology is leveraged in the industry.

9.6.1 Manufacturing

In manufacturing, the Industrial Internet of Things (IIoT) enables the generation of sensor data such as temperature, pressure and vibration readings from machines. Production systems log detailed metrics on batch performance, defects and assembly line activity. Supply chain data, including shipment tracking and inventory levels, is collected and stored along-side maintenance logs that predict equipment failures. Design processes generate large files from Computer Aided Design (CAD) software, including 3D models and simulations. Manufacturers store this data using industrial data lakes that integrate operational and business data, edge storage for real-time monitoring and cloud services for scalability.

9.6.2 Transportation and Logistics

In transportation and logistics, big data comes from GPS systems that track vehicle locations, traffic conditions and optimised routes. Vehicle telematics data includes fuel consumption, engine performance and maintenance status. Shipping companies collect data on package tracking, delivery times and weather conditions that might affect schedules. Customer interactions, such as delivery preferences and complaints, also contribute to the data pool. Autonomous vehicles generate sensor and image recognition data to support navigation and decision-making processes. Storage solutions include high-throughput pipelines for fleet monitoring, federated cloud systems for global logistics and secure repositories for compliance with transportation regulations.

9.6.3 Energy and Utilities

The energy and utilities sector generates substantial data through smart meters, which track electricity and gas usage patterns in real-time. Sensor networks monitor grid conditions, including voltage levels and load distribution, while predictive maintenance systems analyse data from turbines and pipelines to identify potential issues. Renewable energy sources rely on weather data to optimise operations at solar farms and wind turbines. Geospatial mapping data supports infrastructure planning, such as pipeline layouts and power distribution networks. This data is stored in Supervisory Control and Data Acquisition (SCADA) systems for real-time management, time-series databases for historical analysis and cloud-native platforms to handle scalability.

9.6.4 Retail

In retail including e-commerce, customer transaction data is generated through purchases, payment methods and abandoned carts. Online retailers collect browsing data, customer demographics and activity logs to build detailed profiles. Social media interactions, including reviews and recommendations, contribute to unstructured data. IoT devices like smart shelves in physical stores generate data on inventory levels and in-store traffic patterns. Retailers store this data in customer data platforms for personalisation, data lakes for combining structured and unstructured information and scalable databases that support fast retrieval and analysis.

9.6.5 Healthcare

In healthcare, data generation comes from multiple sources, including electronic health records (EHRs), which capture patient history, diagnoses and treatments. Medical imaging technologies like X-rays, MRIs and CT scans produce vast amounts of high-resolution image data. Wearable devices and IoT sensors continuously generate real-time health metrics such as heart rate and glucose levels. Genomic data from sequencing efforts contributes another layer of complexity, enabling personalised medicine. Pharmaceutical companies also produce detailed datasets from clinical trials and drug development processes. This data is stored in cloud-based repositories, genomic data platforms, or on-premise data centres to meet compliance and privacy regulations.

9.6.6 Finance

The finance and banking industry generates vast amounts of transaction data from ATMs, credit cards and digital payment systems. Additionally, market data from stock exchanges and trading volumes is collected continuously. Financial institutions maintain customer data, including credit histories, spending patterns and account information. Fraud detection systems generate logs of suspicious activity, while unstructured data like financial news and social media sentiment are also analysed for market insights. Storage solutions in finance include data lakes for real-time analytics, distributed ledgers like blockchain for secure transactions and regulatory-compliant storage systems for sensitive customer data.

9.6.7 Entertainment and Media

The entertainment and media industry generates data from streaming services, where viewing times, preferences and user interactions are tracked continuously. Social media trends, such as hashtags and audience engagement metrics, produce unstructured text data. Advertisers collect data on impressions, clicks and conversions to optimise campaigns. Platforms like YouTube and Spotify store user-generated playlists, subscriptions and recommendations, while live streaming services handle latency metrics and real-time user activity data. Content delivery networks (CDNs) store and distribute this data for seamless user experiences, while distributed databases support personalisation at scale.

9.6.8 Public Sector

In the public sector, data comes from census surveys, which capture demographic, housing and income information, as well as from smart city sensors monitoring traffic, air quality and waste management. Public safety agencies generate emergency call logs, surveillance footage and crime statistics, while government departments maintain policy data, legislative histories and project records. Disaster management relies on weather alerts, flood maps and evacuation plans. Governments store this data in federated data centres, hybrid cloud systems for disaster recovery and open data platforms accessible to the public.

9.7 CONCLUSION

Big Data has transformed the way organisations store, process and analyse massive amounts of information, allowing them to extract valuable insights and make informed decisions. This chapter explored key aspects of Big Data, including its characteristics, storage solutions, processing techniques, project lifecycle and its impact across various industries.

The evolution of data management has seen a shift from traditional relational databases to more scalable NoSQL databases, data lakes and cloud-based architectures. The implementation of parallel and distributed processing frameworks, like Hadoop and Spark, has made it possible to process data efficiently at scale. Additionally, the growing adoption of ELT alongside ETL methods reflects the increasing reliance on cloud computing for flexible and powerful data transformations.

Understanding Big Data's lifecycle, from data acquisition to storage, processing and analytics, ensures businesses can effectively manage and extract value from their data. As industries continue to embrace Industry 4.0, integrating Big Data with advanced technologies such as artificial intelligence, machine learning and the Internet of Things (IoT) will further drive innovation.

The growth potential of Big Data is substantial. According to forecasts from Statista.com, the global total of created, captured, copied and consumed data will reach 180 zettabytes in 2025. This exponential growth emphasises the importance of data-related services and infrastructure in the ever-evolving digital age.

Ultimately, the ability to harness Big Data effectively is becoming a critical competitive advantage in today's digital economy. Future developments in data governance, security and real-time analytics will continue to shape the landscape, making it essential for professionals to stay informed and adapt to new advancements in the field.

SUMMARY

- Big Data is a field dedicated to the storage and processing of large amounts of data
- Types of data can be
  - Structured – stored in table format
  - Unstructured – text, image, audio, video
  - Semi-structured – XML, JSON
- Big Data characteristics (5Vs)
  - **Volume:** The vast amount of data generated from different sources.
  - **Velocity:** The rate at which data is generated and processed.
  - **Variety:** The various data types – structured, semi-structured and unstructured.
  - **Veracity:** The trustworthiness of data, ensuring quality and consistency for analysis.
  - **Value**: Extracting meaningful insights for decision-making.
- Big Data Storage
  - **RDBMS Database:** Organises data into structured tables and uses SQL for management.
  - **NoSQL Databases:** No fixed schema, ideal for handling unstructured/semi-structured data.
    - o **Key-value Stores:** Simple key-value pairs (e.g., Redis, DynamoDB).
    - o **Document Stores:** JSON/BSON-based (e.g., MongoDB, CouchDB).
    - o **Column-Family Stores:** Organises data in flexible columns (e.g., Apache Cassandra, HBase).
    - o **Graph Databases:** Stores data as nodes and edges, suited for relationship-based applications (e.g., Neo4j, Amazon Neptune).
  - **Data Warehouses:** Centralised storage optimised for query and analysis.
  - **Data Marts:** A smaller subset of a data warehouse, tailored for specific departments within an enterprise.
  - **Data Lakes:** Stores vast amounts of raw data in its native format.
- **OLTP** focuses on real-time transactional data processing, ensuring speed and consistency, while **OLAP** is designed for complex queries and data analysis, supporting decision-making through multidimensional analysis.
- **ETL (Extract, Transform, Load)** transforms data before loading it into a structured data warehouse, ensuring high data quality and compliance but requiring preprocessing time. **ELT (Extract, Load, Transform)** loads raw data first into a data lake or cloud warehouse, leveraging scalable compute for on-demand transformations, making it ideal for big data, real-time analytics and AI/ML workloads. ELT is more flexible and faster for large datasets, while ETL is better for strict regulatory and structured environments.
- Big Data Processing involves handling massive datasets using Parallel and Distributed Processing. **Parallel processing** utilises multiple cores within a single machine, while

distributed processing spreads tasks across multiple machines to enhance scalability and fault tolerance. Techniques such as **MapReduce**, **Hadoop** and **Apache Spark** enable large-scale data processing. Workloads are categorised into **batch processing** for large historical datasets and **real-time (stream) processing** for instant insights.

- Big Data implementation approach
  - Business Case Evaluation
    - o Define the project's purpose and expected outcomes.
    - o Engage stakeholders to set measurable success criteria.
    - o Assess feasibility, including data availability, technical requirements and compliance with regulations (e.g., GDPR, HIPAA).
  - Data Identification
    - o Identify internal and external data sources.
    - o Internal data is typically structured; external data can include public datasets, third-party providers, social media and IoT sensors.
  - Data Acquisition and Ingestion
    - o Extract data from structured (databases), semi-structured (JSON, XML) and unstructured (images, videos, text) sources.
    - o Data ingestion methods:
      - Batch ingestion – periodic collection for historical analysis.
      - Streaming ingestion – real-time data flow for immediate analysis.
    - o Technologies: Apache Kafka, Flink, Spark Streaming.
  - Data Storage
    - o Store raw data in data lakes (e.g., HDFS, Amazon S3, Azure Data Lake Storage).
    - o Data warehouses (e.g., Amazon Redshift, Snowflake, Google BigQuery) for structured data.
    - o Add metadata for governance and discoverability.
  - Data Processing and Transformation
    - o ELT (Extract, Load, Transform) for flexibility in analytics and ML applications.
    - o ETL (Extract, Transform, Load) for compliance and structured analytics.
    - o Steps include:
      - Data cleansing (handling missing values and duplicates).
      - Normalisation and aggregation for business needs.
      - Schema reconciliation for integrating multiple sources.
    - o Distributed computing tools: Apache Spark, Hadoop MapReduce.
  - Descriptive Analytics and Data Visualisation
    - o Validate processed data through statistical analysis and visualisations.
    - o Identify anomalies before further analysis.

- Deployment, Monitoring and Maintenance
  - o Deploy using cloud-based (AWS Glue, Google Dataflow, Azure Data Factory) or on-premises (Kubernetes, Hadoop) solutions.
  - o Monitor key metrics like ingestion rates and pipeline execution times.
  - o Implement real-time alerts for failures and inefficiencies.
  - o Maintain data governance through compliance audits and access controls.
  - o Adapt infrastructure for new data sources, regulations and technologies.

EXERCISE

Multiple Choice Questions (1 Mark)

1. Which of the following is NOT a characteristic of Big Data?

(a) Volume
(b) Variety
(c) Velocity
(d) Validation

2. Big Data is typically stored in which of the following systems?

(a) Relational Databases
(b) Data Lakes
(c) Flat Files
(d) Local Disk Storage

3. Which of the following is used for processing large-scale data efficiently across multiple nodes?

(a) MySQL
(b) Hadoop
(c) SQL Server
(d) MongoDB

4. The term 'NoSQL' refers to:

(a) A database model used only for structured data
(b) A type of structured database
(c) A non-relational database system
(d) A cloud-based storage solution

5. Which of the following describes the ELT method in data processing?

(a) Data is extracted, transformed and loaded into storage
(b) Data is extracted, loaded into storage and then transformed

(c) Data is analysed and then transformed
(d) Data is aggregated and loaded into a cloud-based system

6. What is the main advantage of using cloud-based storage for Big Data?
   (a) Limited scalability
   (b) High performance with low latency
   (c) Greater scalability and flexibility
   (d) Requires on-premise infrastructure

7. Which of the following frameworks is used for parallel processing of Big Data?
   (a) Apache Spark
   (b) Microsoft Excel
   (c) Google Sheets
   (d) Amazon DynamoDB

8. The first step in the Big Data project lifecycle is:
   (a) Data storage
   (b) Data processing
   (c) Business case evaluation
   (d) Deployment and monitoring

9. In the data identification stage of the Big Data lifecycle, what is the primary focus?
   (a) Acquiring data from external sources
   (b) Cleaning and normalising data
   (c) Analysing data patterns
   (d) Identifying which datasets are required

10. Which method of data ingestion is designed for real-time applications?
    (a) Batch ingestion
    (b) Stream ingestion
    (c) ETL process
    (d) Data cleansing

11. Why is metadata important in Big Data storage?
    (a) It stores the original dataset
    (b) It helps in data governance and discoverability
    (c) It transforms data into structured formats
    (d) It is not important in Big Data management

12. What is the purpose of continuous monitoring in the deployment phase of Big Data projects?
    (a) To track key metrics and detect inefficiencies
    (b) To transform raw data into structured formats

(c) To upload data to external cloud providers
(d) To cleanse the data for better performance

13. What does 'Defense in Depth' refer to in the context of Big Data security?
(a) Using a single security layer for data protection
(b) Implementing multiple layers of security measures
(c) Storing data in a single, centralised location
(d) Encrypting data once and for all

14. A healthcare organisation wants to use Big Data to enhance patient care by integrating Electronic Health Records (EHRs), medical imaging and real-time health data from wearable devices. What storage solution would best support the integration of structured and unstructured data?
(a) Data Warehouse
(b) Data Lake
(c) SQL Database
(d) Blockchain

15. A transportation company is analysing GPS data, fuel consumption and vehicle health to optimise its fleet performance. Which data processing model should the company use to analyse this data in real-time?
(a) Batch Processing
(b) Stream Processing
(c) Cloud Data Storage
(d) OLAP Cubes

16. A retail company wants to deliver personalised recommendations based on customers' browsing and purchase data. Which of the following Big Data technologies would be best for combining large volumes of structured and unstructured data?
(a) Data Warehouse
(b) Data Lake
(c) OLTP Database
(d) Blockchain

17. A pharmaceutical company is conducting a clinical trial and needs to store and analyse genomic data, patient health records and treatment outcomes. Which storage solution would best support this diverse dataset?
(a) SQL Database
(b) Data Lake
(c) Data Mart
(d) OLAP System

18. A smart city initiative collects data from various sensors monitoring traffic, air quality and waste management. Which storage solution would best support the integration of this sensor data for analysis and visualisation?

(a) Data Warehouse
(b) Data Lake
(c) Relational Database
(d) Blockchain

19. A manufacturing plant uses IoT sensors to detect abnormalities in machine performance. Which approach should the plant use to analyse these sensor data streams in real-time to predict equipment failure?
(a) Batch Processing with Hadoop
(b) Stream Processing with Apache Kafka
(c) Data Warehouse with OLAP
(d) Relational Databases for Querying

20. In healthcare, which of the following is a major source of Big Data?
(a) Social media activity
(b) Electronic health records (EHRs)
(c) Consumer purchases
(d) Marketing campaigns

Short Answer Type Questions (2 Marks)

1. What are the primary differences between OLTP and OLAP systems?
2. How does a data lake differ from a data warehouse in terms of data storage and processing?
3. What are the key components of the ETL process in Big Data?
4. Why is ELT preferred over ETL in modern Big Data architectures?
5. What is the role of schema-on-read in Big Data storage?
6. How does stream processing differ from batch processing in handling Big Data?
7. What is an example of a NoSQL database and why is it used in Big Data?
8. What is the significance of a data mart in Big Data storage?
9. How does Big Data analytics support predictive maintenance in manufacturing?
10. What are some challenges of implementing Big Data solutions in an enterprise?

Long Answer Type Questions (5 Marks)

1. Explain the key characteristics of Big Data and how they influence data storage and processing strategies.
2. Compare and contrast traditional RDBMS with Big Data storage solutions such as NoSQL and Data Lakes.

3. Explain the importance of ETL and ELT processes in Big Data workflows. How do they differ?
4. Discuss the role of parallel and distributed computing in Big Data processing with examples.
5. Describe the significance of stream processing in Big Data. How is it different from batch processing?
6. What are the challenges of implementing Big Data solutions in an organisation?
7. What is the significance of metadata management in Big Data environments?
8. Discuss the advantages and disadvantages of cloud-based Big Data storage solutions.
9. What role does the Internet of Things (IoT) play in generating Big Data across industries?
10. Describe how Big Data is used in fraud detection within the financial sector.

CHAPTER 10

Analytics

OBJECTIVES

Analytics has become an essential component of Industry 4.0, transforming the way businesses operate by providing actionable insights from vast amounts of data. This chapter explores the evolution of analytics, its key methodologies and its role in improving decision-making, efficiency and competitiveness in industrial settings. We categorise analytics into descriptive, predictive and prescriptive approaches, discussing their applications in manufacturing, supply chain optimisation and predictive maintenance. Additionally, we examine industry case studies where analytics has driven cost reduction, improved operational efficiency and enhanced strategic decision-making. The chapter also presents methodologies for implementing analytics projects effectively and highlights cloud-based solutions for handling industrial-scale analytics.

By the end of this chapter, you should be able to:

- Understand the significance of big data analytics in Industry 4.0 and its historical evolution.
- Differentiate between descriptive, predictive and prescriptive analytics and their applications.
- Recognise key techniques such as clustering, regression, classification and anomaly detection.
- Evaluate real-world business cases where analytics has optimised industrial processes.
- Identify key performance indicators (KPIs) and their impact on manufacturing and supply chains and how analytics can help to measure and improve them.
- Apply structured methodologies to implement analytics projects.

10.1 INTRODUCTION

In Chapter 4, we explored the transformative impact of technologies that connect physical assets with computing capabilities, enabling seamless interaction between machines and digital systems. This interconnected ecosystem, known as the Industrial Internet of

Things (IIoT), consists of sensors, smart devices and networked machines that continuously exchange data in real-time.

The widespread adoption of IIoT in industrial settings has led to an explosion of high-volume, high-velocity and complex data streams, commonly referred to as Big Data (covered in Chapter 9). The scale of data generation has increased exponentially over the years, growing from terabytes in 2005 to petabytes in 2010, exabytes in 2015 and zettabytes in 2020. This rapid increase reflects the increasing digitisation of industrial processes, where sensors embedded in manufacturing equipment, supply chain systems and predictive maintenance platforms continuously produce vast amounts of data.

However, raw data alone holds little value unless it is transformed into actionable insights. The sheer volume and complexity of industrial data make it impractical for manual analysis, creating an urgent need for advanced computational techniques. Analytics is the process of examining vast amounts of data using computational techniques to find hidden patterns and insights that can help organisations make better decisions.

10.1.1 Evolution of Analytics

Though the term Analytics may seem new, the concepts are not. In 17th Century, John Graunt, an English tradesman, collected mortality records and analysed the patterns in death rates, leading him to question the reliability of the data, as well as hypothesising the actual death from outbreak of plague from the death numbers of the plague year compared to the previous and subsequent years. He is often touted as the first statistician and first epidemiologist.

In the business context, Fredrick Winslow Taylor, referred to as the 'father of scientific management', introduced a method known as 'time and motion study' in the 1880s. The study involved carefully observing the movements required and measuring the time taken for each step of a task. By analysing this data, the study aims to find and eliminate the unnecessary movements, thereby reducing the time taken for the steps.

Henry Ford's measuring the speed of assembly lines and subsequent analysis is a classic example of applying data analysis to improve industrial processes. In 1913, at the Ford Motor Company's Highland Park, Michigan plant, he added a mechanised belt that moved the material. This innovation significantly increased the speed and efficiency of automobile production from little more than 12 hours to one hour 33 minutes, allowing Ford to produce cars much more quickly and affordably.

Business intelligence is another term that is used interchangeably with analytics. Both describe the general practice of using data in making informed, intelligent business decisions. The term business intelligence was introduced in 1865 to describe how a banker profited from information by gathering and acting on it before his competition.

Data mining is another related term that was coined in the 1990s to denote the process of discovering patterns within large data sets. The usage of data mining arose directly from the advancement of database and data warehouse technology. The new technologies enable firms to store more data while also processing it more rapidly and efficiently.

Though these terms are used interchangeably, we can see from Google search trends (Figure 10.1) the term analytics is gaining popularity in the last decade compared to business intelligence and data mining which were more popular a decade before that. In the trend search, we have used two terms, Data Analytics and Business Analytics, to represent the Analytics domain we are referring to, as a search term, Analytics may cover other unrelated topics too.

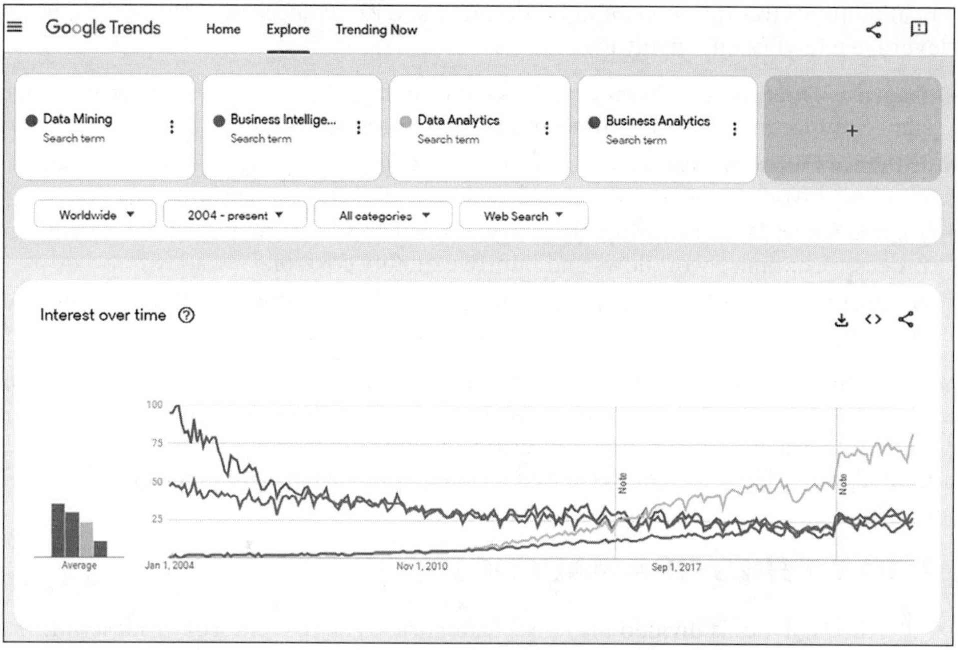

Credit: Google Trends, Google and the Google logo are trademarks of Google LLC.

Figure 10.1 Worldwide Google Search for business intelligence, data mining and analytics (data analytics and business analytics) from 1 Jan 2004 to 31 Dec 2023

All of these would have been impossible without the evolution of big data technologies in parallel. As discussed in Chapter 9, Hadoop was developed in 2005 based on MapReduce algorithm. This enabled the processing of large data in parallel using multiple CPUs and then aggregating them to reduce the processing time drastically. Subsequently, several open-source projects, such as Apache Spark, improved the performance even further. For example, Spark can do a multi-pass map-reduce operation faster than Hadoop since most of the disk I/O of Hadoop is avoided in Spark.

Cloud computing, which allowed customers to 'rent a service' rather than buying the required hardware and software, also helped. An organisation manager could obtain the hardware and software in an affordable, on-demand manner without the need for a lengthy procurement process. Cloud vendors handle all troubleshooting, backups, administration, capacity planning and maintenance. Furthermore, for many corporate tasks, the cloud is just easier and more efficient to use. The cloud now has substantially more storage, is available to several users at the same time, can handle multiple projects and the capacity can be extended or reduced based on demand. Cloud computing is discussed in detail in Chapter 8.

10.1.2 Why Analytics Matters in Industry 4.0?

Industry 4.0 is characterised by the integration of cyber and physical systems through technologies like IIoT and cloud computing into manufacturing and supply chains. The continuous influx of high-volume, high-velocity and high-variety data requires sophisticated analytical methods to extract meaningful patterns and drive decision-making. Organisations that leverage analytics effectively can:

- **Improve Operational Efficiency:** Real-time data monitoring and predictive maintenance reduce machine downtime and increase productivity.
- **Enhance Decision-Making:** Advanced analytics provides insights into supply chain disruptions, inventory management and customer demand forecasting.
- **Enable Smart Manufacturing:** By integrating AI and IIoT, manufacturers can automate processes, optimise workflows and improve quality control.
- **Reduce Costs and Waste:** Data-driven optimisation minimises material waste, energy consumption and inefficient resource allocation.

In the following sections, we will explore key implementation methodologies and the role of descriptive, predictive and prescriptive analytics in industrial decision-making. We will also examine real-world applications in manufacturing and supply chain management, highlighting how analytics drives efficiency, cost savings and strategic improvements in Industry 4.0.

10.2 KEY TYPES OF ANALYTICS

Generally, the analytics is divided into three categories. The first category is descriptive analytics, which summarises the data. It collects information from the raw data to answer the question 'What has happened?' It also tries to explain 'Why did it happen?' Sometimes this part of descriptive analytics is also called diagnostic analysis.

The second category is Predictive Analytics which is used for prediction or forecasting. It combines the descriptive analytics output with machine learning (ML) algorithms and techniques to build models that predict the future. It answers the questions 'What will happen?' and 'Why will it happen?' in the future.

The final category, prescriptive analytics, goes beyond prediction to provide actionable recommendations that guide optimal decision-making. It is all about providing advice. In general, prescriptive analytics is also predictive analytics that prescribes some courses of action and shows the likely outcome or influence of each action. It provides answers to the questions 'What should I do?' and 'Why should I do it?'

10.2.1 Descriptive Analytics

Descriptive analytics plays a crucial role in manufacturing by providing insights into historical data patterns, enabling organisations to optimise processes and improve decision-making.

Data Visualisation and Exploratory Data Analysis (EDA)

Manufacturing generates vast amounts of data from production lines, quality control systems and supply chains. *Data visualisation techniques* allow engineers and decision-makers to explore patterns, trends and anomalies in an intuitive way before applying more advanced analytics. Tools like Grafana, Power BI and Tableau can be used to create *Production Monitoring Dashboards* that help visualise key metrics such as machine uptime, defect rates and cycle times. Heatmaps and box plots can reveal outlier patterns in sensor data, helping manufacturers pinpoint irregular temperature fluctuations in metal forging or pressure inconsistencies in plastic moulding. In quality control and process optimisation, understanding the distribution of defects, production times, or cycle efficiencies is essential. Analysing histograms and probability distributions provides insights into variability and process stability. For example, in Quality Control in Automotive Manufacturing, Histograms can help analyse the distribution of defect rates per 1,000 units, identifying whether process adjustments are needed. Many manufacturing inefficiencies follow the *Pareto Principle*, where 80% of production issues stem from 20% of causes. Pareto analysis helps manufacturers focus on high-impact areas for process improvements. By ranking the most common types of product failures, manufacturers can prioritise process changes that address the top 20% of defect causes, leading to a significant reduction in overall defects.

Association Rule Mining

Association rule mining helps uncover meaningful relationships between different data points (Figure 10.2), enabling manufacturers to identify correlations between process parameters, quality defects and machine performance. This method is particularly useful for predictive maintenance, defect detection and process optimisation.

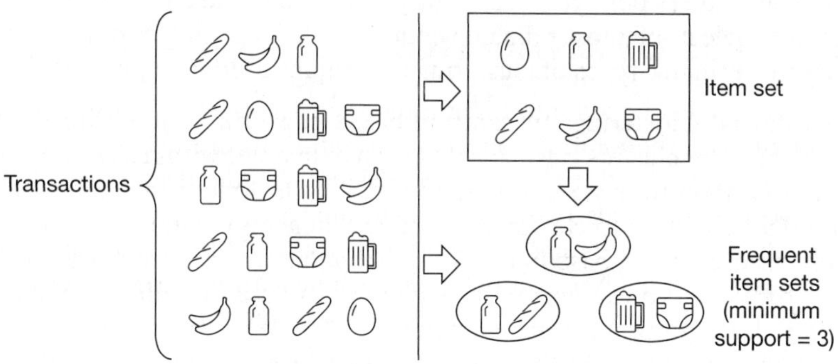

Figure 10.2 Association rule mining to identify frequent itemsets

For instance, in a predictive maintenance use case, association rules can analyse sensor data from industrial machinery to identify conditions that precede equipment failure. If historical data shows that high vibration levels combined with increased motor temperature often lead to motor failure, maintenance teams can be alerted in advance to perform preventive servicing.

Another example is quality control in automotive manufacturing. If certain defects in engine assembly frequently occur when specific suppliers provide a batch of components, association rules can help pinpoint potential quality issues. This allows manufacturers to take corrective action, such as adjusting supplier selection criteria or refining production processes.

Traditional methods for discovering associations include the Pearson correlation coefficient, Chi-square statistics and regression analysis. However, these approaches were designed for small datasets with manually cleaned records and limited variables, making them impractical for modern manufacturing environments, which generate trillions of records and millions of variables from IoT sensors, Enterprise Resource Planning (ERP) systems and production logs. The Apriori algorithm was developed to improve efficiency handling large data. It works by reducing the number of possible combinations it needs to check when looking for patterns. Later, other methods like Frequent Patter Tree (FP-Tree) and Equivalence Class Clustering and bottom-up Lattice Traversal (ECLAT) made the process even faster by organising data in ways that make frequent patterns easier to find.

Clustering

While association rule mining focuses on finding relationships between variables, clustering groups similar objects into meaningful categories based on shared characteristics. This is particularly useful in customer segmentation, process optimisation and anomaly detection within the manufacturing industry.

A typical application of clustering in manufacturing is grouping defective products based on failure patterns. By analysing defects in semiconductor wafer production, clustering algorithms can categorise failures into distinct groups, helping engineers identify the root causes, whether they stem from material inconsistencies, machine calibration issues, or environmental factors like temperature and humidity.

Another practical use case is optimising supply chain efficiency. Manufacturers can cluster suppliers based on delivery performance, quality consistency and cost efficiency, enabling procurement teams to select the most reliable vendors while balancing cost and risk.

There are three primary types of clustering techniques used in manufacturing analytics:

- **Centroid-Based Clustering:** In centroid-based clustering, the entire dataset is partitioned to different clusters, ensuring that each data point within a cluster is more similar to each other than to those in other clusters. This is done by allocating the datapoints based on its proximity to cluster centroid and hence the name. K-Means algorithm is a widely used centroid-based clustering technique. In manufacturing, K-Means clustering can be used to categorise production lines based on efficiency metrics, such as cycle time, defect rate, and machine downtime. However, the centroid-based clustering method assumes clusters have convex shapes, which may not be suitable for complex datasets like irregularly shaped product defects or dynamic supply chain behaviours.
- **Density-Based Clustering:** Density-Based Spatial Clustering of Applications with Noise (DBSCAN) addresses the limitations of centroid-based methods by identifying clusters based on density rather than distance. This method is effective in anomaly detection, such as identifying rare but critical failures in aerospace manufacturing, where specific defects may appear in highly localised conditions but are crucial for safety compliance. DBSCAN is particularly useful when working with non-uniform data distributions, such as irregular customer order patterns in seasonal production cycles.
- **Grid-Based Clustering:** Grid-based clustering divides the entire data space into grids and assigns objects to these grids, making it computationally efficient for large datasets. This technique is beneficial in real-time process monitoring, where production data streams continuously. For example, in food processing plants, temperature and pressure sensor readings can be grouped in real-time to detect anomalies in product quality.

10.2.2 Predictive Analytics

Predictive analytics involves constructing a mathematical or statistical model that defines the relationship between a target (dependent) variable and one or more explanatory (independent) variables. The primary objective of predictive modelling is to minimise error, which is the difference between the model's predicted values and the actual observed values. By refining the model, businesses can make accurate forecasts and data-driven decisions. A predictive modelling schema is represented in Figure 10.3.

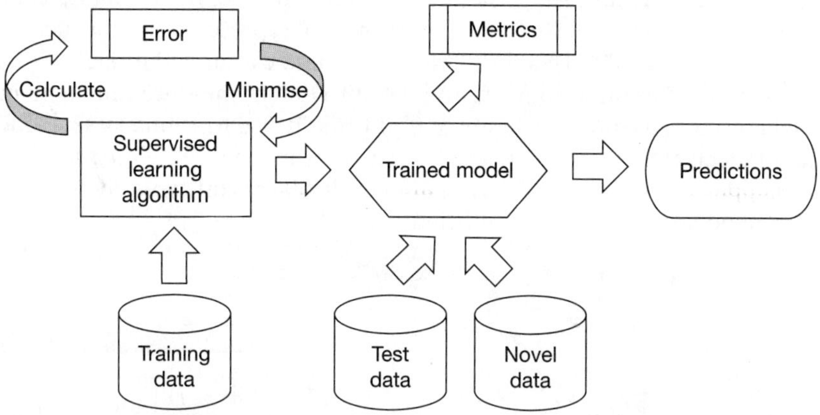

Figure 10.3 Predictive modelling process flow

A predictive model consists of a structured framework incorporating parameters, which include variables, operators and constants that define how input data influences the target outcome. The process of developing a predictive model involves training these parameters using a supervised learning algorithm, ensuring that the model generalises well across unseen data. The dataset used for this learning process is known as the training dataset, which contains historical examples that help the model identify patterns and relationships.

Predictive models can follow two structural approaches. Some models have a static, predefined structure, such as linear regression models, where the mathematical relationship between variables is set before training. Others, such as decision trees or neural networks, develop dynamically during training, adjusting their internal parameters based on the data they encounter. This adaptability allows them to capture complex patterns and improve predictive accuracy over time.

Predictive analysis can primarily be categorised into two key streams: regression and classification. Regression is used when the goal is to predict a continuous numerical value, such as forecasting sales revenue, estimating house prices, or predicting temperature trends. In contrast, classification is applied when the objective is to categorise data into discrete labels, such as determining whether an email is spam or not, identifying customer churn, or diagnosing a medical condition based on symptoms. Both approaches rely on patterns in data to make informed predictions, with regression focusing on numerical precision and classification emphasising accurate categorisation.

Regression

Regression analysis is a fundamental predictive modelling technique used to understand relationships between variables and predict continuous numerical outcomes. At its core, regression models estimate the relationship between one or more *independent variables (predictors)* and a *dependent variable (outcome)*. The simplest form, linear regression, assumes a straight-line relationship, represented as:

$$Y = \beta_0 + \beta_1 X + \varepsilon$$

where Y is the predicted value, X is the independent variable, β_0 is the intercept, β_1 is the coefficient and ε represents the error term. The graphical representation for such a regression model is shown in Figure 10.4. This linear regression line captures the relationship between a car's fuel efficiency measured in MPG to its displacement measured in cubic inches. This public dataset is obtained from University of California Irvine Machine Learning Repository. Displacement reflects the size of the engine, that is, how much air-fuel mixture an engine can draw in during one complete engine cycle. A larger displacement typically means more fuel/air burned; more power but lower fuel efficiency.

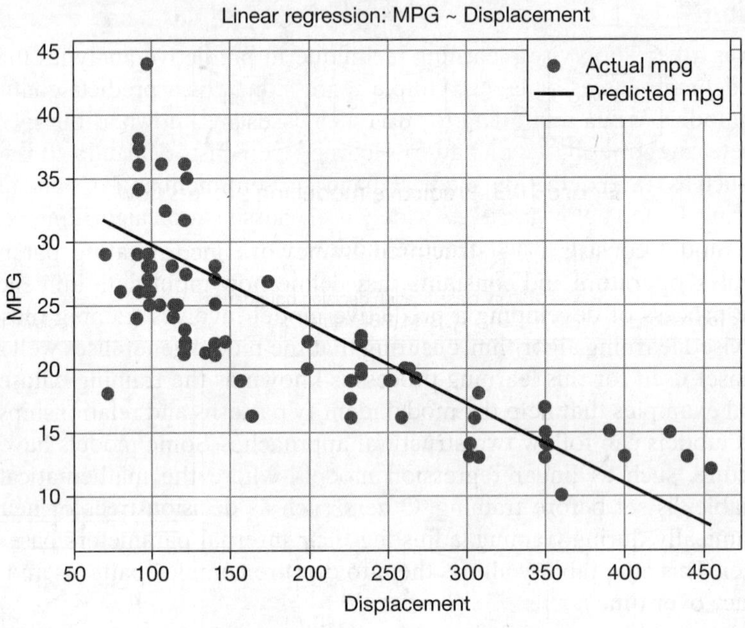

Figure 10.4 Linear regression line capturing the relationship between a car's fuel efficiency to its displacement.

For more complex relationships, *polynomial regression* captures non-linearity. Regularisation techniques such as *Ridge and Lasso regression* are used to improve model stability and prevent overfitting.

Regression analysis plays a crucial role in improving quality control and process optimisation in smart manufacturing. By analysing variables such as raw material properties, production speed, machine settings and environmental factors, multiple regression models can determine which parameters have the most significant impact on product defects. Manufacturers use these insights to fine-tune production processes, reducing defects and ensuring consistent product quality. This data-driven approach helps minimise waste, improve efficiency and enhance customer satisfaction.

Another important application of regression in Industry 4.0 is energy consumption optimisation. Smart factories generate vast amounts of data from IoT-enabled sensors, monitoring energy usage across different machines and processes. Regression models can analyse operational conditions, machine workloads and external factors like ambient temperature to predict energy consumption patterns. By understanding these patterns, manufacturers can optimise energy usage, implement demand-response strategies and reduce operational costs, contributing to sustainability efforts and overall efficiency.

Regression is also extensively used in supply chain and logistics optimisation, where accurate predictions are essential for reducing costs and improving reliability. By analysing historical delivery data, weather conditions, traffic patterns and fuel prices, regression models help estimate delivery times and transportation costs. These insights enable businesses to optimise logistics routes, schedule deliveries more efficiently and reduce supply chain disruptions.

Classification

Classification is a key supervised learning technique in predictive analytics that categorises data into predefined classes or labels. Unlike regression, which predicts continuous values, classification models assign input data to two (such cases are known as binary classification) or more discrete categories based on patterns learned from historical data. It is widely used in applications such as fraud detection, medical diagnosis, sentiment analysis and quality control in manufacturing. Listed below are a few widely used classification algorithms and Figure 10.5 shows the decision boundary's each of these models.

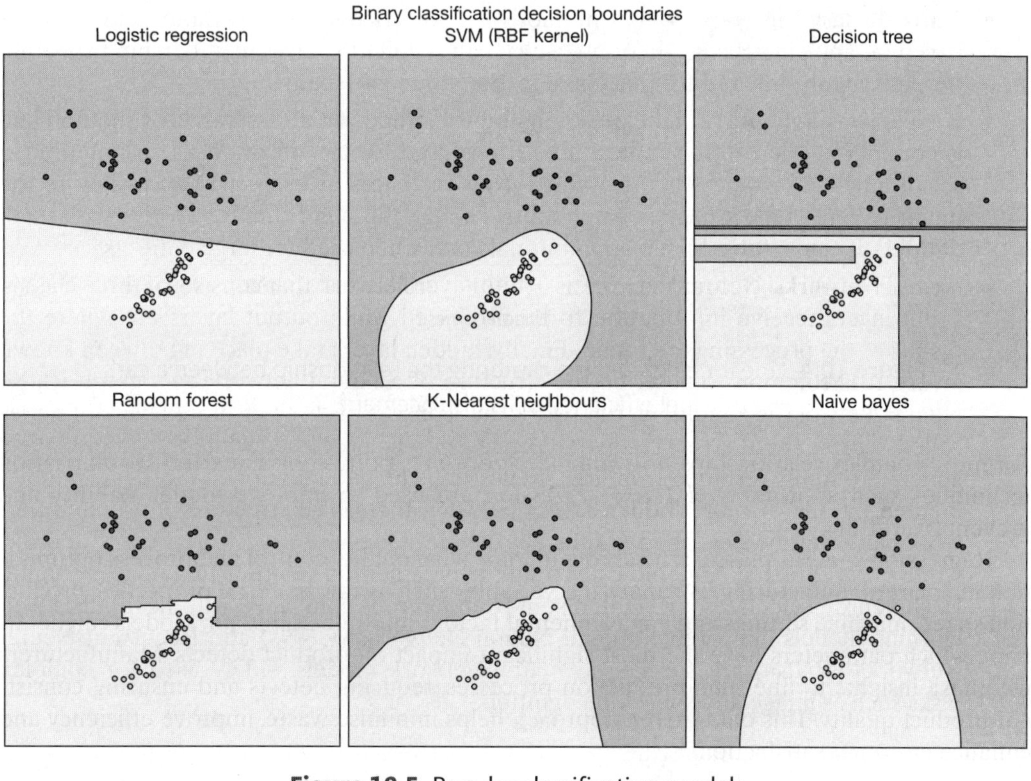

Figure 10.5 Popular classification models

- **Logistic Regression:** Logistic regression is a classification algorithm that predicts binary outcomes (e.g., pass vs. fail, malignant vs. benign, spam vs non-spam, etc). From the input features, it predicts the probability of success (one of the outcomes, e.g., pass, malignant or spam in the example quoted above) using the sigmoid function to restrict the values between zero and one. Then, based on a threshold value, say 0.5, it is assigned to one class (pass, malignant, or spam) if the probability is greater than or equal to 0.5. or the other class (fail, benign, non-spam) if the probability came less than 0.5.
- **Decision Tree:** Decision trees classify the data based on a series of rules organised in an inverted tree-like structure that splits into branches based on conditions of an input feature. They are easy to construct and interpret the results. They can handle both categorical and continuous inputs. These advantages made it among the most popular predictive algorithms. However, they tend to suffer from overfitting problem, unless corrected by regularisation or pruning.
- **Random Forest:** Random Forest is an ensemble technique that constructs multiple decision trees and aggregates their results to enhance accuracy and minimise overfitting. It is robust against noise and works well with large datasets.
- **Support Vector Machine:** Support vector machine (SVM) determines the support vectors of a hyperplane which separates the observations of two or more classes with maximum margin. It provides a kernel function that measures the similarity of observations in a higher dimensional feature space, thus working well for both linear and non-linear classification.
- **Naïve Bayes:** The Naïve Bayes methods use knowledge of probability and statistics based on applying Bayes' theorem, which can predict the class membership probabilities. It is highly efficient for text classification tasks like spam filtering.
- **K Nearest Neighbour:** The nearest neighbour method is a non-parametric method and is considered one of the simplest algorithms used for classification. In this method, a set of labelled instances is maintained. New instances are classified according to the majority vote of the k nearest neighbours. Though this algorithm is effective with small datasets, it can involve high computational costs when used for large datasets.
- **Neural Networks:** Neural Network is a multilayer network that consists of three classes. Input layers receive information to be processed while output layers are where the results of the processing are found. Finally, hidden layers take place in between known layers. It is common to describe the structure of a neural network as a graph whose Neural Network is a multilayer network that consists of three or more layers of nodes called neuron. Input layers receive information to be processed while the output layers generate outputs resulting from this processing. The two mentioned layers are supplemented by one or more hidden layers between them. The structure of a neural network can often be described graphically, where nodes are used to represent neurons while every directed edge indicates a link from a neuron's output to a neuron's input. Each neuron calculates a weighted sum from neurons connected with its inputs. Deep learning models like Convolutional Neural Networks (CNNs) and Recurrent Neural Networks (RNNs) are specialised neural network models widely used for complex tasks such as image and speech recognition.

Classification techniques are a key component of Industry 4.0, helping businesses to automate decisions, optimise processes, and improve overall productivity by classifying information into various buckets. Following are some of the most critical uses of classification in Industry 4.0:

1. **Predictive Maintenance and Equipment Monitoring:** Classification algorithms are also used to monitor industrial equipment health in real-time by classifying sensor data into one of different types like 'normal,' 'at risk,' or 'faulty.' This allows manufacturers to predict equipment failure before it occurs, and maintenance can be planned ahead of time. Classification models like decision trees or support vector machines (SVM), for example, can identify the likelihood of coming equipment failure by classifying temperature or vibration data from machines, reducing downtime and repair costs.
2. **Quality Control and defect Detection:** Classification in smart factories is applied for identifying products as 'defective' or 'non-defective' depending on sensor measurements and visual inspections. Computer vision techniques and classification models (e.g., convolutional neural networks) can analyse images of products to detect defects, scratches, or other marks. This facilitates automatic quality inspection, reduces the likelihood of human error, and ensures high-quality production levels.
3. **Supply Chain Optimisation:** Classification models can classify logistics information such as 'on-time,' 'delayed,' or 'late delivery.' Past delivery performance can be coupled with machine learning algorithms to predict probable supply chain operation disruptions, classify them by degree of risk, and recommend the necessary course of action. It improves delivery accuracy, reduces operational cost, and enhances customer satisfaction.
4. **Energy Consumption Optimisation:** Classification procedures are used to categorise energy consumption patterns, and industries can analyse when energy usage is under a 'high consumption' or 'optimal' state. The categorisation of patterns in energy data based on factors like equipment loading and weather helps companies prevent energy wastage proactively and reduce operational costs, heading towards sustainability.

Time Series Forecasting

Time series forecasting plays a crucial role in Industry 4.0, where interconnected cyber-physical systems and real-time data streams drive industrial automation and smart decision-making. Time series forecasting, which predicts future values based on historical data, is a key enabler for optimising production efficiency, minimising downtime, and improving supply chain operations. Unlike conventional predictive models that analyse static data, time series forecasting explicitly accounts for temporal dependencies, recognising that past observations influence future trends.

In the context of Industry 4.0, time series data is continuously produced by sensors embedded in machinery, industrial processes, and smart grid systems. Such datasets often exhibit patterns with trends and seasonality. We illustrate the same with electricity consumption data, a public dataset obtained from the University of California Irvine Machine Learning Repository, in Figure 10.6. A trend indicates long-term changes in system behaviour, such as increased energy consumption in a factory due to increased production levels. Seasonality occurs in industrial environments where demand fluctuates based on external factors, such as higher electricity usage during peak operation periods or increased machine usage during certain manufacturing seasons. There could be multiple seasonality in the time series data, for example, based on the hour of the day and month of the year. In addition, time series data can include noise, random variations caused by unforeseeable external factors, calling for the creation of robust forecasting models with the ability to filter out extraneous fluctuations and increase predictive power.

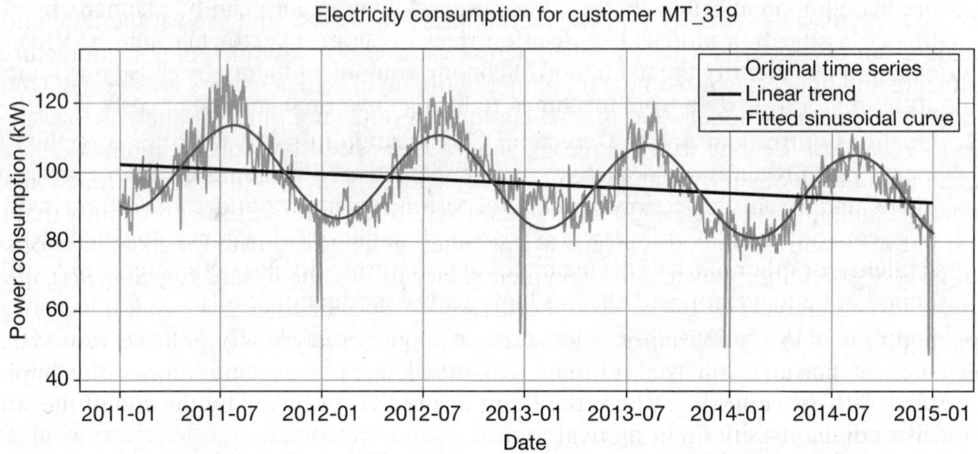

Figure 10.6 Time series data for electricity consumption indicating trend and annual seasonality

Various forecasting models are widely utilised in the context of Industry 4.0. *Autoregressive Integrated Moving Average* (ARIMA) model is a traditional statistical method that accurately detects trends and seasonal swings through the method of autoregression, differencing, and moving averages, thus making it applicable to stable industrial operations with evidently defined patterns. Exponential smoothing methodologies, including the Holt-Winters model, are able to handle seasonal demand and production patterns by assigning higher importance to more recent observations. On the other hand, in the circumstance of intricate and non-linear industrial processes, advanced deep-learning structures, including *Long Short-Term Memory* (LSTM) networks, a specialised recurrent neural network (RNN), possess superior performance. LSTMs are especially proficient in handling sensor information in the context of predictive maintenance, where the task involves the recognition of long-term associations and the identification of minute degradation trends in machines.

The prediction of time series data has widespread use in the era of Industry 4.0, enabling greater efficiency and stimulating innovation. Predictive maintenance is one of the key areas impacted by this approach, where prediction models analyse sensor data, vibration levels, and temperature logs to anticipate probable machine failures before they happen. This proactive approach reduces unscheduled downtimes, reduces maintenance costs, and increases the working life of industrial equipment. Time series models in the context of smart manufacturing are utilised to optimise production schedules by predicting fluctuations in the availability of raw materials, machinery capacity and labour requirements. Similarly, prediction techniques in supply chain management help in anticipating changes in demand, thereby avoiding excess inventory while ensuring timely order fulfilment. The energy industry benefits greatly from time series prediction as well, where smart grids use predictive analysis to optimise electricity generation, in turn reducing operating expenses and enhancing sustainability. Quality control in manufacturing also utilises time series models to detect anomalies in the production process, ensuring uniform product quality and minimising defects.

Survival Analysis

Survival analysis is a predictive analytic method used to analyse and predict the time until a specific event occurs. It is widely applied across various domains, including manufacturing, healthcare and finance, where understanding the timing of critical events is essential. Unlike traditional regression models that predict continuous outcomes, survival analysis focuses on time-to-event data, dealing with unique challenges such as censored observations.

The foundation of survival analysis lies in key concepts such as the survival function and the hazard function. The survival function, denoted as $S(t)$, represents the probability that the event has not occurred by a given time t. This function helps estimate the likelihood of survival or system reliability over time. Complementary to this, the hazard function, $h(t)$, measures the instantaneous risk of the event occurring at a specific time, given that it has not yet happened. One of the most challenging aspects of survival analysis is handling censored data, which occurs when an event's exact timing is unknown due to study limitations or incomplete observations. Right-censoring, where the event has not yet occurred by the end of the study period, is a common scenario in survival studies.

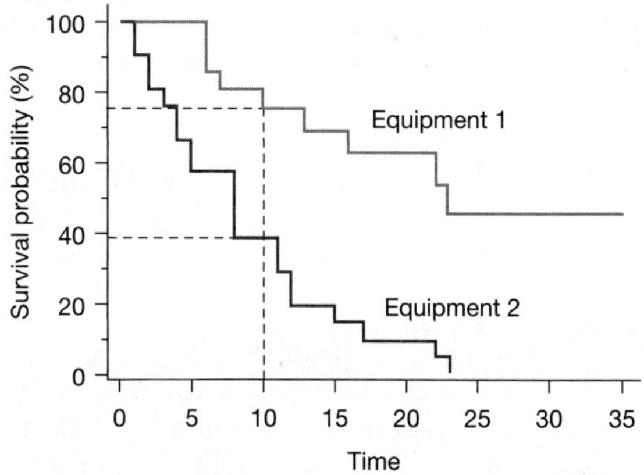

Figure 10.7 Survival analysis. In 10 months, equipment 1 has 75% survival chance, whereas equipment 2 has 40%

Several techniques are used to conduct survival analysis. The *Kaplan-Meier estimator* is a widely used non-parametric method for estimating the survival function from observed data. It provides a stepwise survival curve, offering a visual representation of how survival probabilities evolve over time (Figure 10.7). This method is particularly useful in medical research and reliability studies where event occurrences vary across time. Another powerful approach is the semi-parametric *Cox Proportional Hazards model*, which examines the relationship between survival time and explanatory variables. Unlike Kaplan-Meier, it allows for the inclusion of multiple predictors and estimates the impact of different factors on survival while making the assumption that the hazard ratios remain constant over time. For cases where parametric modelling is required, the Weibull and Exponential models are often used, especially in reliability engineering and failure prediction. These models assume specific probability distributions for survival times and are highly effective in industrial applications where machine failures follow predictable patterns.

Survival analysis finds extensive applications across various industries. In manufacturing and predictive maintenance, survival analysis plays a crucial role in forecasting the remaining useful life of machines and components. Depending on sensor readings and operating status, failures can be forecast ahead of time to enable timely intervention and reduce unplanned downtime. In customer retention and churn, survival models are also used to forecast probability and time until customer loss. Subscription businesses, for example, model drivers for churn and use forecasted customer behaviour to deploy retention programs. Healthcare and medicine use it for forecasting patient survival rates for cancers, comparing new treatments' efficacy, and determining risk factors associated with patient death. Through patient histories and treatment outcome analysis, survival models help physicians with disease prognosis and treatment selection. Banking also benefits from survival analysis in terms of credit risk modelling, where it forecasts time until loan defaults, and investment risk analysis, where it examines longevity of financial products.

Survival analysis remains a critical component of predictive analytics with a well-organised approach to modelling time-to-event data. The capacity to effectively handle censored data and provide reliable estimates of survival probabilities creates essential information not possible through regular regression procedures. Survival analysis enhances decision making by forecasting significant events and enabling the optimisation of resource allocation.

Anomaly Detection

Anomaly detection is a critical predictive analytics approach identifying unusual patterns not observed within regular behaviour. These anomalies, otherwise known as outliers, can be rare yet critical situations such as fraud, systems crashes, or cyber threats. Through machine learning methods and statistical methods, anomaly detection enables organisations to respond preemptively to threats before they turn into major problems.

Anomalies are grouped into *point anomalies*, where a single instance differs considerably from the average (a rare burst of credit-card purchases, for example), *contextual anomalies*, where something in a certain context alone is out of sorts (a summer-months, rather than winter-months, electricity bill, for instance), and *collective anomalies*, where a group of data points individually are not necessarily out of line, yet collectively display an abnormal trend (a coordinated cyber attack, for example).

There are different data type- and properties-dependent anomaly detection techniques. Statistical methods like Z-score and Inter-Quartile Range (IQR) rely upon distributional properties to identify outliers, while ML methods like isolation forests, auto-encoders, and one-class Support Vector Machines (SVM) improve detection in high-dimensional data. Anomaly detection for time series data applies models like ARIMA, LSTMs, and Prophet to identify abnormal patterns in sequential data.

Anomaly detection use cases span multiple industries. It's applied in manufacturing for predictive maintenance through sensor anomaly detection. Its uses in finance include fraud detection through alerting unusual transactions. It applies in healthcare for medical anomaly detection for detecting diseases even before symptoms are apparent. In cybersecurity, it's applied for network intrusion detection and abnormal access patterns.

By integrating anomaly detection into predictive analytics, organisations can identify impending critical events, make more informed decisions, and minimise losses, making it a key instrument in risk management and operational efficiency.

10.2.3 Prescriptive Analytics

While predictive analytics focuses on forecasting future trends and likely outcomes, prescriptive analytics goes a step further by not only forecasting them, but also prescribing how best to achieve them. It achieves this by taking into account potential situations, limitations, and trade-offs for the best possible means to achieve a particular objective. Prescriptive analytics benefits greatly in manufacturing where business organisations need to optimise cost, efficiency, resource allocation, and operating performance within strict budget, manpower, and time constraints.

Prescriptive analytics evolved from operations research, a mathematical applied specialty focused upon optimisation techniques for decision making. In practice, it overlaps with decision science and management science in helping organisations resolve complex problems in supply chain management, production planning, workforce scheduling, and maintenance optimisation.

Prescriptive analytics techniques are broadly classified into the following categories.

Convex Programming (Mathematical Optimisation)

Convex programming is used when a problem has a convex structure, meaning that an optimal global solution can be mathematically determined. This class of methods includes:

- **Linear Programming (LP):** Used for optimising production schedules and resource allocation (Figure 10.8).
- **Second-Order Cone Programming (SOCP):** Applied in quality control optimisation, ensuring that variations in product specifications remain within acceptable limits.
- **Semidefinite Programming (SDP):** Helps in optimising complex supply chain networks, where multiple constraints and dependencies must be considered.
- **Geometric Programming:** Useful for cost minimisation in energy-intensive manufacturing, such as reducing electricity consumption in metal processing plants.

Heuristic Search Methods

Not all problems have a convex structure, meaning an exact solution is impractical or impossible to compute within a reasonable timeframe. In such cases, heuristic search techniques help find a near-optimal solution efficiently. Some of the examples are:

- **Simulated Annealing:** Mimics the process of metal cooling and recrystallisation to find a low-cost, high-efficiency solution in production optimisation,
- **Genetic Algorithm:** Inspired by natural selection, this method is used for multi-objective optimisation, such as balancing production speed, energy consumption, and product quality,

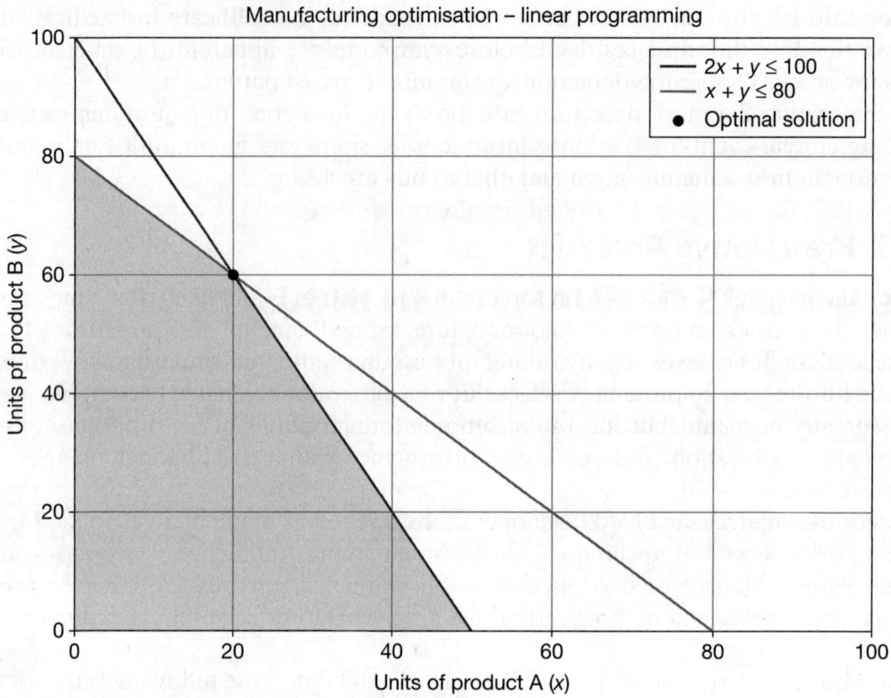

Figure 10.8 Convex optimisation - linear programming

- **Tabu Search:** Helps in routing optimisation for logistics, ensuring that raw materials and finished goods are transported using the most efficient paths.

Reinforcement Learning

Reinforcement learning (RL), a subset of machine learning, enables systems to learn optimal decision-making strategies by interacting with the environment and maximising rewards over time. Unlike traditional optimisation techniques, RL continuously adapts to changes and can handle dynamic, uncertain environments. RL can be used in smart factories to continuously adjust machine parameters (e.g., speed, pressure and temperature) for optimal production efficiency. For example, in steel manufacturing, RL can dynamically modify furnace temperature to minimise energy consumption while ensuring product quality. In assembly lines, reinforcement learning can train collaborative robots to improve task efficiency by adjusting movement patterns based on real-time sensor feedback. RL can optimise warehouse management by learning how to minimise delays, reduce excess inventory and maximise throughput in distribution centres.

Simulation-Based Optimisation

In complex manufacturing environments, real-world testing can be expensive and time-consuming. Simulation-based optimisation allows manufacturers to test different scenarios and determine the best operational strategies before implementing them in production. A digital twin can simulate different factory layouts, machine configurations and material flows to identify the most efficient production settings. For instance, in automotive manufacturing, a digital twin can help redesign assembly lines to reduce cycle time and minimise worker fatigue. In supply chain logistics, simulation-based optimisation can model delivery routes and transportation schedules, ensuring that goods are delivered on time while minimising costs. We will discuss digital twins and simulation in Chapter 11.

10.3 BENEFITS OF ANALYTICS IN INDUSTRY 4.0

Advanced analytics maximises operations performance through three main applications – decreasing downtime, improving productivity and optimising complex production networks. Figure 10.9 shows the key benefits of Analytics in Industry 4.0.

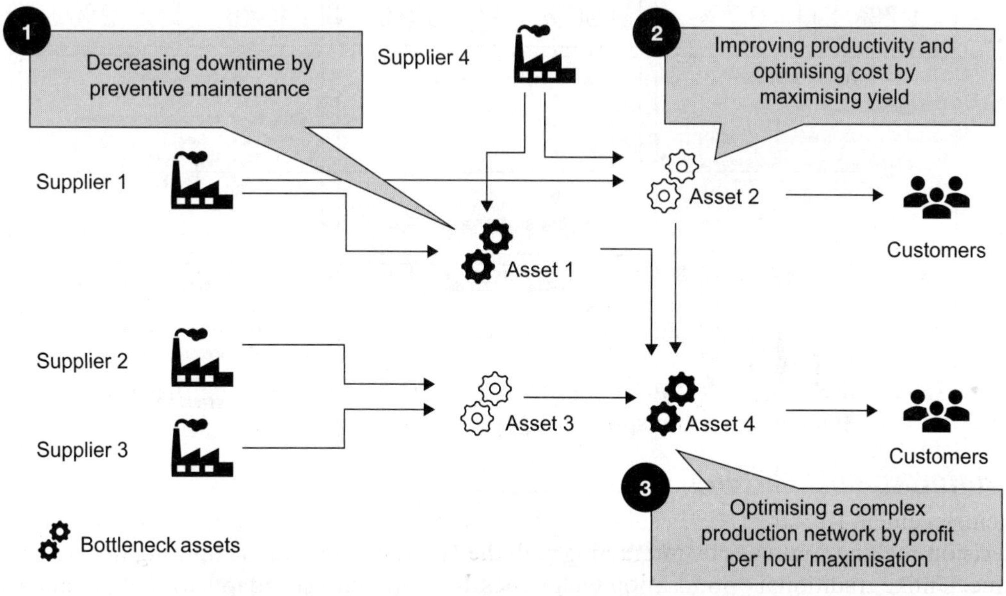

Figure 10.9 Three key benefits of analytics in Industry 4.0

10.3.1 Decreasing Downtime

Equipment failure is one of the most significant challenges in manufacturing, often leading to costly disruptions and lost productivity. Traditionally, machine maintenance depended on operator experience and routine inspections, where skilled workers detected early signs of mechanical issues and intervened before a complete failure occurred. However, with modern factories operating thousands of machines under high efficiency and uptime pressures, relying solely on human observation is no longer feasible.

With predictive maintenance, manufacturers can now anticipate failures before they happen. Advanced analytics systems collect and analyse historical sensor data from machines

DID YOU KNOW?

Key Performance Indicators in Industry 4.0

Business executives utilise analytics to identify possible problems that may influence important Key Performance Indicators (KPI) of their business, both in their factories and throughout their supply chains. A single KPI cannot accurately reflect the performance of a plant or manufacturing organisation. Furthermore, some KPIs, such as on-time delivery, represent not only the success of a single factory but the entire supply chain. Leading manufacturers employ analytics to help managers understand the challenges that underpin each of these KPIs, as well as how they relate to one another. Figure 10.10 shows an example KPI dashboard which can give a glimpse of the health of the organisation.

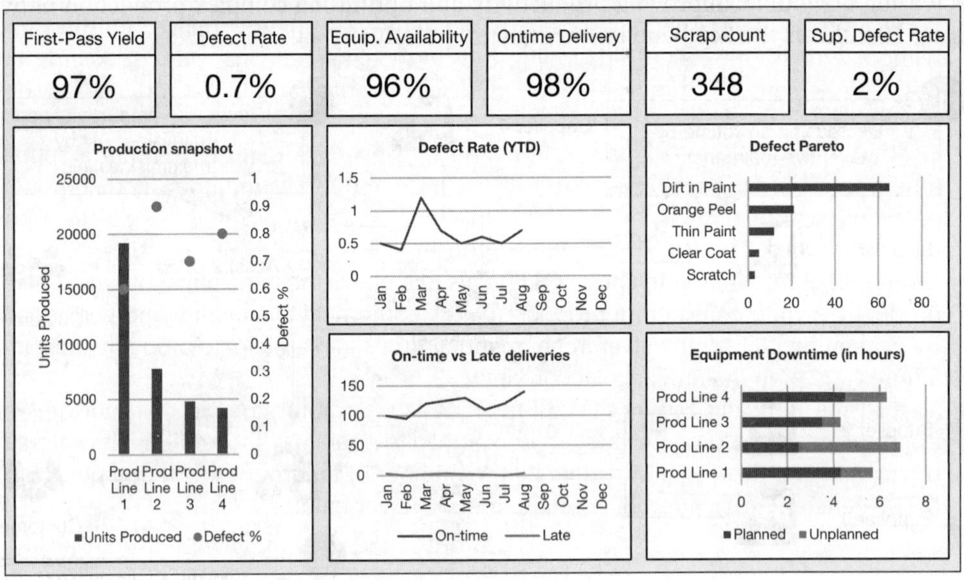

Figure 10.10 Example KPI dashboard

The most common KPIs include the following:

- **Throughput:** This measures the efficiency of a plant or manufacturing based on the total quantity of goods produced during a specific time period. Continuous monitoring of this type of data allows producers to discover potential equipment inefficiencies, manage resource backlogs and change production schedules to meet targets.
- **Yield:** This evaluates the efficiency with which things are produced and is calculated by dividing the number of units produced to standard specifications by the total number of units produced.
- **Profit Per Hour:** Profit per hour is a simple calculation that determines the profitability generated by a specific piece of equipment or a production line over a certain period.

- **Cycle Time:** Cycle Time refers to the average time required to convert raw materials into a finished product, encompassing all stages of production from the initiation of processing to the final output.
- **Cycle Time Variability:** Cycle time variability refers to the variation in the time it takes to complete a specific task or process. In manufacturing, this can impact productivity and output. Reducing cycle time variability can lead to more consistent and predictable production rates.
- **Capacity Utilisation:** This indicates how successfully a manufacturer matches capacity to demand and is measured by dividing total capacity used in a particular timeframe by total available production capacity, typically presented in a percentage form.
- **Scrap Rate:** This is the amount of material that must be scrapped after a task is completed. It is a matter of wastage and hence the lower the rate, the better.
- **Perfect Order Rate:** This measures a manufacturer's ability to deliver finished goods without errors. This includes shipping the correct number of goods, packaging them correctly and ensuring documentation matches the quantity shipped and invoiced at the correct price.
- **On-time Delivery:** This refers to the percentage of items delivered within a specified time limit as promised to the customer. This research assists in understanding potential order fulfilment delays and determining the specific cause, whether it is due to supplier delivery concerns or order management constraints.
- **Overall Equipment Effectiveness (OEE):** This calculates the percentage of time a plant is productive while considering product quality, equipment availability and performance. By evaluating OEE at any given time, manufacturers can anticipate probable equipment failures and plan maintenance accordingly.
- **Mean Time Between Failures (MTBF):** MTBF is a reliability metric that indicates how long a piece of equipment or system is expected to operate without failing. It's calculated by dividing the total operating time by the number of failures during that time. A higher MTBF generally indicates better reliability and fewer failures.

to detect patterns and anomalies that indicate potential breakdowns. By continuously monitoring critical performance parameters, such as temperature, pressure, vibration and energy consumption—these systems predict when a machine is likely to fail, allowing companies to perform proactive repairs and avoid costly unplanned downtime.

A compelling example is from the oil and gas sector, where offshore platforms utilise high-performance compressors. One of the major oil producers experienced repeat failures of the compressors, which cost $1 to $2 million per platform daily. The cause was unknown despite intensive engineering intervention. Through advanced analytics on sensor data from hundreds of compressors, analysts concluded that failure was correlated with high temperature, high pressure and a number of other factors. Based on this revelation, they built an AI-based predictive model that could predict breakdowns well in advance, enabling maintenance crews to intervene before they occurred. Downtime decreased from 14 days to only six days as a consequence, saving the company millions of dollars with every incident.

10.3.2 Improving Productivity and Optimising Cost

Aside from minimising downtime, analytics could also make machines run more efficiently and maximise throughput, resulting in greater profitability even with a small improvement in operations. A very useful tool in this regard is yield, energy and throughput (YET) analysis, which simplifies the tradeoff between output, utilisation of resources and cost savings at every phase of the production process.

A global manufacturer of chemicals had unreliable throughput and generally low production at one of its European plants. Its monomer furnace, a key element in its production process, had variable production rates and was running below its optimum efficiency. To troubleshoot the problem, the company collected sensor data for 615 days, comprising 600,000 samples of 63 operating variables, creating around 40 million data points.

Using sophisticated data analysis, the company discovered subtle interactions among important process variables that affect furnace performance. By creating an optimisation model, engineers learned how small changes would greatly increase output. From this revelation, the company conducted a set of controlled experiments to optimise the production parameters. The outcome was astounding – the output was boosted by 18–30%, which meant an added $5 million in net contribution by the furnace alone. Extending the analysis plant-wide, they discovered a potential boost of $30 million in profitability, showing the tremendous power of analytics in uncovering opportunities for efficiency enhancement and cost reduction.

Furthermore, analytics is also instrumental in cost management because it enables manufacturers to identify and eliminate hidden inefficiencies. For instance, businesses over-procure raw materials as a safeguard against shortages, therefore accumulating unnecessary inventory carrying costs. Analytics software can suggest the ideal inventory level after examining historical procurement and demand trends, thereby eliminating wastage without disrupting production.

10.3.3 Optimising Complex Production Networks

While YET analysis concentrates on single machines, profit per hour (PPH) maximisation takes analytics to the systems level, making whole production networks – from procurement and production through logistics and sales – run at optimal profitability. Such optimisation requires sophisticated process model algorithms that can assess thousands of interdependent variables and constraints, far exceeding what previous human planners could handle. It assists in deciding the procurement mix, product mix and sales mix so that maximum profits are gained for each time period. A multinational chemicals producer with several plants and products and highly volatile raw material prices could not optimise its production, pricing and supply strategy. The company had a complicated combination of spot and long-term contracts with interregional and interdepartmental transfer prices, so its production decisions were often not optimal with much potential profit unrealised.

To solve this problem, the company used a mixed-integer model featuring 500+ decision variables and 3,000+ operational constraints on contracts, non-linear cost functions of finished goods, raw material cost structure, production steps and capacities and transportation logistics.

This prescriptive analytics model simulated various production and distribution scenarios, helping the company identify bottlenecks, redirect resources and optimise pricing models.

The impact was dramatic. Short-term tactical benefits included the shifting of production to underutilised lines and reducing raw material costs, which yielded significant cost savings. It also enabled the company to identify strategic opportunities for capacity expansion and revenue growth. A counterintuitive suggestion led to the elimination of polyvinyl chloride (PVC) production, which, in turn, improved overall profitability by allowing the company to focus on higher-margin products. These changes yielded an increase in earnings before interest and taxes (EBIT) of over 50%, demonstrating the immense value of analytics in high-complexity production environments.

10.3.4 Other Benefits

Support New Business Models

Traditionally, manufacturers have emphasised making and selling completed products, earning their revenues from single transactions. But most are now moving into service-based business models, where rather than selling products outright, they are offering them as part of a continuous service. This shift, better known as product-as-a-service (PaaS), enables manufacturers to make their products available on a pay-per-use, subscription, or outcome-based basis, with continuous value realisation for the customer and a stable source of revenue. For instance, Rolls-Royce has pursued a 'power-by-the-hour (PBH)' approach in its aircraft engine business. Under a PBH contract, airlines are able to pay a fixed fee per flying hour, instead of taking on the complete cost of engine ownership and maintenance Analytics lies at the heart of enabling such novel business models, enabling manufacturers to take advantage of real-time IoT data, predictive maintenance and usage analytics to make the product more reliable and service delivery more optimal.

For instance, Rolls-Royce's highly sophisticated engine health monitoring system drives its PBH service, which offers real-time performance feedback to enhance operational availability, reliability and efficiency. In contrast with conventional sales, where the manufacturer relationship often concludes at the sale, the services-based approach facilitates long-term customer interaction. Through insights derived from analytics, manufacturers are in a position to provide personalised upgrades, efficiency improvements and service contract renewals, thereby affecting customer satisfaction and loyalty.

Monitor Supplier Performance

Producers leverage analytics to assess and streamline their supply chain by determining the suppliers that regularly deliver on schedule. Advanced analytics software interfaces with real-time GPS tracking and logistics data to monitor shipments in transit and enable proactive decisions in case delays occur. Apart from punctuality, analytics can also be used to measure product quality, price competitiveness and adherence to labour and environmental laws. This fact-based measurement assists companies in handling more reputable suppliers while identifying loopholes in their procurement process to enhance.

Gain Supply Chain Visibility

Supply chain performance is vital for manufacturers with thousands of suppliers and orders to manage. Analytics provides real-time monitoring of inventory with visibility into raw material inventory, goods in transit and inventory available across various plants. Manufacturers have visibility and hence can strategically redistribute inventory to counteract shortages and order optimised fulfilment so that production continues uninterrupted even in the event of supply chain failure.

Optimise the Sequencing of Production

Analytics assists in informing manufacturing teams about what production runs and work orders to address first, versus a series of factors like customer urgency, supply chain constraints and inventory levels. By tying work order information to sales orders and inventory levels, production sequencing can be made smarter by supervisors. For instance, a plant manager can rush an order from a high-volume or premium customer that has to be filled right away and put off a less important order with a longer lead time. This meets critical shipments with maximum use of resources on the production lines.

Reduce the Extent of Product Recalls

Through the tracking of granular real-time information, analytics improves quality control by detecting defects at the point of origin. By analysing production reports, machine logs and quality assurance data, manufacturers can monitor when and where a defect happened and thus limit the extent of probable product recalls. This selective recalling process is economical, avoids wastage and avoids loss of reputation by recalling only the affected batches and not whole product lines. Moreover, manufacturers can utilise these insights to enhance quality control procedures and defect minimisation initiatives.

Enhance Worker Productivity

Analytics serves an important function in minimising non-productive idle time by scheduling maintenance routines according to production calendars. Historically, maintenance crews are subjected to inefficiencies because of inadequate scheduling, which results in delays and underutilisation. Through predictive analytics, organisations are able to schedule maintenance during downtime, which ensures that machines as well as employees are fully utilised. Additionally, analytics is used to streamline shift scheduling through the plotting of worker availability in relation to material delivery windows, production cycles and other exogenous events. This leads to improved workforce management, less idle time and greater operational efficiency.

Reduce Employee Attrition

A properly managed workforce is the recipe for long-term prosperity. Analytics enables manufacturers to pinpoint workplace safety risks, overloading and underuse of workers, the principal drivers of job dissatisfaction and worker turnover. Through workplace analytics, companies can get ahead of the curve to enhance working conditions, work schedules and career development possibilities. Furthermore, skills-matching analytics enables companies to reassign workers to jobs that are more aligned with their skills, which enhances job satisfaction and retention.

Ensuring Accurate Financial Data

The majority of manufacturers still rely on manual spreadsheets and disconnected financial systems, which result in reporting errors and inconsistencies. By embedding analytics-integrated financial reporting, businesses are able to automate data extraction from various sources, including ERP systems, production logs and machine data. This delivers real-time, error-free financial reporting with little chance of miscalculations or falsified figures. With credible financial insights, manufacturers are able to improve budgeting, forecasting and overall financial decision-making, leading to greater transparency and operational efficiency.

10.4 IMPLEMENTATION STRATEGIES

Implementing analytics in an industrial environment requires a structured approach to ensure that data-driven insights translate into actionable business value. In this section, we will first discuss the best practices to follow for the step-by-step implementation of an analytics project in Industry 4.0. We will then move to a few established methodologies for the same and map those steps. Lastly, we will look at the most frequent implementation scenario.

10.4.1 10 Steps to Implement Analytics for Industry 4.0

Implementing a successful analytics project requires alignment with business goals, effective data management and scalable analytical capabilities. The following *10 best practices* outline the key steps to developing a robust analytics strategy (Figure 10.11). Many of these steps overlap with big data implementation (as discussed in Chapter 9). If an organisation has already established a big data infrastructure, some of these steps can be skipped or streamlined.

1. **Treat It as a Business Project, Not Just an IT Initiative:** The first and most critical step in ensuring the success of an analytics initiative is treating it as a business project rather than just an IT implementation. One of the most common reasons analytics projects fail is that they are perceived as technology-driven rather than as business enablers. To gain traction and deliver meaningful value, these projects must be closely aligned with business objectives and priorities. This means engaging key stakeholders from the very beginning, including executives, operations managers and frontline employees who will ultimately rely on the insights generated. A good analytics project should be framed in terms of solving significant business problems, such as an on-time delivery goal, manufacturing downtime reduction or supply chain efficiency streamlining.

   Another way analytics can be made into a business-driven initiative is by demonstrating early success. This is done through making reports about those metrics first that have a direct bearing on important performance metrics such as customer satisfaction, order fulfilment rates and cycle times for production. An example would be when companies align operations data into IT infrastructures and then find out about the late deliveries' effect on customer retention or how machine breakdown causes disruptions in supply chains. Through experiencing early tangible results, analytics organisations are then able to secure leadership sponsorship and continue investing in analytics broader use.

2. **Inventory All Data Repositories Including Operational Data:** Defining business priorities, the next challenge is an inventory of all available data. Most organisations have data located in multiple systems, departments, or legacy applications, leading to inefficiency and inconsistency in referencing and making use of the data. A comprehensive inventory of the data involves discovering all potential sources of data, classification of data and evaluation of existing datasets' access and governance framework. These are structured data, typically located in relational databases such as PostgreSQL, MySQL,

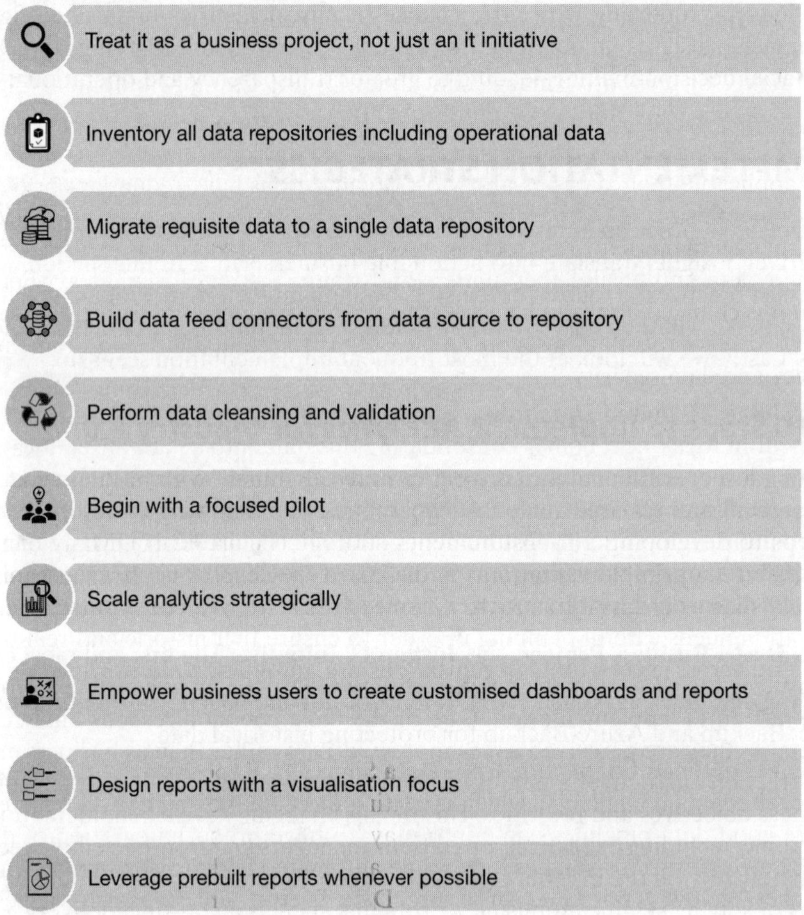

Figure 10.11 Ten steps of implementing a successful analytic program

or SQL Server; unstructured data, such as IoT logs, video streams and reads from sensors, typically located in NoSQL databases such as MongoDB or Elasticsearch; and semi-structured data, such as JSON or XML files, which may come from APIs or from streaming platforms such as Apache Kafka.

To acquire the most complete and accurate analysis, manufacturing management application data should be fused with factory equipment and manufacturing process data. With real-time machine performance, sensor and production information integrated into ERP systems, manufacturers will have visibility into order fulfilment and operation efficiencies like never before possible. For instance, matching work order information from an ERP system with cycle time information from the manufacturing floor is an effective metric for determining if an order will be completed on time or not. Visibility is critical because manufacturing delays translate directly into customer satisfaction and revenue, enabling companies to remove bottlenecks proactively, optimise resource use and enhance overall supply chain reliability.

Concurrently, the target state has to be defined by determining what metrics and analytics results the organisation is seeking to achieve. They could be worker safety metrics, quality, or preventive maintenance metrics like Overall Equipment Effectiveness (OEE), Mean Time Between Failures (MTBF), or cycle time variance.

While listing data is vital, organisations need to ascertain quality, completeness and how applicable their datasets are. Most organisations have redundant or old information, which results in skewed analytics results and misguided conclusions. With metadata management platforms such as Apache Atlas, organisations can track the lineage of their data and ensure they understand where the data originates, when and how often it is changed and who has access to it. This is required for setting up the building blocks for effective data integration, as well as for compliance requirements for data privacy and data security.

3. **Consolidate Required Data into a Single Repository:** With a vision for the data environment in focus, next comes consolidating this data into a centralised location. A single location of truth is required for any analytics initiative since it eliminates silos for data and allows all departments to work from the same and up-to-date information. Based upon organisational requirements, data will be located in a highly defined, highly structured data warehouse such as Snowflake, Google BigQuery, or Amazon Redshift, or a less defined, highly flexible data lake such as AWS S3, Azure Data Lake, or HDFS. Migration needs careful planning in order to ensure that historic and real-time data is migrated across successfully. Organisations use automation aids for loading and transforming data, such as Apache NiFi, Talend, or dbt and backup automation tools such as AWS Backup and Azure Backup for protecting historical data.
4. **Design Data Feed Connectors from Data Source to Repository:** With centralised data comes the second challenge, which is getting the data moving between systems. That's where good data pipelines come into play. To have proper integration of data, workflows between various sources have to be automated and provide updates in real-time. Apache Airflow, AWS Glue, or Azure Data Factory are standard engines for such workflows, from extraction all the way through to storage. To enable real-time analytics, organisations can use platforms like Apache Kafka or Apache Flink which can ingest a continuously arriving stream of data from IIoT sensors, factory sensors and trading systems. With appropriately architected data pipelines, organisations can get analytics models fed accurate and timely data, providing real-time visibility into business operations.
5. **Validate and Clean the Data:** Data quality is another factor that heavily contributes to analytics programs' success. Poor-quality data will put businesses back into making bad decisions based on spurious information, which could lead to financial losses or operational inefficiency. To avoid this from happening, businesses should have good quality data cleansing and validation procedures in place. This will include removing duplicate records, repairing inconsistencies and normalising across datasets. AI-based data validation engines like Great Expectations, Deequ and Apache Spark MLlib will ensure accuracy in data through the use of pre-defined rules for detecting and repairing anomalies. Schema validation platforms like JSON Schema and Avro Schema Registry, however, will ensure incoming data is in the expected format, leaving minimal room for error. Let's take an example of the data validation and cleansing. IIoT sensors on a

production line may intermittently lose connection, leading to missing temperature or pressure readings. This can be imputed using interpolation or forward/backward filling to estimate missing values.

6. **Begin with a Focused Pilot:** To properly demonstrate value for analytics, it is wise to go small and targeted rather than enterprise-wide from day one. A good starting point would be to collect and analyse data from a relatively small group of machines, ideally bottlenecks or business-critical machines. This way, organisations can keep costs low, provide tangible results within a matter of weeks and generate momentum for further analytics adoption.

   To yield maximum benefit from an initial analytics project, collecting data must be aligned with clearly defined business goals. When top priority is given to lowering equipment downtime, then sensor readings should be collected from pieces of equipment that require a high amount of maintenance. To optimise production output, analytics should be geared toward tracking production volume and time-series information so that one can precisely quantify how much product is being produced within a given span of time.

   A practical approach is to start out with analytics applied to a single production asset for which an established use case exists. An acceptable starting point would be predictive maintenance where ML algorithms can be trained from historical and real-time inputs and use these inputs to predict imminent failure prior to its occurrence. TensorFlow, PyTorch, or AWS SageMaker can be employed for building predictive models and real-time telemetry from sensors embedded in IoT devices can be ingested through Message Queuing Telemetry Transport (MQTT) brokers and analysed through Elasticsearch for detecting anomalies.

7. **Scale Analytics Strategically:** Once the analytics value has been established, the next step is to apply these initiatives across other supply chain operations and manufacturing lines. Growth has to be targeted toward areas where present data coverage is low. This can be done by introducing additional sensors, new sources, or enhancing existing systems for full coverage. While businesses scale up analytics scope, they need to accelerate analysis sophistication levels, from mere performance tracking to more complex order quality and effectiveness metrics.

   The disciplined way to scale begins with fundamental production metrics such as throughput and capacity utilisation. Once this fundamental knowledge has been acquired, businesses can introduce secondary quality metrics such as accepted units to total produced units. With this knowledge, manufacturers are able to calibrate their processes, remove waste and maximise overall effectiveness.

   To handle increasing amounts and complexities of data, businesses can employ distributed machine learning libraries such as Spark MLlib, Dask, or Ray for scalable prediction and data handling in large datasets. Digital twin technology on platforms such as Siemens Mindsphere or NVIDIA Omniverse can also be employed in such a way as to simulate manufacturing configurations, potential failures and optimise processes even before becoming problems in real life.

8. **Empower Business Users to Create Customised Dashboards and Reports:** Giving business users the capability to develop reports and dashboards themselves is a component of democratising analytics in an organisation. Equipping them with suitable

tools, organisations can make all employee levels capable of probing and analysing data without requiring IT assistance. Amongst such modern business analytics tools are Tableau, Power BI and Looker, which offer intuitive, self-service interfaces for users to derive insights, visualise trends and track key performance indicators in real-time. SQL-on-Hadoop engines like Presto or Trino offer ad hoc analysis for large datasets and allow users to query and infer insights without requiring special knowledge about data infrastructure.

Except for routine reporting, no-code ML functionality inbuilt into analytics solutions brings it into a higher gear as far as ease of use is concerned. With such functions, factory floors can tap into useful patterns in past data, such as inventory bottlenecks, machine failure forecasts, analysis of waste in resources and impact of production slowdowns on profit margins and revenues. With such advanced tools in the hands of business users, organisations are well-placed to create a data-led organisation, accelerating decision-making and opening up new avenues for process enhancement and expansion.

9. **Design Reports with a Visualisation Focus:** For information to drive action, graphically formatted reports need to be authored and visualisation comes first, leveraging real-time graphs, heatmaps and trend analysis instead of tables. Well-defined graphical reports enable the user to visually discern anomalies at a glance, expose trends and spot patterns, making complex information that much simpler for them and act accordingly. Advanced data visualisation platforms like Grafana and Kibana allow organisations to develop dynamic, web-based dashboards for real-time measurement of important parameters. Advanced solutions provide simplicity through visually appealing interfaces for presenting insights, allowing business users to make quicker data-driven choices without requirements for advanced analytics skills. Advanced visual analytics provide organisations potential for lowering complexity, enhancing predictive insights for effectiveness and enabling well-informed strategic planning.

10. **Utilise Canned Reports Whenever Available:** Finally, industry benchmarks and prebuilt reports are crucial in order to accelerate insight creation and data-informed decision-making. Most analytics platforms have prebuilt reports and dashboards aligned to industry-standard KPIs, enabling organisations to compare themselves to industry averages or leading competitors.

Tools such as SAP BusinessObjects, IBM Cognos and Oracle Analytics provide out-of-the-box solutions for tracking key operational metrics like machine uptime, production efficiency and overall equipment effectiveness (OEE). By leveraging these preconfigured reports, businesses can streamline analysis, identify performance gaps and make data-backed improvements without the need for extensive custom report development. This approach not only saves time and resources but also ensures that organisations stay competitive by aligning their analytics strategies with industry best practices.

Implementing an analytics program in Industry 4.0 is an iterative process that begins with a small-scale project and gradually expands in scope. Manufacturers can leverage analytics-driven insights by integrating data from sales, inventory, fulfilment, production systems, customer experience and third-party sources to enable real-time decision-making and dynamic production adjustments. By systematically incorporating analytics into operations, organisations can unlock valuable insights for predictive maintenance, quality enhancement and operational efficiency. This data-driven approach not only optimises current processes but also lays the groundwork for advanced analytics adoption and full-scale digital transformation, positioning manufacturers for greater agility, innovation and competitiveness in an evolving industrial landscape.

10.4.2 Established Process Models

To ensure a structured and systematic approach to implementing analytics projects, various formal process models have been developed. These models serve as frameworks that guide organisations in executing data-driven projects, ensuring that each phase, ranging from data preparation to model deployment, is effectively managed. In this section, we explore three widely recognised methodologies: Knowledge Discovery in Databases (KDD), Sample Explore Modify Model Assess (SEMMA) and CRoss-Industry Standard Process for Data Mining (CRISP-DM). While these methodologies have distinct stages, they share core principles that focus on data selection, preprocessing, modelling and evaluation before final deployment.

KDD: Knowledge Discovery in Databases

The Knowledge Discovery in Databases (KDD) process, introduced in 1989, is a structured methodology for extracting valuable knowledge from large datasets. It is designed to apply data mining techniques while ensuring that preprocessing and transformation steps are carried out efficiently. KDD consists of five key stages that allow businesses to systematically uncover patterns and insights from their data (Figure 10.12):

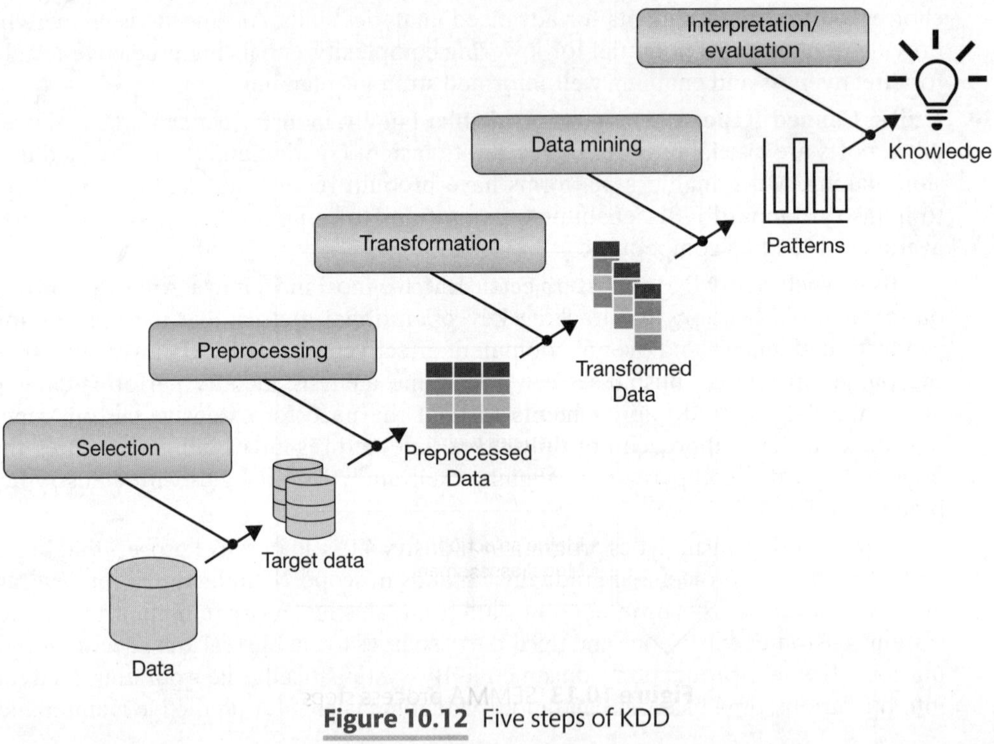

Figure 10.12 Five steps of KDD

- **Selection:** This phase involves defining a target dataset and selecting relevant variables or data samples that will be used for pattern discovery.
- **Preprocessing:** This stage focuses on cleaning and preparing the data, ensuring consistency, handling missing values and removing redundant information.
- **Transformation:** Here, dimensionality reduction or feature transformation techniques are applied to make data more manageable for analysis.
- **Data Mining:** This phase involves applying pattern recognition and predictive modelling techniques to extract meaningful insights from the dataset.
- **Interpretation and Evaluation:** The final stage focuses on validating mined patterns, ensuring they align with business objectives and refining models where necessary.

The KDD process is highly interactive and iterative, requiring domain expertise to make informed decisions at each stage. Before applying KDD, organisations must understand the business problem, incorporate prior knowledge and establish clear goals to ensure that insights can be effectively translated into actionable improvements.

SEMMA: Sample, Explore, Modify, Model, Assess

Developed by SAS Institute, SEMMA is a structured approach for implementing analytics projects, designed to facilitate data exploration and modelling through five essential stages (Figure 10.13). The methodology provides a practical roadmap for structuring data mining projects and is particularly useful for businesses looking to refine and improve predictive analytics models.

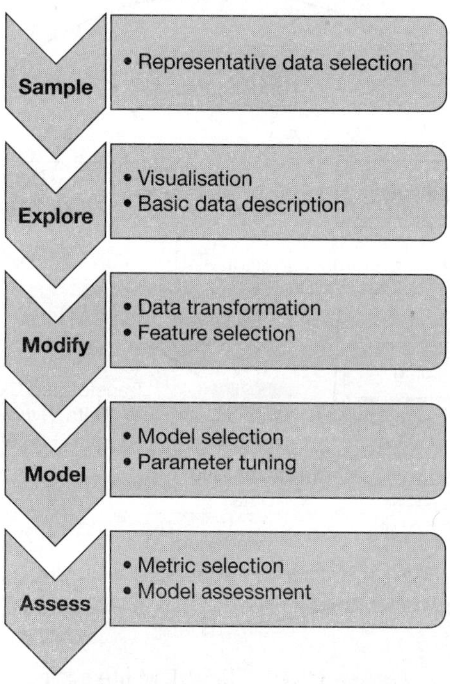

Figure 10.13 SEMMA process steps

- **Sample:** Extract a representative subset of the data that is large enough to capture essential patterns but small enough for efficient processing.
- **Explore:** Conduct exploratory data analysis (EDA) to identify patterns, anomalies and trends in the dataset.
- **Modify:** Refine the dataset by creating, selecting, or transforming variables to improve model performance.
- **Model:** Apply statistical and machine learning models to extract insights and generate predictions.
- **Assess:** Evaluate the model's reliability and effectiveness, ensuring it meets business objectives.

SEMMA provides a structured yet flexible framework, allowing businesses to iteratively refine data mining projects. The methodology ensures that analytics initiatives are organised, repeatable and aligned with business goals, making it a preferred approach for organisations deploying data-driven solutions at scale.

CRISP-DM: Cross-Industry Standard Process for Data Mining

CRISP-DM is one of the most commonly adopted methodologies in analytics, developed as a cross-industry initiative involving Daimler Chrysler, SPSS and NCR. It provides a structured yet adaptable six-phase process (Figure 10.14) that helps businesses translate raw data into actionable insights while ensuring that analytics models align with organisational objectives.

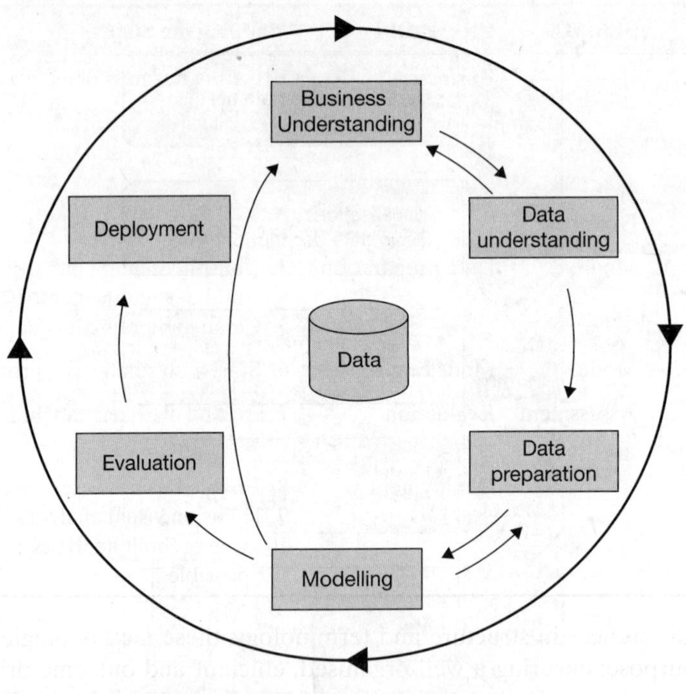

Figure 10.14 CRISP-DM life cycle

- **Business Understanding:** The first phase focuses on defining the problem, identifying key business challenges and converting them into a data mining problem statement.
- **Data Understanding:** This phase involves collecting initial data, exploring its structure and identifying potential quality issues.
- **Data Preparation:** Constructing the final dataset by integrating, cleaning and transforming data for analysis.
- **Modelling:** Applying statistical and machine learning techniques to discover patterns or make predictions.
- **Evaluation:** Validating the model's performance to ensure that it meets business objectives.
- **Deployment:** Implementing the model into a real-world production environment, ensuring users can leverage insights for decision-making.

CRISP-DM is highly adaptable, making it suitable for both small-scale experiments and enterprise-wide analytics implementations. It ensures that analytics projects remain focused on business objectives while providing a structured approach to data-driven decision-making.

Though we see there is a difference between the processes at a conceptual level, we observe more similarities than differences. Table 10.1 gives a summary of the corresponding stages in these three methods.

Table 10.1 Summary of the Corresponding Stages in KDD, SEMMA and CRISP-DM

KDD	SEMMA	CRISP-DM	Mapping the Steps
Pre KDD	-------------	Business understanding	1. Define business objectives.
Selection	Sample	Data Understanding	2. Inventory current data.
Pre-processing	Explore		
Transformation	Modify	Data preparation	3. Consolidate data into a single repository. 4. Build data connectors. 5. Clean and preprocess data.
Data Mining	Model	Modelling	6. Start with small-scale projects.
Interpretation/ Evaluation	Assessment	Evaluation	7. Expand analytics capabilities gradually.
Post KDD	------------	Deployment	8. Enable business users to create reports. 9. Focus on visual analytics. 10. Use prebuilt analytics models where possible.

Despite their differences in structure and terminology, these methodologies serve the same fundamental purpose: ensuring a well-organised, efficient and outcome-driven approach to analytics implementation. By leveraging these frameworks, manufacturers can successfully extract value from data, optimise decision-making and drive innovation in Industry 4.0 and beyond.

10.4.3 From Device to Dashboard – Data Flow

The true value of an IIoT system does not lie merely in the collection of individual sensor events or massive data archives but in the interpretation, analysis and decision-making derived from that data. A vast network of interconnected devices is beneficial only if the data it generates can be processed, analysed and acted upon in a meaningful way. The cloud plays a pivotal role in enabling scalable storage, real-time analytics and deeper data processing, ensuring that manufacturers can uncover patterns, predict failures and optimise operations through data science and machine learning. Figure 10.15 below shows a general data flow from the device to the dashboard in the cloud.

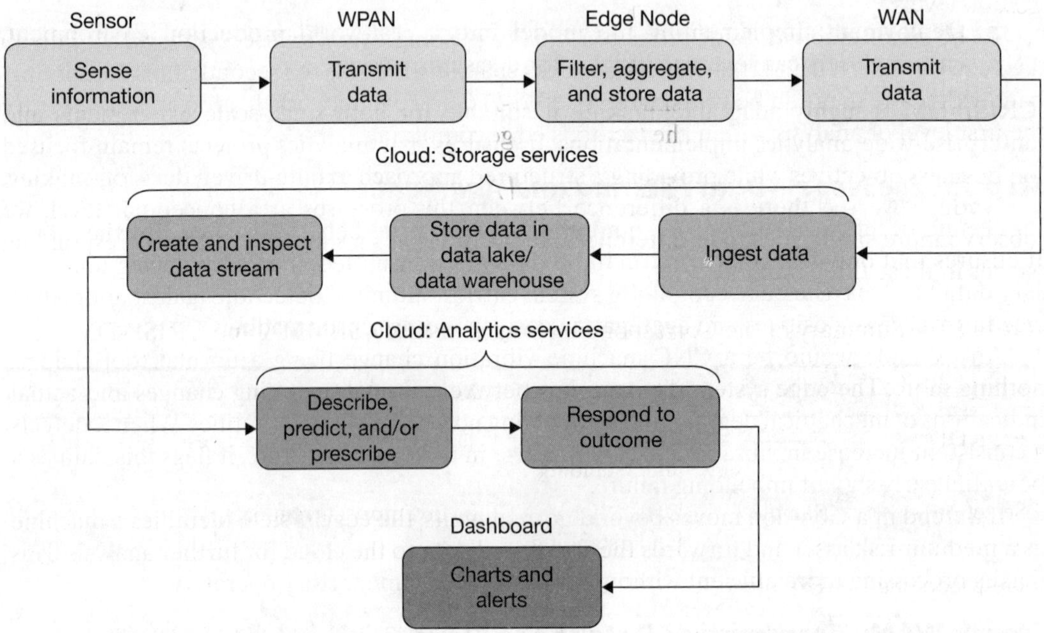

Figure 10.15 Device to dashboard data flow

Let us consider predictive maintenance for Computer Numerical Control (CNC) machines in a smart factory as one example. They are core to precision manufacturing and unwanted failures can result in huge downtime, production schedule delays and loss. Predictive failures can be prevented by deploying a sensor-to-cloud analytics pipeline, allowing for scheduled proactive maintenance, while downtime is kept at a bare minimum. The following are the steps by which data moves from sensor detection through the seven steps.

Step 1: Sensor – Sensing Information at the Edge

The first stage occurs within the CNC machine itself, where onboard acoustics, temperature and vibration sensors constantly monitor the machine condition. The sensors stream real-time data related to mechanical stress, heat generation and efficiency, providing a comprehensive picture of machine condition. A sensor can, say, sense increased frequency deviation from its operating baseline, suggesting wear and tear in a spindle bearing. Meanwhile, a sensor can sense unusual motor heating to a degree never seen before, suggesting lubricant problems or friction. The data at this stage is uninterpreted and unprocessed and must be transmitted and interpreted further to provide useful information.

Step 2: WPAN – Local Communication and Data Transfer

Once CNC machine sensor data has been generated, it must be transmitted to a local gateway or factory network. It accomplishes this over a Wireless Personal Area Network (WPAN) connection or industrial Ethernet, depending on factory deployment. A Zigbee gateway can gather data from multiple machines and offer transparent communications to the rest of the network. Here, the objective is to get sensor readings across with low latency and data congestion levels.

To be more efficient, aside from continually pushing every millisecond-level vibration, data can be collected and transmitted in blocks spanning a whole ten seconds, thus conserving bandwidth. The data can be transmitted to a Wi-Fi 6 or a 5G network, leading to a transfer to the first level of analysis within the factory's edge computing.

Step 3: Edge Node – Data Filtering and Aggregation

Processing occurs at a local edge computing server before being transmitted to the cloud. It ensures that only value-added, structured data are transmitted, thereby reducing unnecessary data burdens. The edge computing system carries out initial detection against anomalies, weeding out false positives and aggregating corresponding sensor readings.

An example would be a CNC machine vibration change being a normal tool change, nothing more. The edge system distinguishes between normal operating changes and actual indications of mechanical degradation by applying noise cancellation routines. When it detects a consistent increase in vibration level with a rise in motor temperature, it flags this data as a potential early sign of impending failure.

If a trend in a vibration moves beyond normal limits, the edge system identifies a machine as a medium-risk asset and forwards the structured data to the cloud for further analysis. This makes processing more efficient without impacting real-time factory operations.

Step 4: WAN – Transmitting Data to the Cloud

After preprocessing, structured data is transported through a Wide Area Network (WAN) to a cloud analytics platform. Typically, it happens over 5G, fibre-optic cable, or IIOT messaging protocols like MQTT. Since manufacturing installations require data to be communicated securely with minimal latency, WAN technologies provide filtered data to cloud servers in a timely fashion.

Here, data from vibrations and temperature anomalies from a variety of CNC machines are transmitted to a cloud repository within AWS, Azure, or Google Cloud. No sensor data is transmitted, although aggregate indicators by machine in terms of average frequency of vibrations, peak temperature and risk level are transmitted. This preserves network bandwidth without sacrificing data integrity and availability for later analytics.

Step 5: Cloud – Data Ingestion, Storage and Inspection

It flows into the ingesting pipeline once it arrives in the cloud, where it gets processed for real-time monitoring and long-term data storage. The cloud platform feeds arriving data into structured formats in order to put it into a standardised form for storing. It forwards data to a data repository such as HDFS, Amazon S3, or Azure Data Lake, where data is kept for historical trend analysis and to train machine learning models.

Meanwhile, a real-time analytics engine (such as Apache Kafka or AWS Kinesis) is continually monitoring data for deviations. If a CNC machine experiences a sudden boost in its vibration frequency over a 24-hour window, a cloud platform marks it a high-risk device and schedules it for further analysis by AI.

Step 6: Cloud Analytics Service – Predict, Infer and Respond

During this stage, advanced machine learning algorithms analyse historic data and sensor data in real-time to create predictive trends and predict potential failures. The system evaluates multiple sensor readings through deep learning platforms such as TensorFlow or PyTorch to identify impending failures. For instance, if past data from comparable CNC machines indicate that vibrating frequencies above 40 Hz, along with temperature increases, lead to spindle bearing failure, then the model forecasts a probability score for impending failure. If a probability level above 90% is reached, then the system automatically recommends rescheduling a check within a span of 48 hours. Additionally, prescriptive analytics evaluates multiple approaches to maintenance and suggests the most cost-effective. If, for example, replacing a spindle bearing at this point in time prevents a costly future failure, this approach puts this recommendation foremost for maintenance engineers.

Step 7: Dashboard – Visualisation and Alerts

The final step in the process involves presenting predictive insights to factory managers and maintenance teams through an interactive dashboard. The processed data, risk scores and recommended actions are visualised on platforms like Power BI, Tableau, or Grafana, allowing decision-makers to take proactive measures.

The dashboard displays the live health status of all CNC machines, with colour-coded alerts indicating their risk levels. Machines classified as low risk remain green, those in medium-risk zones appear yellow and those in high-risk categories turn red, signalling urgent action. Additionally, the system automatically schedules preventive maintenance for flagged machines and notifies technicians via email or SMS alerts.

In this case, the factory manager sees that three CNC machines are in a moderate-risk category, while one machine is at high risk of failure. The system has already scheduled a preventive maintenance check for tomorrow morning, preventing unplanned downtime.

10.5 BUSINESS USE CASES

As mentioned in Section 10.4, manufacturers leverage data analytics to enhance the efficiency of their operations and supply chains, gaining valuable insights into KPIs. This section explores several manufacturing use cases where analytics played a crucial role.

- **Manufacturer of Refractory Products:** This multinational manufacturer is the largest supplier of refractory products in the United States. The company is over 150 years old and has close to 20 manufacturing plants in the US and Canada and sourcing centres spread across North America, Europe and Asia. However, their data collection and analysis were challenging due to acquisitions and various applications used over the years. The company overcame these challenges by consolidating data and applications into a single cloud ERP system. This move resulted in improved forecast accuracy, reduced worker overtime, optimised inventory levels and enhanced on-time delivery, achieving a rate of over 90%.

- **Leader in Digital Storage Solution:** After acquiring two large companies in the last decade, this manufacturer faced issues with data reporting workflows due to the different data and workflow platforms used by these companies. Data came from over 2,000 applications deployed on these diverse platforms, resulting in data warehouse updates taking in excess of eight hours. By the time reports were available, data were already one to two days old, essentially depriving business users of real-time analytics. To address this, the organisation standardised its data and workflows on a new cloud-based system, providing business leaders with analytical data in about 20 minutes through pre-configured reports. This consolidation streamlined workflows and ensured consistency in data and reports for all leadership teams.
- **Mechatronic Device Manufacture:** This privately held Italian manufacturer of mechanical and electronic devices for home appliances, HVAC (heating, ventilation and air conditioning), EV (Electric Vehicle) charging, smart meters and automotive has 18 plants in different parts of Italy. They created an analytics dashboard as a single destination for insights. They used Oracle cloud analytics to eliminate silos, allowing managers to easily create the reports they need. This simplifies data aggregation and helps manage manufacturing processes more effectively. This significantly reduced the time executives spend searching for data and enabled more data-driven decisions than instinct-driven ones.
- **Manufacturer of Aluminium Extrusion:** A premium manufacturer of custom-fabricated aluminium extrusions serving industries such as building and construction, automotive and industrial equipment, faced interoperability issues with data across its five manufacturing plants. Plant managers were manually combining data from spreadsheets into a homegrown ERP system and attempting to integrate it with HR and Finance data, leading to inconsistencies and poor detection of global supply shortages, underperforming suppliers and inefficient prioritisation of client orders. By implementing a new cloud-based ERP and analytics platform, the company improved purchasing and inventory decisions (for example, which products are in the highest demand), identified process bottlenecks (such as supplier delays) and made necessary adjustments (such as reassignment of resources and personnel) to meet customer demands more effectively.

10.6 CONCLUSION

The integration of analytics into manufacturing has fundamentally reshaped how organisations optimise operations, improve efficiency and make data-driven decisions. By leveraging structured data collection, predictive modelling and visualisation tools, manufacturers have transitioned from reactive problem-solving to proactive, strategic decision-making. This shift has resulted in measurable improvements in operational efficiency, reduced downtime, enhanced quality control and supply chain optimisation, positioning data-driven manufacturers at a significant competitive advantage.

Market forecasts further underscore the growing importance of analytics in manufacturing. In 2022, the global industrial analytics market was valued at approximately $32.1 billion, with projections suggesting it could reach around $117 billion by 2032, reflecting a compound annual growth rate (CAGR) of over 12%. Some estimates suggest even faster expansion, predicting the market will grow from $40.42 billion in 2023 to $150.15 billion by 2032, with a CAGR of 15.8%. These figures highlight the increasing reliance on advanced analytics to optimise manufacturing processes, reduce costs and improve decision-making.

Looking ahead, several emerging trends will continue to drive the evolution of manufacturing analytics. AI and ML will become even more integral, powering predictive and prescriptive models that enable automated decision-making and real-time process optimisation. Edge computing will gain prominence, allowing manufacturers to process and analyse data closer to the source, reducing latency and enabling real-time operational insights. The adoption of digital twin technology will further revolutionise industrial analytics by creating virtual replicas of physical assets, allowing businesses to simulate different production scenarios, optimise efficiency and detect potential issues before they occur.

Meanwhile, augmented and guided analytics will democratise access to insights, empowering non-technical users to interact with data intuitively and extract meaningful conclusions without requiring deep technical expertise. Manufacturers will also increasingly integrate external data sources, such as weather patterns, geopolitical risks and economic indicators, into their analytics models. This will enable more resilient supply chain management, allowing businesses to anticipate and mitigate disruptions before they impact production.

The future of manufacturing analytics lies in greater automation, smarter decision-making and seamless integration of cutting-edge technologies, ensuring businesses remain agile, competitive and forward-thinking in the rapidly evolving industrial landscape.

SUMMARY

- Analytics is Essential for Industry 4.0. As industrial processes generate vast amounts of data, analytics transforms raw data into actionable insights, improving efficiency, decision-making and competitiveness.
- Types of Analytics
  1. Descriptive Analytics
     - Summarises past data to understand trends and patterns.
     - Uses statistical measures (mean, median, clustering and association rules).
     - Helps in quality control and identifying process inefficiencies.
  2. Predictive Analytics
     - Forecasts future events using historical data.
     - Includes regression (for numerical predictions) and classification (for categorising data). Also, time series forecasting, survival analysis and anomaly detection.
     - Applications: predictive maintenance, energy consumption optimisation, supply chain forecasting.
  3. Prescriptive Analytics
     - Recommends optimal actions based on predictive insights.

- Uses operations research techniques like linear programming and heuristic search.
- Applications: optimising production schedules, minimising costs and improving supply chain efficiency.

- Key Applications in Industry 4.0
  - Predictive Maintenance: Identifies machine failures before they happen, reducing downtime.
  - Quality Control: Uses classification models to detect defects in production.
  - Supply Chain Optimisation: Enhances logistics, inventory management and demand forecasting.
  - Energy Efficiency: Predicts energy consumption patterns to reduce operational costs.
- 10-Step Process for implementing Analytics for Industry 4.0
  1. Define business objectives.
  2. Inventory current data repositories.
  3. Consolidate data into a single repository.
  4. Build data connectors.
  5. Clean and preprocess data.
  6. Start with small-scale projects.
  7. Expand analytics capabilities gradually.
  8. Enable business users to create reports.
  9. Focus on visual analytics.
  10. Use prebuilt analytics models where possible.
- Process Models
  - KDD (Knowledge Discovery in Databases): Focuses on data selection, preprocessing, transformation, mining and interpretation.
  - SEMMA (Sample, Explore, Modify, Model, Assess): A structured approach to analytics project execution.
  - CRISP-DM (Cross-Industry Standard Process for Data Mining): A widely used framework emphasising business understanding, data preparation, modelling, evaluation and deployment.

EXERCISE

Multiple Choice Questions (1 Mark)

1. What is the primary role of analytics in Industry 4.0?

(a) Replacing human decision-making

(b) Automating all industrial processes

(c) Extracting insights from data to improve efficiency and decision-making

(d) Eliminating the need for human intervention in manufacturing

2. Which of the following is not a category of analytics?
   (a) Descriptive Analytics
   (b) Predictive Analytics
   (c) Adaptive Analytics
   (d) Prescriptive Analytics
3. What is the primary question that descriptive analytics aims to answer?
   (a) What has happened?
   (b) What will happen?
   (c) What should I do?
   (d) Why will it happen?
4. Time series analytics is primarily used for:
   (a) Generating historical reports
   (b) Forecasting future events based on past data
   (c) Identifying past inefficiencies
   (d) Manually analysing data trends
5. Which analytics approach focuses on recommending actions for optimal decision-making?
   (a) Descriptive Analytics
   (b) Predictive Analytics
   (c) Prescriptive Analytics
   (d) Diagnostic Analytics
6. Which of the following is a commonly used predictive analytics technique?
   (a) Linear Regression
   (b) Histogram Analysis
   (c) Pareto Charting
   (d) Root Cause Analysis
7. How does predictive maintenance improve industrial efficiency?
   (a) By repairing machines only after they fail
   (b) By preventing unexpected failures through early detection
   (c) By reducing the need for skilled maintenance staff
   (d) By increasing downtime for machine inspections
8. Which of the following is a key benefit of analytics in supply chain management?
   (a) Reducing supplier relationships
   (b) Forecasting demand fluctuations
   (c) Eliminating inventory management
   (d) Removing human intervention from logistics
9. Which of the following best describes clustering in descriptive analytics?
   (a) A technique used to group similar data points together
   (b) A method to predict future failures

(c) A process for automating supply chain operations
(d) A way to manually organise data

10. Which of the following best describes a decision tree in predictive modelling?
(a) A method that randomly selects data points for predictions
(b) A tree-based model that splits data into branches based on conditions
(c) A purely statistical method with no visual representation
(d) A technique used only for unsupervised learning

11. Which machine learning model is commonly used for detecting anomalies in data?
(a) Logistic Regression
(b) Decision Trees
(c) Isolation Forest
(d) K-Means Clustering

12. What is the purpose of survival analysis in analytics?
(a) Predicting the lifespan of equipment or time until failure
(b) Analysing historical sales data
(c) Enhancing customer satisfaction surveys
(d) Measuring website traffic

13. How does cloud computing enhance industrial analytics?
(a) By reducing the need for data collection
(b) By providing scalable data storage and real-time processing
(c) By eliminating security concerns in manufacturing
(d) By reducing the need for trained data scientists

14. Which methodology is widely used in analytics project implementation?
(a) CRISP-DM
(b) SWOT Analysis
(c) Gantt Charting
(d) Pareto Optimisation

15. What is the key goal of anomaly detection in industrial analytics?
(a) Identifying patterns in normal data behaviour
(b) Detecting unusual patterns that could indicate failures
(c) Eliminating noise in datasets
(d) Predicting customer satisfaction levels

16. Why is data quality important in analytics implementation?
(a) Poor quality data leads to incorrect insights and decisions
(b) Data quality has no impact on analytics outcomes
(c) Only structured data matters for analytics
(d) It reduces the number of analytics models required

17. A logistics company is struggling with delivery delays due to weather and traffic. How can analytics help them improve on-time delivery?
    (a) Use historical data to identify patterns and predict potential delays
    (b) Eliminate all human decision-making in route planning
    (c) Reduce the number of drivers to optimise scheduling
    (d) Switch to manual tracking instead of automated systems
18. An automotive manufacturer wants to optimise its supply chain by ensuring the right amount of raw materials is available at the right time. Which analytics approach should they prioritise?
    (a) Predictive Analytics
    (b) Descriptive Analytics
    (c) Heuristic Search
    (d) Clustering
19. A factory producing electronic components wants to minimise product defects and improve quality control. Which combination of analytics techniques should they apply?
    (a) Clustering and Anomaly Detection
    (b) Time Series Analysis and Classification
    (c) Regression Analysis and Association Rules
    (d) Classification and Computer Vision
20. A company wants to determine the most profitable combination of products to manufacture while considering labour, material costs and demand. What technique should they use?
    (a) Clustering
    (b) Prescriptive Analytics with Linear Optimisation
    (c) Predictive Analytics with Neural Networks
    (d) Time Series Forecasting

Short Answer Type Questions (2 Marks)

1. What is the primary goal of analytics in Industry 4.0?
2. Differentiate between predictive and prescriptive analytics.
3. Why is anomaly detection important in manufacturing?
4. What are Key Performance Indicators (KPIs) in manufacturing analytics?
5. What is the significance of survival analysis in industrial settings?
6. What is the role of time series forecasting in smart manufacturing?
7. How does prescriptive analytics support supply chain optimisation?

8. What is the difference between business intelligence and analytics?
9. How do manufacturers use analytics to minimise product defects?
10. What are some common challenges in implementing industrial analytics?

Long Answer Type Questions (5 Marks)

1. Explain the role of analytics in Industry 4.0 and how it contributes to industrial efficiency.
2. Differentiate between descriptive, predictive and prescriptive analytics with examples.
3. Discuss the importance of predictive maintenance and its impact on industrial operations.
4. How does anomaly detection contribute to quality control and cybersecurity in manufacturing?
5. Explain how time series forecasting is used in Industry 4.0 for optimising production and supply chain management.
6. Discuss the challenges in implementing industrial analytics and how businesses can overcome them.
7. Explain the concept of Key Performance Indicators (KPIs) in manufacturing and how they are tracked using analytics.
8. Compare and contrast the KDD, SEMMA and CRISP-DM methodologies for implementing analytics projects.

Digital Twins

CHAPTER 11

OBJECTIVES

The objective of this chapter is to venture into the world of digital twins. A Digital Twin is a virtual representation of real-world entities and processes synchronised at a specified frequency and fidelity. It integrates real time and historical data to depict past and current states while simulating potential future scenarios.

By the end of this chapter, you should be able to:

- Understand the evolution of the digital twin concept.
- Comprehend different types of digital twins and their usage, digital thread and its relationship with digital twins.
- Explore augmented reality, simulations and their alignment with digital twins.
- Understand the roles of digital twins across various sectors and their implementation in the design, build and operation phases of products and processes using IoT, Massive Data, hyper-connective networks, cloud services, AI and Software Applications.

11.1 INTRODUCTION

Invention requires failure and experimentation because, by definition, an invention can only be learned via trial and error. Therefore, until the target level of product quality is reached, the iterative process of design, development, testing, lessons learned from failures and correction will continue.

The Wright Brothers used to dream of building a manned flying machine. They started building the model in 1899. In 1903, they became the first to fly 'Wright Flyer', a power-controlled, heavier-than-air flying machine. Almost 4 years of numerous designs, building models, testing and continuous failures shaped their Flyer. The insights from each failure were used to improve their design. But this was time-consuming and super expensive work. Gradually, industries started making scaled miniature representations or models to visualise the would-be actual product or process. However, that also did not reduce the uncertainty of real-life challenges in a product, a process, or an assembly line until the whole system was built. One can imagine a shop floor where thousands of machines are running and the amount of effort and downtime required to pinpoint the exact issue in case of any failure.

The idea of simulating a physical product or 'twinning' both before it is built and throughout its post-build operations and maintenance phases started shaping up during the early 1960s by NASA for Space Missions. For the Apollo 13 Mission, NASA used 15 simulators to train the Crews, the Ground Engineers and the Mission Controllers. Simulators were adopted to match the conditions of actual spacecraft. A high-fidelity model was used to explore solutions and predict the results. By building the physical replica and using the data from actual spacecraft, the engineers on the ground could work through different scenarios. On its way to the moon on 13 April 1970, Apollo 13 got into an unprecedented problem. Its one oxygen tank ripped apart and damaged another unit inside the spacecraft. Ground Engineers simulated potential fixes using the physical replica of the spacecraft and then sent the instructions to the crew for implementing the workaround. Four days later, Apollo 13 successfully came back to Earth. Today, with hyper-connective networks, IoTs, affordable wireless sensors, high-performance computing capabilities, super large storage availabilities, cloud services, software with modelling and data analytical capabilities and application of Artificial Intelligence (AI), the physical replicas can be digitally simulated fully. The physical and digital replicas could be seamlessly connected and exchange information for near real time analysis of the past, current state and future. This physical replica is called a 'Physical Twin', while its digital replica is termed a 'Digital Twin'.

11.2 WHAT IS DIGITAL TWIN

11.2.1 Definition

Digital Twin Consortium, an industry association working to build the market and recommend standards, came up with the following definition of digital twin on 3 December 2020, considering various use cases of digital twin.

> **'A digital twin is a virtual representation of real-world entities and processes, synchronised at a specified frequency and fidelity'.**

- Digital twin systems revolutionise business processes by expediting a comprehensive understanding of assets like machines, equipment, infrastructure; operation processes and work flows; interactions among different assets and processes, facilitating optimal decision-making and enabling effective action.
- Digital twins utilise real time and historical data to depict the past and current states while simulating potential future scenarios.
- Digital twins are driven by results, customised for specific use cases, supported by integration, constructed upon data, informed by domain expertise and integrated into IT/OT systems.

It should be interesting to explore the fundamental components of this definition as per the Digital Twin Consortium.

- **Virtual Representation:** It is a set of correlated digital models and supporting data that provides consistent and organised information about the subject matter, i.e., real-world entities or processes. Digital models encompass a variety of forms, such as 3D CAD models, mathematical equations, databases, etc. Digital models can be divided into the following categories:

- **Representational Model:** It comprises structured information that depicts the states of entities or processes.
- **Computational Simulation Model:** It is an executable model of a process comprising both data and algorithms.

These can act as digital representations and computational simulations illustrating the past, present and potential future of real-world entities and processes.

- **Real World Entities and Processes:** This implies any physical entity in a real space. It can be any perishable product, i.e., an engine, machine components, medical equipment, spacecraft, buildings, a city, human organs, or human beings as a whole. These entities can connect and interact with each other, creating systems that subsequently execute processes such as satellite launches, automobile production lines, oil refineries, shipping operations, or even the functioning of a medicine on an organ.

  It is essential to understand that physical entities consist of not only the objects themselves but also the environments in which they exist and interact, such as the temperature and humidity of the shop floor. This leads to digital twin technology's third fundamental building block: synchronising the virtual and real worlds.
- **Synchronisation:** Synchronisation aims to create a near resemblance of virtual representation to the real world and vice versa. Every digital twin relies on synchronisation of observation and intervention mechanisms. Observation is capturing real-world changes and reflecting them in the virtual model (e.g., detecting an increase in water level), while intervention implies applying virtual changes to influence the physical world (e.g., opening floodgates).

The speed and frequency of this synchronisation might vary depending on the rate of change of the real-world entities and the methods used for real-to-virtual synchronisation. The frequency of synchronisation can be different, such as real time, daily, or milestone-based updates. For digital twins that utilise multiple synchronisation methods, the synchronisation frequency may differ for each method. Additionally, there might be distinct synchronisation frequencies for observation and intervention, varying from one entity to another. The synchronisation should happen with *appropriate fidelity*. Fidelity refers to the level of precision and accuracy applied to both the virtual representation and its synchronisation methods.

Thus, in the world of digital twins, all products and systems are dual in nature. The system has a physical version and a digital or virtual or information version. Most importantly, these two versions are linked together throughout their lifecycle. This is depicted in Figure 11.1.

11.2.2 Evolution of Digital Twin

David Gelernter, professor of Computer Science at Yale University, in his 1991 book 'Mirror Worlds', anticipated something like a digital twin and called it a Mirror World. In his 'Mirror World', the World Wide Web is not merely a linked network of documents and files but an immersive virtual realm. In this online environment, numerous facets of public life are represented as an extension and a carefully chosen reflection of the physical world. However, in its

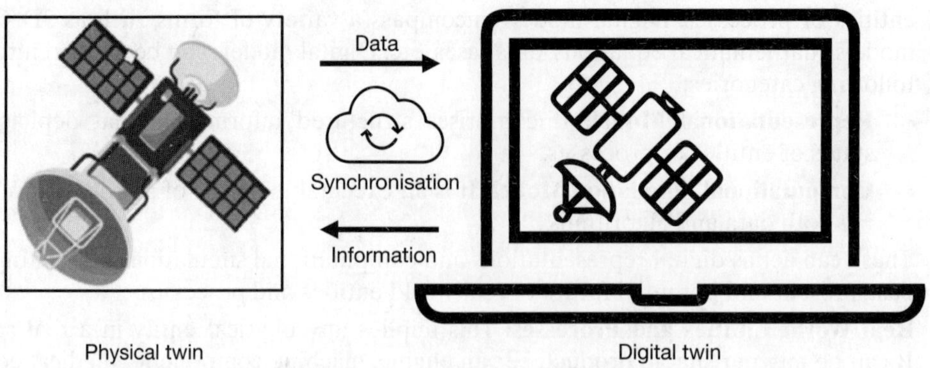

Figure 11.1 Digital twin – a living digital model of the physical twin

current form, the Digital Twin Model was first introduced by Dr. Michael Grieves in 2002 as a conceptual model for Product Lifecycle Management (PLM), although it was not initially given a specific name. Subsequently, it was referred to as the 'Mirrored Spaces Model' (MSM) in 2005 by Grieves, which he later modified to the 'Information Mirroring Model' in 2006. Finally, it became known as the Digital Twin in 2010, a term initially coined by John Vickers of NASA in a Roadmap Report with Grieves. Despite the name change, the core concept and model have remained consistent. The evolution of the name of the digital twin is depicted in Figure 11.2.

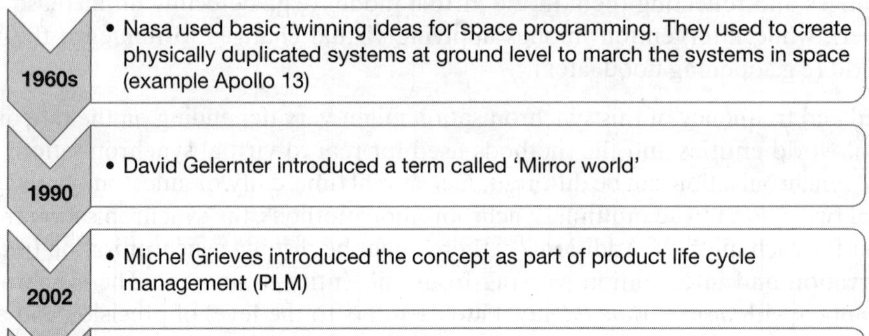

Figure 11.2 How the name 'digital twin' evolved

The digital twin concept has been evolving in terms of an increase in capabilities and fields of application over the last one and a half decades. There is massive potential for further evolution. Today, creating digital replicas involves merging multiple technologies, for example, IoT or sensor integration, robust Big Data systems and advanced processing and analysis tools, such as AI algorithms. We are in the 4th Industrial Revolution and a massive transformation is happening in how things are designed, built and maintained. Digital twins can replicate their real-world counterparts, offering substantial benefits to various professionals, including industrialists, entrepreneurs, urban planners, manufacturers and more. Industry 4.0 provides the ideal setting for digital twins' growth and significant contribution, underscoring their importance.

Let us explore Figure 11.3 below, which demonstrates the roadmap for the evolution of digital twins.

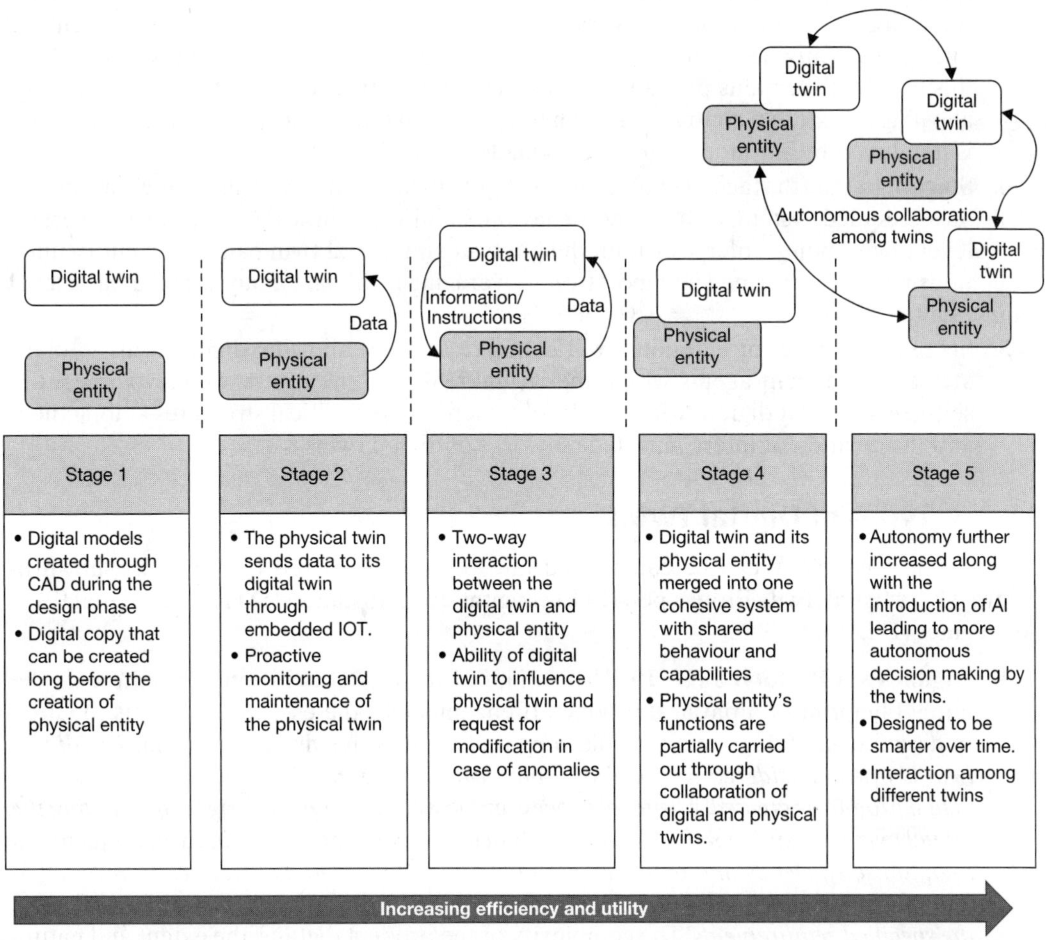

Figure 11.3 Evolution of digital twin

The stages of evolution are as follows:

1. **Stage 1:** Digital models were prepared through CAD tools during the product design phase. CAD had features to keep the revision history. Any previous version could be easily modified and a new design version could be published. Thus, complete traceability of evolution could be established, introducing the 'Digital Thread' concept. We will discuss 'digital thread' later in the chapter.
2. **Stage 2:** The physical entity begins transmitting data to its digital replica, mirroring the current status of the physical entity. This ongoing exchange empowers the manufacturer and individuals involved in operating and maintaining the product to oversee its physical counterpart actively. The alignment of the physical and digital depictions of a product, system, or process is commonly denoted as a Digital Shadow.
3. **Stage 3:** At the 3rd stage of evolution, digital twins enable two-way interaction with their physical counterparts. This transforms digital twins from passive representations into active components that influence the physical entity. However, in most cases today, this capability remains partially realised. The digital twin primarily acts as a gateway, allowing an external controller system to operate on the physical entity. This interaction is not driven by autonomous decisions made by the digital twin itself.
4. **Stage 4:** In the 4th stage of evolution, the digital twin and its physical counterpart merge into a unified system with shared behaviours and capabilities. At this stage, a certain level of autonomy emerges within the physical and digital twin pair. Some entity functions may be performed collaboratively, leveraging local data analysis and a basic level of intelligence.
5. **Stage 5:** The level of autonomy will continue to grow, allowing digital twins to operate as independent agents within the virtual realm. In the future, we may witness the self-generation of digital twins, the development of hierarchical structures among them and the creation of interconnected networks of digital twins.

11.2.3 Types of Digital Twins

Digital Twin technology can be categorised into three types based on usage, as shown in Figure 11.4: Digital Twin Prototype (DTP), Digital Twin Instance (DTI) and Digital Twin Aggregates (DTAs).

1. **Digital Twin Prototype (DTP):** This variant of the digital twin concept represents the initial blueprint of a physical product. It encompasses the essential data sets required to describe and fabricate a tangible copy that mirrors the virtual rendition. *The digital twin is often considered a 'clone' of a physical product, but it usually exists before the actual product is created. The ideal process involves developing and rigorously testing the digital twin to ensure it meets the desired functionalities. Next, we digitally manufacture it to optimise efficiency and manage its simulations throughout its lifecycle. We proceed to manufacture the physical product only when the product design aligns satisfactorily with the specified requirements.* The complexity of the product dictates the extent and nature of the required data and information.

   In essence, products or systems that involve high complexity, substantial risk, a need for continuous monitoring and improvement and where digital simulations can enhance design, performance and maintenance are prime candidates for DTPs.

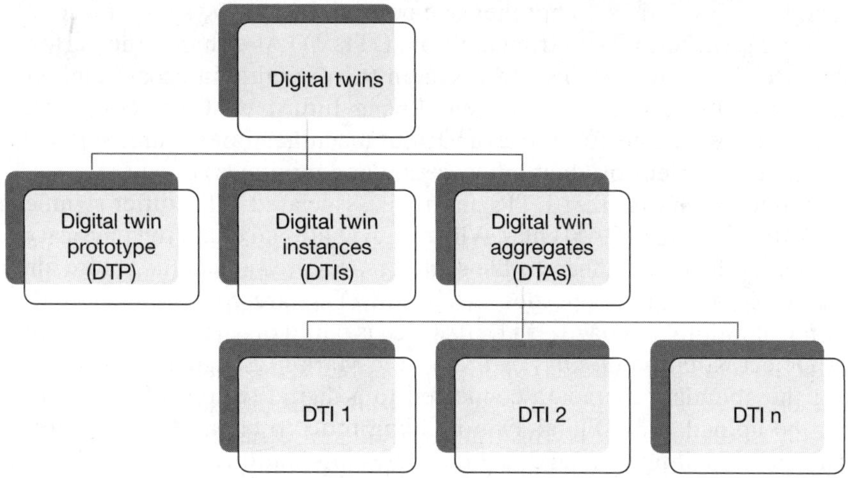

Figure 11.4 Types of digital twins

2. **Digital Twin Instance (DTI):** DTI is a digital twin of a specific product. It reflects the status, condition, and history of the physical product at every stage of the product life cycle. DTIs get integrated with other digital devices like IoT sensors. The real time data collected from these sensors is used to update the DTI so that it accurately reflects the physical product. DTIs contain vast range of information depending on the requirements of the product and industry. DTIs are established exclusively for products where it is fundamental to maintain a comprehensive record of the product's data throughout its lifecycle for operational efficiency, safety, and optimisation. This can include the following:
   - **Geometric Information:** Details about the product's physical shape, size and configuration.
   - **Service Information:** Maintenance schedules, repair records and service history data.
   - **Part Changes:** Information regarding component replacements, upgrades and modifications over time.
3. **Digital Twin Aggregates (DTAs):** DTAs represent a higher-level concept within the digital twin framework, combining data and functionality from multiple DTIs. Unlike DTIs, DTAs may not always exist as standalone data structures; they can be more like computational constructs with access to and interaction with various DTIs. DTAs offer two essential representations:
   - **Longitudinal (Insight):** DTAs analyse historical data from DTIs to identify patterns and correlations. For instance, they can predict component failures based on historical sensor data patterns. This involves learning from past asset behaviour to make future predictions.
   - **Latitudinal (Learning):** DTAs can facilitate collective learning within a system. If certain DTIs within a group acquire knowledge or insights through actions, DTAs can share this learning across all DTIs, contributing to improved system-wide performance and efficiency.

In summary, DTAs provide a comprehensive perspective on a system or a set of assets by aggregating and scrutinising data from multiple DTIs. DTAs enhance the performance and dependability of intricate systems and assets in the digital twin ecosystem. For example, Manufacturing plants, especially those employing Industry 4.0 principles, use DTAs to monitor and optimise the performance of various machinery and equipment on the factory floor. This ensures efficient production processes and minimises downtime.

It is crucial to emphasise that DTPs and their associated DTIs differ significantly from Computer-Aided Drawing (CAD) files. While a CAD file can detail the physical dimensions of a component, it lacks the inherent file structure required to capture, store and maintain comprehensive component properties. DTPs and DTIs are commonly modelled through Digital Twin Definition Language (DTDL), which is based on structured file formats such as JavaScript Object Notation (JSON) or Extensible Markup Language (XML) for the management of the abundant metadata connected to a digital twin. DTDL JSON/ XML files would then be uploaded in Digital Twin building tools to create the DTIs for the assets. DTDL is a data modelling language that facilitates integration with IoT. DTDL was developed by Microsoft and subsequently made available as an open source. Figure 11.5 represents the relationship among digital twin concepts, i.e., DTP, IoT-enabled physical entities, DTI and DTAs for the life cycle management of industrial smart machines.

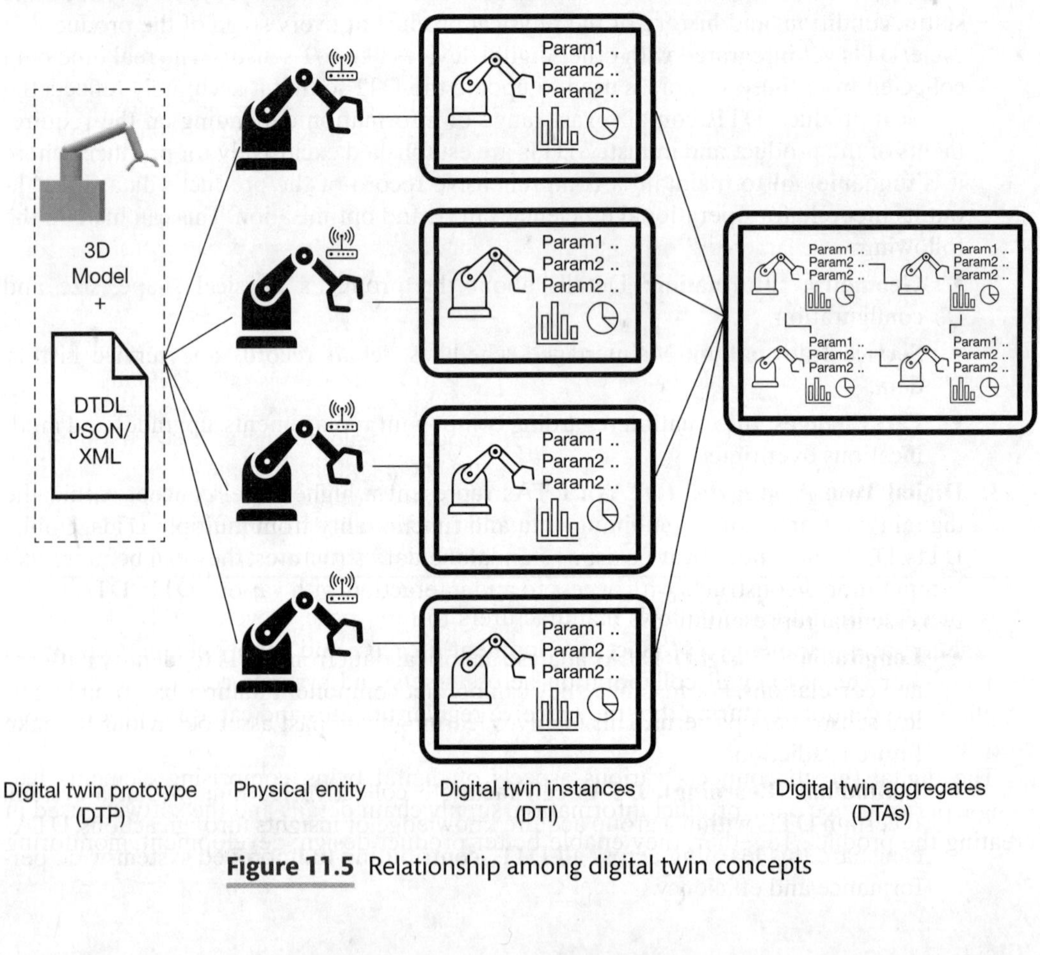

Figure 11.5 Relationship among digital twin concepts

11.2.4 Digital Thread and Its Relationship with Digital Twins

The essence of a digital thread revolves around achieving complete digitisation and traceability throughout the entire product life cycle.

Digital thread uses various technologies, including CAD tools, PLM systems and IoT sensors, to facilitate data flow throughout the product's life cycle. The thread ensures traceability in a multi-party, multi-process composite framework by enabling data sharing, real time monitoring, collaboration and integration. This enhances transparency, communication and efficiency across the product lifecycle, improving decision-making and product quality.

This approach is highly valuable in optimising physical and digital processes across the entire product value chain. The E2E traceability through digital thread generates immense value. For example, in the case of a vehicle accident linked to a system malfunction, such as unintended acceleration, the digital thread, with its thorough lifecycle traceability, can precisely identify the issue. By utilising model-based systems and engineering capabilities, it becomes feasible to conduct a detailed examination of the vehicle's initial design concepts to determine the root cause of the problem.

Let us briefly review the characteristics and benefits of digital twins and digital threads as shown in Table 11.1.

Table 11.1 Digital Twin vs Digital Thread

Category	Digital Twin	Digital Thread
Scalability	Scalable, but to a limited extent. These systems gather real time data from a single source or asset and while they can connect to other digital twins to simulate larger environments (DTAs), their primary utility lies in evaluating specific production environments.	This is highly scalable and can connect to nearly any other enterprise system, *including digital twins*. Therefore, digital threads are better suited for scenarios requiring data collection from various departments, devices, systems and processes.
Data Management	Collects sensor data and compiles it in a readily accessible database.	Collects data not only from sensors but also from digital twins and other sources, centralising data flow across departments and production silos for universal access within the entire organisation. Data is expected to be more accurate due to zero or minimal manual intervention in data transmission among different departments and workflows.

In essence, the digital thread allows manufacturers to repurpose, recycle and track product information across the entire product development lifecycle and supply chain. It enhances inter and intra-departmental collaboration across assets and systems connecting partners, suppliers and vendors, ensuring that they receive regular updates and can trace product data consistently.

The digital thread connects various aspects of digital twins, comprising elements like design, performance data, product information, supply chain details and the software used in creating the product. Together, they enable better product design, development, monitoring and maintenance while ensuring end-to-end data traceability and continuity and, thereby, better regulatory compliance.

11.2.5 Digital Twin vs Simulation

Before trying to discover the differences between digital twin and simulation, let us try to understand briefly what simulation is.

Simulation is an approach that mimics the operation of real-world processes or systems using models. These models capture the fundamental behaviours and attributes of the selected process or system, demonstrating how the model progresses under different conditions over time.

Simulations are digital models used to mimic the behaviour of systems or processes in various real-life conditions without physically building or testing the same. This helps designers, engineers, managers in experiment, decision making and training activities. Let us take the example of a car simulator. It replicates the look, feel and control of an actual vehicle in a virtual environment, offering the driver a sensation similar to being in an actual car. Various driving scenarios, road conditions and weather situations can be simulated to generate a completely immersive experience.

The types of simulation system are described below:

1. **Discrete Event Simulation:** This involves modelling a system's changes over time, triggered by discrete events. For example, traffic flow simulation can model how a car can move through queues or respond to traffic light.
2. **Dynamic Simulation:** Here, the modelling focuses on a system's continuous movement through space using principles of physics and engineering. This helps in testing the movement and response towards forces in a simulated environment. For example, simulation of air flow over a car body to test the drag and improve the design accordingly.
3. **Process Simulation:** This simulation replicates the physical interactions between multiple components or systems over time. This helps in optimising the plan for production and maintenance processes. For example, with this we can model how a product moves in a production line or how an engine part behaves during long distance driving.

While both digital twin technology and simulation use virtual model-based simulations, they are distinct in several ways. The primary differences between them are their scope coverage, real time data interaction and the degree to which they mimic physical systems. Digital twins are dynamic and closely connected to real-world data, whereas simulations often have a narrower focus and may not involve real time engagement of data. Each approach offers unique advantages and is suited to specific use cases. It is essential to highlight that a digital twin establishes a virtual environment capable of running multiple simulations, backed by real time data and a two-way information exchange between the twin and the data-collecting sensors. This significantly improves the precision of predictive analytical models, providing deeper insights for managing and monitoring products, policies and procedures.

These differences can be further described as follows:

- **Elements of Data and Interaction:** Traditional CAD-based simulations are used to model products or processes, enabling the testing of various parameters or design elements. However, they are static and dependent on manual updates by the designer.

In contrast, a digital twin starts as a static simulation model but gradually evolves into dynamic simulation as the real-time data gets integrated. It matures over the product's lifecycle with more and more data and delivers insights and information that cannot be achieved from static simulation.

- **Basis of Simulation:** Simulation tries to model potential scenarios for a product. They are constrained by the parameters and scenarios defined by designers. In contrast, digital twins provide real-time feedback, enabling designers to monitor a product's performance and improve using real-world data. Thus, they can replicate real world behaviour of the specific product. The primary difference is: simulations are theoretical and predictable, while digital twins are real and dynamic.
- **Scope of Use:** Simulations are valuable for testing different scenarios against predefined parameters and primarily used during product design. Digital twins have a broader scope covering all stages of a product's life cycle.
- **Cost Aspect:** Digital twins are expensive than simulations, as they need installation of sensors and to establish two-way communications with analytical tools and user interfaces. Hence it is justified to invest in digital twins for critical assets or processes only.

The differences are summarised in Table 11.2.

Table 11.2 Simulation vs Digital Twin

Category	Simulation	Digital Twin
Elements of Data and Interaction	Static, utilises pre-defined data sets for theoretical analysis.	Active, leverage real time data to update the virtual model continuously.
Basis of Simulation	Interaction takes place before the simulation, involving the adjustment of variables to observe potential outcomes.	Real time feedback. Enables interactive engagement with the model for testing various scenarios dynamically.
Use Cases	Narrow – primarily used for product design, scenario testing, training, education, risk management and market predictions.	Broad scope – encompassing all stages of a product's lifecycle. Primarily used for industrial asset management, predictive life cycle management and process optimisation.
Cost Aspect	Cost-effective due to limited use cases, easy setup and no real time updates.	Expensive due to huge setup costs, integration with massive IIoTs and analytics tools and continuous real time interactions and hence should be planned for critical assets or procedures where the cost can be justifiable for the business.

11.2.6 Digital Twins vs Augmented Reality

In Chapter 7, we discussed how Augmented Reality (AR) augments a real environment with digital components. In the context of Industry 4.0, AR and digital twins are distinct but related concepts, both playing significant roles in various industries. Some key differences are shown in Table 11.3.

Table 11.3 Augmented Reality vs Digital Twin

Category	Augmented Reality (AR)	Digital Twin
Technology	AR overlays digital information, including visuals, sounds, or data, onto the real world in real time. Such information can be viewed through devices like smartphones, tablets, or AR glasses, enhancing the user's perception of the physical environment.	Digital twins are virtual replicas of physical objects, systems, or processes. They are data-driven models that simulate real-world entities and are typically used for design, monitoring, analysis and optimisation.
Use Case	AR enhances the user's immediate environment by providing additional information, context, or interactive elements. It is often used in gaming, education, navigation, marketing, etc.	Primary use cases of digital twins are predictive maintenance, performance optimisation, design validation and simulating complex systems. They find applications in industries like manufacturing, healthcare, aerospace, infrastructure, automotive, etc.
Data Integration Approach	AR typically relies on real time data or information from sensors and cameras to provide context-sensitive experiences. It doesn't always require a comprehensive digital model of the physical world.	Digital twins are built upon extensive data integration, incorporating information from various sources to create an accurate and detailed representation of a physical entity.
Stages of Product Life Cycle	AR is often used during physical entities' operational or usage phase, providing users with real time information or assistance.	Digital twins can be used throughout an entity's lifecycle, from design and development to operation and maintenance. They offer a long-term, holistic view of the physical counterpart.
User Interaction	AR is interactive and user-centric, allowing users to engage with digital elements overlaid on the real world.	Digital twins are primarily used for data analysis, simulation and monitoring. They are also interactive but not in same way as AR.

Despite having differences, AR and digital twins complement each other frequently:

1. AR can provide an intuitive and immersive interface for interacting with digital twins. Users can wear AR glasses or use mobile AR apps to visualise and manipulate digital replicas of physical objects or systems in real time.
2. AR enables experts to provide remote assistance to field technicians or workers using digital twins. By sharing their view through AR interfaces, experts can guide on-site personnel through complex tasks, troubleshooting, or repairs.
3. Designers and engineers can use AR to visualise digital twins of prototypes or planned changes within the context of the physical environment. This allows them to assess the feasibility and impact of design changes more effectively.
4. Through AR, customers can visualise and interact with virtual models before making purchase decisions.

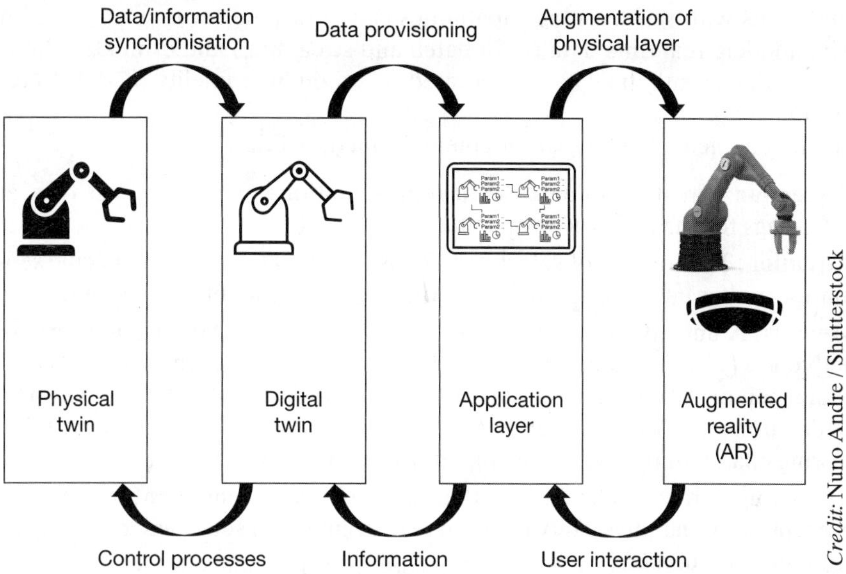

Figure 11.6 Complements - digital twin with AR

As shown in Figure 11.6, the digital twin is connected to the physical world through synchronisation and control processes. Information and functions within the Digital twin can be accessed by applications in the Application Layer. These applications can either actively request data from the digital twin or receive it directly. Finally, AR comes into play by enhancing the physical layer with information sourced from the services within the application layer, effectively augmenting the physical environment with digital data from the digital twin.

11.3 COGNITIVE DIGITAL TWINS

In Section 11.2.2, we have explored the various stages of the evolution of digital twins. Stage 5 of the evolution refers to the increase in the autonomy of digital twins through cognition. In this section, we will understand such cognitive digital twins in more detail.

Cognitive Digital Twin (CDT) is a concept that merges the concept of a digital twin with cognitive computing and AI technologies.

Cognition can be understood as a computational viewpoint that regards the mind as a form of machine. In the context of digital twins, cognition represents a computational method that replicates and imitates assets, machinery and processes in real time using AI technology. Developing cognitive digital twins involves the computation of the status, behaviours and interconnected models relevant to real-world elements within digital environments.

When digital twins can learn in real time and are equipped with AI functionalities and cognitive capacities, they transform into CDTs. This transformation enables organisations to explore new applications for digital twins that were previously not feasible. One of the key advantages of cognitive digital twins is their inherent ability for self-learning and the capability to foresee future events. It introduces cognitive features to enable intricate reasoning and optimising dynamic processes.

Digital twins with self-learning capabilities can anticipate future events and make decisions using models, real time data (both batch and streaming) and estimated data. If there is a shortage of data, synthetic data is generated based on high-fidelity models to replicate the actual system.

CDTs are characterised by six fundamental functionalities:

1. **Perception:** Creating valuable data representations of the physical twin and its surroundings for further processing.
2. **Attention:** The process of selectively focusing on a task, goal, or specific sensory information, either intentionally or in response to environmental signals and circumstances.
3. **Memory:** A unified process encompassing the temporary holding of information during active use (working memory), the recollection of events in the physical twin's life (episodic memory) and the retention of factual knowledge about the environment and its interactions with the physical twin (semantic memory). This process includes encoding, storing, maintaining and retrieving information when necessary.
4. **Reasoning:** Drawing conclusions that align with a starting point, which may include a perception of the physical twin and its environment, a set of assertions, a memory, or a combination thereof.
5. **Problem Solving:** Finding solutions to problems or achieving specified objectives, starting from a given point.
6. **Learning:** Converting the experience gained from physical twin into reusable knowledge for future experiences.

A CDT is capable of performing Front Running Simulation (FRS). FRS takes the present condition of the product and forecasts the future state of the product over a period of time with modelling and behavioural simulation. This functionality helps in foreseeing potential product deterioration or malfunction, identifying probable bottlenecks in production, or pinpointing operation processes of equipment that could lead to unsafe or catastrophic situations.

There are certain schools of thought to embed AI directly into equipment so that it can be able to adapt and 'learn' over time. While this seems reasonable, there is genuine risk of unintended behaviours spiralling out of control, potentially resulting in disastrous consequences. Therefore, before incorporating AI into products, it is important to have a CDT in place. It offers supervision and FRS capabilities, allowing for timely intervention and the restoration of product control.

Below are some examples of CDT Implementation:

1. **Manufacturing:** A CDT of the manufacturing process can gather real time data from Industrial IoT (IIoT) sensors and cameras throughout the factory. Using AI and ML algorithms, it can analyse data to predict maintenance needs, optimise production schedules and detect defects in real time. It can even make autonomous decisions, such as adjusting machine settings to improve efficiency.
2. **Smart City:** CDT can represent the entire urban environment, including traffic systems, public transportation, energy grids and environmental sensors in a smart city. Using AI & FRS, CDTs can analyse data from various sources, such as satellites, drones, traffic cameras, weather stations, public transits and optimise traffic flow in real time, reduce energy consumption and provide residents with personalised recommendations for transportation and energy usage.

11.4 ROLE OF DIGITAL TWINS IN PRODUCT LIFE CYCLE

In Section 11.2, we discuss different types of digital twins and briefly touch upon their usage during the product design, build and operation phases. In this section, let us discuss their role in various facets of the product life cycle, namely in design, build, test and operations phases, in detail.

11.4.1 Role of Digital Twins in Design

In today's highly competitive landscape, there is immense pressure to accelerate the design of innovative products, features and for faster time-to-market. The conventional design approach has several challenges like complex design collaboration and long trial production cycles for prototypes. Instead, there is a growing need to swiftly create precise, functional virtual replicas of upcoming systems and subject them to various scenarios. This is precisely the role played by a digital twin during the design phase.

Over the past few decades, tools available to design engineers have undergone significant advancements.

The invention of AutoCAD in 1982, the first 2D CAD software designed for personal computers, was a revolution in this field as it democratised the field of system and product design. Over the years, CAD software has become increasingly sophisticated thanks to high-performance processors, robust graphics capabilities, big data availability, fast global networks, extensive storage, cloud technology and artificial intelligence. The tools for designing a digital twin extend 3D CAD capabilities with several additional features.

A critical aspect of digital twin modelling is using data to guide and improve design of physical products and processes. Historical data can be fed into the model to simulate its performance under the new design. Hypothetical or simulated data are used for what-if analysis, to predict future behaviour of the physical system.

Advanced design processes are currently reinforced by generative design. Designers set parameters like material requirements, cost constraints and manufacturing methods, while the software examines various design options. Simulations assess each alternative, to identify the effective design choices. Here comes the benefit of DTPs as we discussed in Section 11.2.3.

We are also witnessing the usage of digital twins in the design phase for higher-quality 3D printing. Simulating the 3D manufacturing process can yield superior outcomes as it provides insight into complex design and material choices in advance.

In summary, a digital twin in the design phase can more accurately predict the performance of a future product or system. It verifies and validates design decisions, giving design engineers confidence that the physical result will meet expectations, avoid failures and mitigate risks.

Let us evaluate some examples here:

1. **Building Construction Design:** Traditionally, when designing such a building, we would build it first and then use sensors and an Internet of Things (IoT) system to determine its efficiency. However, an alternative approach is to create a virtual representation or digital twin of the building before construction begins. This digital twin is a detailed, computer-generated building model, including its architectural and engineering aspects. Designers and engineers can experiment with various design choices and configurations with the digital twin. For instance, they can test the impact of using double-pane glass

windows, altering the building's orientation to the sun, or adjusting insulation materials. Also, the structural aspects like stress and vibration i.e., how the building physically responds to various forces like earthquakes and wind, as well as behaviours related to software and control systems i.e., building's internal systems, like HVAC (Heating, Ventilation and Air Conditioning) and lighting function can be tested.

The key advantage here is the ability to explore 'what-if' scenarios in a virtual environment. Design teams can make changes and adjustments to the digital twin to see how they affect the building's performance without the need for physical construction.

2. **Smart Vehicles Design:** Let us now discuss how digital twins are getting used in the design of new cars. Historically, taking a car from concept to production has taken anywhere from two to five years. However, digital technologies have significantly reduced this timeframe, often by more than 50%. The highly competitive nature of the automotive market demands shorter development cycles.

The term 'car digital twin' is becoming common, involving data, sensors, artificial intelligence and collaboration. The automotive industry is undergoing substantial changes, focusing on sustainable designs, electrification, cost control and adding innovative features, including autonomous capabilities.

Car designers and engineers now prioritise virtual simulations. Each innovation and the entire vehicle system must undergo rigorous testing under numerous scenarios. Digital twin technology enhances this process by enabling diverse simulations, generating high-quality data and predicting real-world outcomes. It supports multi-physics simulations, considering various interacting physical properties like electrical, mechanical, fluid and chemical. Additionally, it can also test the vehicle within its operational environment, considering factors like weather, altitude, road conditions and more. Figure 11.7 shows the design of a smart vehicle with digital twin with data synchronisation between a real asset and its twin. The design can be moderated with advanced 3D CAD tools, analytics and predictive (What/If) simulations to choose the desired model.

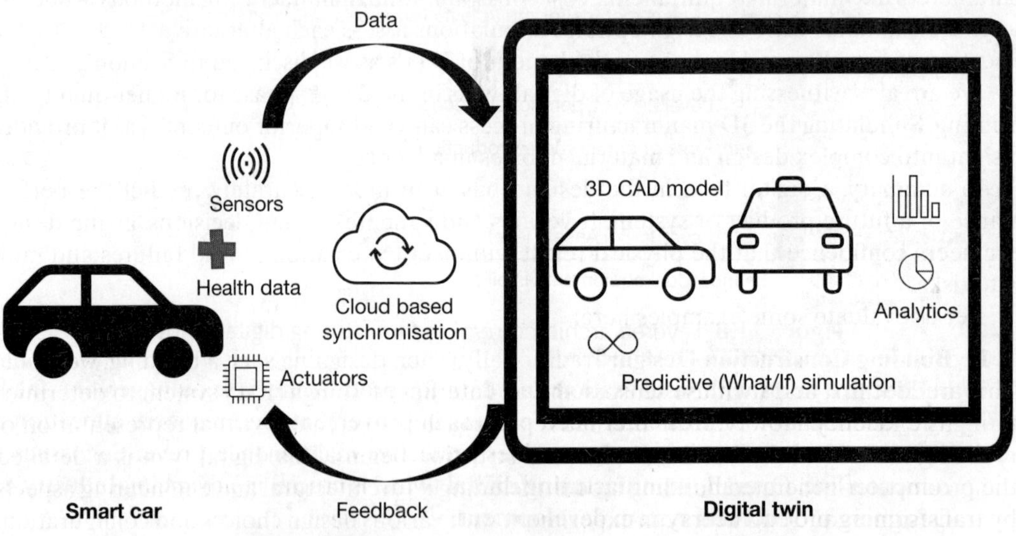

Figure 11.7 Smart vehicle design with digital twin

Designing self-driving or autonomous vehicles is even more complex, requiring advanced simulators and digital twin technology. These vehicles rely heavily on sensors to understand their surroundings and ensure safety. Digital twins can simulate the placement and functioning of these sensors and how the vehicle processes the data it collects. They can also address challenges such as how artificial intelligence handles unpredictable driving situations and decision-making processes. You can relate this with the concept of CDT, which we have learnt in Section 11.3.

11.4.2 Role of Digital Twins in Production and Testing Phase

Smart factories, generally integrate digital twins across the manufacturing life cycle, from design, production, testing, to operations and maintenance. During the production phase, advanced digital twins can define manufacturing specifications, including production line setups, monitoring metrics for individual components, final products and the entire production line. It can monitor the performance of machines and workflows in real time during production. It can also identify anomalies, inefficiencies in the processes and causes of down time, by analysing real time data generated by IIoT. It can do testing for quality control by comparing the production data against the ideal configurations to detect deviations. It can also track actual usage or performance of the delivered products. It can use 'what-if' analysis to simulate product or system performance under different stress conditions like excessive heat, extreme weather conditions and improper handling.

The digital twin architecture in manufacturing can be structured into layers tailored to specific business needs. At Level 1, the Assets level offers visibility into individual equipment; at Level 2, the Operational Process level provides insights into one or more production lines while at Level 3, the Enterprise Business level, stakeholders focus on high-level metrics for multiple factories or a single factory, as shown in Figure 11.8.

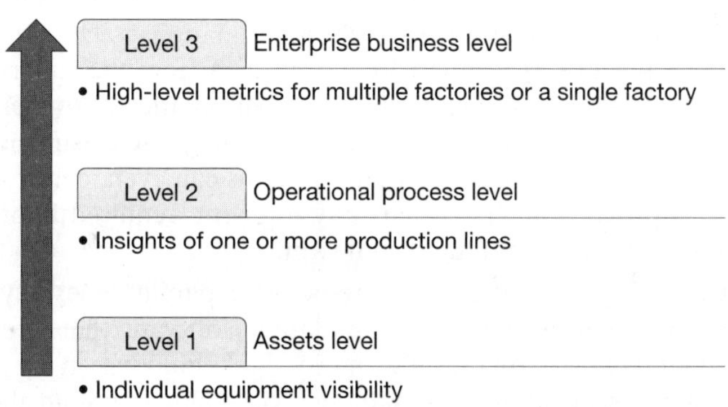

Figure 11.8 Layered architecture of manufacturing digital twins

This architecture helps with visualisation and data for product lifecycle management, facilitating the identification of areas that need process improvements and resource allocation in real time or for future planning. It also enables predictive maintenance, preventing issues in the production line. Overall, 'manufacturing digital twins' is crucial in advancing Industry 4.0 by transforming product and system development.

For example, an engineering firm was tasked with enhancing maintenance capabilities for a company's production line. The main objective was to evaluate maintenance strategies for robots equipped with welding guns, using digital twins to improve performance and cost savings. Three maintenance approaches were considered: scheduled, condition-based and predictive maintenance. Initially, the entire production line was modelled, but the project was focused on specific parts of certain robots prone to wear and tear. IIoT sensors were deployed to collect relevant data, then a digital twin was created to represent these physical parts. These digital twins served as intelligence sources for a 'what if' application, enabling stakeholders to manipulate parameters, conduct simulations and assess maintenance impacts. The project involved scenario comparisons and cost-benefit analysis. The project was expected to yield several positive outcomes, including educating management on digital technologies' role in a factory, expanding the project to include more assembly line components, exploring return on investment opportunities and recognising the potential of digital twins in reducing maintenance costs.

11.4.3 Role of Digital Twins in the Operational Phase

A digital twin brings significant advantages during the operational and maintenance phases. While used in operations, these digital twins are commonly referred to as '*supervisory twins*'. They support three major operational processes i.e., supervision, diagnosis and control, and prediction, as described below.

- **Supervisory Capability:** During the operational phase, digital twin can accurately replicate and helps visualising the behaviour of the physical object. As an example, if the physical twin is in running state, the digital twin provides data or graphical representations of its parameters, like position, speed, and temperature. It generates real time characteristics based on data from sensors within the actual device. Often, this data is presented alongside the performance targets.
- **Diagnosis and Control Capabilities:** Real time data from the supervisory digital twin can be fed into analytical systems, which can automatically trigger alerts when certain parameter thresholds are exceeded. They can diagnose existing issues and, in some cases, predict potential future issues. Some systems can even identify opportunities for performance improvement. In a smart environment, required improvements can be digitally initiated directly from the digital twin.
- **Predictive Capability:** A digital twin's predictive capability goes beyond performance enhancement. It involves predictive simulation software that works alongside the supervisory digital twin to run future scenarios. Stakeholders can examine the results of 'what-if' simulations, enabling informed decision-making without the associated risks or costs of making changes to the physical counterpart. Positive outcomes from predictive capabilities can trigger investment decisions for both processes and systems.

Aerospace Giant Boeing has revolutionised aircraft inspection and maintenance with the implementation of an AR application that utilises a digital twin of one of its aircraft. This innovative approach has empowered Boeing to produce more than 100,000 synthetic images with the help of digital twins, significantly enhancing the training of machine learning algorithms employed within the AR application.

11.5 USE CASES OF DIGITAL TWIN APPLICATIONS IN INDUSTRY 4.0

So far, we have learned about the evolution of digital twin concepts, various types of digital twins as per industrial needs, and their role in the design, production, testing, and operation phases. In this section, we will explore how all these concepts come together and real-life use cases are implemented using digital twins in various Industry 4.0 setups.

11.5.1 Use of Digital Twins in Manufacturing

Manufacturers are already experiencing the advantages of implementing digital twins, ranging from quicker product development to enhanced productivity among frontline workers. More than 80% of companies that adopted these technologies reported enhancements in their capacity to innovate and collaborate during their production, manufacturing and operational processes.

The following are some use cases of digital twins in the manufacturing sector:

1. **Complex Machine Design:** Building digital prototypes (DTPs) using 3D models enables designers to visualise and iteratively test product designs in a simulated environment before creating physical prototypes. This minimises the requirement for time and investment in expensive physical prototypes, expediting the design process.
2. **Factory Planning and Arrangement:** Optimise machine placements, assembly sequences, employee interactions and other aspects by creating spatial maps of factories.
3. **Simulation of Robotics:** Highly realistic simulation of the robot's behaviour and interactions with its surroundings. This simulation includes various scenarios, environmental conditions and potential challenges that the robot may encounter during its tasks, thereby reducing risks associated with accidents or errors during real-world operations. The performance and safety of using robots could be rigorously validated within a digital twin environment before actual deployment.
4. **Supervision, Guided Maintenance and Repairs:** Streamline routine and time-consuming tasks with AR technologies that enable remote assistance. Consistent sensor-driven information gathering and analysis also helps in decision-making for end-of-life and replacement of machine and spare parts.
5. **Training for Operators:** Digital twins can serve as tools for employee training and simulating real-life situations, enabling them to acquire expertise and understanding without potential physical harm or equipment damage. For instance, by creating a digital twin of a production line, employees can practice operating the equipment, do complex troubleshooting and improve productivity with low training costs.

11.5.2 Use of Digital Twins in Infrastructure

Digital twin technology assists urban developers, planners and administrators globally in gaining a deeper understanding of public spaces and enhancing their utilisation. Through the utilisation of advanced, interactive models and real time IoT data, various stakeholders can replicate and assess factors such as traffic dynamics, impacts of climate changes

DID YOU KNOW?

A prominent manufacturing and compressed air solutions provider has successfully implemented digital twins to transition from a product-oriented business model to a service-based one. Through its digital twin system, the company continuously receives operational data throughout the equipment's lifecycle, with a primary focus on air consumption rates. This data-driven approach has enabled it to bill its clients based on their actual air consumption rather than fixed rates, leading to a substantial 30% reduction in commodity costs. Notably, this innovative strategy has attracted nearly half of the industry's major vendors to adopt digital twins for similar benefits.

and evolving landscapes near to critical infrastructure like airports, highways and transport centres. Whether individual facilities or entire smart cities, digital twins empower proprietors, administrators and policymakers to effectively manage vast datasets, equipping them with the insights needed to optimise the infrastructure for future requirements.

The following are some use cases of digital twins in infrastructure:

1. **Real time Traffic Flow Monitoring:** Digital twins can be used to monitor traffic flow in real time. Sensors, and cameras can collect data and feed it into the digital twin model of road networks. This helps in analysing traffic patterns, congestion detection and dynamic route optimisation to improve overall traffic management.
2. **3D City Planning and Expansion Simulation:** Digital twins enable urban planners to create 3D models of cities. These models can simulate the effects of various development and expansion scenarios and help in optimising the city plan. Planners can assess the impact of new infrastructure, buildings, before implementing them in real world.
3. **Real Time Parking Monitoring:** Digital twins can help manage parking spaces efficiently, reducing congestion and improving the parking experience. Sensors on parking meters are integrated with the digital twin of the parking space. Sensors will share the real time occupancy data with the digital twin. Drivers can access the information about available parking spots in real time.
4. **Real Time Safety Monitoring:** Digital twins play a vital role in enhancing public safety. They can integrate data from various sources, such as surveillance cameras and IoT sensors, to monitor public spaces. Suspicious activities or safety threats can be detected in real time, that can trigger rapid responses from law enforcement agencies.
5. **3D Visual Incident Reporting and Analytics:** Digital twins can provide a 3D visual representation of the scene, when any incident occurs in public places. This helps in incident reporting and analysis. Investigators can reconstruct events, identify critical details and gather evidence more effectively.

11.5.3 Use of Digital Twins in Aerospace

Aerospace projects are highly complex, particularly when working on aircraft and spacecraft, which are expensive to design and build. They demand exceptional precision to prevent costly delays and ensure the safety of astronauts and pilots.

DID YOU KNOW?

There were recurring flash flooding and land management challenges in densely populated Singapore. Hence, the Singapore Land Authority (SLA) initiated a 3D mapping project in 2012. They launched Virtual Singapore, a digital twin project powered by Bentley Systems' software, in collaboration with GPS Lands Singapore. A significant accomplishment was reached in February 2022 as Bentley Systems and GPS Lands Singapore collaborated to develop the world's inaugural digital twin of an entire nation, Singapore. This digital twin was developed using aircraft equipped with laser scanning technology and aerial imagery. Additionally, in June 2022, VIZZIO Singapore utilised satellite data, numerous photographs, and advanced 3D modelling technology to produce a comprehensive 3D reconstruction of the entire city.

In India, the city of Amaravati in Andhra Pradesh is undergoing development as a 'smart city' with a total investment of $6.5 billion. This ambitious project uses a digital twin technology incorporating over 1,000 datasets. These datasets serve various purposes, such as facilitating the permitting process, tracking construction progress and assessing the viability of design plans, particularly in the city's challenging climate conditions.

The primary focus in the aerospace sector is on enhancing production efficiency and engineering processes, spanning from parts manufacturing to aircraft design. Having said that, maintenance and repair expenditures have continued to rise, despite efforts to build more durable engines and employ innovative maintenance techniques. Hence, the involvement of digital twins, spanning from design to maintenance, is providing immense benefits for this industry.

In specific scenarios, machines in the developmental phase may be tasked with unprecedented challenges or destined for locations where direct tracking is unfeasible, such as space missions. By employing digital twins, companies can digitally monitor these machines with an impressive 147% improvement in accuracy compared to traditional methods. An aerospace giant has achieved approximately 40% improvement in first-time-right quality of parts by using digital twins.

As this technology continues to evolve, its potential applications are expanding. It might be possible that in just a few years, we could see fully functional digital twins of entire airports.

The following are some use cases of digital twins in the aerospace industry:

1. **Simulation for Training:** Digital twins can simulate real-world scenarios for training purposes. Pilots, astronauts, and ground crew can train in a safe and controlled virtual environment, and become proficient in operating complex aerospace systems.
2. **Predictive Maintenance:** Aerospace companies leverage digital twins to monitor the health and performance of aircraft and spacecraft in real time. By collecting sensor data and comparing it to the digital twin's model, maintenance teams can predict when components may fail and proactively schedule maintenance.

11.5.4 Use of Digital Twins in Automotive

The automotive industry leverages digital twins to simulate and evaluate new design ideas, enhance production procedures and forecast a vehicle's performance in diverse scenarios before physical production. The primary advantage for automotive Original Equipment Manufacturers (OEMs) in adopting digital twins is the capacity to optimise time and resources by detecting and resolving potential problems pre-emptively. According to market research, the digital twins' market size in the automotive sector was assessed at $2.2 billion in 2022 and it is projected to reach approximately $34.6 billion by 2032, exhibiting a Compound Annual Growth Rate (CAGR) of 32.6%.

The following can be major use cases of digital twins in the automotive sector:

1. **Interactive User Experiences:** Digital twins enable the creation of interactive 2D and 3D user experiences, particularly useful for in-vehicle infotainment (IVI) systems and digital cockpits, enhancing the user interface and experience within vehicles.
2. **Autonomous Driving Simulation:** Digital twins are used to simulate various scenarios and provide real time visualisation of outcomes. This is crucial for testing autonomous driving systems safely within virtual environments before deploying them in the physical world.
3. **Training and Guidance for Technicians:** Interactive and immersive experiences offered by digital twins are employed to train and guide frontline workers. These experiences have been proven to enhance knowledge retention and increase the overall productivity of the workforce.
4. **Sales and Marketing:** Leveraging existing 3D data, digital twins are used to create photorealistic renders and interactive 3D configurators. These tools are valuable for sales and marketing purposes, allowing potential customers to visualise and customise the vehicles of their choice.

Figure 11.9 depicts the application of digital twins in the automotive industry (design, build, maintenance, marketing) and is duly complemented by AR support.

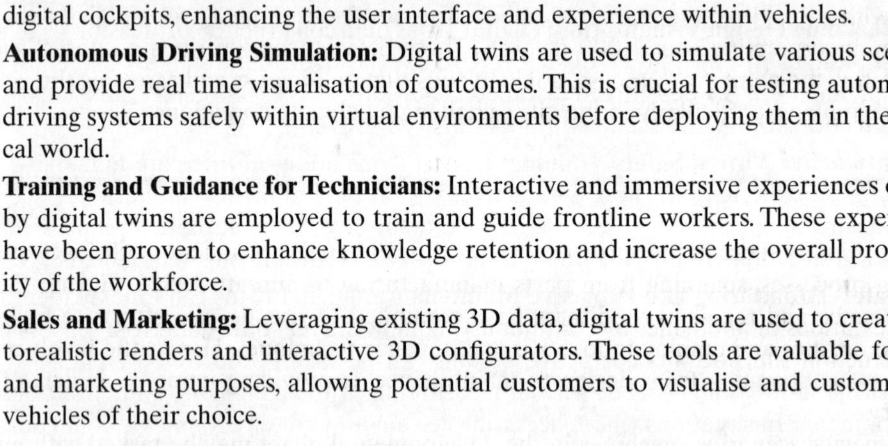

Credit: Charlie Hutton / Shutterstock; Mikhail Bakunovich / Shutterstock

Figure 11.9 Application of digital twin in automotive industry

11.5.5 Use of Digital Twins in Construction

Construction activities comprise various stakeholders like architects, engineers, contractors, building owners, and local authorities. Digital twins manage and share data among these diverse groups and enhance decision-making process. In construction projects, digital twins are employed to create precise replicas of real-world environments through 3D Models. These 3D models allow construction teams to interact virtually with physical assets during the design and planning phases, providing real-time access to essential building information. By incorporating digital twins into the Building Information Modelling (BIM) process, construction companies, along with architectural and engineering firms, can address some of the industry's most significant challenges, including low productivity, profitability concerns, performance issues and high error rates.

The following can be major use cases of digital twins in construction:

1. **Real time Design Visualisation:** Digital Twins help construction professionals to visualise designs in real time. Utilising AR to overlay BIM data enables teams to swiftly examine both the design and installed conditions. This functionality aids in early issue detection, mitigating the need for expensive rework.
2. **Interactive Virtual Safety Training:** Digital twins are used to create immersive safety training experiences. These interactive, virtual environments simulate key scenarios, transforming job-critical learning and training. Workers can better prepare for real-world situations, improving safety awareness and reducing on-site accidents.
3. **Safety Monitoring and Proactive Maintenance:** Digital twins can quickly identify any deviations or anomalies by continuously comparing real time data from sensors in construction site, so that corrective actions can be taken proactively. Digital twins are crucial in providing precise virtual measurements of challenging and hazardous environments. In situations where accessing certain areas or environments could pose risks to employees, remote inspection through digital twins is a safer alternative.
4. **Improved Project Closure and Maintenance:** Digital twins maintain the critical model and project data. This ensures that operations and maintenance can be optimised in long run.

11.5.6 Use of Digital Twins in Oil and Gas

Digital twins, equipped with advanced simulations, AR and ML are extensively getting utilised in oil & gas sector. Exploration, processing and maintenance, in all processes, digital twins are enhancing overall planning, efficiency and safety.

The following can be major use cases of digital twins in oil and gas:

1. **Offshore Drilling Operations:** Offshore drilling operations are complex and require constant monitoring. A digital twin of an offshore well allows operators to continuously assess its structural integrity, identify potential issues and plan maintenance or repairs proactively. This minimises the risk of costly downtime or environmental incidents.
2. **Cost-Weight Estimations Simulation:** Oil and gas projects involve massive structures and equipment and accurately estimating costs and weight is crucial. A digital twin

with simulation capabilities can help in refining these estimates by considering various design alternatives and their associated costs and weights.

3. **Remote Monitoring:** The digital twin of a pipeline can be used to monitor the flow of oil and make adjustments to the pumping rate as needed.

11.5.7 Use of Digital Twins in Health Care

Let us imagine a day when each one of us has a personal digital twin that maps our body, including our organs, nervous system, immune system, etc. For example, we know the electric, mechanical, or chemical behaviour of our heart, how our immune system will react in case of transplant, how a particular drug will work on us, or even detect an early sign of anomalies in the body and can take immediate actions to avoid any life-threatening disease. Though with the current technologies, this still looks like a sci-fi movie, a lot of work is going on for twinning a part of the whole entity; for example, the digital twin of the heart's electrical behaviour that would help in setting up the pacemaker more accurately and for a particular individual.

The following are major use cases of digital twins in the healthcare sector:

1. **Diagnostic Support:** The patient's digital twin utilises diverse health data sources such as imaging records, physical measurements, lab reports and genetic information to aid in the diagnostic process. This digital representation of the patient will mimic their health condition using the clinical data at hand and deduce any missing variables through statistical models.
2. **Patient Monitoring:** Wearable sensors and health trackers can be used to continuously provide real time data to a patient's digital twin. This data, which includes biometric, behavioural, emotional, cognitive and psychosocial aspects, can help understand disease progression and develop software for early symptom detection. This empowers doctors and users to diagnose patients before the onset of illness and assess the efficacy of treatment. The digital twin can be nourished with data from various sources such as medical records, lab test results, pharmaceutical data, wellness and disease management records, data from well-being devices and social determinants like location, local weather and purchasing habits.
3. **Development of Drugs and Dosage Optimisation:** *As per statistics, developing a single drug can take approximately ten years with a 90% probability of failure in clinical trials.* A digital twin can be extensively analysed with thousands of drugs to pinpoint the most suitable ones for a particular case. Importantly, this process might not be limited to existing drugs. We can construct a digital group comprising patients with diverse phenotypes exhibiting similar symptoms. With this, we can assess new candidate drugs to forecast their likelihood of success and determine the ideal dosage. Enhancing this initial selection process can significantly reduce the need for extensive clinical trials. Such digital twins will also pave the way for the development of personalised medicine.
4. **Medical Device Design and Optimisation:** Here we will talk about the possible convergence of two realms of digital twins. One represents the patient with their unique characteristics and the other portrays the design of the medical device. These models can be synchronised to assess the outcomes when a specific device is implanted in a particular patient. This approach is particularly valuable for populations that cannot undergo clinical investigations without posing potential harm.

5. **Simulation of Surgery:** Surgery is carefully adapted to meet the individual needs of the patient. Personalisation is extremely important in enhancing the success of interventions while minimising patient risks. Digital twins play a crucial role in this context, as they can simulate invasive clinical procedures, allowing medical professionals to predict outcomes before selecting the most appropriate procedure.

11.6 KEY IMPLEMENTATION CONSIDERATIONS

So far, we have discussed how the digital twin concept has revolutionised various industries with all its use cases and is among the top enablers of Industry 4.0. However, it is important to understand the key implementation aspects before its wide-scale adoption. Let us discuss some of the most critical ones here.

11.6.1 Cybersecurity Threats

A significant risk associated with digital twins is that they can give cybercriminals a new avenue to infiltrate and exploit sensitive data. For instance, attackers can gain access to detailed information about the facility's layout, machinery, and operational processes through the digital twin of the manufacturing facility.

The capability of establishing two-way communication between the physical twin and its digital twin significantly amplifies the potential for wrongdoing. If an external malicious entity gains control of factory systems, it can manipulate them to disrupt or distort machines, robots, sensor data, potentially resulting in catastrophic accidents.

Possible Mitigations

1. As mitigations, organisations should adopt a proactive approach to safeguard their digital twins. The initial measure to ensure the security of digital twins is to secure the underlying system. It is advisable to consider implementing a zero-trust security program at this stage. This robust 'trust no one' security framework, continuously validates all users, regardless of whether they are within or outside a company's network. It is required to use of multi-step authentication protocols to ensure secured access. Even lateral movements within the network are subject to continuous reassessment and reauthorisation under this framework.
2. Regular penetration testing is key for assessing and mitigating vulnerabilities in digital twin systems. This involves simulating cyberattacks on the digital twin systems to validate their response and uncover any weaknesses that require attention. Additionally, it's essential to properly isolate digital twin systems from other networks and systems to enhance security and minimise potential risks.

11.6.2 Choosing Right Data

The widespread adoption of digital twins leads to substantial transmission of data from the system and its components. The more sensors and digital twins are employed, the greater the volume of data collected and transmitted.

It is essential to choose the right set of data for the business and system and then convert it into meaningful information that can facilitate informed decision-making about the system and its usage.

11.6.3 Standardisation Issue

The increasing usage of digital twins could potentially result in varied representations of their data, depending on the system's manufacturers. Even components within the same system might possess incompatible formats and information due to a lack of collaboration and standardisation among different manufacturers. The possible mitigation should be defining a standard for data representation and that should be followed by different manufacturers.

11.7 IMPLEMENTING A DIGITAL TWIN

With all the above concepts in mind, it is important now to understand how to implement a digital twin. Implementing a digital twin involves several steps and the process may vary depending on the system's complexity or the object we wish to replicate. The organisation should create a blueprint to clearly articulate the objectives and purpose of creating digital twins, the sequence in which they will be constructed to optimise their worth, the approach by which their functionalities will progress and the ownership and governance frameworks in place. The following are some key questions that an organisation should ask before making the decision to implement the digital twins:

- **a.** Why do we need to create digital twins?
- **b.** Which digital twins offer the most significant advantages for our enterprise?
- **c.** What is the value associated?
- **d.** Which use cases should be prioritised to give the biggest value?
- **e.** What is the data source and how is it gathered and stored?
- **f.** How do we choose the right data based on requirements?
- **g.** What Models should be developed based on the data?
- **h.** What Key Performance Indicators (KPIs) and Operational Performance Indicators (OPIs) should be monitored and what actions should be taken based on the KPIs?

Here is a general step-by-step guide for building the digital twins solution based on the blueprint created:

1. **Select the Target System:** We should first choose the physical system, object, or process for creating a digital twin. This can be anything from a machine, or entire shop floor operations to a building.
2. **Data Processing:** This includes data collection and integration. First, the relevant data from the physical system should be gathered. This data may come from various sources, including sensors, IoT devices, video cameras, historical records and manual input. The collected data should be integrated into a centralised database or platform. It is to be ensured that the data is accurate and consistent. The appropriate cloud platform

(public, private or hybrid cloud on hyper-scale cloud providers) should be used to handle the data.

3. **Data Transformation:** The data collected from different assets and systems should be in a format suitable for analysis. The source data can be of heterogeneous format and hence, appropriate ETL (Extract-Transform-Load) tools should be used to transform the data and store it in a meaningful format.
4. **Model Creation:** Develop a digital model that mirrors the physical system. This model should capture the key attributes, behaviours and functionalities of the real-world object or process. 3D CAD Models, 3D Point Cloud Method and DTDL would be used to create models and upload them to the digital twin tools to create DTIs.
5. **Real time Data Integration:** Integrate the digital twin into your existing IT/OT systems as needed for seamless operation. Set up mechanisms to continuously feed real time data from the physical system into the digital twin. This ensures that the digital twin remains synchronised with its physical counterpart. Big Data frameworks like Kafka could be used to continuously influx huge data from sensors and other sources. For lower volumes of record, a separate data management process can be used.
6. **Visualisation through Digital Twin Application/ GUI:** Create a user-friendly interface to visualise the digital twin and its data. Visualisation tools help users interact with and gain insights from the digital twin. The prominent Digital Twin Tools from big organisations like Microsoft, Siemens and SAP provide these features of visualisation of the twin along with the KPI dashboards (like energy management, security alerts, analysis of machine temperature, vibration, accuracy, oil leakage, maintenance/ fault diagnosis, etc).
7. **Data Analysis and Simulation:** Implement algorithms and simulation techniques to analyse the data and simulate the behaviour of the digital twin. This step helps to predict and assess various scenarios. Sophisticated analytical tools and AI/ML-based algorithms help in what-if simulations, predictive analysis and diagnosis.
8. **Implementation of Control and Decision Support Process:** Incorporate control systems that allow you to interact with the physical system through the digital twin. Additionally, use the digital twin to support decision-making processes.
9. **Problem Detection and Feedback:** Based on steps 6 and 7, the digital twin should be able to detect problems and automatically communicate back to the management interfaces for necessary decision-making. Also, the targeted commands or suggestions can be transmitted to the digital twin for necessary optimisation.
10. **Adjustments and Iteration:** If any diagnosis is incorrect, it can be returned to the engine as negative training data. The AI algorithms then make necessary adjustments. This iterative process ensures the digital twin system continually learned and improved, providing accurate and actionable insights for operations.
11. **Security:** Implement robust data security measures to protect sensitive information within the digital twin. In the previous section, we discussed the 'zero trust' policy towards security for digital twins.
12. **Use Case-Specific Applications:** Tailor the digital twin to address specific use cases or objectives, whether for predictive maintenance, process optimisation, or any other application.

A few key steps are depicted in Figure 11.10 below for an understanding of the E2E solution.

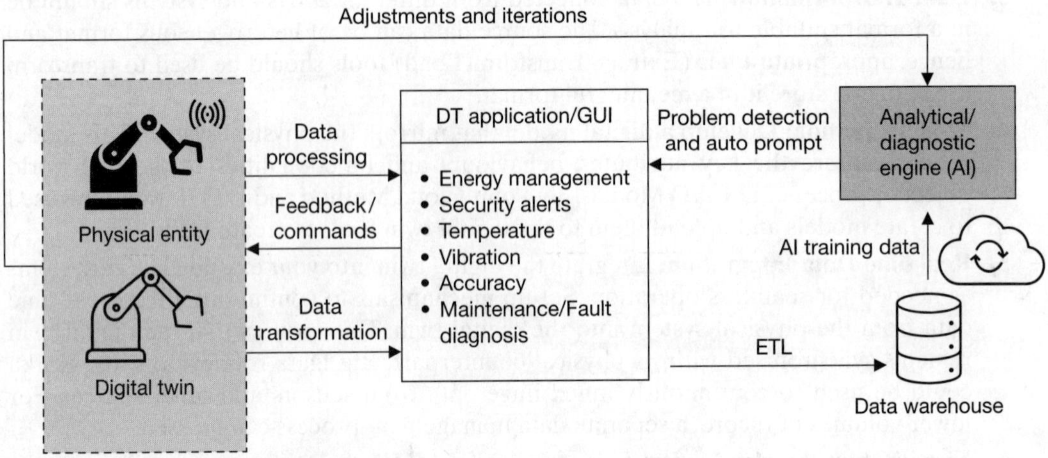

Figure 11.10 Steps to create a digital twin

DID YOU KNOW?

The 3D Point Cloud Method involves capturing the shape and surface details of real-world objects or environments by recording millions of precise points in three-dimensional space (X, Y and Z coordinates). They are generated through technologies like LiDAR, photogrammetry, or 3D scanning. 3D Point clouds provide the foundational data needed to build detailed 3D models, digital twins and virtual environments. This method enables highly accurate digital representations without physically altering the original object. Common software tools used for creating and processing point clouds are AutoDesk ReCap Pro, MeshLab, Bentley ContextCapture and CloudCompare.

11.8 CASE STUDY – DIGITAL TWIN IMPLEMENTATION FOR A SMART HOSPITAL BUILDING[1]

Let us explore a real-life case study on how a complete digital twin ecosystem has been implemented for a smart building.

11.8.1 Introduction

The need for efficient daily management and operational safety becomes more critical with the complexity and volume of newly constructed buildings. Managing numerous rooms with

[1] Peng, Yang, et al., 'Digital twin hospital buildings: an exemplary case study through continuous lifecycle integration.' *Advances in Civil Engineering* 2020.1 (2020): 8846667.

diverse requirements, intricate mechanical and electrical systems and unpredictable human behaviours can be daunting. Apart from normal buildings, hospital buildings face unique challenges as they are typically equipped with advanced facilities, specialised medical equipment, stringent security protocols and integrated business systems. This case study explores the successful implementation of Digital Twin technology in a hospital building, showcasing the benefits of this cutting-edge approach.

11.8.2 Background

A reputed hospital in China, constructed a new clinical centre building, offering 500 beds and covering an area of approximately 83,000 square meters. Earlier, the hospital's mechanical, electrical systems, medical systems, and sewage treatment operated independently. Their existing management systems were not integrated and working cohesively to meet medical service demands. This resulted in inefficiencies, making it challenging for hospital management to make timely decisions and placing a heavy burden on the Operations and Maintenance Department. Traditional manual management methods were outdated, as they struggled to monitor and analyse the extensive data generated from various sources.

11.8.3 Innovative Approach

This case study introduced the 'continuous lifecycle integration method' as a key technical innovation. This method involves a detailed schedule for gradually and punctually integrating relevant content throughout the entire lifecycle of a building project. The building project typically includes phases like Architecture, Engineering and Construction (AEC) and Operation and Maintenance (O&M). Among these phases, the O&M phase is crucial, given its significance in the building's longevity. However, O&M is interconnected with AEC, as critical static information is determined during the design phase and intelligent hardware devices are installed during construction.

Another innovative practice highlighted in this case study is the concept of 'early movement' by general contractors, which ensures data consistency within the digital twin. Traditionally, general contractors are mainly involved in the construction phase. In this case, the general contractor's digital twin consultant team participated in the planning and design phase, collaborated with owners and designers on data interfaces and collected essential design metadata. During construction, the general contractor coordinated contracts with various system suppliers, set up smart device networks and tracked changes. This integrated project delivery approach ensured data consistency throughout the entire building lifecycle, not just during construction.

11.8.4 Steps for Continuous Life Cycle Integration

The initial step was the digitisation of the hospital and it was achieved through geometric modelling. This process relied on CAD drawings from the design phase, although more complex mechanical rooms were digitised using the 3D point cloud method, employing laser scanning or photogrammetry techniques.

Following geometry modelling, the integration of BIM was a pivotal step. The dynamic aspects of the BIM emerged as processes were monitored through business systems and sensor devices. Subsequently, AI models were integrated to process the incoming monitoring data.

The final step involved object mapping through on-site investigations. This step was essential for establishing a connection with the real-world hospital, ensuring that all functional modules and diagnostic suggestions were accurately synchronised with the physical hospital building based on precise object mapping.

11.8.5 Digital Twin Development Process

A unified Digital Twin system was developed, encompassing desktop client, smartphone app and tablet app interfaces, based on continuous integration. In Section 11.7, we have seen how a digital twin system is developed. Similar steps were followed in this case.

Here is an overview of the major steps within this system:

1. **Data Processing:** Raw data from the physical hospital were initially processed. Some of this data was handled through the Apache Kafka engine, which manages high-density stream data.
2. **Data Transformation:** Two significant data transformation methods were employed: the Apache Flink engine and scheduled ETL processes. These continuously converted the analysis data regarding the building's status into a structured data warehouse format. The data warehouse served as the source for training data required by AI models.
3. **Diagnosis System:** The AI-driven diagnosis engine regularly receives data for diagnosis from the data warehouse. Once properly trained, these AI models formed the core of a diagnosis engine. When the diagnosis engine detects an issue or anomaly, it immediately communicates this information to management interfaces so that necessary actions can be taken.
4. **Feedback and Instructions:** Targeted commands or suggestions were transmitted to the digital twin for necessary optimisation.
5. **Adjustments and Iteration:** If any diagnosis was found to be incorrect, it was sent back to the engine as negative training data, to train the models for further fine-tuning.

Big data services were implemented using free, open-source engines to address budget constraints. Kafka and Flink, two big data frameworks, handled the continuous influx of quick and dense stream data originating from various sensors.

Conversely, there was a considerably lower volume of records to process for systems related to repair or medical equipment repositories. Therefore, a more concise approach was adopted, employing Python ETL scripts with Pandas for data management. This tailored approach ensured efficient handling of the less intensive data streams.

Building Automation (BA) hardware networks comprised over 1900 intelligent sensors, distributed across 13 subsystems, including the elevator system. Approximately 230 smart meters were also sourced from two energy monitoring systems (electricity and water supply), along with 100 sensors serving three medical gas systems. These devices collectively generated a staggering 2 million records per day, accumulating over 10 terabytes of recorded data in total. Despite the substantial volume, the data format was relatively straightforward and continuous integration was facilitated by robust network infrastructure. The primary challenge in this context was needing a database backup every three months.

In contrast, the records sourced from systems related to repairing medical equipment or outpatient information systems were significantly more complex due to their complex properties and this required the development of flexible ETL scripts.

Security measures were put in place, such as security monitoring videos and patient visit records. Digital twin systems were restricted to access private cloud HTTP APIs within the building's firewalls. This posed considerable challenges during the development phase. Finally, a private cloud was deployed in the digital twin control centre and the hospital's information department to securely store sensitive business data and address these safety concerns.

An AR application was developed specifically for use on an iPad tablet to assist in verifying object mapping.

The core algorithms for diagnosis were developed using the Python programming language. AI models were implemented using well-known frameworks such as TensorFlow, wrapped with Keras and PyTorch for deep learning purposes.

11.8.6 Digital Twin Application

The control centre was well-equipped to ensure reliable operation, with features like a server room, network infrastructure, power backup, private offices, security access including facial recognition, communication systems and environmental controls. This centre had significant authority and access to data, making it a central hub responsible for coordinating daily activities in the hospital.

The following areas were closely monitored and necessary actions were taken.

- Space management
- Room management
- Energy management
- Water management
- Security alerts
- Humidity/ Temperature
- CO_2 level
- Maintenance/ Fault diagnosis

The system collected requests from automatic fault diagnosis modules and assigned tasks to the nearest worker in the hospital building, either manually or automatically.

The system gathered requests from automatic fault diagnosis modules and allocated tasks to the nearest worker within the hospital building through manual intervention or automated processes. The security platform seamlessly integrated all security cameras by linking them with the hospital's video monitoring platform, providing features such as real-time monitoring of critical areas. Facial recognition cameras were strategically placed at all entrances on each floor, facilitating real time management of guest flow. The system calculated the total flow every five minutes using ETL processes. A warning signal would be transmitted to the control centre if the average flow strength surpassed a threshold of 0.25.

Systems were built to detect abnormal electrical usage, frequent repair patterns and poor-quality maintenance.

11.8.7 Key Takeaways

The digital twin project in this Smart Hospital Building in China has been running for more than 4 years and is considered to have reached the desired performance:

1. Survey results indicated a 10% improvement in management satisfaction compared to the previous clinical building.
2. The energy savings were estimated at about 1% every year.
3. The utilisation of digital twin diagnosis prevented over 10% of faults in the facility and requested repairs.

This case of the 'continuous lifecycle integration' concept is not confined to any country or region. Given common management requirements and similar design, this case offers valuable problem-solving insights for other modern buildings worldwide.

11.9 CONCLUSION

The adoption of Digital Twin technology is rapidly accelerating across industries, driven by advancements in IoT, AI, cloud computing and real-time data analytics. As industries seek greater efficiency, predictive capabilities and innovation, digital twins have emerged as a strategic enabler, offering real-time insights, enhanced design accuracy and optimised operational performance.

According to market research, the global digital twin market is projected to grow from approximately $7 billion in 2022 to nearly $110 billion by 2028, reflecting an impressive CAGR of 58.3%, as shown in Figure 11.11. North America is expected to lead this growth, followed by Europe and Asia-Pacific, with particularly strong adoption anticipated in the healthcare and manufacturing sectors.

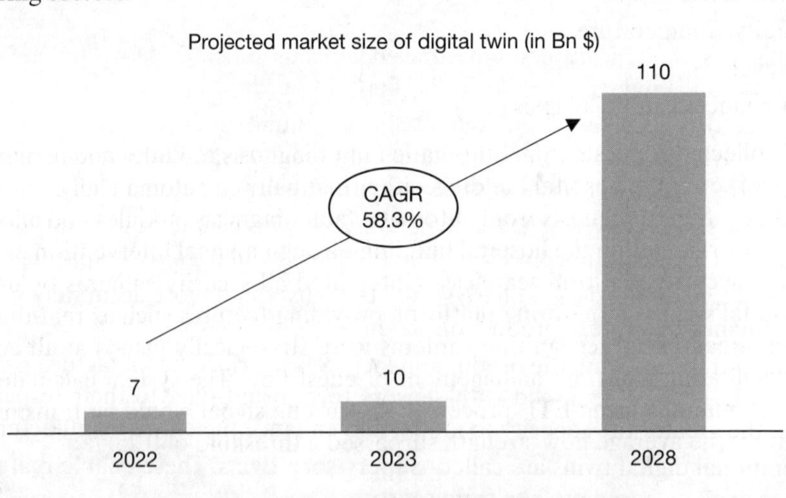

Figure 11.11 Projected digital twin market size

Looking ahead, the future of digital twins is set to be shaped by several transformative advancements. Integrating AI and ML will enable more autonomous, self-learning digital twins to make predictive decisions with minimal human intervention. Edge computing will allow for faster, decentralised data processing, enhancing the real-time responsiveness of digital twin models. Blockchain technology is also expected to strengthen data security and traceability within digital twin ecosystems. Emerging areas such as cognitive twins, which incorporate behavioural models and decision-making logic, are poised to redefine the future industrial and business landscapes. These advancements and the increasing emphasis on sustainable and resilient systems highlight the vital role that digital twins will play in the next generation of Industry 4.0 innovations. As organisations increasingly embrace digital twins to bridge the physical and digital worlds, this technology will become a foundational pillar for driving smarter, safer and more adaptive industries.

SUMMARY

- **Digital Twin:** It is a virtual representation of real-world entities and processes, synchronised at a specified frequency and fidelity.
- Digital Twin technology can be categorised into three types based on usage, i.e.,
  - **Digital Twin Prototype (DTP):** This variant of the digital twin concept represents the initial blueprint of a physical product. It encompasses the essential data sets required to describe and fabricate a tangible copy that mirrors the virtual rendition.
  - **Digital Twin Instance (DTI):** This variant of the digital twin concept pertains to a particular real-world product with a continuous, real time connection to its digital twin over the entire lifespan of the physical product.
  - **Digital Twin Aggregates (DTAs):** DTAs represent a higher-level concept within the digital twin framework, combining data and functionality from multiple DTIs.
- **Digital Thread:** The thread ensures traceability in a multi-party, multi-process composite framework, by enabling data sharing, real time monitoring, collaboration and data integration.
- Simulations are typically computer-based, utilising software-generated models to assist managers and engineers and for training purposes.
- The main difference between digital twins and simulations – Digital twins are comprehensive, dynamic and intimately connected to real-world data. In contrast, simulations often have a narrower focus and may not involve real time data engagement.
- **Cognitive Digital Twins (CDTs):** When digital twins can learn in real time, they transform into CDTs, equipped with AI functionalities and cognitive capacities.
- A digital twin in the design phase serves the purpose of more accurately predicting the performance of a future product or system.
- The digital twins within the production facility are termed 'manufacturing digital twins'. The more devices equipped with sensors to transmit data to their respective digital twins, the greater the capacity to model and enhance the entire manufacturing process.
- Operational digital twins are called 'Supervisory Twins'. These enable real time performance analysis based on data from sensors.

- Key implementation considerations for digital twins:
  - Cybersecurity threats
  - Choosing the right data
  - Standardisation issue

EXERCISE

Multiple Choice Questions (1 Mark)

1. What key technology supports the functionality of a Digital Twin?
  (a) Machine Learning
  (b) Real time data integration
  (c) 3D Printing
  (d) Blockchain

2. Which of the following is NOT a benefit of a Digital Twin?
  (a) Predictive maintenance
  (b) Process optimisation
  (c) Enhanced physical durability
  (d) Fault detection

3. What is the importance of standardising data in Digital Twins?
  (a) To increase sensor performance
  (b) To ensure compatibility across systems and manufacturers
  (c) To reduce production costs
  (d) To enable 3D modelling

4. What is a key difference in the use cases for Augmented Reality (AR) and Digital Twins?
  (a) AR is primarily used for predictive maintenance, while Digital Twins are used for enhancing user experiences.
  (b) AR overlays real time data on the physical world, whereas Digital Twins simulate entire systems or objects for optimisation.
  (c) AR relies on batch data, while Digital Twins rely solely on pre-recorded models.
  (d) AR is designed for lifecycle monitoring, whereas Digital Twins are focused on user interaction.

5. Which tool can manage real time high-density data streams in Digital Twins?
  (a) Python ETL scripts
  (b) SAP HANA
  (c) TensorFlow
  (d) Apache Kafka

6. What is the first step in building a Digital Twin?
  (a) Model creation
  (b) Selecting the target system

(c) Data visualisation
(d) Implementing control systems

7. Which platform is commonly used to store large-scale Digital Twin data?
(a) Local databases
(b) USB storage devices
(c) Cloud platforms
(d) Local servers

8. Which is not a use case of Digital Twin simulation?
(a) Predictive maintenance
(b) Visualising the system design
(c) Enhancing physical component durability
(d) Process optimisation

9. What distinguishes a Digital Twin Instance (DTI) from a Digital Twin Prototype (DTP)?
(a) DTI focuses only on the design phase, while DTP is used for maintenance.
(b) DTI represents a digital mirror continuously synchronised with its real-world physical counterpart, while DTP represents the initial virtual design.
(c) DTI aggregates data from multiple prototypes, whereas DTP aggregates multiple systems.
(d) DTI focuses on geometric modelling only, while DTP focuses on lifecycle data.

10. Which of the following represents the third stage in the evolution of digital twins?
(a) Development of CAD-based digital models
(b) Bi-directional interaction between the physical and digital twins
(c) Complete autonomy of digital twins in decision-making
(d) Simulation of product behaviour using historical data

11. Which of the following is not a capability of a 'supervisory twin'?
(a) Real time visualisation of a physical asset's status
(b) Running predictive simulations for future scenarios
(c) Replacing physical assets with digital counterparts
(d) Diagnosing existing and potential issues in real time

12. Which key functionality do digital twin aggregates (DTAs) provide?
(a) Storing raw data collected from IoT sensors.
(b) Predicting future component failures using insights from multiple Digital Twin Instances.
(c) Real time monitoring of a specific physical product.
(d) Designing virtual prototypes of high-risk machinery.

13. What is the main advantage of integrating the Digital Thread into the product lifecycle?
(a) Minimising costs by replacing all physical sensors.
(b) Enhancing scalability by focusing only on individual components.

(c) Achieving end-to-end traceability and collaboration across assets and systems.
(d) Limiting data collection to the early stages of product design.

14. What unique capability does a Cognitive Digital Twin (CDT) offer compared to traditional digital twins?
   (a) Static modelling of physical systems for design purposes.
   (b) Real time synchronisation without decision-making capabilities.
   (c) Self-learning and reasoning abilities to predict future events and optimise processes.
   (d) Data aggregation from multiple independent systems.
15. An urban planner needs to simulate traffic flow changes based on a new flyover design. What characteristic of a digital twin would make this possible?
   (a) Representational modelling of real-world entities
   (b) Physical replica construction for hands-on testing
   (c) Bi-directional synchronisation with existing traffic cameras
   (d) Computational simulation of scenarios in the virtual twin
16. An engineering firm implements digital twins for a production line.
   Which of the following best demonstrates the use of a 'what-if' scenario in their project?
   (a) Visualising the production line layout in 3D
   (b) Simulating the impact of different maintenance strategies on performance and costs
   (c) Tracking the lifecycle of each machine part
   (d) Implementing AR-assisted training modules for operators
17. An aerospace company aims to design a new aircraft with enhanced safety features by utilising digital twin technology.
   At which stage of digital twin evolution should they aim for two-way interaction between the physical and digital entities?
   (a) Stage 2
   (b) Stage 5
   (c) Stage 4
   (d) Stage 3
18. How can Augmented Reality (AR) complement Digital Twins?
   (a) By replacing digital twins in data-heavy systems.
   (b) By providing immersive visualisation and interaction with virtual models.
   (c) By eliminating the need for data synchronisation in digital twins.
   (d) By enabling virtual replicas to function without real time data.
19. A vehicle manufacturer wants to test new designs for autonomous vehicles in different weather conditions.
   Which digital twin feature is most relevant?
   (a) Condition-based monitoring of vehicle sensors
   (b) Multi-physics simulation using virtual environments

(c) AR-assisted visualisation of manufacturing defects
(d) Real time diagnostics of fuel efficiency

20. A manufacturing company wants to train new operators without disrupting ongoing production.
What role could digital twins play in this scenario?
(a) By creating interactive training modules with digital replicas of the production line
(b) By enabling condition-based monitoring of production equipment
(c) By optimising production schedules based on historical data
(d) By providing a digital inventory of all production materials

Short Answer Type Questions (2 Marks)

1. What are the two main components of a digital twin and how are they connected?
2. What are the two primary representations offered by Digital Twin Aggregates (DTAs)?
3. Why was the concept of 'digital thread' significant in the evolution of digital twins?
4. How does fidelity impact the synchronisation between a physical twin and its digital counterpart?
5. What is the primary purpose of a Digital Twin Aggregate (DTA)?
6. How are 'what-if' simulations used in maintenance strategies in the build phase?
7. What role does a supervisory twin play in optimising operations?
8. How does Digital Twin Definition Language (DTDL) enhance the functionality of Digital Twins?
9. How does Augmented Reality (AR) enhance the usability of Digital Twins in industrial applications?
10. Why is synthetic data important in Cognitive Digital Twins (CDTs)?

Long Answer Type Questions (5 Marks)

1. Briefly explain the concept of Digital Twin with examples.
2. Describe the differences between a Digital Twin Prototype (DTP), a Digital Twin Instance (DTI) and Digital Twin Aggregates (DTA) and their respective role.
3. How do digital twins differ from simulations and what advantages do digital twins provide in real-world applications?
4. Explain the significance of high-fidelity synchronisation in the context of digital twins. Provide examples where low-fidelity synchronisation might lead to inefficiencies.
5. Discuss the role of digital twins in healthcare, particularly in patient monitoring, drug development and surgery simulation.

6. Describe the six fundamental functionalities of Cognitive Digital Twins (CDTs) and explain how these functionalities enable advanced decision-making.
7. Explain the security challenges of digital twins and the measures to mitigate these risks.
8. A real estate developer plans to implement a digital twin for a smart building project to optimise energy efficiency and improve tenant comfort. Outline how they should approach the implementation.
9. Discuss the cybersecurity risks associated with digital twins and outline possible mitigation strategies to address these risks effectively.
10. Compare and contrast the role of real time data in Digital Twins and traditional simulations. Why is real time data critical for Digital Twins?

Blockchain

CHAPTER 12

OBJECTIVES

Until a few years ago, cryptocurrencies were the primary use case for Blockchain. However, rapid development in this field has added capabilities to blockchain technology, which is gaining traction across almost all industries today. So, it is essential to understand this trending technology and its use cases, especially in the context of Industry 4.0.

By the end of this chapter, you should be able to:

- Understand the role of Blockchain in Industry 4.0 implementation
- Describe the fundamental concepts of blockchain technology
- Analyse the potential applications of blockchain
- Evaluate the benefits and challenges of blockchain in the Industry 4.0 context
- Recognise real-world implementations of blockchain

12.1 INTRODUCTION

What is Blockchain?

The term 'blockchain' has become widely recognised, largely due to the rise of discussions around cryptocurrencies. It's often confused with Bitcoin, but blockchain is far broader than just Bitcoin and other cryptocurrencies. Before diving into blockchain and its associated components, it's important to first grasp a few fundamental concepts that have made blockchain a groundbreaking technology: centralised vs. decentralised systems and mutable vs. immutable systems.

Centralised and decentralised systems refer to how control, decision-making and operations are structured, particularly concerning who has authority and how data is managed across the network. Refer to Figure 12.1. A centralised system is controlled by a single entity or group, leading to a more structured but potentially less transparent and resilient model.

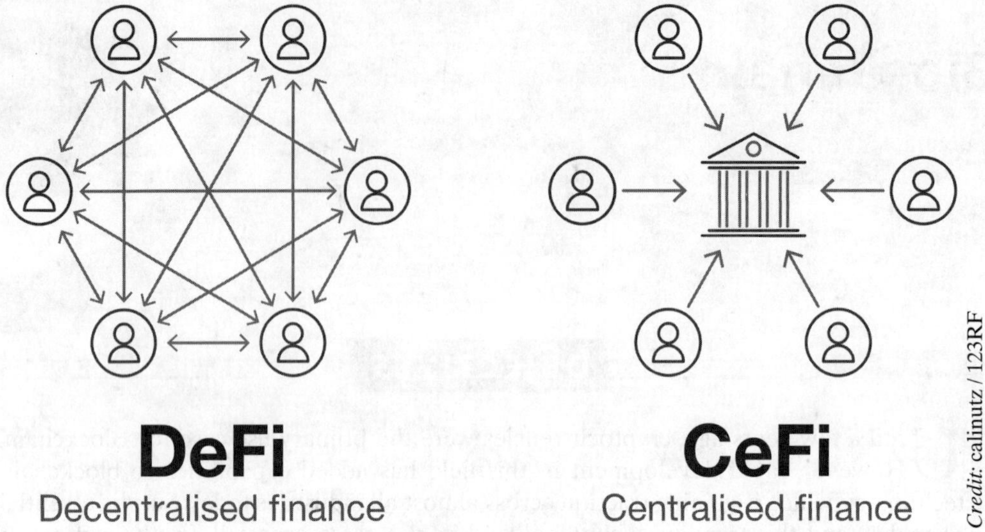

Figure 12.1 Centralised vs decentralised system

As the control and decision-making are concentrated in the hands of a single entity or a few entities, the central authority has full control over the data and operations. For example, banks are financial institutions that centralise the monetary transactions between multiple entities. They act as intermediaries that set up the rules and manage the financial transactions. On the other hand, decentralised systems distribute control across many participants, which enhances transparency, trust and security while making the system more robust and resistant to failures or attacks. As the control is distributed among many independent participants or nodes, rather than being concentrated in one central authority, there is no single point of control or failure and decision-making is done collaboratively by the network participants. This distribution increases the security and transparency of the system. We will discuss later in this section how blockchain implements a decentralised system.

Mutability or immutability denotes the ability (or inability) to change data once it has been recorded on a system. Mutability means that the records can be modified or changed after they have been written into the system of records. Most of the centralised systems are mutable, as a small number of participants can make a decision to change a past record. Immutability refers to the characteristic of data being unchangeable once written. When a record is immutable, it cannot be modified, deleted, or tampered with after it has been confirmed and added to the system of records. This concept is crucial because it influences how blockchain systems are perceived in terms of trust, security and data integrity. The blockchain systems are generally immutable due to their decentralised nature.

We can define Blockchain as a decentralised, immutable ledger that can record transactions and track assets in a peer-to-peer network.

An asset can be either tangible, such as a house, car, cash, or land, or intangible, like intellectual property, patents, copyrights and branding, that has some value. Almost any asset can be recorded and exchanged on a blockchain network, helping to minimise risk and lower costs for all parties involved. Blockchain functions as a decentralised ledger or database that is distributed across thousands of computers worldwide. It keeps an ever-expanding series of records known as blocks. These computers are called nodes, which enable the blockchain network to operate by maintaining and verifying transactions in a secure and connected system.

While discussing the types of blockchains, we will see that a blockchain network can also be set up as a centralised system or mutable system based on need.

Industry 4.0 has led to a surge in connected devices, assembly lines and factories, with supply chains, logistics and data exchange now extending across a global network. Because blockchain offers a secure, transparent, and decentralised method to track and verify transactions in real time among multiple parties, it has a crucial role to play in this connected industrial landscape. As all stakeholders have access to a single, tamper-proof record, it minimises the risk of fraud and errors. The traceability and accountability of blockchain networks not only result in greater trust but also simplify processes by removing intermediaries. Moreover, advanced blockchain technologies like smart contracts drive faster decision-making, lower operational costs and enhanced efficiency in interconnected systems by automating transactions.

12.2 KEY CONCEPTS OF BLOCKCHAIN TECHNOLOGY

12.2.1 How Do Blockchains Work?

To comprehend the core parts of a blockchain, it is important to understand its workflow. Let us take a simple example. Two friends, John and Peter, are involved in some monetary transaction. Both have a notebook in their hands to record the transactions. John gives Peter 200 dollars and both of them have noted it in their corresponding notebooks. But John does not trust Peter and thus he asks Peter to show his notebook. Surprisingly, he finds that Peter wrote the amount received as 150 dollars! So, in this trustless environment, they have two options to establish trust. One option is to conduct transactions through a trusted third-party intermediary, such as a bank. Or they can invite five more independent people who also bring their notebooks to record the transactions. After the transaction is completed, each member will show his/her notebook and the group will reach a consensus on the transaction that has happened. So, this group now possesses distributed records of the transaction which can't be modified and thus, the process brings back trust to this environment. This decentralised collaborative and consensus-based ledger system that allows peer-to-peer transactions in a trusted way without involving any third-party intermediaries (like a bank) is a simplified example of the principle of blockchains.

If we ponder on the critical issue that can happen in any digital transaction is the *double spend problem*, where the same digital currency could be spent more than once. In a centralised system, banks resolve this issue as intermediaries. In 2008, Satoshi Nakamoto, a pseudonymous person or group of persons, released the *Bitcoin whitepaper* which solved the double spend problem in a peer-to-peer transaction and also provided a framework to establish security, transparency and trust in such a transaction. The technology is generically known as the blockchain.

In a blockchain, every transaction is recorded in a block, which is linked to the previous block, creating a secure and immutable chain of records. Once a transaction is verified and added to the blockchain, it becomes part of the public ledger and is nearly impossible to alter or duplicate. To prevent double spending, blockchain uses a consensus mechanism (like Proof of Work in Bitcoin) where participants in the network (miners or validators) work together to verify the legitimacy of each transaction. If the same cryptocurrency is attempted to be spent more than once, then it would require altering past transactions, which is computationally infeasible due to the decentralised storage of the blockchain. This ensures that once a transaction is confirmed, it is final and cannot be reversed or duplicated, effectively preventing double-spending.

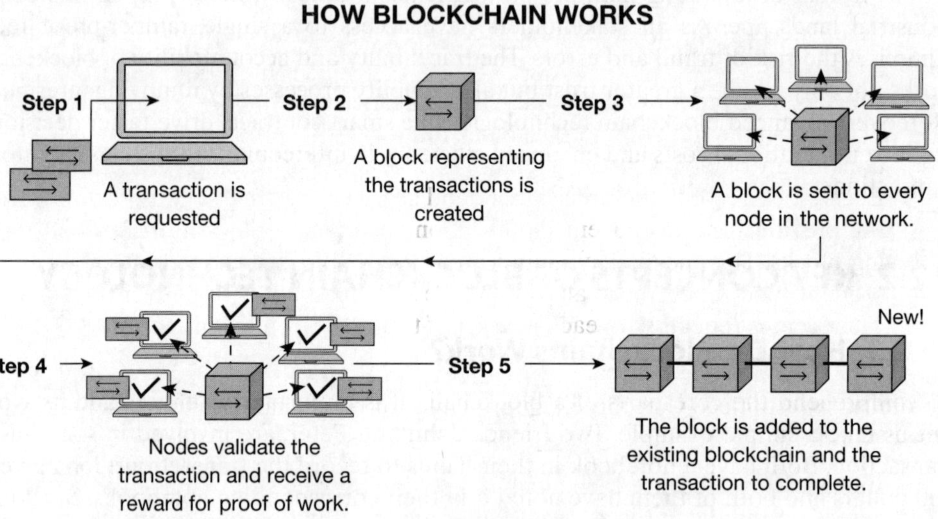

Figure 12.2 Blockchain transaction workflow

Here's how it works step by step (refer to Figure 12.2):

1. **Transaction or Action is initiated:** Transactions, in a blockchain, refer to the exchange of information, assets, or data between participants within a network. A few examples of transactions are the transfer of digital currency or cryptocurrency, the execution of a smart contract, the journey of a batch of goods through different traceable checkpoints, etc. Those transactions can track the movement of either a tangible asset (a product) or an intangible asset (intellectual). As a transaction is initiated, all the details like the sender, the receiver, the amount, the digital signature, etc., are captured within the transaction record.
2. **The Transactions Are Broadcast and Validated:** The transactions are broadcast to a network of computers (also called nodes) that are part of the blockchain. These computers check the transactions to make sure they are valid (for example, checking that the senders have enough funds to send).

3. **Transactions Are Inserted into a Block:** The validated transactions are then bundled together and added into a *block*. Transactions are recorded using encrypted data to ensure that they remain safe and secured during transmission over the network. A data block can store any chosen information, including details like who, what, when, where and the amount involved. It can also capture specific conditions, such as the temperature of a food shipment. Recall our previous example of five friends recording a transaction in their notebooks. A block can be imagined as a digital page in the notebook, with a group of transactions written on it.
4. **Consensus and Verification Completed:** Once the computers/nodes in the blockchain network agree about the legitimacy of the block, the block is ready to be inserted into the blockchain. This method of agreement is called *consensus.* In some blockchains (like Bitcoin), this involves solving complex math problems (called proof of work) to confirm the block.
5. **Block is Inserted into the Chain:** After the block is verified, it's appended to the chain of blocks in the right order. Each block references the previous block, creating a continuous 'chain' of blocks. As an asset moves or changes ownership, this data chain expands. Each block records the precise time and sequence of transactions, securely linking to the previous one to prevent modifications or unauthorised insertions between existing blocks. This means any modification in any block will need alteration in all subsequent blocks and those alterations are possible only with network-wide consensus. This process repeats with each new set of transactions, allowing the blockchain to grow continuously.

You must have noticed that three key principles of blockchain are implemented within the above workflow:

- **Immutability:** Because each block is linked to the one before it and the whole network agrees on the data, it isn't easy to alter any information once it's added to the blockchain. The information is considered permanent and public. Everyone can see it, but no one can change it. This makes the blockchain secure and tamper-proof. If someone tries to change a block, the network will notice and reject it.
- **Decentralisation:** Every computer (node) on the network has a copy of the entire blockchain. This makes the system decentralised. No single computer has control; it's spread across many. To tamper with the blockchain, we would need to change every single copy on the network, which is practically impossible.
- **Transparency:** The ledger (the block) is updated for everyone simultaneously, ensuring everyone has the same information.

12.2.2 Components of Blockchain

To support the workflow outlined in the previous section, a blockchain architecture must include several key components. Understanding the design and role of each of these components is essential if we intend to implement a blockchain solution for our needs. Refer to Figure 12.3 for an overview of these components.

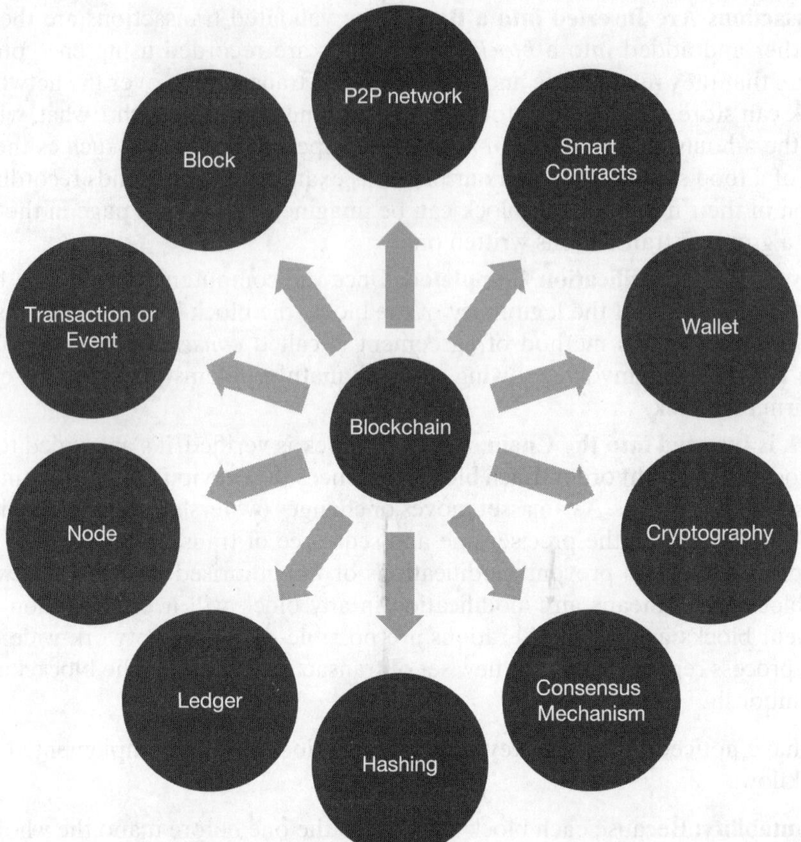

Figure 12.3 Components of blockchain

Peer-to-peer Network

A blockchain functions on a decentralised Peer-to-Peer (P2P) network, allowing participants to interact and conduct transactions directly and thus does not need intermediaries or central authorities in the whole process. In a P2P network, interconnected computers or nodes communicate, execute transactions and securely store records in a transparent, tamper-proof and immutable manner (see Figure 12.4). Note that this concept of distributing the ledger or record across multiple nodes ensures that no single node has control of the entire system. It makes the system more resilient to failures and malicious attacks, as no single point of failure can compromise the entire network.

Node

Each participant of the P2P network functions as a blockchain node and plays an important role by processing transactions and verifying new blocks in order to maintain the distributed ledger. Every node is distinct and identified by a unique identifier (see Figure 12.4). A higher number of nodes enhances the network's decentralisation and security. Blockchain networks consist of different types of nodes, each serving a specific purpose. In this book, we will focus on the three most commonly used types.

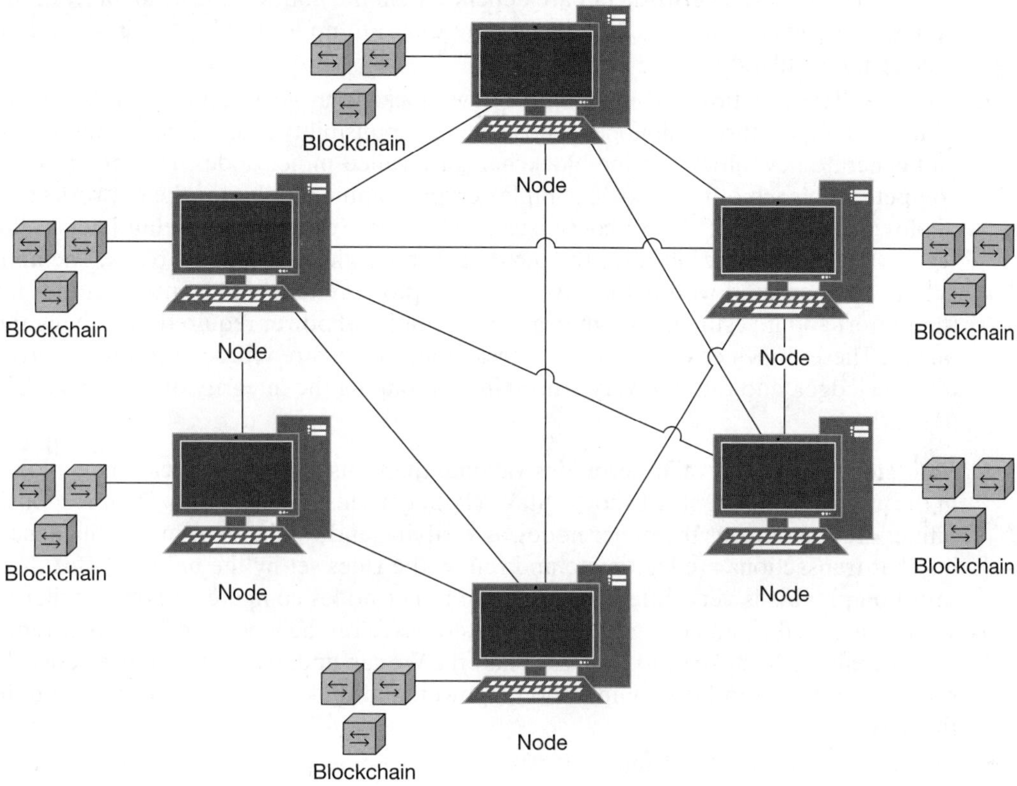

Figure 12.4 BlockchainP2P network with nodes

1. **Full Nodes:** These nodes are critical for maintaining the stability and security of blockchain networks. They download and preserve every transaction record and block, and thus can independently verify the entire blockchain history, whether it conforms to the network's rules. Because these nodes store and maintain the complete copy of a blockchain ledger, they are called full nodes. These nodes form the backbone of the network as they perform the crucial tasks such as transaction verification, block validation, data storage, consensus participation and data propagation. As we can understand, the full nodes need extensive resources, in terms of data storage and computational power.
2. **Light Nodes:** All the nodes can't dedicate high processing and computational resources, especially if they use devices with limited processing power and storage capacity, such as smartphones, IoT devices, tablets, etc. So, the concept of light node evolved. The light nodes are structurally and responsibility-wise similar to the full nodes, but are optimised for limited resources. These nodes do not store the entire copy of the blockchain, but store only the essential header part where the metadata of the transactions is stored. They contact the full nodes if they need to retrieve the full transaction details. Thus,

light nodes can quickly verify the authenticity and integrity of the blockchain without requiring extensive computational resources. They enhance accessibility as more users can participate in the network, but are dependent on the honesty and availability of full nodes, which makes them slightly less secure than full nodes in terms of independent transaction validation.

3. **Miner Nodes:** Creation and insertion of new blocks in a blockchain is also known as Mining. The specialised full nodes, whose main responsibility is to validate transactions and generate new blocks on the blockchain, are called miner nodes. The miner nodes compete with each other to solve complex cryptographic puzzles for the networks that deploy Proof of Work (PoW) consensus mechanisms (described in detail later in the section). The first miner to solve the puzzle gets the right to create the block. The miners are compensated with the newly minted cryptocurrencies and transaction fees for their effort and investment in significant computational power required for solving the puzzle. These nodes ensure that the transactions are securely processed and inserted into the ledger, and thus are very crucial in maintaining the integrity of the blockchain network.
4. **Validator Nodes:** The validator nodes validate the transactions and create new blocks in the networks that deploy Proof of Stake (PoS) consensus mechanism. Their responsibilities are the same as the miner nodes, as verifying and validating transactions, ensuring that transactions are legitimate and follow the rules set by the network. But their selection process is very different. While the miner nodes compete with each other to solve a puzzle, the validator nodes are selected based on the amount of cryptocurrency they are willing to stake or lock up as collateral. We can understand that these nodes do not need to invest in high computational power, and thus, it is easier to participate in the network.
5. **Observer Nodes:** As the name suggests, observer nodes play a passive role in blockchain networks. They do not participate in the validation of transactions or the creation of new blocks, but their key role is to monitor and maintain an accurate and up-to-date version of the blockchain ledger. Similarly, they do not stake any cryptocurrency or receive any rewards or transaction fees. Various purposes for which observer nodes are maintained are auditing, analysing transaction data, and providing access to blockchain information. They are useful for managing decentralisation, transparency, and external monitoring of the network.

Other types of nodes, such as pruned full nodes, master nodes, staking nodes, authority nodes, super nodes, etc. are used based on specific categories and needs of the blockchain.

Transaction or Event

In blockchain, a transaction is an action or a transfer of data or value that takes place between two or more parties, recorded and verified using the blockchain network. It typically involves a transfer of cryptocurrency (like Bitcoin or Ethereum), but it can also involve other types of data, such as smart contract interactions, asset transfers, or voting results, based on the blockchain's type. Transactions are signed by the sender with a private key, safeguarding authenticity and preventing tampering. This digital signature proves that the sender authorised the transaction. Transactions are recorded with a timestamp, indicating when the transaction was initiated. Often, transactions include a small fee that is paid to validators or miners to process and confirm the transaction. This fee is usually paid in the blockchain's native currency.

Block

A block is a group of transactions recorded at a particular time. The most important set of information contained in a block is a group of transactions, a timestamp, a reference to the previous block (via a hash) and other metadata. The structure of a block is explained in detail in the next section.

Ledger

A blockchain ledger digitally records transactions and data and stores them in a decentralised manner across multiple computers worldwide, making it a type of distributed ledger. This ledger consists of a series of blocks and every block contains a set of transactions. Every computer in the blockchain network preserves a copy of these blocks. This ensures transparency and data accuracy. For instance, when a user purchases something using cryptocurrency, a monetary exchange takes place between the buyer and the seller. Before the transaction is finalised, a majority of blockchain nodes must validate its legitimacy through consensus.

Hashing

Hashing is the process of using a cryptographic hash function to encode information. This function performs a mathematical operation that converts a block of data into a unique, fixed-length sequence of hexadecimal characters. The resulting hash value acts as a cryptographic fingerprint to ensure the integrity of the block's contents by detecting any alterations.

For example, Bitcoin, the leading cryptocurrency, utilises the Secure Hash Algorithm (SHA-256). Irrespective of the size of the input data, this algorithm consistently generates a 256-bit hash output. A completely different hash value will be generated even with the slightest modification, such as changing a letter's capitalisation, as demonstrated in the examples below:

Input	Hash Output
my blockchain	9e64dda3c88541ca5d3c201479b365c45581ae987268a831a908a3626b636723
My blockchain	b90ec79655b9311b999ee95bee70a0c186768b85c23d5c80de7b7ac23145c54b

Consensus Mechanism

Traditional financial institutions depend on central authorities to safeguard transactions and ensure system stability. In contrast, decentralised systems face challenges in reaching consensus, raising concerns about maintaining blockchain security and preserving transaction records. To address this, blockchain networks use a set of protocols and systems that are used by the nodes to collectively approve the validity of any transaction and the integrity of the blockchain. In return, they collect incentives. This decentralised governance model ensures both security and functionality and forms the foundation of cryptocurrencies like Bitcoin, Cardano, Ethereum, etc., by verifying transactions and protecting blockchain integrity.

Based on the consensus mechanism implemented by a blockchain, nodes can independently verify transactions and update the ledger. These mechanisms vary, with some requiring nodes to contribute computational work, others relying on stake-based validation and some operating through authority-based verification. Below are some common consensus mechanisms implemented in blockchain, with the potential for new models to emerge in the future.

1. **Proof of Work (PoW):** Proof of Work is a consensus mechanism where miners try to solve complex mathematical puzzles and the first person who solves the puzzle and validates transactions can add those validated transactions to the blockchain. This process is highly resource-intensive as it demands significant computational power and energy. The miner who successfully solves the puzzle first receives the newly minted cryptocurrency (such as Bitcoin) as the reward along with transaction fees. PoW is a highly secure consensus process as it would require a huge amount of computational effort if the blockchain is attempted to be altered, making attacks extremely difficult. However, the high energy consumption and resource intensity create an entry barrier. This led to the development of alternative consensus mechanisms.
2. **Proof of Stake (PoS):** In this consensus mechanism, the validator nodes are selected based on how much cryptocurrency they can lock up or 'stake' as collateral, instead of the nodes competing to solve complex puzzles. This approach significantly lowers the computational requirement and energy consumption and is thus more scalable. So, it is a popular choice for new blockchain projects. However, PoS can have one drawback that it may potentially favour the wealthy participants with large stakes, which could eventually centralise the control of the network.
3. **Proof of Capacity (PoC):** This is another consensus-building mechanism where the 'capacity' or the available hard drive disk space becomes the deciding factor in determining the miner of the new block. Miners create and store 'plots' or files that contain pre-computed solutions for the cryptographic puzzles and quickly search through them to provide a valid proof when a new block is proposed. Thus, a miner who can allocate more space can store a larger number of plots and stands a higher chance of being selected to mine a block and win rewards. The initial setup to create and store plots will take some effort, but after that, no high-energy processing is required. A few advantages of this innovative consensus process are the low energy consumption, faster processing and lower entry barrier, still keeping the overall process transparent and decentralised.
4. **Proof of Authority (PoA):** This consensus mechanism relies on a group of individuals' reputation or identity staked in the network to create a small pool of trusted 'authorities'. The authorities are responsible for validating the transactions and mining blocks. The network users who want to contend as the validators go through a formal notarisation process. In this step, they submit documentation to prove their identity and link it with the network identity to stake their digital reputation. The existing validators can vote to add new users to the group of authorities. As this consensus process works based on credibility and accountability of the authorities and does not require any computational processing (PoW) or cryptocurrency investment (PoS), the process is faster and energy-efficient. This mechanism is well-suited for permissioned networks. One flip side of PoA is that it may result in the centralisation of power in the hands of selected individuals, introducing bias in the transaction handling.
5. **Byzantine Fault Tolerance (BFT):** This consensus algorithm is designed to ensure that the reliability and security of a distributed system are maintained, even if faulty or

malicious nodes are present in the network. Byzantine Generals' Problem (BGP) is a classical distributed computing problem, where the challenge is to reach a consensus in the presence of 'traitor' nodes who may send conflicting messages to the other nodes. BFT algorithm solves this, using techniques such as cryptographic verification and redundancy introduced in the network. BFT assumes that there are more honest nodes than faulty ones and thus relies on intercommunication between each node and a voting mechanism to arrive at consensus. Though this algorithm is highly effective in maintaining network integrity despite the presence of conflicting messages from the faulty nodes, the communication overhead increases multi-fold as the network size increases. Practical Byzantine Fault Tolerance (pBFT) has overcome this problem by introducing a leader node in the network. It handles large networks efficiently and thus is used in popular blockchain platforms like Hyperledger Fabric.

In real life, the consensus mechanism should be selected based on the composition of the blockchain network and the business needs. We will elaborate on the process of selecting a consensus mechanism in the latter section.

Cryptography

Cryptography employs computer algorithms to protect data, ensuring that only authorised recipients can access its contents. This technology encrypts messages through mathematical algorithms and makes them unreadable during transmission through communication channels.

Most leading blockchains and cryptocurrencies function on public, decentralised networks which means that they won't have centralised servers or an advanced setup of firewalls. One might assume that their peer-to-peer networks can be easy prey for attackers and hackers. However, blockchain has proved to be one of the most secure and groundbreaking innovations of the 21st century, largely due to cryptography along with other advanced security mechanisms they use.

Wallet

Cryptocurrency, through blockchain technology, has introduced an alternative solution to the traditional centralised global financial system. Blockchain wallets, also known as digital wallets, allow users to securely store, manage and conduct financial transactions with cryptocurrencies. These transactions are more secure than traditional banking methods as they are cryptographically signed and also safeguard user data and privacy.

Each user's private key is stored within their wallet, serving as the crucial access point to their cryptocurrency holdings. While the digital assets remain on the blockchain, the wallet securely holds the user's access credentials and transaction data.

DID YOU KNOW?

Not all blockchains support smart contracts. Blockchains such as Ethereum, Avalanche and Cardano are designed to be able to execute smart contracts, while the core Bitcoin blockchain does not provide native smart contract functionality. The main distinction between these blockchains lies in their ability to store custom logic and execute it as needed.

Popular blockchain wallets include Bitcoin Wallet, Coinbase Wallet, Bitcoin Core, Electrum, Ledger Wallet, CoinPayments, etc.

Smart Contracts

A set of self-executing programs that run on a blockchain network is known as Smart Contracts. They contain code that defines predefined conditions, which, when met, automatically trigger specific actions. Unlike traditional systems that rely on centralised servers, smart contracts operate on a decentralised blockchain, allowing multiple parties to achieve a shared outcome in an accurate, timely and tamper-resistant manner. These tamper-proof programs function based on logical rules such as: 'If/when event X occurs, then execute action Y'. A single smart contract can incorporate multiple conditions and several smart contracts can work together within an application to facilitate a series of connected processes. Various programming languages are used to develop smart contracts, with Solidity being the most common for Ethereum-based contracts.

12.2.3 Structure of a Block

An individual block within a blockchain contains four main parts – block header, block body, block height and block hash as shown in Figure 12.5.

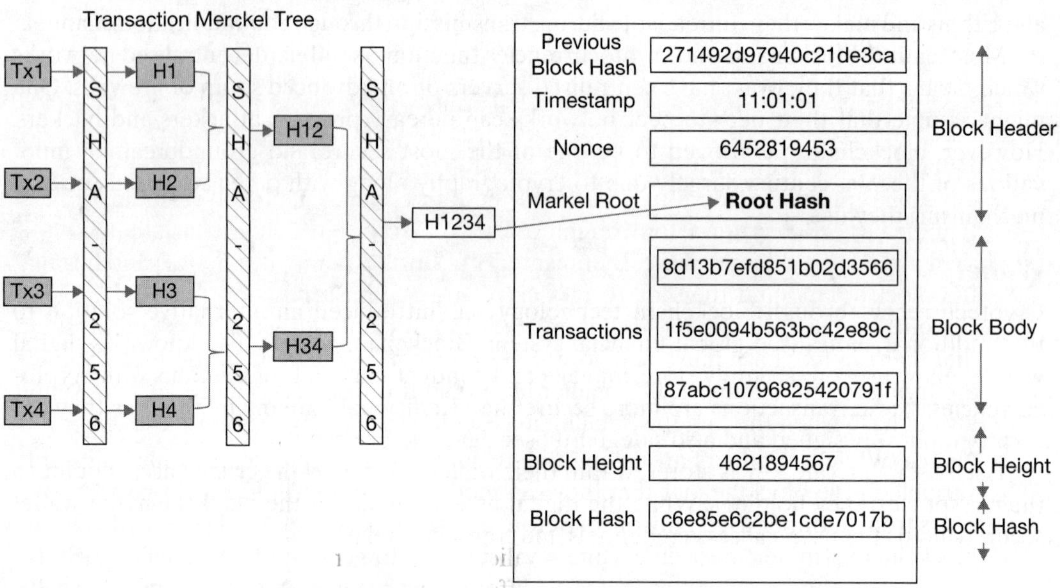

Figure 12.5 Structure of a block

Block Header

The block header contains the following:

- **Version:** The version field signifies the version of the blockchain software used. This is a 4-byte field, where the number specifies the software version used to create the block. The software used by the blockchain may be upgraded due to the incorporation of new features. However, as the blockchain is a decentralised system, there is no central admin to push the software to all the nodes. So, a miner will signal the readiness of the upgrade through the version number field. The network participants can vote through this field whether the proposed upgrades to the software should be used.

- **Timestamp:** The timestamp indicates when the block was created or added to the blockchain, providing a chronological order to the blocks.
- **Previous Block Hash:** This is the cryptographic hash of the preceding block in the blockchain. It establishes a link between the current block and the one before it, maintaining the sequential integrity of the blockchain. Refer to Figure 12.6 on how this chain of blocks is maintained.

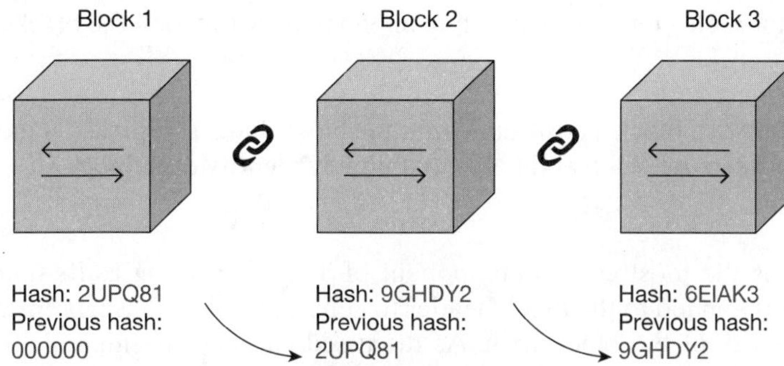

Figure 12.6 Chain of blocks

- **Difficulty Target**: In PoW-based blockchains like Bitcoin, miners must solve a computational puzzle to add a new block to the blockchain. This involves finding a hash that meets certain criteria, for example, the goal can be to produce a hash that is lower than a predefined target value. By repeatedly altering a number (a nonce) and hashing the block's data, and then applying a cryptographic hash function (e.g., SHA-256 in Bitcoin) on these, the target hash is created. The difficulty target defines the threshold that the resulting hash must be below to be considered valid. It is a number and miners are trying to find a numerically smaller hash than this target. It is more difficult to find a valid hash when the target is smaller, making the mining process more challenging.
- **Nonce:** While the miners are attempting to solve the hash puzzle, the only adjustable variable in the block header that they can modify is the nonce. The term 'nonce' originates from 'number used only once'. Before a new block is attempted to be added to the blockchain, a miner must determine a valid nonce to generate that specific block. By altering the nonce, miners can generate different hash values and explore possible solutions. As this nonce value is increased with 1 continuously till the correct value is found, you can understand that this number can be used only once. The first miner who finds a valid hash meeting the network's difficulty target earns a reward. The block is then broadcast to the network's other participants and finally inserted into the blockchain. This mining process repeats. Due to the vast number of possible nonce values, finding the correct combination demands significant computational power.

- **Merkle Root:** A Merkle root links the block header to the transactions within the block. It is generated by hashing pairs of Transaction IDs, creating a concise and unique fingerprint for all the transactions in that block. This Merkle root is included in the block header to ensure the block's contents remain secure and cannot be altered. If someone attempts to add/update/delete a transaction, the newly generated Merkle root of the transactions in the block will lose alignment with the Merkle root in the block header, indicating tampering.

Block Body or Transactions

The body of a block holds a collection of transactions, recording the data and information captured on the blockchain. These transactions may include digital asset transfers, smart contract executions, or other relevant data, depending on the specific function the blockchain provides.

Block Height

Block height refers to a block's position within the blockchain. It represents the total number of preceding blocks in the chain and helps to uniquely identify each block.

Block Hash

The block hash is the most critical component of a block, serving as its unique identifier. A hash function is applied to the block header to generate a block hash. It ensures the immutability and integrity of the blockchain. As discussed earlier, a hashing algorithm (such as SHA-256 in Bitcoin) processes the block header's fields to produce a fixed-length hash value that uniquely represents the block's data.

Any modification to the transaction data within a block alters the Merkle root, which subsequently changes the block hash. Similarly, updates to the block header also result in a different hash. Since each block contains its own hash as well as the hash of the previous block, any tampering disrupts the chain, making unauthorised changes immediately detectable by network nodes. Table 12.1 provides a view of the size of important components of a block and when they are updated.

Table 12.1 Size and update frequency of block components

Field	Purpose	When updated	Size (Bytes)
Version	Version of the Blockchain software	Software is upgraded	4
PrevBlockHash	256-bit hash of the previous block	A new block is created	32
MerkelrootHash	256-bit hash based on all the transactions in the block	Transactions are verified and accepted	32
Timestamp	Current block timestamp as seconds since 01-01-1970T00:00 UTC	Every few seconds	4
Bits	Current target is compact format	The difficulty is adjusted	4
Nonce	32 bit number	Incremented as a hash is tried	4

12.2.4 Types of Blockchain Networks

Blockchain networks can be designed in various ways, including public, permissioned, private, or consortium-based models. Understanding these different types is essential for selecting the right network structure for Industry 4.0 applications.

Public/Permissionless Blockchain Networks

In this type of blockchain network, any participant can view, access and contribute without needing any central authority's permission. The transaction data is visible to any peer-to-peer network of connected nodes that access the blockchain network and can participate in the validation mechanism. Though this fosters inclusivity and makes the ledger data immutable once committed, the challenge may arise due to the lack of transaction privacy, as the sensitive data is open to all participants. These networks generally deploy the PoW consensus mechanism, which can lead to slower processing and higher energy consumption as the network grows.

Private Blockchain Networks

The main difference between a public blockchain and a private one is that in the latter, access is restricted to a group of participants, who are often from a single organisation. A central entity controls participation, maintains the distributed ledger and ensures that the consensus protocols are followed. Thus, a private blockchain is not decentralised. The main focus is to maintain the immutability, security and efficiency, and they are suitable for enterprise applications.

Permissioned Blockchain Networks

These networks adopt characteristics of both public and private networks. They allow anyone to participate in the network after a suitable verification process, but the permissions to specific activities, such as traversing the full chain or validating transactions, are restricted to only authorised entities. This hybrid approach maintains transparency while ensuring accountability.

Consortium blockchains

These blockchains are also called Federated Blockchain, as more than one organisation manages this partly private blockchain. PoA is the common consensus mechanism used in these networks. Similar to the private network, consortium networks can limit user participation with different levels of data access and validation rights. But the governance is managed collectively by multiple organisations, distributing the decision-making power equally among members. Thus, the security and trust are enhanced. Large enterprises that depend on consortia or partnering arrangements to run their business usually form this type of blockchain network to build trust in transactions among them.

12.2.5 Blockchain Protocols Relevant for Industry 4.0

Now that we have understood the key components of blockchains, it is time to understand how different blockchain protocols implement these components to provide a secure and transparent platform for Industry 4.0 implementation. A blockchain protocol refers to a set of rules and standards that define how a blockchain network functions and how its participants interact. These protocols make sure that the network functions in a secure, transparent and decentralised manner.

Ethereum

Ethereum is a decentralised, open-source blockchain proposed by Vitalik Buterin in 2013 and launched in 2015. It expanded the concept of blockchain beyond cryptocurrencies to create a platform for smart contracts, enabling developers to build decentralised applications (dApps).

- **Consensus Mechanism:** Ethereum originally relied on PoW like Bitcoin, but it has transitioned to PoS through the Ethereum 2.0 upgrade. In PoS, validators stake a portion of their ETH as collateral to propose and validate new blocks.
- **Security:** Ethereum's network is highly secure, leveraging strong cryptography and an evolving security model.
- **Scalability:** Ethereum has faced scalability challenges, but Ethereum 2.0 and Layer 2 solutions such as Optimistic Rollups and ZK-Rollups aim to significantly improve transaction throughput.
- **Applications:** Ethereum supports a flourishing ecosystem, particularly for decentralised finance (DeFi), non-fungible tokens (NFTs), decentralised autonomous organisations (DAOs) and more. It is considered one of the leading platforms for building dApps.

Hyperledger

Hyperledger is an open-source initiative hosted by the Linux Foundation, focused on advancing blockchain technologies for cross-industry use. Unlike public blockchains like Ethereum and Bitcoin, Hyperledger is designed for private, permissioned blockchains, tailored for enterprise solutions where businesses can create customisable blockchain systems. It is not a cryptocurrency platform but a set of frameworks and tools to build permissioned networks. The Hyperledger project has incubated several DLT frameworks that can be used to build business applications. A few such Hyperledger frameworks are described below.

- **Hyperledger Fabric** is a modular DLT framework that supports smart contracts in a permissioned network. The identities of the participants are authenticated, and the ledger is exposed only to these authenticated members. Due to its high privacy, security, and scalability level, it is used to build enterprise-level applications in industries like finance, supply chain, and healthcare.
- **Hyperledger Sawtooth** is another open-source Hyperledger project that provides a flexible and modular framework for building blockchain applications according to business needs. It separates application logic from the core blockchain functionality, which enhances flexibility for customisation and extension. It deploys consensus mechanisms, such as pBFT or a more innovative mechanism like Proof of Elapsed Time (PoET), which is a lottery-based approach to reach consensus while validating transactions and creating new blocks.
- **Hyperledger Iroha** was originally developed by Soramitsu Co., Japan. It works on a permissioned network and offers smart contracts. Using its built-in tools and commands, applications can be created simply and quickly.

- **Hyperledger Indy** project offers a blockchain-based digital identity management solution. Through this platform, an individual can control how much information from his/her identity needs to be shared with an organisation based on the need. The identity is verifiable through independent validators in the blockchain network.
- **Hyperledger Burrow** platform is created to support smart contracts for financial institutions and investors in a permissioned network. It uses the Oracle consensus algorithm, which is an advanced version of PoS. This is a fast and reliable platform, and thus is suitable for financial institutions and large corporations.
- **Security:** It offers rigorous security features, such as access control, cryptographic signatures, and enhanced privacy.
- **Scalability:** Hyperledger networks are a permissioned network and thus do not have the overhead of PoW. They can support high transaction volume and can scale more efficiently than public blockchains.
- **Applications:** Hyperledger is primarily used in industries that require high privacy and compliance standards, like supply chain management, finance, healthcare, and identity management.

Corda

Corda, developed by R3, is a blockchain platform designed specifically for enterprise use cases, particularly in financial services. Unlike traditional blockchains, Corda focuses on privacy, scalability and the integration of existing systems. Initially developed for finance, Corda has expanded to industries such as healthcare, insurance, trade finance and supply chain management.

- **Consensus Mechanism:** Corda uses a notary-based consensus model, ensuring transactions are not duplicated and are valid. This approach is more efficient for businesses, as it reduces the computational and storage requirements compared to traditional blockchain systems.
- **Security:** Corda ensures that transaction data is only visible to the parties involved, providing high levels of privacy and safeguarding sensitive business data.
- **Scalability:** Corda is designed for scalability in enterprise applications, reducing network load by sharing transactions only among relevant participants, unlike Ethereum's public broadcast system.
- **Applications:** Corda is extensively used in industries where privacy, security and regulatory compliance are critical, like trade finance, settlement of securities, cross-border payments, insurance and supply chain management.

Cardano (ADA)

Cardano is a blockchain platform designed to provide a more secure and sustainable infrastructure that supports smart contracts and dApps. It emphasises peer-reviewed research to develop its protocols, focusing on scalability and sustainability.

- **Consensus Mechanism:** Cardano utilises PoS protocol, which is both energy-efficient and more scalable than PoW.
- **Security:** PoS offers a strong theoretical foundation, ensuring robust security and resistance to potential attacks.

- **Scalability:** Cardano addresses scalability through PoS and additional technologies like Hydra, which enables higher transaction throughput.
- **Applications:** Cardano is designed for use cases in decentralised finance, identity management and supply chain solutions, positioning itself as a competitor to Ethereum in the smart contract space.

Solana (SOL)

Solana is a blockchain platform built to efficiently support decentralised applications and cryptocurrencies. This high-performance platform is designed for scalability, offering fast transaction speeds and low costs, capable of processing thousands of transactions per second (TPS).

- **Consensus Mechanism:** Solana combines Proof of History (PoH) with Proof of Stake (PoS) to enable faster transaction processing while maintaining security. PoH is a unique pre-consensus algorithm that verifies duration between two blocks by the ensuring that the timestamps are tamper-proof. This pre-consensus step drastically reduces the transaction validation and block finalisation time.
- **Security:** Solana maintains robust security through its consensus mechanism but has faced network outages and some concerns over security vulnerabilities.
- **Scalability:** Solana excels in scalability, supporting thousands of TPS with low transaction fees. It leverages parallel transaction processing for enhanced performance.
- **Applications:** Solana is widely used in the DeFi space, NFT and other dApps that require low latency and high throughput.

DID YOU KNOW?

We did not include Bitcoin in the above discussion of different blockchain protocols because our focus is on the protocols relevant to industry 4.0 applications. But it is important to know about this protocol that started the idea of a blockchain. Bitcoin is the first and most well-known blockchain protocol, introduced in 2008 with the goal of creating a decentralised, peer-to-peer electronic cash system. Bitcoin is primarily a cryptocurrency protocol. Its main use is as a digital currency and store of value. It uses a PoW consensus mechanism (we discussed this mechanism earlier in the chapter) and inserts them into the blockchain. While Bitcoin is primarily used for transferring value, it has also gained traction as a store of value (often referred to as 'digital gold').

12.2.6 Blockchain as a Service (BaaS)

Blockchain as a Service (BaaS) is a cloud-based service offering that enables businesses to develop, host and manage their own blockchain applications, smart contracts and other related functions. While businesses benefit from blockchain technology, the cloud-based IT provider or service partner takes care of the complex tasks of infrastructure management, ensuring the system remains operational.

The advent of BaaS is a major advancement and a key milestone in the blockchain ecosystem, offering a solution to accelerate the widespread adoption of distributed ledger in businesses across various industries.

Importance of BaaS

As more organisations recognise the strategic benefits of adopting blockchain technology, they often face challenges such as the technical complexities involved, the scarcity of domain-specific expertise and the significant operational overhead of developing and maintaining a blockchain network. These challenges slow down the adoption process. BaaS provides a solution as it simplifies blockchain deployment. The ready-to-use products help businesses integrate blockchain technology with their core business processes.

By collaborating with the right BaaS provider, businesses can leverage expert blockchain developers, governance and process expertise and a fully managed cloud infrastructure, eliminating the need for upfront investment and ongoing operational expenses.

BaaS Model

A BaaS service provider offers end-to-end blockchain management services:

- **Initial Setup:** A BaaS service provider is responsible for deploying all the necessary infrastructure to set up the blockchain network for the customer. This includes:
- **Infrastructure Setup:** The BaaS provider sets up the physical and virtual infrastructure in order to support the blockchain network. The principal considerations while setting up the cloud-based servers are that they should have sufficient storage, processing power, and network connectivity to handle the blockchain nodes.
- **Blockchain Node Setup:** As we discussed earlier, nodes are the key components of a blockchain network as they perform the critical activities like validation of transactions, mining of blocks, etc. The BaaS provider configures and maintains these nodes on behalf of the business client. The availability, communication and synchronisation among the nodes are to be ensured for the seamless functioning of the distributed network.
- **Scalability:** The transaction volume and demand for network capacity may vary for any business. So, the provider must ensure that the infrastructure is scalable to accommodate such changes. For example, adjusting the storage and processing capabilities of the blockchain network based on real-time usage may be necessary.
- **Blockchain Configuration:** Once the initial setup is done, the service provider need to configure the blockchain based on the business need. The BaaS models allow for flexibility in setting up the right blockchain network, and thus, the service provider customises the blockchain network based on the needs.
- **Blockchain Protocol:** Depending on the nature of the business and its specific requirements (such as speed, scalability, privacy, or consensus mechanism), the BaaS provider will configure the network to adopt one of several distributed ledger protocols. We discussed the important protocols earlier in this chapter.
- **Network Topology:** This includes configurations like, number and types of nodes required, consensus mechanism (PoW, PoS, etc.) and smart contract templates depending on the business needs.

- **Ongoing Maintenance and Management:** Regular maintenance and management of the deployed blockchain network to ensure reliable and uninterrupted operations are crucial. The BaaS service provider takes full responsibility for this.
- **Performance Monitoring:** The BaaS provider regularly checks the blockchain network for any performance bottlenecks that can slow down the performance. A few examples are connectivity, network speed, and transaction processing delays.
- **Upgrades and Patches:** New upgrades and patches for introducing new functionalities, enhancing security, and improving performance are unavoidable as blockchain technologies evolve. The BaaS provider ensures that the network is updated with the latest patches and versions. They may also modernise the core components, like smart contract updates or blockchain architecture modifications, if required.
- **Security Management:** It is important to continuously monitor the blockchain network for potential vulnerabilities and keep it updated with the latest security protocols. The BaaS provider ensures that the nodes, the transaction and the network are secure from external threats and malicious attacks.
- **Incident Management:** Disruptions can occur even after continuous monitoring and patching. For example, a node failure, data inconsistency, or system breakdown are the common failures in blockchain networks. The BaaS provider is responsible for ensuring minimal downtime by addressing and resolving the incident promptly.

Support Services BaaS providers often offer a comprehensive range of support services that go beyond just infrastructure management. These services include:

- **Bandwidth and Resource Management:** Blockchain networks, especially the public blockchains where transaction volumes can be high, demand high network bandwidth. The BaaS provider monitors and optimises bandwidth usage to ensure seamless processing of transactions and maintains optimal network latency during high-demand periods.
- **System Health Monitoring:** The health monitoring of the blockchain network involves activities such as ensuring all nodes are properly synchronised and functioning, no critical errors have compromised the network's integrity and the state of the overall system is stable. This is one of the key responsibilities of the BaaS provider.
- **Proactive Security Surveillance:** Security is a major concern in blockchain networks. A BaaS provider ensures proactive surveillance to prevent security breaches, hacking attempts, or unauthorised access to the network. By deploying advanced encryption, multi-signature authentication, and other security protocols, sensitive data and assets are protected.
- **Incident Response and Recovery:** Incidents are the unwanted events or disruptions, like security breaches, network congestion, etc. The BaaS provider puts a response plan in place to quickly address and resolve the issue. This could involve isolating affected nodes, rolling back transactions, or initiating a recovery process to restore normal operations.

Advantages of BaaS

- **Quick Implementation:** BaaS providers often offer ready-to-use, customisable templates and plug-and-play modules that can be set up quickly by users with basic programming knowledge. This helps organisations avoid lengthy and expensive research and planning phases, enabling them to integrate blockchain with existing systems without the complexity of developing the technology from scratch.

- **Use of Expertise:** By using BaaS organisations can focus on their main business operations while entrusting the blockchain infrastructure to a service provider. This results in a more efficient and secure management of distributed ledger technologies, enabling businesses to execute fault-tolerant, high-performance workloads.
- **Cost Savings:** Traditional on-premises blockchain setups come with a high Total Cost of Ownership (TCO), including costs for infrastructure, hardware, software, personnel, licensing and consulting, as well as ongoing management, monitoring and bandwidth expenses. In contrast, BaaS offers a more flexible and scalable pricing model based on usage, such as volume of transaction, number of channels and concurrency, making it a cost-effective alternative.
- **Customisation Focus:** While the technical infrastructure is handled by the BaaS provider, the in-house team can focus on tailoring the blockchain solution to their specific business needs. This ensures that the blockchain service complements its operations and business strategy, with the provider maintaining and managing the infrastructure.
- **Staff and Resource Optimisation:** BaaS eradicates the need for large on-premises IT infrastructure, reducing physical overheads such as equipment space, power and cooling. This leads to lower staffing needs and allows the internal IT team to focus on more critical, business-centric tasks and strategic projects.

Potential Issues with BaaS

Despite the many advantages, there are some challenges to consider when using a BaaS solution:

- **Lack of Visibility and Control:** In most BaaS models, the service provider owns and manages the infrastructure, which can lead to limited visibility for the customer. When technical issues arise, businesses often have to rely on the information provided by the BaaS provider, which could result in communication breakdowns and a lack of control. In the case of a service failure or security breach, businesses may feel disconnected from the issue, leading to potential trust issues with the provider.
- **Data Compliance Challenges:** Blockchain technology, by its very nature, involves the global distribution of data. This can raise compliance concerns, particularly when data is shared across borders or when the BaaS provider hosts its servers in regions that are subject to economic sanctions or differing regulations. Ensuring data privacy and compliance with jurisdictional laws becomes more complex when blockchain networks are involved.

By offering businesses a managed blockchain solution with cloud-based infrastructure and expert support, BaaS provides a cost-effective and efficient way to adopt blockchain technology while mitigating some of the complexities of traditional blockchain implementations. However, organisations must judiciously evaluate potential risks, including data compliance issues and the challenges of working with third-party providers.

Well-known BaaS Solution Providers

Though there are a large number of BaaS services hitting the market every day, with the growing popularity of blockchain use cases, we will discuss a few reputed and enterprise-scale solutions here.

Microsoft Azure Blockchain Service This is a fully managed blockchain platform that supports multiple blockchain protocols, including Hyperledger Fabric, Ethereum, and Corda. It is used by organisations to set up their permissioned networks, and the integrated tools are useful for enabling governance, security and scalability. The platform facilitates the creation and management of consortia between partners.

If an organisation is already using the other Azure services, such as Azure Active Directory, Azure SQL Database, and Azure Logic Apps, then they can create a unified enterprise system as the Azure BaaS seamlessly integrates with those services. Some of the key features of the platform are that the built-in governance tools help define participant roles, manage consensus mechanisms, enforce network rules and ensure compliance.

IBM Blockchain Platform It is a permissioned blockchain platform built on the Hyperledger Fabric protocol. This is a flexible platform that allows high customisation to ensure that enterprises can tailor their blockchain networks to meet specific operational requirements. A rich set of developer tools, including APIs, SDKs, and a web-based interface for managing networks.

Some of the key enterprise features that it offers are private channels, multiple consensus options, and identity management. IBM BaaS integrates effortlessly with IBM Cloud, AI services, and data analytics, enabling advanced applications in supply chain management, financial services, and healthcare. It focuses on meeting the needs of industries requiring transparency, scalability, privacy, and security.

Amazon Managed Blockchain This fully managed blockchain service can use either Ethereum for public decentralised applications or Hyperledger Fabric for permissioned private blockchains. A big advantage of this platform is that it is deeply integrated with AWS-based infrastructures and services, such as AWS Identity and Access Management (IAM), Amazon Simple Storage Service (S3), and AWS monitoring service (CloudWatch), making it easy to incorporate blockchain at the enterprise level.

Some of the key benefits offered are the built-in security with advanced encryption, comprehensive compliance certifications, flexible pricing model and scalability. So, this is a popular and cost-effective BaaS solution for businesses of various sizes that need support for both public and private blockchain networks.

Oracle Blockchain Service Oracle's Cloud BaaS provides a robust platform for developing and deploying permissioned blockchain networks with a focus on high performance and security. Built on Hyperledger Fabric, it includes enterprise-grade features such as secure smart contracts, access control and private channels.

Part of Oracle Cloud, it integrates seamlessly with other Oracle applications like ERP, supply chain management and databases, enabling businesses to extend their existing infrastructure into the blockchain. Oracle also provides powerful analytics tools to derive actionable insights from blockchain data, which helps businesses optimise operations and make data-driven decisions.

Security features include end-to-end encryption, identity management and compliance with industry regulations. Oracle's platform also integrates advanced AI features for better operational decision-making and automation.

Enterprises already using Oracle Cloud services or looking for a high-performance, secure blockchain platform tightly integrated with ERP and supply chain management systems will find Oracle Blockchain an ideal fit.

Based on the above discussion, Table 12.2 shows a comparative overview of the leading BaaS platforms.

Table 12.2 Comparative analysis of different BaaS platforms

Feature/ Service	Microsoft Azure Blockchain Service	IBM Blockchain Platform	Amazon Managed Blockchain	Oracle Blockchain Cloud Service
Supported Protocols	Ethereum, Hyperledger Fabric, Corda	Hyperledger Fabric	Ethereum, Hyperledger Fabric	Hyperledger Fabric
Enterprise Integration	Strong integration with Azure services	Integration with IBM Cloud, AI and data analytics	AWS ecosystem integration	Deep integration with Oracle Cloud
Developer Tools	Azure SDK, Visual Studio integration	Extensive developer tools, APIs, SDKs	AWS SDKs, integration tools	Blockchain APIs, Oracle Cloud SDKs
Security Features	Advanced encryption, compliance tools	Encryption, identity management, compliance	Built-in encryption, AWS compliance	End-to-end encryption, identity management
Pricing Model	Consumption-based pricing	Subscription and usage-based	Pay-as-you-go pricing	Subscription-based pricing
Best For	Businesses using Azure, need flexibility	Enterprises needing Hyperledger Fabric and deep customisation	AWS users looking for scalable blockchain	Oracle Cloud customers needing blockchain for ERP, supply chain

12.3 BLOCKCHAIN USE CASES IN INDUSTRY 4.0

Blockchain is highly relevant to Industry 4.0, which focuses on the integration of digital technologies (like the Internet of Things, artificial intelligence, automation and big data) into manufacturing and industrial processes. Blockchain, with its secure, decentralised and transparent architecture, can significantly enhance the capabilities of Industry 4.0 by improving several key areas. So, let us understand these use cases well.

12.3.1 Supply Chain Transparency and Traceability

- **Problem:** Traditional supply chains are often intricate and involve multiple stakeholders, making it challenging to track the origin and movement of products throughout the supply chain.
- **Blockchain's Role:** Blockchain offers real-time transparency and traceability, enabling secure and immutable tracking of products as they move through the supply chain. Every stage of a product's journey is logged on the blockchain so that the data becomes reliable and tamper-proof.
- Blockchain is used to track a product's entire journey from raw materials to the end user, especially in sectors prone to tampering, like food, pharmaceuticals, and manufacturing. The aim is to ensure authenticity and minimise the risk of fraud. Let's see an example. In the food industry, it is critical to trace perishable goods from the source to the consumption points. Blockchain is transforming the way this can be automated. Companies like Walmart and Nestle leverage blockchain to monitor food items from farm to consumer. First, a unique QR code is assigned to the produce at harvest, which captures important information such as farm location, certifications, and grower details. This data is logged on the blockchain. The product is traced at every stage of the supply chain, and a new block is added by each of the supply chain partners, like the processor, the transporter, the distributor, and the retailer. Suppose a product, traced in this way, had a customer complaint; then the associated blockchain would facilitate quick identification of affected batches, which reduces costs and waste. Also, the consumers can scan the QR code to access detailed information about the products, which improves transparency, trust and accountability. Refer to Figure 12.7.

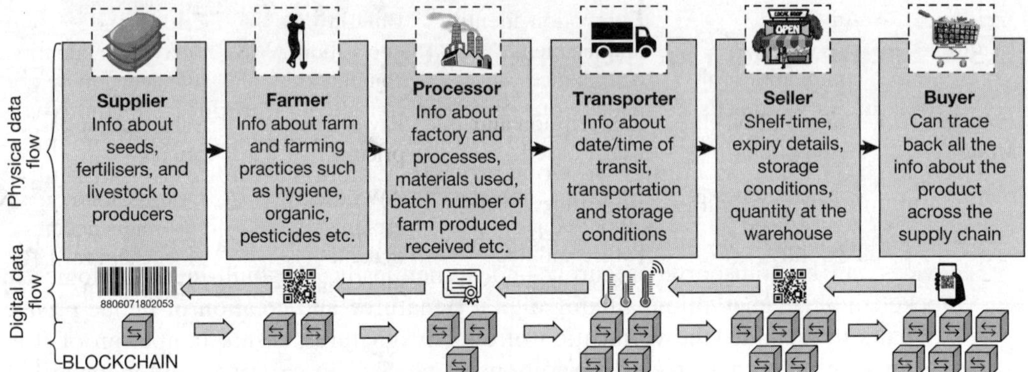

Figure 12.7 Blockchain in managing the food supply chain

12.3.2 Secure and Efficient Tracking of Physical Assets

- **Problem:** Expensive physical assets such as machinery, equipment, or raw materials need to be tracked closely during transfer of location or ownership. This often relies on paper-based tracking systems and centralised databases, which are vulnerable to errors and fraud.

- **Blockchain's Role:** Blockchain can facilitate smart asset management in these scenarios. The physical assets are transformed into digital representations on the blockchain, called tokenisation. For example, in industrial environments, businesses can tokenise assets like machinery or even intellectual property. The associated blockchain can streamline managing the transfer of location and ownership by tracking various activities like issuance, payment, exchange, escrow, and retirement. By enabling secure, transparent and efficient tracking of ownership and transactions, blockchain in these cases improves asset management and reduces the risk of fraud.

12.3.3 Reform Predictive Maintenance

- **Problem:** Predictive maintenance requires accurate, real-time data from multiple sources to foresee potential equipment failures, but managing and ensuring the integrity of this data can be challenging. Additionally, manually coordinating maintenance actions and sharing data across stakeholders can be inefficient and error-prone.
- **Blockchain's Role:** Blockchain enhances predictive maintenance by ensuring immutable, secure and transparent data storage. It facilitates seamless data sharing among stakeholders and automates maintenance actions through smart contracts. Blockchain also tracks equipment parts' origin, ensuring quality and authenticity while offering a transparent, auditable maintenance history.
- **Example:** In an industrial environment, blockchain securely records sensor data from machines, ensuring its integrity. When predictive analytics signal a potential failure, a smart contract automatically schedules maintenance or alerts the team, ensuring timely and accurate interventions. Additionally, the machine's maintenance history and source of parts are easily accessible, improving efficiency and reducing downtime.

12.3.4 Secure System for Industrial IoT Data

- **Problem:** As we discussed in previous chapters, IoT devices produce enormous volumes of data, and managing this data in real-time while ensuring its security and integrity can be challenging. Though the cloud provides a solution for storage, it is difficult to maintain the integrity and authenticity of data.
- **Blockchain's Role:** By integrating blockchain with IoT, the data generated by IoT devices can be transported securely and authentically. The built-in technologies in blockchain, like, encryption, affixing digital signatures, and creation of public-private key pairs to facilitate the verification of digital signatures, ensure tamper-proof storage through immutable records. Additionally, blockchain can track and manage IoT devices, ensuring that data is transmitted securely and accurately between devices and systems.

For example, in a smart factory, machines automatically log their production data, performance, working parameters and status onto the blockchain. This allows managers to have a transparent and reliable record of machine activity, to monitor production in real-time, to ensure timely equipment maintenance, and, as a result, to reduce downtime and improve productivity. Refer to Figure 12.8.

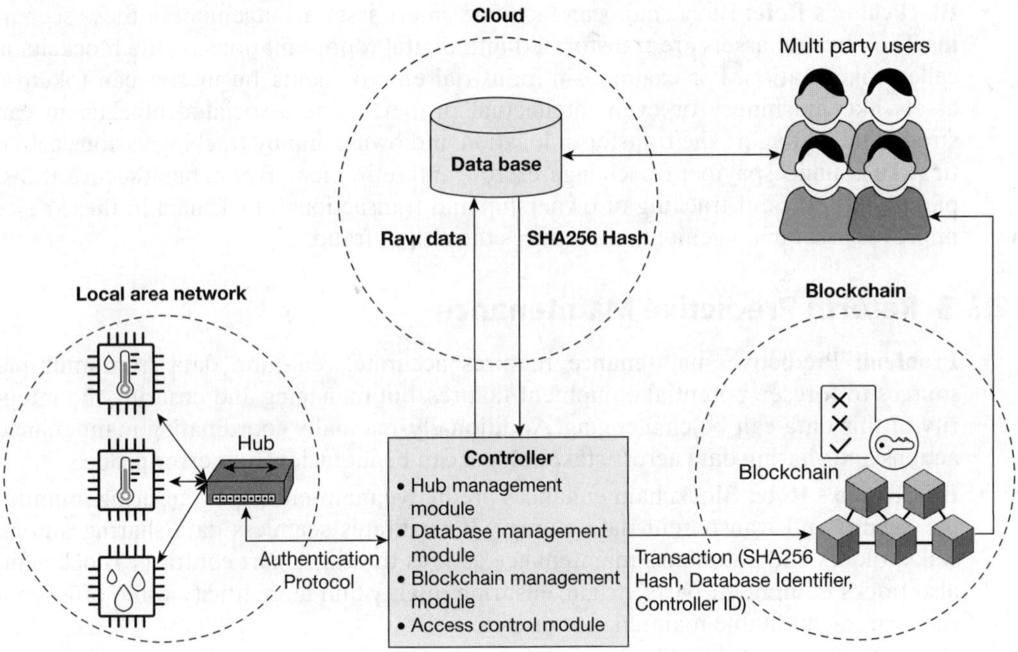

Figure 12.8 Blockchain in managing the IoT data security

12.3.5 Improve Collaboration Between Industry Players

- **Problem:** In traditional industries, collaboration between suppliers, manufacturers and distributors is often inefficient, as each party operates with separate systems and records.
- **Blockchain's Role:** Blockchain facilitates seamless collaboration by providing a shared, secure ledger where all participants in the supply chain or manufacturing process can access the same real-time information.
- **Example:** Suppliers, manufacturers and partners can securely share production schedules, inventory data and quality checks on a blockchain, ensuring that everyone involved has accurate, up-to-date information, leading to improved decision-making and streamlined operations.

12.3.6 Demonstrate Compliance with Social and Environmental Regulations

- **Problem:** Industries are expected to comply with social and environmental regulations, such as ethical sourcing, carbon footprint reduction, fair labour practices and sustainability standards. However, tracking compliance across supply chains and operations is complex, prone to errors and difficult to verify. So, proving compliance to regulators, consumers and stakeholders is also challenging.

- **Blockchain's Role:** As blockchain creates a tamper-proof record of transactions, actions and data points, it ensures that any steps taken to meet social or environmental regulations (e.g., sustainability efforts, ethical sourcing, labour practices) are permanently logged and can be easily accessed by regulators or auditors, guaranteeing transparency and authenticity. Blockchain is used to track the complete lifecycle of a product, from the sourcing of raw materials to the manufacturing and distribution of the finished product. This is useful for verifying claims like the use of ethically sourced raw materials or adherence to environmental standards, such as carbon emissions or water usage.
- **Example:** In the food industry, blockchain helps track sustainable farming practices, carbon footprints and certifications such as organic or fair trade, allowing companies to demonstrate compliance with environmental regulations and social initiatives.

12.4 BLOCKCHAIN BENEFITS FOR INDUSTRY 4.0

Based on the use cases outlined above, we can identify several benefits of incorporating blockchain into an Industry 4.0 setup. With the rise of IoT, the volume of transactions has surged, leading to inefficiencies such as redundant record-keeping within internal systems and third-party validations. Traditional record-keeping systems are also vulnerable to cyber-attacks and fraud and limited transparency can delay data verification. All of these issues slow down business processes, harm the bottom line and highlight the need for a more effective solution.

- **Enhanced Trust:** Blockchain ensures that network members receive accurate and timely data. Confidential records are only shared with authorised network participants, providing a high trust level.
- **Increased Security:** Blockchain requires consensus on data accuracy from all network participants. Once validated, transactions are immutable and permanently recorded, making them tamper-proof. Even system administrators cannot alter or delete transactions. This ensures secure IoT integration and robust data protection.
- **Improved Efficiency:** Blockchain eliminates the need for time-consuming record reconciliations by utilising a distributed ledger shared across the network. Additionally, smart contracts, sets of pre-defined rules, are stored on the blockchain and executed automatically, speeding up transactions. Real-time, traceable records of every transaction enhance transparency and operational efficiency.
- **Enhanced Collaboration:** Blockchain facilitates the seamless, secure sharing of information between partners and suppliers, ensuring accurate and reliable data management.

All these help to optimise the entire supply chain, improving trust among stakeholders, reducing inefficiencies and enabling quicker response times to disruptions.

12.5 IMPLEMENTATION STRATEGY

We have discussed the fundamentals of blockchain and its applications in various domains. Let's explore the step-by-step approach to how organisations can implement it. A particularly promising area is end-to-end supply chain tracking, where blockchain enables transparency and collaboration among the participating organisations. We will assume that BaaS is used to accelerate the implementation process. Refer to the workflow diagram in Figure 12.9.

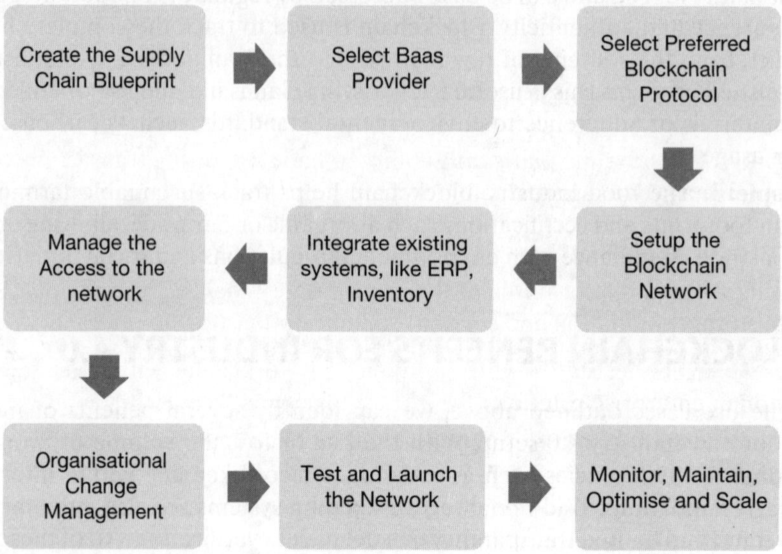

Figure 12.9 Steps to implement blockchain in supply chain

Step 1: Supply Chain Blueprint Creation

The initial phase involves understanding your supply chain's specific needs and defining your blockchain integration goals.

- **Business Process Mapping:** Document the current supply chain processes and identify areas where blockchain can add value and streamline operations. This is a key step that helps focus the implementation to the most impactful part of the business.
- **Stakeholder Mapping:** Identify key players in your supply chain (suppliers, manufacturers, distributors, retailers).
- **Data Inventory:** Determine the data to be recorded on the blockchain (product details, timestamps, inventory, shipment information, quality checks).
- **Access Control Design:** Define data access levels for different stakeholders, particularly for sensitive information.

Step 2: BaaS Provider Selection

Choosing the right BaaS provider is crucial. BaaS platforms allow rapid blockchain network deployment without extensive infrastructure management. Consider the following factors:

- **Protocol supported:** If you have already decided on the blockchain protocol to be used, then you may be constrained in choosing a platform. As we discussed earlier, each BaaS platform supports a limited set of protocols. So, check the available options.

- **Security Assessment:** Evaluate the platform's security measures for protecting sensitive data and transactions.
- **Integration Capabilities:** Ensure easy and efficient integration with existing enterprise systems (ERPs, inventory management tools).
- **User Experience Evaluation:** Select a platform with intuitive features for easy user adoption.
- **Cost-Benefit Analysis:** Assess the platform's pricing and support offerings for long-term viability.
- **Scalability:** Consider the platform's ability to handle growing transaction volumes and data storage as your supply chain grows.

Step 3: Protocol Selection

After choosing a BaaS platform, select the appropriate blockchain protocol. This protocol governs data storage, validation and sharing. Popular options for supply chain include:

- **Hyperledger Fabric:** Suitable for permissioned blockchains with strict access controls, common in enterprise solutions.
- **Ethereum:** Ideal for public or permissioned blockchains requiring smart contracts and DApps.
- **Corda:** A permissioned platform well-suited for financial services and adaptable to supply chains needing secure, auditable transactions.
- **Consortium:** Determine if a consortium blockchain (shared among a group of organisations) is the best fit for your supply chain.

We discussed these protocols in detail in the previous section.

Step 4: Network Construction

With a protocol selected, set up your blockchain network. Key activities include:

- **Node Creation:** Each participant (supplier, manufacturer, distributor) has a node. BaaS simplifies this process. Assign roles and permissions.
- **Consensus Protocol Configuration:** Choose a mechanism (e.g., PoW, BFT) to ensure agreement on the blockchain's state.
- **Smart Contract Development:** Create self-executing contracts to automate processes (payments, inventory updates) based on predefined conditions. Use the BaaS provider's tools or the protocol's SDK.
- **Data Governance:** Establish clear guidelines for data ownership, access and management within the blockchain network.

Step 5: Systems Integration

Blockchain's value is maximised through integration with existing systems (ERP, inventory management), as you should be able to execute the use cases like below:

- **Real-time Tracking:** Integrate IoT sensors for live updates on goods in transit.
- **Automated Payments:** Use smart contracts to trigger payments upon meeting specific conditions (e.g., delivery).

- **Data Synchronisation:** Ensure real-time updates of product status, order details and shipment information across all nodes.
- **API Development:** Create APIs to facilitate communication between the blockchain network and existing systems.

Step 6: Identity and Access Management (IAM) Implementation

For multi-organisational supply chains, robust IAM is essential:

- **Digital Authentication:** Use digital certificates to verify participant identities.
- **Role-Based Access Control (RBAC):** Define data access permissions for each stage, ensuring authorised access.
- **Key Management:** Implement secure key management practices for cryptographic keys used in the blockchain.

Step 7: Training and Change Management

Being able to use the blockchain-integrated supply chain requires considerable organisational change management.

- **User Training:** Provide comprehensive training to all stakeholders on how to use the blockchain system.
- **Change Management Strategy:** Develop a plan to address any resistance to change and ensure a smooth transition to the new system.
- **Documentation:** Create clear and concise documentation for users.

Step 8: Network Testing

The system is aimed at bringing trust to the full supply chain workflow and thus, thorough testing is crucial before going live:

- **Simulation Testing:** Simulate various scenarios to ensure the blockchain handles different conditions effectively.
- **Smart Contract Validation:** Verify smart contract performance.
- **System Integration Testing (SIT):** Confirm existing systems could be seamlessly integrated with the blockchain system.
- **User Acceptance Testing (UAT):** Involve users in testing to ensure the system meets their needs.

Step 9: Network Launch

After successful testing, deploy the network:

- **Participant Onboarding:** Ensure all partners are set up with credentials and roles.
- **Go-Live Execution:** Begin recording transactions. Monitor closely for smooth operation.
- **Phased Rollout:** Consider a phased rollout to minimise disruption and allow for adjustments.

Step 10: Monitoring and Maintenance

Continuous monitoring and maintenance are vital:

- **Transaction Monitoring:** Track transactions for accuracy.
- **Performance Monitoring:** Monitor for issues like slow processing.
- **Security Audits:** Regularly audit the blockchain and smart contracts.
- **Incident Response:** Have a plan in place to remediate any issues that arise.

Step 11: Scaling and Optimisation

Scale and optimise the blockchain as the network grows:

- **Node Expansion:** Add nodes as new stakeholders join.
- **Smart Contract Optimisation:** Optimise contracts for efficiency.
- **Continuous Improvement:** Regularly review and update the blockchain system to meet evolving business needs.

By following this roadmap, organisations can effectively implement blockchain in their supply chains, leading to increased transparency, faster transactions, reduced fraud and a more efficient and reliable network.

12.6 CHALLENGES AND CONSIDERATIONS FOR BLOCKCHAIN ADOPTION

While blockchain technology offers numerous benefits, such as enhanced security, transparency and efficiency, there are quite a few challenges and considerations that organisations and industries must address before adopting blockchain. These challenges range from technological and regulatory hurdles to issues related to scalability and integration. Below are the key challenges and considerations for blockchain adoption:

1. **Scalability Issues**
   - **Challenge:** As we discussed earlier in this chapter, some of the consensus mechanisms, like POW may consume substantial computational resources and time for validating and inserting the transactions into the blockchain. If the transaction volume increases, this may significantly slow down the performance of the network, making the corresponding Blockchain architecture (like Bitcoin) inefficient.
   - **Consideration:** Scalability is crucial for blockchain to be viable for large-scale applications in the finance, healthcare and supply chain industries. Alternative solutions are being explored to improve blockchain scalability.
2. **Energy Consumption**
   - **Challenge:** Some blockchain networks, predominantly those that use PoW (like Bitcoin), consume massive amounts of energy. This environmental impact can make blockchain adoption less attractive, especially for businesses that prioritise sustainability.

- **Consideration:** Transitioning to consensus mechanisms that are more energy-efficient, like PoS, or exploring hybrid models, can help mitigate the environmental impact. Additionally, utilising blockchain in sectors like renewable energy may help reduce its carbon footprint.

3. **Regulatory Uncertainty**
   - **Challenge:** Blockchain technology, especially cryptocurrencies, exists in a grey area in terms of regulation. Different countries have different rules regarding the use of blockchain and in many cases, regulations are either unclear or non-existent.
   - **Consideration:** Governments are still figuring out how to regulate blockchain technologies, particularly regarding issues such as taxation, data privacy, KYC (Know Your Customer) requirements and AML (Anti-Money Laundering). Businesses need to stay well-versed in evolving regulations and ensure their blockchain use cases are compliant with local and international laws.
4. **Integration with Legacy Systems**
   - **Challenge:** Many industries still rely on legacy systems (e.g., older databases, ERP software) that may not be compatible with blockchain. Integrating blockchain with these systems can be complex, time-consuming and costly.
   - **Consideration:** Blockchain adoption may require significant changes to existing workflows and technologies. Companies may need to develop middleware solutions or adopt hybrid systems that combine traditional and blockchain-based approaches.
5. **Data Privacy and Confidentiality**
   - **Challenge:** Blockchain's public ledger ensures transparency, but this transparency can raise concerns regarding data privacy. Information stored on a blockchain is often permanent and visible to all participants in the network.
   - **Consideration:** Technologies that can enhance privacy, such as zero-knowledge proofs (ZKPs) and private blockchains (permissioned blockchains), can address data privacy concerns by allowing private transactions without compromising the integrity of the ledger. In simple terms, ZKP is a cryptographic method where one party proves to another that they hold the knowledge about the authenticity of a data without actually revealing the data.
6. **Complexity and Skill Shortage**
   - **Challenge:** Blockchain is still a relatively new technology and there is a shortage of skilled professionals who are experienced in blockchain development and implementation. This creates a barrier to entry for many organisations.
   - **Consideration:** Companies looking to adopt blockchain must invest in training their employees or hire external consultants. Additionally, blockchain development protocols (such as Ethereum, Hyperledger and Solana) are evolving, so professionals need to stay up to date with the latest tools and technologies.
7. **Interoperability**
   - **Challenge:** Different blockchain protocols (e.g., Ethereum, Hyperledger) may not be able to easily interact with one another, creating an issue of interoperability. This lack of communication between blockchains can limit the ability to transfer data and assets between systems.

- **Consideration:** The development of cross-chain communication protocols and interoperable blockchain solutions is essential for the broader adoption of blockchain, especially in multi-party ecosystems such as supply chains, finance and healthcare. Breakthroughs in this area are occurring with projects like Polkadot and Cosmos, which utilise relay chains and hub concepts to enable blockchain platforms to communicate with each other. This would ensure more scalable cross-chain applications as these decentralised systems can connect, communicate and cooperate with each other.

DID YOU KNOW?

The Polkadot protocol is designed to enable secure communication between unrelated blockchains, allowing value and data to flow seamlessly between networks like Ethereum and Bitcoin without the need for intermediaries. It is also fast and scalable as it utilises multiple parallel blockchains (known as 'parachains') to offload much of the processing workload of the main blockchain to these parachains. The Polkadot network consists of a primary blockchain, called the 'relay chain,' along with various user-created parallel chains (parachains). Additionally, it features a 'bridge' layer that facilitates the transfer of value and data between most blockchains and can even connect to non-blockchain databases.

8. **Adoption and Trust**
   - **Challenge:** Blockchain is a disruptive technology and many businesses may be resistant to change or sceptical of the technology. The transition to blockchain often requires a shift in mindset, business models and operations, which can be difficult.
   - **Consideration:** Building trust among stakeholders is essential. Blockchain's inherent transparency and security features can help overcome scepticism, but organisations need to educate and demonstrate the benefits of blockchain to internal and external stakeholders. Also, successful pilot projects and proof of concept (PoC) implementations can help prove the value of blockchain.
9. **Cost of Implementation**
   - **Challenge:** While blockchain offers long-term benefits, the initial cost of implementation can be high, especially when integrating with existing systems, developing custom solutions and training staff.
   - **Consideration:** Organisations must weigh the long-term benefits of blockchain, such as reduced transaction costs, greater transparency and improved security, against the upfront investment. In some cases, the adoption of blockchain may lead to significant cost savings and operational efficiencies in the long run.
10. **Governance and Accountability**
    - **Challenge:** Blockchain networks, particularly decentralised ones, may lack clear governance structures, making it difficult to regulate who is responsible for decision-making, upgrades and resolving disputes.
    - **Consideration:** Developing clear governance models, especially for permissioned blockchains, can help define the roles of participants and ensure accountability. In

public blockchains, decentralised governance models (such as DAO – Decentralised Autonomous Organisations) are emerging as a solution.

11. **Network Effect and Adoption by Ecosystem Players**
    - **Challenge:** Blockchain technology's value increases as more participants join the network, creating a network effect. However, achieving this critical mass of participants can be challenging, especially in industries that are traditionally slow to adopt new technologies.
    - **Consideration:** Successful adoption often depends on gaining widespread support from industry players. Collaboration between businesses, governments and other stakeholders is key to accelerating blockchain adoption across the ecosystem.

While blockchain adoption offers significant advantages, businesses must carefully consider the associated challenges. Overcoming barriers such as scalability, integration, regulatory uncertainty and governance is essential for successful implementation. By addressing these issues thoughtfully, organisations can reveal the transformative potential of blockchain in various industries, driving efficiency, transparency and trust in their operations.

12.7 CONCLUSION

The demand for blockchain technology is rapidly increasing as businesses and industries recognise the need for transparent, secure and decentralised solutions for transaction and data management. Among its many applications, smart contracts are gaining significant traction across various sectors, offering automation, efficiency and trustless agreements. According to market research, the Blockchain market is expected to grow globally from USD 20 billion in 2024 to USD 250 billion by 2029, underscoring its growing influence. A particularly exciting and transformative area where blockchain is poised to have a major impact is in the development of Web3. Web3 represents the vision for a decentralised internet, built on blockchain technology and distributed ledgers that are collectively governed by participants rather than central authorities. In Web3, control and access are decentralised and governed by community-driven networks, in contrast to the Web2 model, where a handful of powerful corporations hold dominance. Key technologies supporting the Web3 architecture include blockchain, smart contracts and digital assets/tokens, which together enable secure, transparent and autonomous interactions in this next-generation internet. This shift has the potential to transform industries, empowering users and creators by providing more control over their data, digital identities and financial assets.

SUMMARY

- Blockchain is a secure, decentralised, immutable ledger used for recording transactions and tracking assets. It differs from Bitcoin, as blockchain technology has applications beyond cryptocurrencies.
- Key principles of blockchain include decentralisation, immutability and transparency.
- Transactions are recorded in blocks, verified through consensus mechanisms and linked in a tamper-proof chain.

- Blockchain eliminates intermediaries, reducing costs and increasing trust among participants.
- Types of blockchain networks:
  - Public (Permissionless) – Open to all (e.g., Bitcoin, Ethereum).
  - Private – Controlled by a single entity (e.g., enterprise blockchains).
  - Permissioned – Restricted access for approved participants.
  - Consortium – Governed by multiple organisations.
- Consensus mechanisms ensure transaction validity, including Proof of Work (PoW), Proof of Stake (PoS) and Byzantine Fault Tolerance (BFT).
- Smart contracts are self-executing programs that automate agreements without intermediaries.
- Major blockchain platforms include Ethereum, Hyperledger, Corda, Cardano and Solana, each suited for different use cases.
- Blockchain as a Service (BaaS) empowers businesses to adopt blockchain without managing infrastructure (offered by Microsoft Azure, IBM, AWS and Oracle).
- Use cases in Industry 4.0:
  - Supply chain transparency – Tracks product origin and movement.
  - Asset management – Secure tracking of equipment and resources.
  - Predictive maintenance – Automates machine maintenance using sensor data.
  - IoT security – Protects industrial IoT device data.
  - Regulatory compliance – Verifies environmental and ethical standards.
- Blockchain benefits for Industry 4.0:
  - Enhanced trust – Ensures data integrity and authenticity.
  - Stronger security – Prevents fraud and cyberattacks.
  - Greater efficiency – Eliminates manual record-keeping and intermediaries.
  - Seamless collaboration – Acts as a single source of truth for all stakeholders.
- Challenges in blockchain implementation:
  - Scalability challenges arise as blockchain networks grow, requiring alternative solutions.
  - Energy consumption concerns, especially with PoW systems like Bitcoin, drive a shift towards energy-efficient alternatives like PoS.
  - Regulatory uncertainty affects blockchain adoption as governments define legal frameworks for compliance.
  - Integration with legacy systems is complex, requiring middleware solutions and APIs for smooth adoption.
  - Interoperability issues prevent different blockchains from communicating, but solutions like Polkadot and Cosmos address this.
- The future of blockchain includes Web3, a decentralised internet model where users have more control over their data, digital assets and interactions.

EXERCISE

Multiple Choice Questions (1 Mark)

1. What is a blockchain?
   (a) A centralised database
   (b) A decentralised, immutable ledger
   (c) A type of cryptocurrency
   (d) A cloud computing service
2. Which consensus mechanism does Bitcoin use?
   (a) Proof of Stake (PoS)
   (b) Practical Byzantine Fault Tolerance (PBFT)
   (c) Proof of Work (PoW)
   (d) Proof of Authority (PoA)
3. Which of the following is not a type of blockchain network?
   (a) Public
   (b) Private
   (c) Hybrid
   (d) Secured
4. Which blockchain platform is primarily designed for smart contracts?
   (a) Bitcoin
   (b) Ethereum
   (c) Hyperledger Fabric
   (d) Corda
5. What is the function of a Merkle root in a blockchain?
   (a) Store user passwords
   (b) Link transactions within a block
   (c) Create new blocks
   (d) Encrypt data
6. Which consensus mechanism is most energy-efficient?
   (a) Proof of Work (PoW)
   (b) Proof of Stake (PoS)
   (c) Proof of Capacity (PoC)
   (d) Byzantine Fault Tolerance (BFT)
7. What is the purpose of hashing in blockchain?
   (a) Speed up transaction processing
   (b) Encrypt data for security

(c) Store large amounts of data
(d) Generate public keys

8. Which of the following is not a real-world use case of blockchain?
   (a) Supply chain management
   (b) Predictive maintenance
   (c) Online shopping reviews
   (d) Digital identity management
9. What does 'immutability' mean in blockchain?
   (a) Data can be edited anytime
   (b) Transactions are irreversible
   (c) Only admins can modify data
   (d) Data is deleted after a certain period
10. Which blockchain network is primarily used in enterprise solutions?
   (a) Solana
   (b) Ethereum
   (c) Hyperledger Fabric
   (d) Bitcoin
11. What is the main advantage of decentralisation in blockchain?
   (a) Increased central authority control
   (b) Reduced dependency on a single entity
   (c) Higher operational costs
   (d) Faster government regulation
12. What does BaaS stand for?
   (a) Blockchain as a System
   (b) Blockchain as a Service
   (c) Business as a Software
   (d) Bitcoin as a Service
13. Which blockchain protocol is designed for financial institutions?
   (a) Ethereum
   (b) Hyperledger
   (c) Corda
   (d) Solana
14. Which industry benefits from blockchain-based predictive maintenance?
   (a) Retail
   (b) Manufacturing
   (c) Social media
   (d) Online gaming

15. What role do miner nodes play in a blockchain network?
    - **(a)** Store user credentials
    - **(b)** Validate transactions and add new blocks
    - **(c)** Track real-world asset ownership
    - **(d)** Create private keys for users
16. Which of the following is a challenge for blockchain adoption?
    - **(a)** High scalability
    - **(b)** Strong regulatory support
    - **(c)** Energy consumption
    - **(d)** Low transaction costs
17. Which organisation developed Hyperledger?
    - **(a)** Microsoft
    - **(b)** IBM
    - **(c)** Google
    - **(d)** Linux Foundation
18. What type of blockchain is Bitcoin?
    - **(a)** Private
    - **(b)** Consortium
    - **(c)** Permissioned
    - **(d)** Public
19. Which of the following is not a component of a blockchain block?
    - **(a)** Timestamp
    - **(b)** Previous block hash
    - **(c)** User login details
    - **(d)** Merkle root
20. What is the primary function of a smart contract?
    - **(a)** Execute agreements automatically
    - **(b)** Store user credentials
    - **(c)** Mine new cryptocurrencies
    - **(d)** Encrypt transaction data

Short Answer Type Questions (2 Marks)

1. What is the difference between public and private blockchains?
2. What is a blockchain node?
3. What problem does blockchain solve in supply chain management?

4. What is the role of a hash function in blockchain?
5. What are smart contracts?
6. Why is blockchain called 'immutable'?
7. What are the key benefits of blockchain for Industry 4.0?
8. What is the role of cryptography in blockchain?
9. What is the main difference between Hyperledger and Ethereum?
10. How does blockchain prevent double spending?

Long Answer Type Questions (5 Marks)

1. Explain the working of blockchain step by step.
2. Discuss the role of blockchain in Industry 4.0.
3. What are the challenges of blockchain adoption and how can they be addressed?
4. Describe the structure of a blockchain block in detail.
5. How does blockchain enhance security and trust in transactions?
6. What is Blockchain as a Service (BaaS) and what are its advantages and disadvantages?
7. What is the role of smart contracts in blockchain applications?
8. How does blockchain improve supply chain management?
9. What is the significance of consensus mechanisms in blockchain?
10. What are the different types of blockchain networks and where are they used?
11. How does blockchain help in regulatory compliance?
12. What are the environmental concerns of blockchain and how can they be mitigated?

Cybersecurity

CHAPTER 13

OBJECTIVES

We explored how Industry 4.0 has revolutionised the industrial landscape by introducing advanced connectivity and automation, which also brings significant cybersecurity challenges. This chapter focuses on the principles of cybersecurity in Industry 4.0, emphasising the need to protect interconnected operational technologies (OT) in addition to information technology (IT) from evolving threats. By examining established security frameworks, the chapter provides a detailed guide to developing robust cybersecurity programs for smart factories. It also highlights key challenges such as employee awareness and resource constraints.

By the end of this chapter, you should be able to:

- Understand the cybersecurity principles and challenges of Industry 4.0.
- Design layered security approaches using the Defence in Depth model.
- Develop and implement ICS cybersecurity programs using established frameworks.
- Identify and address key challenges in securing smart factories.

13.1 INTRODUCTION

Advancements in technology areas like Big Data, Analytics, Cloud Computing, IIoT have transformed us into this new digital age of Industry 4.0. However, this also created new challenges and one of the major challenges is security. In today's cyberspace, it is called cybersecurity.

You might have also heard the term information security and some of the larger companies even have a CXO member called Chief Information Security Officer or CISO. So, are information security and cybersecurity all one and the same? Let's understand with an example.

Imagine you're in charge of storing sensitive customer data for your company. You decide to keep a backup copy of this data on an external hard drive. Now, if you lock that hard drive in a secure cabinet, disconnected from any computer or network, you're still protecting the information. This protection, keeping physical access restricted and preventing unauthorised hands from getting to that hard drive, falls under *information security* but not under *cybersecurity*.

Now, suppose you plug that same hard drive into a company laptop that's connected to the internet. Suddenly, the data becomes vulnerable to online threats like malwares or unauthorised remote access. Protecting the hard drive from these digital dangers falls under the domain of *cybersecurity* (and, also, under *information security*).

So, what's the difference? Information security is the practice of protecting all forms of information, whether digital or physical, whereas cybersecurity is a subset of information security that focuses specifically on protecting digital data and systems from cyber threats like hacking, malware, and phishing.

The graph in Figure 13.1 depicts the popularity of 'cybersecurity' and 'information security' as search terms in Google searches over the years. We can clearly observe that the cybersecurity term is gaining popularity and becoming more popular since the year 2017, whereas the term information security is on the decline.

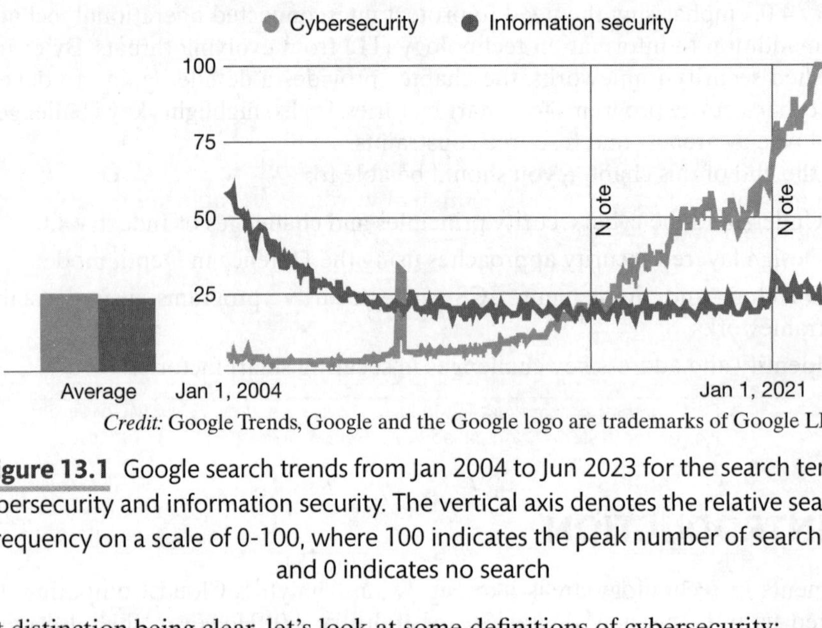

Credit: Google Trends, Google and the Google logo are trademarks of Google LLC.

Figure 13.1 Google search trends from Jan 2004 to Jun 2023 for the search term cybersecurity and information security. The vertical axis denotes the relative search frequency on a scale of 0-100, where 100 indicates the peak number of searches and 0 indicates no search

With that distinction being clear, let's look at some definitions of cybersecurity:

- 'Things that are done to protect a person, organisation, or country and their computer information against crime or attacks carried out using the internet' (Cambridge Dictionary)
- 'The state of being protected against the criminal or unauthorised use of electronic data, or the measures taken to achieve this'. (Oxford Online Dictionary)

So, in simple terms, cybersecurity looks at securing the connected devices from unauthorised use. In subsequent sections, we will look at various types of cyber-attacks, as well as different security measures employed to protect the systems from such attacks.

13.1.1 Evolution of Cybersecurity

In the late 1960s, computers were proliferating among the researchers. However, the computers were expensive and thus people tried to find a way to connect to other computers to utilise unused computer time. From this requirement, the ARPANet project was born and funded by the Advanced Research Projects Agency (ARPA). The first link was established between Stanford University and the University of California at Los Angeles in 1969. However, it quickly grew to link 40 nodes by 1972. Soon other universities started joining this network or creating their alternative network. The challenge was the underlying protocols were different and computers in one network could not talk to computers in another. In the 1970s, ideas of a 'network of networks' were floated and the term 'internet' was used as a shorthand for 'inter-network'. In 1982, TCP/IP was standardised and that permitted the worldwide proliferation of the internet.

Creeper, the first computer virus (non-self-replicating and benign) that moved across the computer network leaving a message trail (I am the creeper, catch me if you can), was written in 1971. The first antivirus program, Reaper, which followed and removed Creeper was written in 1973.

However, this was not before the need for the security of computer systems was conceived. The Department of Defence in the United States of America published a report titled 'Security Controls for Computer Systems' (Ware, 1970) indicating hardware, software, communication lines and personnel controls to secure the computer systems in the network. In 1977, Ruthberg and McKenzie (1977) introduced the concept of the CIA triad, Confidentiality, Integrity and Availability. These are the three basic principles of cybersecurity. In the 1980s, ARPAnet became the internet, and from then onwards, as technologies became more sophisticated, both cyber attacks as well as defence mechanisms leveraged those advancements. See Figure 13.2 for a quick reference to the evolution of the internet and cybersecurity.

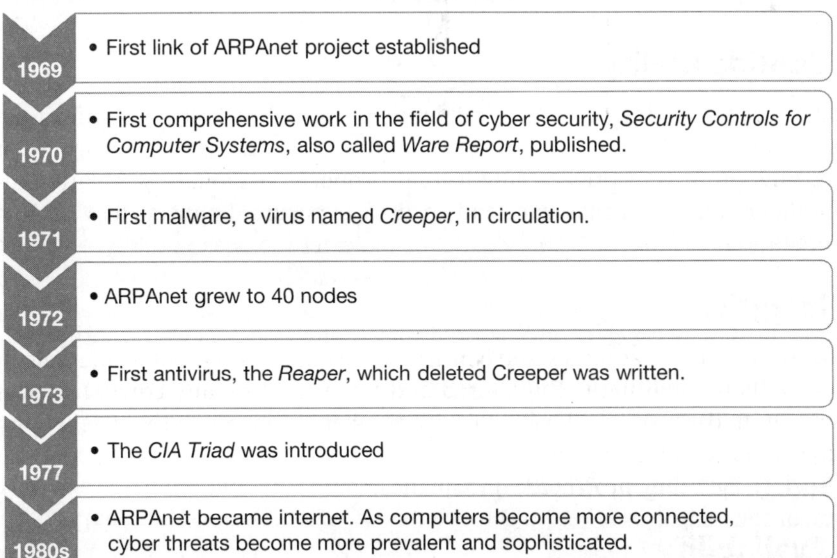

Figure 13.2 Evolution of internet and cybersecurity

13.2 CYBERSECURITY PRINCIPLES

The CIA Triad – Confidentiality, Integrity and Availability – is the guiding model in cyber security. This triad forms the bedrock of protection against a myriad of cyber threats, providing a structured framework for evaluating and implementing security measures. See Figure 13.3 for a quick reference to the CIA triad.

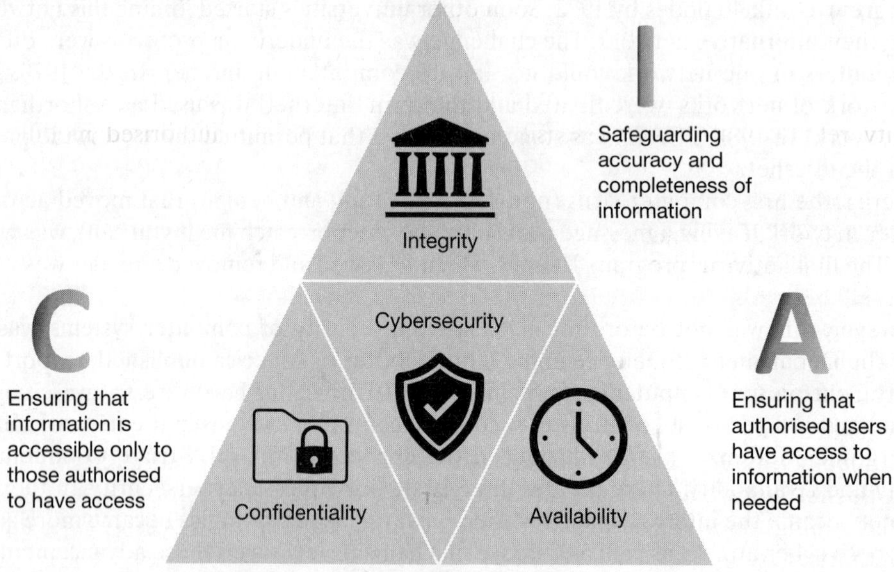

Figure 13.3 CIA Triad

13.2.1 Confidentiality

This aspect refers to limiting access to confidential information solely to authorised persons. The major aim of confidentiality is to prevent unauthorised persons from securing confidential documents, thus protecting privacy, intellectual property, personal information, and other aspects. Confidentiality breaches can lead to data exposure, financial loss, and harm to a person's reputation.

13.2.2 Integrity

Integrity forms the core of accuracy and consistency in data throughout its lifecycle. It ensures data preserves its original nature, unaltered and unaffected by any corruption while being handled, transmitted, or stored. Data integrity is essential in order to protect it from illicit modifications, intentional tampering, or errors with potentially severe consequences in sectors critical to society, including healthcare and finance.

13.2.3 Availability

Availability refers to data and resource access and use according to need. It ensures effective and reactive system and service operation, thus allowing authorised users to access information and use activities without delays. Downtime or non-availability instances can cause disruption to processes, financial loss, and, in critical sectors like emergency services, risk to individuals' safety.

13.2.4 Balancing the Elements for Security

The CIA triad's effectiveness lies in its holistic approach to security with its understanding that emphasising one may compromise others. Placing a suitable balance among confidentiality, integrity, and availability requires conscious thought and strategic implementation.

- **Confidentiality and Integrity:** Achieving a balance between confidentiality and integrity requires safeguarding sensitive data while allowing authorised modifications. Encryption can be a great security measure that ensures both aspects. Data can be encrypted while being stored, as well as while in transit, ensuring confidentiality, while using cryptographic hashing or digital signatures ensures data integrity by ensuring its authenticity.
- **Availability and Integrity:** The provision for data availability in conjunction with data integrity requires implementing mechanisms for redundancy and data backup. Regular data backup, in addition to failing over systems in critical operations, prevent data loss while keeping backup data from being modified, thus maintaining data integrity.
- **Confidentiality and Availability:** Confidentiality can be preserved while ensuring availability by deploying controlled access. Role-based access control (RBAC) makes it possible to provide access solely to persons for data and information required in performing their assigned roles, thus preventing unauthorised access while allowing authorised users to execute assigned tasks without interference.

Practically speaking, an organisation's security architecture must consider how confidentiality, integrity, and availability are interconnected. For instance, while strong access controls (confidentiality) are essential for keeping sensitive financial data secret, it is essential to balance them against systems for quick and convenient data retrieval (availability) along with ones preventing unauthorised alteration (integrity). Within complex systems, a balance among these components can occur. For example, applying more secure encryption methods (confidentiality) can lead to a decreased data processing speed, which then affects the availability of a system. Finding a suitable balance requires a careful determination of the unique requirements and vulnerabilities of a system, its data's nature, and the potential consequences of a breach.

13.3 TYPES OF CYBER THREATS

13.3.1 Malware

Malware, short for malicious software, is one of the biggest threats to a computer, tablet, phone, or any other connected device. Malware is unwanted software that gets secretly installed into a device. Once the malware is installed on the device, it can steal personal information, demand payment to unscramble files on the device, send unwanted ads, or make the device vulnerable to further attacks. Malware structurally can be classified as a virus, worm, or trojan

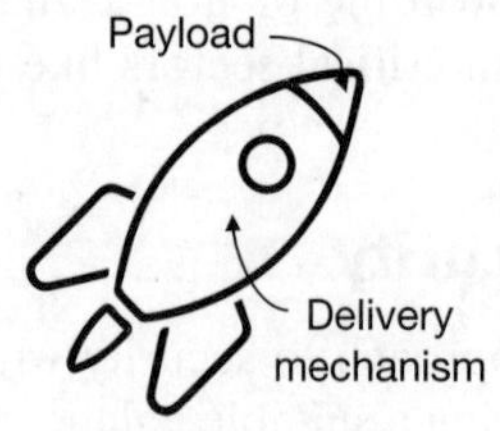

	Payload	Delivery
Worm		✓
Virus	✓	✓
Trojan	✓	

Figure 13.4 Malware categorisation based on structure

horse based on whether they contain payload and delivery mechanism or not (Figure 13.4).

Based on the work they carry out, they could belong to one or more categories, such as spyware, adware, or ransomware.

Worms

Worms are designed to proliferate through a network. Once it gains access to a device, it replicates and spreads into additional machines (delivery), typically remaining active on all the infected machines. Though worms do not change any other programs in the infected machine (no payload), they cause harm by taking up space in the host machines, and choking the network with bandwidth demand.

Virus

Viruses attach themselves to other programs, called hosts. It lies dormant within the host files till a trigger mechanism, like opening the host file, is initiated. Once triggered, they self-replicate and spread from one host to another (delivery). They are also designed to have malicious intent (payload), typically corrupting valuable data or files and rendering them unusable. Thus, they can cause considerable operational problems to organisations.

Trojan

Trojans are disguised as legitimate software programs. But once downloaded, it can cause direct harm (payload) by modifying or deleting files, or indirectly by downloading another malware. Unlike viruses and worms, trojans do not self-replicate (no delivery).

Ransomware

Ransomware is malicious software that encrypts files after gaining access to the system, rendering them useless to the user and then demands a financial payout for the files to be decrypted. When the attacker receives payment, the data is decrypted.

DID YOU KNOW?

Per Greek mythology, during the war with Troy, the Greeks were unable to break into the city of Troy. They placed a large wooden horse near the gate and pretended to sail off. The Trojans believed the horse was a peace offering and they brought it into the city. However, there were Greek soldiers inside the hollow horse, who in the night were able to open the gates from inside and the Greek troops entered and conquered Troy.

Spyware

Spyware is malicious software that runs secretly on a computer and reports back to a remote user. Spyware is typically used to steal sensitive (financial or personal) information, by relaying it to the attacker, rather than simply disrupting a device's operations. For example, keylogger is a specific type of spyware that records all keystrokes made by a user and relays them to a third party. Suppose you go to a bank website and then type your user ID and password. A keylogger will record and relay the website URL as well as the user ID and password, making your bank account vulnerable.

Adware

Adware is malicious software that generates revenue for the malware developer by displaying advertisements to the user of the system. Adware collects data on your computer usage and provides personalised advertisements. However, typically the user is bombarded with advertisements, making working on the infected system very inconvenient, as well as slowing down the system noticeably. Adware can also redirect your browser to unsafe sites.

13.3.2 Phishing

In Phishing, the attacker sends fraudulent communications (usually through email) that appear to come from a legitimate and reputable source. A phishing attack relies on social engineering efforts and exploits the vulnerabilities of human nature – trust, curiosity and other emotions. Unsuspecting users take actions like downloading malware, going to a fake website and divulging personal information, or transferring money.

Phishing attacks originally were very low-risk and low-cost for the attacker, with email being zero-cost medium and bulk email lists available for purchase. So, the attacker sent out mass emails and counted on a few of them to fall victim. However, with the advance of technologies, phishing has evolved in terms of complexity and become one of the largest and most costly cybercrimes on the internet. The phishing can be targeted to an individual or specific group of people. The information of the target(s) is harvested, typically from social media, so these messages may address the target personally, use familiar language, or appear to come from a colleague or friend. This type of targeted phishing attack is also known as *spear phishing*. As spear phishing communication is much more personalised, it looks more legitimate and hence very dangerous. *Whaling* is also a type of spear phishing attack that targets high-value individuals (whales) whose credentials or access to resources have the ability to compromise an organisation. These often lead to business email compromise (BEC) or email account takeover (ATO). For example, a cybercriminal gains access to a company's vendor list and sends an email to the finance department, pretending to be a known supplier. The email looks legitimate and includes a PDF invoice for a large payment, along with updated banking details. The company wires the money to the attacker's account, believing it's paying the vendor. The fraud isn't discovered until the real vendor follows up about the unpaid invoice weeks later.

The delivery mechanism of the phishing attack may also vary from email. It can be delivered via text message (also known as *smishing*) and phone call or voice message (also known as *vishing*).

13.3.3 Denial of Service

In a Denial of Service (DoS) attack cybercriminals flood a server with traffic, making the machine or the network resource unavailable to the users. In *Distributed Denial of Service*

(DDoS) attack, the attack is coordinated from multiple source machines to achieve the same purpose. A distributed denial of service attack is performed by a *botnet*, a group of interconnected devices that can be controlled by command-and-control software by the attacker. A large botnet allows hackers to take down much larger targets than a single machine can.

The first known distributed denial of service attack occurred in 1996 when Panix, one of the oldest internet service providers, was brought down and its service was not available for several days. Even the largest companies can be targeted for DDoS attacks. For example, in 2020, Google reported that it was attacked using three Chinese Internet Service Providers (ISP) which lasted for six months and the peak volume of traffic measured at 2.5 terabytes per second. Amazon was also hit by a similar attack in 2020. The attack lasted for three days and peaked at 2.3 terabytes per second.

The Denial-of-Service attacks are very costly, as the target organisation loses revenues and brand trust.

13.3.4 Data Breach

A data breach happens when information is taken from a system without the owner's permission or knowledge. Any organisation, irrespective of their size or type, can experience a data breach. However, the impact is especially serious for organisations in healthcare, finance, and the public sector. These groups handle highly sensitive data, like government documents, medical records, bank details, and login credentials. Because of the value of this data and strict regulations, breaches in these fields tend to be more costly.

Most data breaches are caused by hacking or malware. But there are other common causes too:

1. **Unintended Disclosure:** Sensitive data is shared by mistake, like when an employee emails private information to the wrong person.
2. **Insider Leaks:** Employees, especially those who are angry, laid off, or tempted by bribes, might steal data.
3. **Payment Card Fraud:** Credit or debit card details are stolen using skimming devices or skimmers. Skimmers are designed to be inconspicuous and blend in with the legitimate equipment. They are often attached over the card reader or inserted into the card slot. When a card is swiped or entered, the skimmer captures the card number and other data.
4. **Loss or Theft:** Physical items like laptops, USB drives, or files are lost or stolen.

As you can see, not all data breaches are caused by online attacks. Some happen through physical theft or devices that don't rely on the internet.

13.3.5 Insider Threat

Insider threats are cybersecurity risks that come from people within an organisation, such as employees, contractors, or business partners, who misuse their authorised access, either on purpose, by accident, or because their accounts are taken over by cybercriminals. Insider threats can be divided into three types:

1. **Negligent Insiders:** Negligent insiders create security threats through ignorance or carelessness. They can expose confidential data often as a result of social engineering or

phishing, or via a lost or stolen device. They may inadvertently send emails/files to an incorrect recipient. However, they do not have malicious intent.

2. **Compromised Insiders:** Threat actors who have stolen a legitimate access credential pose as authorised users, utilising their accounts to extract sensitive data. Typically, it is the negligent employees who are compromised and often do not know that they have been compromised.
3. **Malicious Insiders:** Malicious Insiders are typically disgruntled (or greedy) users who intentionally misuse their access for revenge (or financial gain). Some malicious insiders can also be a *mole*, who works on behalf of a malicious outsider to disrupt the business.

13.4 SECURITY AT DIFFERENT LAYERS

Since cyber-attacks can occur at many levels, it's important to have layered security measures that provide multiple lines of defence to counter the threat at each layer.

13.4.1 System Level Security

The first step in formulating an effective cybersecurity strategy is to ensure that the proper system-level security is configured and maintained. This includes assigning administrative roles for the system, managing user privileges, implementing strong password policies, and patching vulnerabilities.

13.4.2 Network Level Security

To secure your internal network resources from illegal access and infiltration, you must make sure that your network has sufficient network-level security mechanisms in place before connecting it to the internet. The most common means for providing network security is a firewall. Also, the ISP plays an important role in your network security plan, defining elements such as filtering rules for the ISP router connection and public Domain Name System (DNS) precautions.

13.4.3 Application Level Security

Application level security controls how users interact with specific applications. Though you should specify security options for each application, extra care needs to be taken to secure the apps and services that can be accessed over the Internet. These applications and services are vulnerable to exploitation by unauthorised individuals seeking access to your network systems. The security methods you choose must address both server-side and client-side security vulnerabilities.

13.4.4 Transmission Level Security

Transmission level security protects data communications within and across networks. When communicating across the internet, your data flows through a number of systems that you cannot control and without adequate security measures, it is available for everyone to see and use. Transmission level security measures protect your data as it flows through the internet. Secure sockets layer and virtual private network connections are examples of transmission level security measures.

Secure Sockets Layer

The secure sockets layer (SSL) protocol is an industry standard for safeguarding client-server communication. SSL was originally designed for Web browser applications, but it is increasingly being used by a growing variety of other apps. SSL uses digital certificates to authenticate communication partners and establish secure connections. Digital certificates lay the groundwork for employing SSL for secure communications and enhanced authentication.

Virtual Private Network

Virtual private network (VPN) connections can be utilised to create a secure channel of communication between two endpoints. The data that passes between the endpoints, like an SSL connection, can be encrypted, ensuring both data confidentiality and data integrity. VPN connections, in addition, allow you to restrict the type of traffic that can utilise the connection and limit the traffic flow to the endpoints that you choose. As a result, VPN connections offer some network security by assisting you in protecting your network resources from illegal access. A VPN can be seen as an extension of a company's intranet over the existing public or private network and can aid in private and secure communication within your firm.

13.4.5 Cloud Security

With more and more organisations using cloud computing, cloud security has become an important part of the cyber security plan and secures an organisation's cloud-based services and assets. In general, cloud security follows the shared responsibility model: the cloud provider is responsible for securing the services they provide and the infrastructure used to deliver them, while the customer is responsible for protecting their data, code and other assets stored or run in the cloud. Cloud security approach for safeguarding IIoT systems in Industry 4.0 setup is explained in detail in Chapter 8, Section 8.7.

13.4.6 Mobile Security

With organisations allowing mobile devices to connect to the organisation network, Mobile security plays an important role. Mobile security comprises a variety of technologies specific to smartphones and other mobile devices, such as mobile application management (MAM) and enterprise mobility management (EMM). Recently, mobile security has become available as part of unified endpoint management (UEM) systems, which offer configuration and security control for all endpoints (including mobile devices) from a single console.

13.5 CYBERSECURITY IN INDUSTRY 4.0

Prior to Industry 4.0, industries mainly relied on physical security. The premises were access controlled and people were checked physically or using automation for entry and/or exit. The industrial systems were predominantly digitally isolated and hence the likelihood of the cyber threat was minimal. As Industry 4.0 integrates the digital world to the physical world through smart autonomous technologies to enable advanced manufacturing and smart factories, it also brings with it new cybersecurity risks.

Here are the most prevalent vulnerabilities in the Industry 4.0 scenario:

- Numerous devices in various industrial plants continue to operate for extended periods, often weeks or even months, without receiving essential security updates or utilising antivirus tools.
- A significant vulnerability lies in the controllers employed within Industrial Control System (ICS) networks. These controllers can be easily disrupted by either malformed network traffic or excessive volumes of properly formatted traffic. This susceptibility stems from their design during an era when these were isolated from internet and hence cybersecurity was not a paramount concern.
- Many ICS networks expose themselves to cybersecurity threats through multiple entry points. These vulnerabilities can circumvent existing cybersecurity safeguards. For instance, laptops regularly move in and out of facilities and USB sticks are shared across various computers without undergoing proper malware checks.
- A common security flaw is the continued use of large, flat ICS networks, lacking both physical and virtual isolation between unrelated networks. This configuration leaves them susceptible to malware propagation, even extending to remote plant sites.
- Most organisations lack a comprehensive view of the OT devices within their smart factories. Many of these devices are older and were built at a time when cybersecurity wasn't a design priority. As a result, they often lack built-in security features, making them vulnerable to modern threats. One effective way to improve this is by conducting regular system risk assessments, which help identify potential vulnerabilities before attackers can exploit them. However, many organisations do not perform these assessments consistently. A common reason is the inability to scan or assess machines during operational hours, as any downtime could disrupt production and lead to significant costs.

In the subsequent sections, we explore the connected devices, digital supply chain and smart factories, focusing on the unique cyber risks faced by each.

13.5.1 Connected Devices

The number of IoT devices worldwide is estimated to be at 15.1 billion at the start of 2023 and is forecasted to almost double to more than 29 billion by 2030. IoT devices are utilised in both industrial and consumer markets, with the industry segment accounting for around 40 per cent of all IoT devices in 2023. This proportion is expected to remain mostly unchanged for the next 10 years.

As manufacturing facilities integrate and deploy IoT devices, it becomes increasingly critical to examine the security risks these devices offer to manufacturing, production and business networks. Production downtime, equipment, or facility damage, including catastrophic equipment failure and, in extreme circumstances, loss of life are all security implications of compromised IoT devices. Furthermore, possible monetary costs are not limited to production downtime and incident remediation but can include fines, litigation expenditures and revenue loss from brand damage that might last for months or even years after the occurrence.

In 2016, the Mirai malware launched a DDoS attack on Dyn, a company that manages significant portion of the internet's Domain Name System (DNS). This attack caused major websites like Twitter, Netflix, Reddit, CNN, and The Guardian to go down across Europe and the US. The attack worked by using a botnet, a network of infected devices, that overwhelmed Dyn's servers with too much traffic, causing them to crash. Unlike older botnets made up mostly of computers, the Mirai botnet used IoT devices, such as digital cameras and DVRs. Because there were so many of these devices, around 100,000, the attack reached a traffic peak of 2.1 megabits per second, which was more than double any known DDoS attack at that time. These IoT devices were easy targets because they still had default factory passwords (or hardcoded credentials that couldn't be changed), and had not received security updates or patches.

Mapping from the security at different layers, as mentioned in Section 13.4, we could argue that the system level security needs to be considered during the manufacturing of the device itself. The manufacturers should keep the provision of changing the password and the policy should be to change the password during deployment and if possible, at regular intervals.

Similarly, the *update paradox* is another challenge for manufacturing plants. Many industrial production facility networks are rarely upgraded because it is expensive for firms to schedule the necessary production downtime. The manufacturer of the devices should follow a secure software development life cycle to address the vulnerabilities early in the design phase and minimise the need for upgrade patches. They should also minimise the patches that need the device to be restarted.

13.5.2 Smart Factory

The smart factory is a cyber-physical system that analyses data, drives automated operations and learns as it goes, using modern technologies such as AI and ML. Smart factories deploy an ICS, a management and control system that ensures the automatic operation of industrial technical equipment. From an operational perspective, modern ICS environments allow engineers to deploy unmanned sites while maintaining high efficiency and resource control. The term ICS is broad and refers to a variety of control systems, including Supervisory Control and Data Acquisition (SCADA) systems and Distributed Control Systems (DCS).

Contrary to traditional cyber risk which focuses on the confidentiality of information, the smart factory focuses on the availability and integrity of the physical process. Thus, ICS security focuses on the industrial process, unlike IT security where the focus is the information.

As suggested in the network layer of security in Section 13.4.2, we need to ensure that the ICS network is not exposing any unnecessary elements to the internet. Setting up a perimeter network/firewall may provide another layer of security by reducing the attack area. Isolating unrelated networks will also help in containing the propagation of malware.

Despite the growing trend of integration and interconnection, smart factory systems remain relatively isolated due to challenges of integration with legacy systems, lack of interoperable formats, and high implementation costs. This 'closed world' security architecture has the advantage of being less susceptible to conventional mass attacks. However, the downside is that a single security flaw could allow an attacker to gain access to a factory machine. Since the network operates under the assumption that 'what's inside is trusted', an attack can spread throughout the network. Here are some potential entry points for cyber threats in a smart factory:

1. **Malicious Industrial Add-in:** Even when industrial software applications are built with robust security, the add-ins developed to extend functionalities can often introduce vulnerabilities.
2. **Digital Twin Compromise:** A digital twin file, which is a 3D model along with the code that defines the dynamic properties of the modelled object (e.g., weight, velocity), can be altered. Any changes to the digital twin file will be reflected in the physical object it models. If anyone modifies the 3D model definition or the code, the digital twin will represent a different object. Often, cloud-based digital twin catalogues do not enforce any integrity checks of the digital twin files, leading to security lapses. The cybersecurity threats in the Digital Twin system and the possible mitigations are detailed in Chapter 11, Section 11.6.
3. **Compromise via 3rd-party Firmware:** The ease of quickly provisioning custom sensors and actuators using pre-made components opens up entry points for third-party firmware. These may contain compromised libraries, creating cyber risks.
4. **Mobile Human Machine Interface (HMI):** Mobile HMIs are the most common interfaces to monitor and control IIoT devices. If not designed and developed securely, HMI apps can facilitate attacks and thus increase the overall risk. For instance, a vulnerability could allow an attacker to obtain the secret credentials to connect to and interact with the industrial robot.
5. **Data Manipulation on the MES:** Manufacturing Execution Systems (MES) are common targets of attack due to the implicit trust relationship between the MES and its database (or ERP). Most of the MES used in smart manufacturing systems trust any data coming from its database, without performing strong validation before taking action. This can lead to security issues.

It might be worthwhile to mention here that there are recent security frameworks which challenges the assumption of 'what's inside is trusted'. Zero trust framework is a security framework that operates on the principle of 'never trust, always verify'. It shifts away from traditional perimeter-based security models by assuming that no user, device, or application should be inherently trusted, regardless of their location or previous verification. Instead, each access request is rigorously validated and authorised before access is granted.

13.5.3 Digital Supply Chain

Industry 4.0 goes beyond automating the manufacturing plant and looks into automating the entire supply chain process. It introduces intelligent, connected platforms and devices throughout the ecosystem, resulting in a digital supply chain capable of capturing data from points throughout the value chain and informing one another. As a result, there may be better material management and flow, more efficient resource usage and supply that better satisfy consumer needs.

For a digital supply chain to function effectively, it requires open data sharing among all participants in the network. However, this creates a challenge for organisations: they must be fully transparent with some data while still protecting other sensitive information.

For example, imagine Supplier S provides parts to both Manufacturer A and Manufacturer B. When sharing data with A, Supplier S should only reveal information relevant to A and not disclose any details about its business with B.

To manage this securely, organisations typically use a combination of authorisation controls, network segmentation, and intermediary systems. Authorisation Controls ensure that only approved users or systems can access specific data, based on roles, identities, or permissions. Network Segmentation divides the network into isolated sections, limiting access to sensitive data and reducing the impact of potential breaches. Intermediary Systems act as secure gateways or data brokers that collect, filter and share information selectively between supply chain participants, enforcing both security and privacy policies.

13.6 DEFENCE IN DEPTH MODEL

As ICS networks are rarely directly connected to the internet, most ICS attacks happen in two stages. In stage 1, it is to gain access to the business or enterprise network of an organisation that is connected to the internet and then find an entry point to the ICS network. In stage 2, secure access to the ICS network and launch the actual ICS cyber-attack. For example, it could be a phishing campaign, to click a link to install the malicious software on the PC. The malicious software then scans the network for the ICS workstation, which is connected to the ICS network. By exploiting vulnerability in the ICS workstation, it gains access to the ICS network and halts the production line. To counter the same, the methodology behind securing an ICS is also done with a layered defence approach, the defense-in-depth model.

13.6.1 ICS Security vs Traditional IS Security

Traditional IS security may not be directly applicable to most ICS systems as:

1. Most of these devices are embedded devices with just enough memory and CPU to get the specific job done. It may not have any room for anything extra (e.g., antivirus).
2. The ICS network relies heavily on real-time communication. Extra traffic from a Network Intrusion Detection System (NIDS) can introduce latency or delay, which may bring down the entire process.
3. The operator should be able to login to the machine quickly. This is a safety requirement, as in case of emergency the operator should be able to communicate with ICS to manipulate the process. For this reason, the password cannot be too complicated (chances are that they may not be able to remember complicated passwords under stress) or you may not apply a policy of restricting user login after several failed login attempts.
4. Many ICS run production systems with extremely high uptime requirements. For example, in 99.9% uptime, there will be 1 minute of downtime every 1000 minutes. Some of the systems may have even higher uptime requirements. There is simply no time to do any security-related activities like maintenance or patching on systems with these kinds of uptime requirements.

> **DID YOU KNOW?**
>
> DMZ stands for Demilitarised Zone, which is an area where military activities are prohibited, typically based on international treaties or agreements. DMZs are created to prevent military conflict by creating distance between opposing military powers.
>
> Similarly, DMZ in cyber security is a network segment that acts as a buffer zone between two networks with differing levels of trust, such as the OT and IT networks. The purpose of a DMZ is to isolate critical systems and provide controlled access to resources, minimising the risk of cyberattacks spreading across network boundaries.

On the other hand, there are many advantages in securing ICS too.

1. As the ICS systems are stagnant it is easier to detect anomalies. For example, in an ICS network it is relatively easy to determine the traffic pattern and also to find any deviation from that pattern.
2. As the ICS does not change very often, it might be easier to secure it physically. For example, once a Programmable Logic Controller (PLC) is programmed and commissioned, it need not be changed and can be kept in a locked cabinet.

13.6.2 DMZ Layers in ISA-95 Framework or Purdue Model

As discussed in Chapter 2, the Purdue Model for Control Hierarchy is a widely recognised framework used in industrial control system (ICS) environments. It divides systems into hierarchical layers, ranging from Level 0 (physical processes) to Level 5 (enterprise-level operations). To enhance cybersecurity, the concept of Demilitarised Zones (DMZs) is introduced within this model to securely bridge the gap between IT networks and OT networks (Figure 13.5).

Figure 13.5 Purdue/ISA95 model with DMZ layers

Industrial DMZ

The Industrial DMZ is a secure network layer that acts as a buffer between the OT network (covering Levels 0–3 of the Purdue Model) and the IT network (Levels 4 and 5). Its core purpose is to enable secure, monitored communication between OT and IT systems, while protecting sensitive ICSs from unauthorised access, malware, or disruptions originating from the IT side.

To prevent direct communication between IT and OT, all data exchange passes through the DMZ, where it is strictly controlled and filtered. This setup ensures OT systems, like SCADA or PLCs, can safely share data with enterprise tools such as ERP or MES.

Communication through the Industrial DMZ uses secure protocols such as OPC UA, MQTT, or HTTPS. Application gateways or proxies in the DMZ manage and translate these data flows, converting OT-specific protocols (e.g., Modbus, BACnet) into IT-compatible formats (e.g., HTTPS, REST APIs). Data historians like OSIsoft PI or Ignition are typically placed in the DMZ to collect and store operational data, giving enterprise users access to insights without direct contact with the OT layer.

Security is maintained using dual firewalls – one separating IT from the DMZ and another isolating the DMZ from OT. These firewalls enforce strict access controls, only permitting approved protocols and performing deep packet inspection to analyse ICS-specific traffic. Intrusion Detection and Prevention Systems (IDS/IPS), such as those from Nozomi Networks or Dragos, actively monitor traffic for anomalies and threats unique to OT environments.

To further enhance isolation, network segmentation is implemented using VLANs or Software-Defined Networking (SDN), while Network Access Control (NAC) limits which devices can connect to the DMZ. For remote maintenance, jump servers located in the DMZ provide secure access to OT systems, logging all activities for security auditing purpose.

In high-availability or high-security environments, the Industrial DMZ may include redundant configurations with load balancers to ensure continuous operations. Additionally, unidirectional data diodes can enforce one-way data flows (typically from a high security network to a low security network)allowing OT data to reach IT systems while preventing any reverse communication. For time-critical operations, Time-Sensitive Networking (TSN) can be used to maintain deterministic and secure communications within the DMZ.

Enterprise DMZ

Enterprise DMZ is situated where the IT network is accessed from the exterior, that is, a partner network or the internet. Its primary goal is to create a communications channel from the interior IT systems inside an enterprise to the outside organisations and then protect the interior systems against direct exposure to possible dangers.

The Enterprise DMZ is designed to shield information technology resources from outside attack, providing a secure zone for offsite users, vendors, or business partners to access certain resources of the enterprise. The DMZ is designed to segregate critical business systems, databases, and file servers from internet exposure. Publicly accessible services like web servers, email servers, and VPN gateways are housed in the DMZ, segregating the services from the inside network to mitigate impact from any security intrusion. Reverse proxies and dynamic firewalls are among the security controls incorporated in the Enterprise DMZ, screening the incoming and outgoing network data, checking for malware or weakness prior to forwarding the data to the inside network. Network segmentation is employed to bar unauthorised lateral movement within the information technology environment.

Multi-factor authentication and robust access controls for outside users accessing the enterprise systems considerably enhances the security infrastructure of the Enterprise DMZ. Strong access control and multi-factor authentication for outside users accessing enterprise systems ensures access only for right users to sensitive resources, excluding unauthorised access.

The Enterprise DMZ is used as a security buffer between the information technology (IT) network and its exterior, with the use of key technologies for the purpose of good security. Essentially, two firewalls define the DMZ; one for internet access and the other for possible inside-based threats within the DMZ. Most firewalls use Unified Threat Management (UTM) features like virtual private network (VPN) security, antivirus security, and content filters. Reverse proxy servers like NGINX or HAProxy in the DMZ are used to secure access to web applications inside the network through SSL/TLS encryption, load balancing, and Web Application Firewall (WAF) protection from standard threats like SQL injections and cross-site scripting attacks.

DMZ uses VPN concentrators to provide secure access for remote and external cooperators with access to business systems through either SSL-based Virtual Private Networks (VPNs) or IPsec with the inclusion of multi-factor authentication (MFA). Public services like websites and mail systems are typically accompanied with Content Delivery Networks(CDN) like Cloudflare and Akamai, protecting against Distributed Denial of Service (DDoS) and enhancing performance simultaneously. Log streams from devices inside the DMZ are sent to a Security Information and Event Management (SIEM) platform, represented through Splunk and QRadar, in order to facilitate centralised monitoring and provide real-time alerting. Email and DNS security solutions lay the foundation for hardening the Enterprise DMZ, with the use of SMTP gateways to block phishing and including DNS filtering solutions for blocking resolution of hostile domains. Advanced deployments include deceptions like honeypots, aimed at drawing attackers into controlled zones, and zero-trust principles with authentication and verification for all communications within the DMZ.

13.6.3 Layered Security Approach

From Level 5 to Level 0 in ISA-95 model, devices have progressively greater access to critical processes but less security capabilities. As a result, lower-level devices are more dependent on the security of the upper-level devices as well as on network and architectural defences, to protect them. Key security considerations for each level include:

- **Levels 0 and 1:** Implement physical security controls (e.g., locks, restricted access zones), device-level authentication and encryption to secure communication between sensors and actuators, device hardening by disabling unused ports and services.
- **Level 2:** Implement strong role-based access control for human-machine interfaces, regularly update software to patch known vulnerabilities and secure remote access.
- **Level 3:** Secure data exchange with encryption and use firewalls to control data flow between level 3 and level 4.

- **Level 4 and 5:** Implement standard IT security practices, including endpoint security solutions like antivirus tools, patch management and security monitoring.

Physical Security

Physical security focuses on preventing unauthorised individuals from accessing critical equipment or sensitive areas like control rooms, server rooms and network panels. It emphasises creating barriers such as secure walls, locked doors and cabinets and installing surveillance systems like CCTV and alarms to deter or detect intrusions. Policies for visitor management and handling guests are also key components to ensure controlled physical access.

Device Security

Device security revolves around safeguarding ICS devices through practices like patching, hardening and applying both physical and logical access restrictions. It also involves managing the device lifecycle, which includes acquisition, configuration, maintenance, upgrades and proper disposal. These steps are essential to ensure devices remain secure and operational throughout their use.

Network Security

The goal of network security is to limit access to specific parts of the ICS network by implementing logical divisions called security zones. Tools like industrial DMZs, firewalls, access control lists and IDS/IPS help monitor and restrict unauthorised traffic between these zones. IDS passively monitors network traffic for malicious activity and generates alerts when suspicious patterns or attacks are detected. It does not actively block or prevent attacks; it relies on human intervention for the same. IPS actively monitors and analyses network traffic. It takes actions to block or prevent detected attacks in real-time. Effective zoning allows organisations to detect and mitigate anomalies in sensitive network segments.

Computer Security

Computer security aims to prevent breaches in computing systems through measures like timely software updates, installation of antivirus programs and host-based IDS/IPS. Another vital control is disabling unused communication ports, such as USB ports, to prevent unauthorised data transfer or malware infections.

Application Security

Application security ensures that only authorised users interact with applications and services running on a computer system. Key techniques include authentication (verifying user

DID YOU KNOW?

The Defense in Depth model concept came from the architecture of the medieval castle architecture. The castle will be on an open field or a high ground. Which enables the ability to spot the enemy from a distance. The second layer of defence is a moat around the castle, sometimes infested with crocodiles. Tall and smooth perimeter wall acted as a third layer of defence with soldiers guarding from watchtowers and parapets.

identity), authorisation (defining user privileges) and auditing (maintaining logs of user activities). These measures collectively help safeguard application integrity and trace any unauthorised actions.

13.7 CYBERSECURITY IMPLEMENTATION FRAMEWORK

Till now we understood the cybersecurity-related challenges and how to architect the layered security needed for Industry 4.0 factories. In this section, we will look into the framework from a process perspective.

The NIST Cybersecurity Framework (CSF)[1] has been adopted by many public and private-sector organisations to guide cybersecurity efforts and assess cybersecurity risks. The Framework is made up of five concurrent and continuous Functions: Identify, Protect, Detect, Respond and Recover (Figure 13.6).

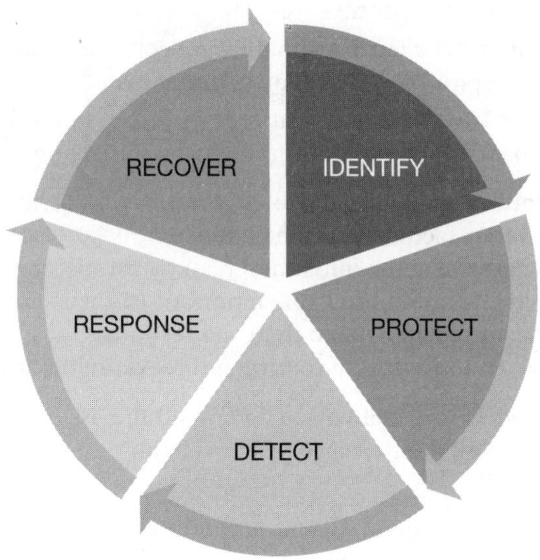

Figure 13.6 Governance model

13.7.1 Identify

The Identify Function's goal is to create an organisational understanding of how to manage cybersecurity risks to systems, people, assets, data and capabilities.

---

[1] In this context organisations also refer to ISA/IEC 62443: Cybersecurity for Industrial Automation & Control Systems, ISO/IEC 27001: Information Security Management Standard. In India, NCIIPC Guidelines govern Security for Critical Information Infrastructure, Digital Personal Data Protection Act (DPDP Act) governs personal data management and CERT-In Compliance for mandatory reporting for breach covering incident response, logging.

A risk assessment evaluates what vulnerabilities exist in the system under consideration, the likelihood that these vulnerabilities will be exploited and the impact of such successful exploitation on the system, the organisation and the environment as a whole. The result of the risk assessment is a risk score for the identified vulnerabilities. The risk score can be given by the following equation:

$$\text{risk score} = \frac{\text{severity} + (\text{criticality} \times 2) + (\text{likelihood} \times 2) + (\text{impact} \times 2)}{4}$$

where the parameters are as follows:

- **Severity:** a number between 0 and 10. National Vulnerability Database (NVD), a US Government repository, maintains a database of vulnerabilities and provides a Common Vulnerability Scoring System (CVSS) score.
- **Criticality:** a number between 1 and 5, that reflects the importance of the system under consideration to the overall process.
- **Likelihood:** a number between 1 and 5, that reflects the chance that the vulnerability will be successfully exploited and become a threat event.
- **Impact:** a number between 1 and 5, that reflects direct and indirect (health, safety and environmental impact, brand reputation) financial impact on the company.

Please note that different organisations may consider different formulas to calculate the risk score but should have all four factors mentioned above.

The risk score helps determine the mitigation plan by prioritising which issues need to be addressed first. Since budgets are limited, it's important to focus mitigation efforts on high-risk items to make the best use of available time and resources.

During a risk assessment exercise, the following three activities are undertaken:

- **Asset Identification:** This step involves finding all the assets of the system under consideration and assessing their *criticality*. The outcome of this step is a list of potential targets.
- **Vulnerability Identification and Threat Modelling:** This step involves identifying any potential vulnerabilities within the assets discovered in the asset identification stage and their associated CVSS score for *severity*. This step also does the threat modelling for the identified vulnerabilities and assesses the *likelihood* of this vulnerability becoming a threat event.
- **Risk Calculation:** This step involves assessing the total *impact* of a threat event for the identified vulnerabilities and combining all the parameters to calculate the risk score.

13.7.2 Protect

The Protect Function serves to limit or contain the impact of potential cybersecurity events. This encompasses various areas, including Awareness and Training, Identity Management and Access Control, Data Security, Maintenance and Information Protection Processes and Procedures.

Security awareness plays a crucial role in preventing OT incidents. Organisations must identify, document and provide training for all personnel, with tailored content based on their roles and responsibilities. Training should cover basic knowledge of social engineering techniques, recognising anomalous behaviour in the OT environment, knowing when and how to connect or disconnect the OT environment from external security domains, understanding password complexity and management requirements and adopting appropriate reporting practices.

Data security involves safeguarding the confidentiality, integrity and availability of data at rest and in transit, as well as preventing data leaks. Cryptography, including encryption, digital signatures, hashing and other cryptographic functions, can support data security requirements by preventing unauthorised access or data modification.

Maintenance tracking solutions enable organisations to schedule, track, authorise, monitor and audit maintenance and repair activities in OT. This ensures proper documentation of maintenance logs or changes, creating an audit trail that aids in cybersecurity-related troubleshooting, response and recovery activities. Maintenance tracking also offers visibility into scheduled maintenance for OT devices, assisting in end-of-life decisions.

Policies, processes and procedures should be maintained to manage the protection of information systems and assets. Countermeasures and outcomes must be in place to handle configuration changes throughout the life cycle of the component and system. Regular backups should be maintained and response and recovery plans should be prepared and tested. A comprehensive plan for vulnerability management should be developed and implemented throughout the life cycle of the components.

13.7.3 Detect

The Detect Function ensures the timely discovery of cybersecurity events through the development and implementation of appropriate activities.

To establish an effective detection capability, organisations need to comprehend various events, anomalies and their potential impacts on systems and the environment. To minimise false positives and unnecessary alarms, organisations should set OT alerting thresholds based on baselines of normal network traffic, data flows and human/OT process behaviour. Considering that OT components are often physically remote and not continuously staffed, alerting thresholds must account for the response time associated with the alert. For instance, a temperature alert threshold might need to trigger earlier based on the expected response time to rectify the situation and prevent an incident.

As part of the organisational risk management strategy, organisations should implement continuous monitoring to assess the effectiveness of protective measures. In an OT environment where active scanning may not be feasible, continuous monitoring can be achieved through passive scanning or manual monitoring at a frequency proportional to the risk. Examples of continuous monitoring processes include network monitoring, system use monitoring, malicious code detection and vulnerability scanning.

13.7.4 Respond

The Respond Function supports the ability to undertake the necessary actions and activities for containing a cybersecurity incident as it unfolds.

When addressing the event, organisations should adhere to their documented response plan, previously prepared as part of the Protect Function. Responding to a cybersecurity incident involves communication with both internal and external stakeholders. Hence, the response plan generally incorporates a communication plan, which should also be implemented. Enhancements to organisational response activities can be achieved by integrating insights gleaned from detection and response activities and these improvements should be incorporated into the response plan.

13.7.5 Recovery

Timely recovery to normal operations after a cybersecurity incident is critical. The Recover Function addresses developing and implementing activities to maintain system resilience and ensure the timely restoration of capabilities and services affected by a cybersecurity incident.

When recovering from events, organisations should execute their documented recovery plans (prepared in the Protect Function). As a recovery effort is ongoing, the recovery steps taken should be documented to identify lessons learned. These lessons can be used to improve recovery plans and processes.

Together, these five Functions provide a high-level, strategic view of an organisation's cybersecurity posture. They help organisations prioritise actions, allocate resources effectively and build resilience against evolving cyber threats. With a solid understanding of each Function, organisations can tailor their cybersecurity programs to align with their specific risks and business needs.

13.8 KEY CHALLENGES IN IMPLEMENTING CYBERSECURITY IN SMART FACTORIES

Implementing cybersecurity in smart factories comes with several challenges:

- **Employee Awareness:** Training employees to handle various aspects of evolving cyber threats is a significant challenge. Employees are often the first line of defence, so they must be aware of the early warning signs of a potential attack to ensure a quick response.
- **Lack of Knowledge to Implement Cybersecurity Systems:** While a wide array of security tools is available, enterprises often lack the necessary in-house knowledge to implement them properly. Misconfigurations are one of the key reasons why breaches occur.
- **Lack of Collaboration Between Cybersecurity Teams and the Executive Team:** Cybersecurity in smart factories at times is evaluated at the plant or regional level and does not receive the global board-level priority given to other functions in the organisation. This lack of collaboration can lead to gaps in security measures.
- **Lack of Budget:** Budgets for smart factories are often allocated from the local/site budget. The absence of dedicated budgets can lead to funds being diverted to other areas of smart factories that seem to require immediate attention, leaving cybersecurity underfunded.

Addressing these challenges is crucial for ensuring the security and integrity of smart factories.

13.9 CONCLUSION

This chapter highlighted the critical importance of cybersecurity in Industry 4.0, particularly in safeguarding Operational Technologies. As manufacturing becomes more connected and automated, the risks associated with cyber threats increase, making robust cybersecurity practices essential. Many organisations are now prioritising cybersecurity efforts due to the rising threat landscape and the need to protect valuable data, assets and operational continuity. The adoption of frameworks like NIST's Cybersecurity Framework and strategies such as Defense in Depth offers manufacturers a structured approach to managing risks. However, challenges such as employee awareness, lack of in-house expertise and budget constraints still persist.

As Industry 4.0 technologies continue to evolve, cybersecurity must be a continuous focus to ensure the integrity, availability and confidentiality of systems. The growing demand for smarter, more connected factories will only amplify the need for comprehensive cybersecurity programs. Manufacturers must recognise that securing their operations is an ongoing journey that requires constant innovation, collaboration and adaptation to new threats. In conservative estimate, cybersecurity revenue is expected to show an annual growth rate (CAGR 2024-2029) of 7.92%, resulting in a market volume of US$271.90bn by 2029. The future of secure smart factories hinges on proactive risk management, employee education and the integration of robust cybersecurity practices throughout the entire lifecycle of industrial systems.

SUMMARY

- Industry 4.0 introduces greater interconnectivity and automation in manufacturing, making robust cybersecurity essential to prevent system vulnerabilities.
- Cybersecurity is built on the principles of confidentiality, integrity and availability, ensuring the protection and accountability of systems.
- Industrial systems face a range of cyber threats such as malware, phishing, insider threats and denial of services, which can lead to operational disruptions, financial loss and damage to reputation.
- Cybersecurity measures must be applied at multiple layers, including physical, network, application and data layers, to safeguard industrial systems against diverse threats.
- In Industry 4.0, IT and OT systems are more integrated, creating unique cybersecurity challenges that require tailored solutions for smart factories and industrial environments.
- The Defense in Depth strategy emphasises multiple layers of security, such as perimeter security, intrusion detection and access controls, to reduce vulnerabilities across various system components.

- The NIST Cybersecurity Framework (CSF) offers a structured approach to managing ICS cybersecurity programs, focusing on the functions of identifying, protecting, detecting, responding and recovering from threats.
- Risk assessments are vital for determining vulnerabilities and calculating risk scores, which help prioritise cybersecurity efforts based on the severity and impact of potential threats.
- Implementing cybersecurity in smart factories presents challenges, including employee awareness gaps, lack of in-house expertise, limited collaboration between teams and insufficient budgets for necessary security measures.
- Employee training and awareness are critical to detecting and mitigating cyber threats before they escalate.
- A collaborative approach between cybersecurity teams and executives is necessary to ensure effective implementation of cybersecurity strategies across all levels of the organisation.
- Limited budgets for cybersecurity in smart factories often lead to underfunded initiatives, leaving systems more vulnerable to attacks.
- Proactive measures, such as regular risk assessments, layered defence strategies and continuous employee training, can help mitigate the impact of cyber threats.

EXERCISE

Multiple Choice Questions (1 Mark)

1. Which principle of cybersecurity ensures that information is not altered during transmission?

(a) Availability
(b) Integrity
(c) Confidentiality
(d) Authentication

2. Which cyber threat involves targeting individuals with fraudulent messages to obtain sensitive information?

(a) Espionage
(b) Phishing
(c) Malware
(d) Denial of Service (DoS)

3. What is the main objective of a Distributed Denial of Service (DDoS) attack?

(a) Steal sensitive information
(b) Encrypt user data for ransom
(c) Overload systems to disrupt services
(d) Modify system settings without detection

4. Which of the following is a challenge unique to cybersecurity in Industry 4.0 environments?
   (a) Limited network connectivity
   (b) High costs of implementing sensors
   (c) Integration of IT and OT systems
   (d) Absence of cyber threat awareness

5. The Defense in Depth model operates on which principle?
   (a) A single layer of strong defence is sufficient
   (b) Combining multiple layers of security to protect assets
   (c) Removing access controls for faster system performance
   (d) Eliminating the need for monitoring systems

6. What does the Identify function in the NIST framework aim to achieve?
   (a) Establishing cybersecurity governance
   (b) Implementing firewalls and access controls
   (c) Developing data recovery mechanisms
   (d) Detecting and mitigating incidents in real-time

7. What is the purpose of risk assessment in cybersecurity programs?
   (a) To eliminate all potential threats
   (b) To identify vulnerabilities and prioritise mitigation efforts
   (c) To design physical infrastructure for secure access
   (d) To focus on cost reduction in security measures

8. Continuous monitoring in OT environments is critical because it:
   (a) Eliminates manual intervention in recovery
   (b) Detects anomalies and alerts administrators in real-time
   (c) Improves sensor functionality in production lines
   (d) Reduces the number of necessary physical devices

9. Why is risk scoring important in cybersecurity programs?
   (a) To determine hardware specifications
   (b) To prioritise vulnerabilities based on their impact and likelihood
   (c) To design user-friendly interfaces
   (d) To monitor network activity in real-time

10. Which of the following is NOT a component of a Defense in Depth strategy?
    (a) Network segmentation
    (b) Role-based access control
    (c) Social engineering
    (d) Continuous monitoring

11. What is one of the biggest challenges in aligning IT and OT systems in Industry 4.0?
(a) The high cost of physical sensors
(b) Ensuring real-time communication without compromising security
(c) Overhauling application development frameworks
(d) Developing entirely new technologies for production systems

12. What is the impact of employee unawareness in cybersecurity?
(a) Faster system performance
(b) Increased vulnerability to social engineering attacks
(c) Lower costs of training programs
(d) Reduced cybersecurity investments

13. Which type of cyber threat involves the insertion of malicious code into legitimate software?
(a) Spyware
(b) Trojan horse
(c) Rootkit
(d) Worm

14. What role do standards and frameworks like NIST play in cybersecurity?
(a) Dictate specific tools for use in enterprises
(b) Provide guidelines for implementing best practices
(c) Enforce legal penalties for data breaches
(d) Reduce hardware costs for security systems

15. Which key principle in cybersecurity ensures authorised users can access data when needed?
(a) Confidentiality
(b) Availability
(c) Encryption
(d) Authenticity

16. An operator in an industrial control system notices unusual spikes in network traffic during non-operational hours. What type of attack might this indicate?
(a) Phishing
(b) Man-in-the-Middle attack
(c) Distributed Denial of Service (DDoS) attack
(d) Data exfiltration

17. After implementing the Defense in Depth model, an organisation still experiences breaches due to human error. What is the most appropriate next step?
(a) Increase the frequency of system audits
(b) Focus on employee training and awareness programs
(c) Add more layers of security to the network
(d) Invest in advanced AI-based threat detection systems

18. A manufacturing facility experiences repeated downtime due to ransomware attacks on its control systems. What is the best initial step to improve their cybersecurity?
    - **(a)** Upgrade all hardware to the latest version
    - **(b)** Implement a comprehensive incident response plan
    - **(c)** Restrict physical access to the control systems
    - **(d)** Conduct a risk assessment to identify vulnerabilities
19. During a cybersecurity audit, it is found that a critical system lacks proper encryption for data in transit. What is the most immediate risk this poses?
    - **(a)** Data breaches from physical theft
    - **(b)** Unauthorised access to transmitted information
    - **(c)** Malware installation on the system
    - **(d)** Service disruption due to power outages
20. A critical manufacturing system is targeted by a phishing campaign and an employee unknowingly clicks on a malicious link. What immediate action should the organisation take to minimise damage?
    - **(a)** Revoke the employee's access credentials
    - **(b)** Shut down the entire network to stop the spread
    - **(c)** Isolate the affected system and initiate incident response protocols
    - **(d)** Notify all employees about the breach

Short Answer Type Questions (2 Marks)

1. What are the three key principles of cybersecurity?
2. Name two types of cyber threats common in smart factories.
3. Why is layered security important in Industry 4.0?
4. What is the formula for calculating a risk score as per the NIST Framework?
5. What role does training play in the 'Protect' function of cybersecurity?
6. Name two key components involved in asset identification during risk assessment.
7. Why is recovery planning critical in smart factories?
8. Mention two challenges in implementing cybersecurity in smart factories.
9. What is the significance of OT-specific alert thresholds?
10. What makes cybersecurity in Industry 4.0 unique compared to traditional IT environments?

Long Answer Type Questions (5 Marks)

1. Explain the significance of cybersecurity in Industry 4.0 factories. How do emerging technologies like IoT and AI influence the security landscape of these factories?

2. What are the main components of an ICS cybersecurity program and why is it crucial to develop a structured approach for cybersecurity in industrial settings?
3. What role does governance play in implementing ICS cybersecurity and how does the governance model influence decision-making in cybersecurity efforts?
4. In the context of ICS cybersecurity, explain how risk assessments are conducted and how they help organisations prioritise mitigation efforts.
5. Discuss the importance of asset identification and threat modelling in the ICS cybersecurity risk assessment process.
6. Explain the importance of the 'Respond' function in the NIST Cybersecurity Framework and how organisations can ensure effective incident response in ICS environments.
7. Discuss the significance of the 'Recover' function in the NIST Cybersecurity Framework and how it helps ICS environments return to normal operations after a cybersecurity incident.
8. How does cybersecurity awareness training play a role in the Protect function of ICS cybersecurity and why is it critical for personnel at all levels?
9. How do continuous monitoring and regular vulnerability assessments support proactive cybersecurity in smart factories?
10. What are some key challenges organisations face when implementing ICS cybersecurity and how can they address these challenges?

Conclusion: Bringing it All Together and Way Forward

CHAPTER 14

OBJECTIVES

After navigating all the core technologies and processes that enable the implementation of Industry 4.0, it is now time to bring them all together, connecting the dots and seeing how these technologies and processes come together to build an Industry 4.0-compatible smart factory.

By the end of this chapter, you should be able to:

- Understand how the different Industry 4.0 technologies are connected.
- Connect the dots by exploring a real-life case study on Smart Aircraft manufacturing.
- Comprehend the challenges in adopting Industry 4.0 and the way forward.

14.1 INTRODUCTION

We have seen that, Industry 4.0 represents a new era of intelligent manufacturing where the true strength lies not in isolated technologies but in their seamless integration. The fundamental aspect of Industry 4.0 is **connectivity**, which enables machines, systems and people to communicate in real-time and in constant interaction. In this book, so far, we have covered key technologies such as IIoT, Automation and Robotics, Additive Manufacturing/3D printing, AR/VR, Cloud Computing, Big Data, Analytics, Digital Twins and Simulations, Blockchain and Cybersecurity. Additionally, we learned the concepts of digital transformation, cyber-physical systems (CPS), horizontal and vertical integration and lean and agile production methodologies. Individually, the efficacy of these tools and processes is less visible. Once they are connected, the collective synergy among these Industry 4.0 tools and processes can create wonders, such as a fully connected, data-driven and automated industrial ecosystem. These technologies interact across the value chain to enable operational efficiency and advanced automation, where decisions are made in real-time and systems operate with minimal human intervention.

Figure 14.1 illustrates the interconnected nature of Industry 4.0 technologies and how they collaboratively drive digital transformation in industrial ecosystems.

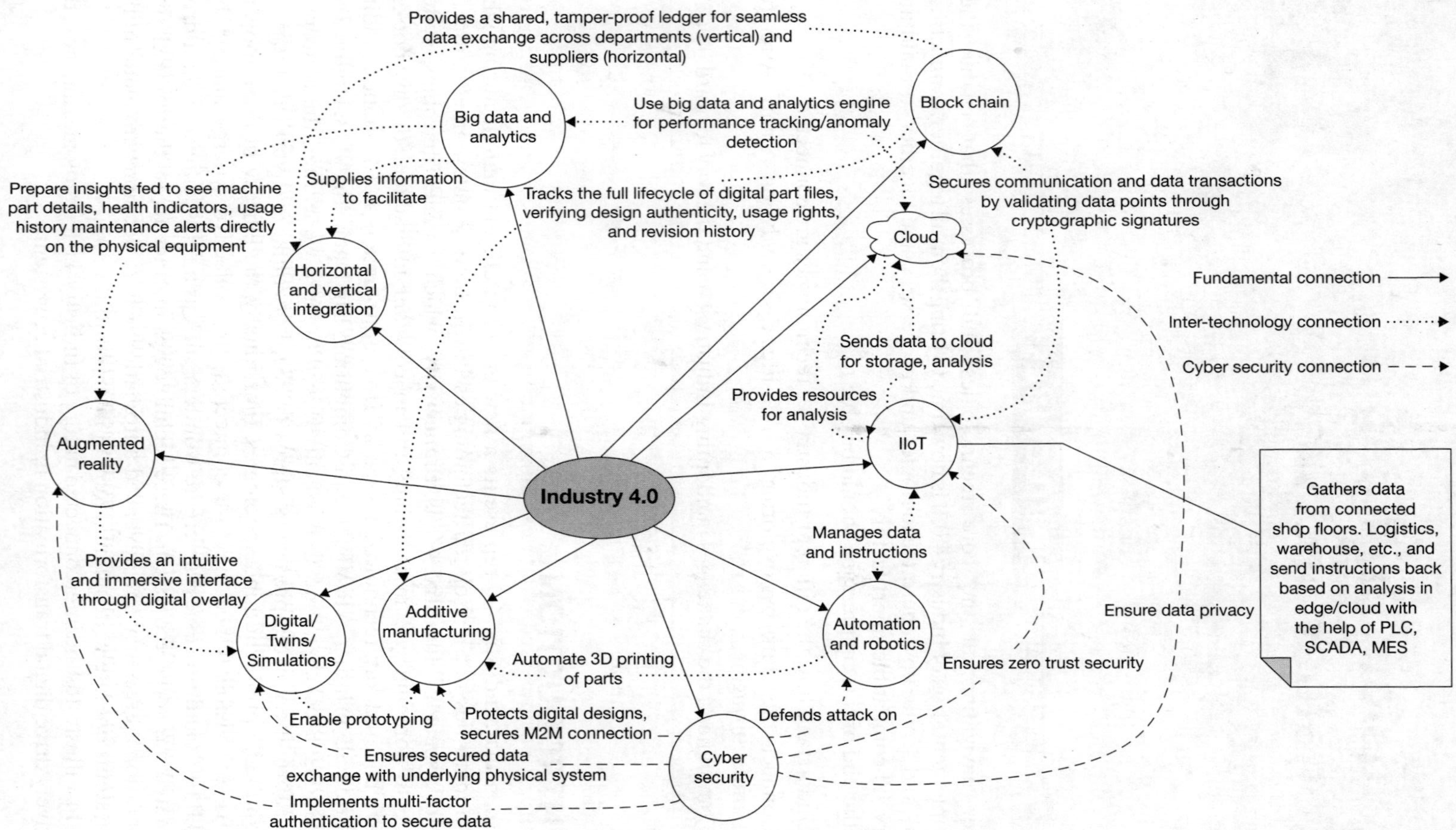

Figure 14.1 Industry 4.0 technologies relationship map

The IIoT is a central enabler that gathers data from connected shop floors, logistics, warehouses and other sources and sends this massive amount of data for deeper analysis and storage in the cloud.

Cloud interacts with the Big Data and Analytics Engine to store this data and perform deep analytics using AI and ML. Big Data and Analytics engines process these vast data streams to detect performance trends and identify anomalies. Based on the feedback, IIoTs trigger actions to end systems with the help of technologies like PLC, SCADA and MES.

These analytics then supply actionable information to enable Horizontal and Vertical Integration. This integration ensures that different departments of an enterprise, from the shop floor to top management and complete supply chains, remain informed and aligned.

Digital Twins and Simulation tools enable virtual prototyping, helping to optimise manufacturing workflows before physical execution. These simulations integrate tightly with AR systems, which provide immersive, real-time overlays on physical equipment, enhancing operator understanding, maintenance efficiency and error reduction. The insights generated through Big Data and Analytics are subsequently delivered to AR-enabled devices, such as smart glasses or tablets, enabling technicians and operators to view contextual information, including machine condition, usage patterns and maintenance alerts, overlaid directly onto the physical equipment in real-time.

3D Printing is powered by advanced robotics, such as Cartesian robots. Digital twins can be associated with 3D printing for fast physical prototyping of the approved digital copy.

Blockchain operates through a shared, tamper-proof ledger that facilitates secure and verified data exchange across departments (vertical) and with suppliers (horizontal). It validates data points using cryptographic signatures to secure communication and data transactions of IIoTs. It also helps in tracking the complete lifecycle of digital part files, including design authenticity, usage rights and revision history.

Cybersecurity is integrated throughout the entire solution, utilising zero-trust frameworks, multifactor authentication and secure system access protocols to protect the entire digital infrastructure. It ensures secure data exchanges and protects against external and internal threats.

This interconnected web of communications among different pillars of Industry 4.0 requires advanced communication technology, such as 5G, to be in place. 5G technology has become a game changer by providing ultra-fast, low-latency and high-reliability wireless communication.

As shown in the above relationship map, there are numerous connections among various technologies of Industry 4.0. In a real-world smart factory setup, the connections become so vast and complex that they cannot be managed through traditional manufacturing processes. In such cases, we need to follow advanced methods, such as lean and agile manufacturing, as discussed in the earlier chapter.

With this knowledge, let us examine a real-life case study of Industry 4.0 implementation in the aircraft manufacturing industry. Aircraft manufacturers are always the front-runners in adopting new technologies. You will notice in the case study how many cutting-edge technologies are interwoven to build this smart production setup.

14.2 SMARTAIR AEROSPACE CASE STUDY: A DIGITAL TRANSFORMATION JOURNEY IN AIRCRAFT MANUFACTURING

In its effort to stay competitive in an increasingly connected and intelligent manufacturing landscape, SmartAir Aerospace, a leading aircraft manufacturer, launched a comprehensive Industry 4.0 transformation initiative. As part of its transformative journey, SmartAir understood that true innovation must start at the very beginning, the design phase, where digitalisation would lay the groundwork for seamlessly connecting and enhancing every stage of the aircraft's lifecycle.

14.2.1 Design and Engineering Transformation

Unlike traditional approaches grounded in static CAD models and manual simulation processes, SmartAir reimagined its design phase by embedding intelligence, connectivity and traceability at every step. Figure 14.2 below illustrates the schematic diagram of Industry 4.0 in operation at an Aircraft Manufacturing Unit.

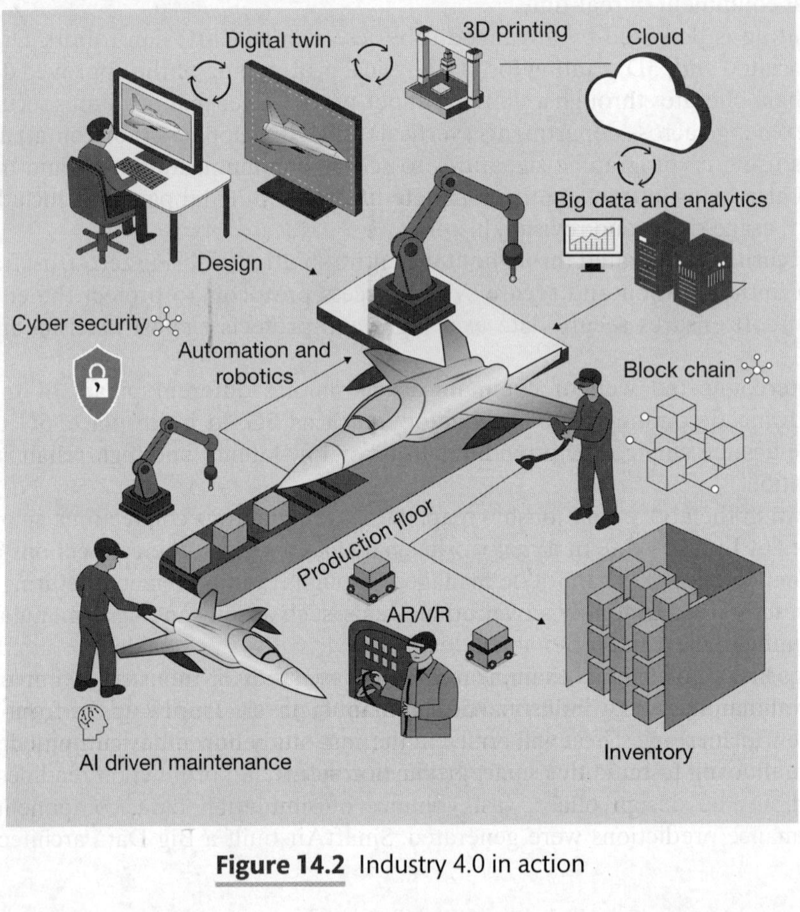

Figure 14.2 Industry 4.0 in action

At the core of this transformation was the adoption of Digital Twin technology, an exact digital replica of each aircraft system, component and structural unit, enriched by real-time feedback from multiple stages of the aircraft lifecycle. SmartAir's engineering teams leveraged high-performance computing environments via Microsoft Azure's cloud infrastructure, deploying digital twin instances that allowed for iterative design testing, load analysis and aerodynamic simulation, all conducted in parallel across globally distributed teams. Every design modification, whether involving a change to the fuselage contour or a new winglet configuration, was logged, versioned and verified through blockchain-based data integrity systems. These immutable records ensured tamper-proof documentation, accessible only through smart contracts that controlled edit rights and ensured traceability, providing regulatory compliance and transparency in collaboration. Figure 14.3 shows a digital twin of the aircraft engine.

Credit: cheskyw / 123RF

Figure 14.3 Digital twin of an aircraft engine

Engineers across geographies collaborated seamlessly using AR and VR platforms. VR-based immersive reviews enabled teams to virtually step inside the aircraft model, inspect ergonomics, wiring pathways and access points. AR headsets were used during prototype validation, overlaying virtual design data on physical mock-ups to ensure precision in component placement. These immersive environments reduced design iteration cycles by nearly 30%, allowing faster convergence between concept and production readiness.

Throughout the design phase, vast volumes of simulation data, component test logs and performance predictions were generated. SmartAir built a Big Data architecture using cloud-hosted data lakes and Apache Spark pipelines to ingest, process and analyse this data. The insights were powered by AI algorithms capable of identifying structural optimisation opportunities, forecasting thermal stress responses and even predicting long-term material fatigue.

To safeguard this digitised intellectual property, SmartAir invested heavily in Cybersecurity infrastructure. A zero-trust model governed all data access, complemented by AI-driven anomaly detection systems and role-based, blockchain-backed authentication. Secure development environments were maintained for each classified aircraft variant, ensuring data sovereignty and compliance across international design centres.

With a fully interconnected, data-driven and simulation-rich design ecosystem in place, SmartAir laid the foundation for the next stages of transformation, where ideas would take shape on the factory floor.

14.2.2 Smart Manufacturing

After building a data-rich, simulation-driven design ecosystem, SmartAir shifted its focus to modernising the production floor. Traditional methods, characterised by rigid assembly lines, isolated processes and paper-based instructions, could no longer meet the rising complexity and speed required in aerospace manufacturing. To stay competitive and scalable, SmartAir was committed to building a smart, modular and fully connected digital factory driven by Industry 4.0 principles.

Introduction of Advanced Robotics and Automated Units

The transformation began with the integration of automated robotic systems into every major phase of assembly. In heavy structural sections such as the fuselage and wing-body joints, articulated robots were deployed to carry out tasks like precision welding, riveting and multi-axis drilling. These are the areas where millimetre-level accuracy and repeatability are critical. Traditionally performed by skilled technicians, these operations were labour-intensive and prone to fatigue-related inconsistencies. By automating them, SmartAir not only improved build quality but also significantly reduced production cycle times. The robots were equipped with tool changers that allowed seamless transitions between drilling, welding and sealing operations along the curved surfaces of fuselage panels. This was especially vital in the assembly of wing-box joints and pressure bulkheads, where precision tolerances are critical for aircraft safety and aerodynamics.

Automated Drilling Units were deployed, which provide high-speed, accurate hole placement in wing structures. Additionally, automated fibre placement (AFP) machines were deployed to lay carbon fibre composites and build fuselage skins. The AFP systems use robotic arms equipped with fibre heads to lay down pre-impregnated composite fibres (prepregs), i.e., resin-coated carbon fibres that cure under heat and pressure to form lightweight, durable structures.

For electronics and avionics assemblies, SCARA and Cartesian robots were strategically deployed. SCARA robots performed high-speed component placements, achieving sub-millimetre alignment in circuit boards, while Cartesian systems handled repetitive bracket installations. Both systems were integrated with vision-guided systems, ensuring accuracy and eliminating the need for post-process corrections.

Vision systems and LiDAR were used extensively for robotic calibration. Cameras above assembly cells continuously scanned work-in-progress parts, comparing them with digital models. If misalignments or surface defects were found, the systems either corrected the issues autonomously or flagged them for human inspection. Cobots were deployed in tasks requiring close interaction with technicians, particularly in final cabin interiors and wiring bays, ensuring safety through force-limited joints and motion sensors.

Usage of AGVs and AMRs

Movement of components across the factory was orchestrated through a dual approach: Automated Guided Vehicles (AGVs) and Autonomous Mobile Robots (AMRs). While AGVs followed RFID-tagged floor paths to transport heavy payloads, AMRs equipped with LiDAR sensors and onboard AI navigated dynamically, adjusting their routes in real-time to avoid congestion or prioritise urgent parts. Advanced conveyor belt systems with RFID were used to track parts through every stage of production. These conveyors fed live data into SCADA systems, allowing line supervisors to adjust the production pace and proactively rebalance workstations. These mobile systems interacted directly with the factory's MES to ensure that parts reached the correct station at the right time, maintaining seamless synchronisation.

AR Glass for Electrical Wiring

One of the most significant shifts was in the collaboration between workers and machines. Technicians assembling electrical wiring in tight fuselage zones were equipped with AR glasses, like Google Glass. These glasses projected digital blueprints over the actual physical wire bundles, providing technicians with real-time guidance and step-by-step instructions. The result was a 25% reduction in wiring errors and a measurable increase in installation speed. AR tools were also integrated with voice commands and quality validation checks, allowing hands-free operation and progress tracking.

Implementation of 3D Printing

SmartAir previously faced several manufacturing challenges, including long lead times, high production costs and inefficiencies in traditional manufacturing methods such as machining and casting. Creating custom tooling was time-consuming, while assembling multiple small parts increased complexity, weight and risk of failure. The need to maintain large inventories and rely on external suppliers resulted in supply chain delays and increased costs. Additionally, machining lightweight materials, such as titanium, resulted in significant material waste. To address these issues, SmartAir built an in-house Additive Manufacturing (3D Printing) lab, using two core technologies. For metallic components, such as engine brackets and cooling ducts, the company utilised selective laser melting (SLM), a process in which a high-power laser selectively melts fine metal powder layer by layer to build complex geometries. SLM is known for producing lightweight, high-strength, aerospace-grade parts with acceptable detail resolution. For polymer tooling and non-structural parts, fused deposition modelling (FDM) was employed. In FDM, thermoplastic filament is melted and extruded through a nozzle, depositing material in layers. This method was ideal for rapidly fabricating jigs, fixtures and enclosures, reducing tooling lead times from weeks to days.

Cloud-based IIoT Backbone: Connected to MES and Secured by Cyber Security and Blockchain

All robotic and additive processes were embedded within a larger IIoT infrastructure. Smart sensors embedded in every machine, from welding robots to conveyor belts, collected real-time data on torque, vibration, temperature and throughput. This data was streamed to a cloud-hosted MES, which monitored key metrics such as station uptime, error rates and tool wear. Predictive analytics triggered maintenance alerts when anomalies, such as abnormal torque in riveting stations, were detected, thereby minimising unplanned downtime and extending the life of assets.

At the systems level, SmartAir's smart factory was supported by a cloud-based digital backbone, orchestrated through containerised microservices hosted on Microsoft Azure. Applications managing inventory, robotics task scheduling and quality assurance operated as independent services within a scalable cloud infrastructure. The Digital Twin of the entire shop floor mirrored all assets in real-time, allowing engineers to simulate layout changes, perform stress testing on production bottlenecks and even rehearse emergency shutdowns before applying any real-world adjustments.

The factory-wide operations were made cyber-resilient through the implementation of secure authentication protocols and network segmentation. Leveraging blockchain-based supply chain management, SmartAir ensured the traceability of critical aircraft components. Data entries, such as inspection results, quality approvals and part origins, were recorded in a Byzantine Fault Tolerant (BFT) blockchain, ensuring tamper-proof audit trails and real-time authentication of parts received from third-party suppliers. This was particularly crucial for safety-critical components, such as engine parts, where compliance with regulatory documentation is vital for safety.

The integration of IBM's Watson IoT Platform enabled deep analytics across the factory. By processing real-time machine data alongside historical trends, Watson's cognitive engine can suggest schedule optimisations, identify chronic failure points and recommend energy savings strategies. This layer of cloud-based AI made the factory not only automated but also adaptive.

14.2.3 Smart Maintenance

Once the aircraft was rolled out from SmartAir's intelligent production lines, it moved into an equally sophisticated and integrated phase: lifecycle management, predictive maintenance, performance optimisation and supply chain resilience. This phase was powered by a seamless integration of IIoT, Digital Twins, Blockchain, Big Data and Cybersecurity frameworks.

Thousands of IIoT sensors embedded in wings, engines, landing gear, avionics systems and hydraulics continuously streamed data into SmartAir's Watson IoT Platform and cloud-based real-time analytics. This enabled engineers to simulate flight behaviours, monitor part stress in real-world conditions and even digitally validate changes before implementing them physically.

This constant data flow fuels Digital Twin updates in real-time. Every component replacement, software update, or environmental stress was mirrored in the virtual model. The twin was also used during active flight operations to monitor systems such as fuel flow, cabin pressure and wing stress, alerting ground teams instantly to any anomalies.

Maintenance operations benefited significantly from these integrations. Before an aircraft landed, its digital twin might already have flagged a valve nearing fatigue or a filter requiring replacement. When the plane arrived, AR-powered tools guided the technician using visual overlays that showed the exact component, service history and work instructions. Figure 14.4 illustrates the application of AR tools in aircraft maintenance operations.

Credit: Gorodenkoff / Shutterstock

Figure 14.4 Aircraft engine maintenance through AR

Every phase, from design to production and maintenance, was interconnected through digital threads. These digital threads ensured that data from CAD simulations, material certifications, manufacturing processes and maintenance logs flowed seamlessly into a single source of truth. This foundation enabled not just efficient production but also long-term aircraft lifecycle management.

The Digital Thread also optimised Just-in-Time (JIT) Inventory Management. Instead of stocking massive warehouses with surplus parts, SmartAir used real-time demand signals, derived from sensor analytics, service history and predictive models, to trigger precise manufacturing and delivery schedules. This minimised idle inventory, reduced waste and improved sustainability. For example, if a part on Aircraft A showed signs of wear and the same part was due for replacement on Aircraft B within two weeks, the system triggered a consolidated order, manufactured it through additive manufacturing (SLM or FDM methods) and shipped it just in time for both aircraft's servicing cycles.

All these operations, whether data streaming from IIoT, AR-assisted repairs, or JIT ordering, were managed on a cloud-edge infrastructure, allowing for low-latency decision-making and fleet-wide historical analytics. These data streams were massive, requiring data lakes built on Apache Hadoop and NoSQL storage, integrated with AI/ML-based analytics engines that evaluate terabytes of flight and service data.

To ensure data authenticity and secure tracking of aircraft parts across suppliers and service partners, Blockchain technology was embedded into the Digital Thread. Each part had a unique cryptographic ID and all its records, from supplier quality certification to flight cycles, were logged on a blockchain using a BFT mechanism. This ensured that even in the event of cyberattacks or network failures, the record remained tamper-proof and accessible to authorised parties across the ecosystem.

In this setup, Cybersecurity was super critical. With thousands of connected nodes, edge devices and cloud APIs, SmartAir implemented zero-trust architecture, TLS 1.3 encryption, network segmentation and AI-powered anomaly detection systems to safeguard against IP theft, intrusion and supply chain attacks.

14.2.4 Connectivity Using 5G Technology

While all of this was already transformational, the introduction of 5G connectivity became the technological catalyst that unlocked real-time, scalable and ultra-reliable industrial communications. 5G's low latency (as low as 1 Millisecond), high device density (up to 1 million connections per square kilometre) and network slicing capabilities have brought enormous benefits to SmartAir's smart factory ecosystem.

In the design phase, remote collaboration between globally dispersed engineering teams became seamless, with high-fidelity 3D simulations streamed over 5G from the digital twin servers. This enabled concurrent engineering and design adjustments to be made in real-time.

During production, 5G-enabled machine-to-machine (M2M) communication facilitated synchronised multi-robot operations and live equipment diagnostics. For instance, if an articulated arm detected overheating via its sensors, it could instantly alert the digital control system, halt operations and dispatch an AMR to fetch a cooling module, all coordinated via ultra-reliable low-latency communication (URLLC) supported by 5G.

In predictive maintenance, 5G-powered edge devices collect and transmit data from thousands of aircraft parts, supporting real-time condition monitoring, flight analytics and service scheduling without delay. This was crucial for SmartAir's global airline customers, who could remotely monitor aircraft health with zero data loss.

Furthermore, 5G's network slicing ensured that different systems, such as safety-critical, maintenance and logistics, could operate securely and independently over the same infrastructure without bandwidth contention, maintaining system integrity and performance.

14.2.5 Final Outcomes

As SmartAir adopted a holistic Industry 4.0 transformation, encompassing design, production and maintenance, the integration of advanced technologies reshaped every dimension of its operations. The results were not just incremental; they were transformative, cutting across cost, time, quality, traceability and resilience.

The implementation of automation and robotics, including articulated robots for fuselage welding, Cartesian and SCARA robots for precision drilling and electronics assembly and Cobots for human-robot collaboration on cabin interiors, resulted in a 35% increase in assembly line throughput. The use of AMRs and AGVs, integrated with digital twins, reduced intra-facility transport time by over 40%.

In terms of production quality, the use of AI-powered vision systems and AR-guided inspections has achieved defect detection accuracy of nearly 99.8%, particularly in wire harnessing and avionics integration. These improvements directly contributed to a reduction in rework rates and a sharper first-time-right ratio, minimising wastage and delays.

From a supply chain standpoint, the synergy of IIoT, cloud computing and blockchain-backed digital threads ensured complete traceability, vendor accountability and component provenance. With smart JIT inventory systems, SmartAir reduced working capital locked in excess inventory and decreased warehousing needs by 25%, making the process leaner and more sustainable.

SmartAir's journey into Industry 4.0 illustrates how a synergistic, interconnected technology strategy can revolutionise core manufacturing. Each technology, whether it be automation, digital twins, AR, additive manufacturing, cloud computing, big data, or blockchain, was not an isolated solution, but rather a part of a larger, intelligent system.

14.3 FACTORS AFFECTING THE ADOPTION OF INDUSTRY 4.0

In the previous section, we examined how SmartAir adopted Industry 4.0 and transformed all its processes, from design to maintenance. But such transformation is not simple. It is a huge change management encompassing people, processes and technologies. Despite the promising landscape of emerging technologies and the widespread awareness around Industry 4.0, adoption rates remain disappointingly slow across several sectors, particularly in developing economies and traditional manufacturing clusters. A significant proportion of small and medium enterprises (SMEs) have yet to implement even foundational components such as data-driven decision systems or real-time monitoring tools. According to multiple international studies, fewer than 30% of manufacturing organisations worldwide have successfully integrated Industry 4.0 technologies into their operations. This is not due to a lack of awareness or interest but rather to a complex web of technical, organisational, financial, regulatory and cultural barriers. The following subsections delve into these factors in depth, identifying both the bottlenecks and interdependencies that must be addressed to facilitate large-scale Industry 4.0 transformation.

14.3.1 Data Leverage

Data serves as the foundation of Industry 4.0, yet many enterprises still struggle to harness its full potential. In smart factories, sensors embedded in machinery continuously collect operational data, including temperature, pressure, cycle time, torque, vibration and throughput. However, gathering data is only the first step; the real challenge lies in consolidating, contextualising, analysing and acting on it in real-time. Most legacy manufacturing systems operate in silos, where each department or process uses proprietary systems that are not interoperable. This disconnection between Operational Technology (OT) and Information Technology (IT) platforms results in poor data visibility across the organisation.

Even when data is collected, it is often incomplete, inconsistent, or lacks the contextual metadata required for effective analytics. There is also the challenge of unstructured data, such as images, video feeds, or voice commands, which require advanced AI or natural language processing (NLP) systems to be interpreted. Organisations attempting to implement AI or predictive maintenance often find that they lack sufficient, high-quality and labelled historical data to train their algorithms reliably.

Moreover, data privacy and ownership issues often emerge when enterprises rely on external cloud services. This discourages data sharing across the value chain and limits the adoption of shared digital platforms, such as digital twins or blockchain-based traceability systems. Without establishing a central data governance policy and an integrated data lake architecture, companies are unable to fully unlock the value of advanced analytics, machine learning and AI optimisation tools.

14.3.2 Organisational Readiness

One of the most underappreciated challenges in Industry 4.0 adoption is organisational readiness. Implementing smart manufacturing technologies is not just a technical upgrade; it represents a fundamental cultural and operational transformation. Many organisations face internal resistance to change due to fear of redundancy, a lack of understanding of the new technologies, or reluctance to shift from time-tested traditional practices. Effective change management is required, encompassing transparent communication, inclusive planning and strong leadership advocacy.

Another key factor is the readiness of organisational structures. Traditional hierarchical models are often too rigid to support the cross-functional collaboration needed in a digital ecosystem. A booming Industry 4.0 strategy requires real-time coordination between production teams, IT departments, supply chain management and data scientists. This necessitates the redesign of workflows and a flatter decision-making process. Both top-down leadership support and bottom-up innovation input are crucial to sustaining the momentum of transformation.

Moreover, even if the core enterprise is ready, transformation efforts often get derailed due to a lack of preparedness among ancillary units. For example, a smart manufacturing firm may require data integration and digital capability from its Tier-1 and Tier-2 suppliers for effective just-in-time logistics or demand forecasting. If suppliers are still operating with manual records or outdated software, the entire chain suffers. Therefore, Industry 4.0 readiness must be viewed through the lens of the whole ecosystem, not just the anchor company.

Training and upskilling are equally essential. The digital workforce must be proficient in both traditional manufacturing and emerging technologies, such as robotics programming, data visualisation and cloud platforms. Companies often underestimate the time and investment required to develop such a hybrid skill set internally, resulting in execution gaps.

14.3.3 Legal and Regulatory Factors

The legal and regulatory landscape for Industry 4.0 is still evolving. With technologies such as cloud storage, AI-driven decision-making, digital twins and blockchain operating across borders, companies must comply with a complex web of international, national and sector-specific regulations. Data localisation laws often conflict with cloud-hosted services, particularly in regulated industries such as pharmaceuticals, defence, or banking. Manufacturing firms that utilise AI in quality inspection or predictive maintenance must also address issues related to transparency, auditability and bias in algorithmic decision-making.

The lack of globally harmonised standards for IIoT devices, data ownership and digital platforms complicates integration and interoperability. Organisations must also contend with the challenge of intellectual property (IP) protection in digital environments. For instance, sharing CAD files or process data with a third-party supplier over a collaborative cloud platform introduces risks of intellectual property (IP) theft unless secured using encryption and legal safeguards, such as smart contracts.

Emerging compliance requirements, such as ISO/IEC 27001 for information security and the General Data Protection Regulation (GDPR) for the handling of personal data, impose additional responsibilities. Failure to comply not only invites financial penalties but also undermines customer trust, particularly in industries involving high-value products or sensitive data.

14.3.4 Cyber Security

The very features that make Industry 4.0 robust, i.e., interconnectivity, real-time communication and cloud integration, also make it vulnerable to cyber threats. As factories shift from isolated control systems to networked, data-driven operations, the attack surface expands drastically. Legacy PLCs (Programmable Logic Controllers) and SCADA systems, which were not initially designed for internet access, become vulnerable to cyber intrusions when integrated into intelligent networks.

The threats are multifaceted: malware, ransomware, IP theft and even remote control of robotic systems through vulnerable entry points. One of the most alarming risks is the potential manipulation of sensor data, which can lead to incorrect AI predictions or the malfunctioning of autonomous machinery. The consequences are not just financial but also operational and reputational.

While advanced cybersecurity frameworks, such as Zero Trust Architecture and AI-based threat detection systems, are available, many companies lack the skilled personnel to implement and manage them effectively. Industrial organisations must now think beyond basic firewalls and antivirus software. They must deploy intrusion detection systems (IDS), security information and event management (SIEM) platforms and endpoint protection tailored explicitly for industrial environments.

Furthermore, cybersecurity must also extend to the supply chain. A compromised vendor system can serve as a gateway for malware to enter the main production network. To mitigate these risks, comprehensive third-party risk assessments, encrypted communication protocols and continuous vulnerability scans must be institutionalised.

14.3.5 Infrastructure Readiness

Infrastructure forms the backbone of Industry 4.0 deployment. Without a robust physical and digital foundation, even the most advanced technologies cannot operate effectively. Industrial infrastructure encompasses several interconnected layers, including network infrastructure, energy systems, computing resources, smart equipment and the physical layout of the facility.

One of the primary technical challenges is network connectivity. Industry 4.0 requires low-latency, high-bandwidth communication between machines, edge devices, cloud servers and human operators. Technologies such as 5G, industrial Ethernet, TSN (Time-Sensitive Networking) and Wi-Fi 6 are crucial for supporting real-time data transmission, particularly in applications like autonomous robots, AR-guided assembly and remote monitoring of Programmable Logic Controllers (PLCs). In many traditional factories, however, the infrastructure is not equipped to handle such data throughput. Devices may be running on outdated legacy networks that lack the bandwidth, security protocols, or deterministic timing needed for smart automation.

Another key concern is computing infrastructure. While cloud computing platforms like AWS, Azure and Google Cloud offer scalable resources, their reliance on internet connectivity can be a bottleneck for time-sensitive operations. To address this, edge computing devices such as industrial PCs (IPCs), micro data centres and edge gateways are deployed at the machine or cell level to process data locally. For instance, a predictive maintenance application that monitors the vibration and temperature of a CNC machine may use an on-premises edge server to analyse deviations and trigger alerts instantly without relying on the cloud.

Power infrastructure is also crucial. Industry 4.0 environments are equipment-intensive, requiring a stable and clean power supply. High-speed robotic arms, AGVs and 3D Printing machines are susceptible to power fluctuations. Thus, facilities must invest in uninterruptible power supplies (UPS), power conditioning systems and redundant power circuits to ensure operational continuity.

Environmental readiness is often overlooked. Many smart devices and industrial IoT sensors require specific environmental conditions, such as temperature, humidity and vibration tolerance, to function accurately. Harsh environments (e.g., foundries or chemical plants) require equipment with high Ingress Protection (IP) ratings or explosion-proof enclosures, which often increase installation costs.

14.3.6 Technology Integration

Industry 4.0 technologies promise to revolutionise manufacturing and industrial processes; however, their real-world implementation is often hindered by fragmentation, a lack of standards, scalability issues and integration complexity. A smart factory is built upon a constellation of advanced technologies: IIOT, cloud platforms, AI, robotics, digital twins, cybersecurity systems and blockchain networks. However, integrating these disparate systems into a seamless, real-time ecosystem is a major technical challenge.

One issue lies in protocol compatibility. Different systems use different communication protocols, which may not natively interoperate. This creates data silos and prevents effective system-level orchestration. For example, while a robotic arm may use a proprietary controller language, the MES system may only accept standardised data formats, requiring middleware layers for translation and mapping. This adds complexity, cost and latency.

Hardware diversity adds another layer of complexity. Different machines have varying computing capabilities, memory and operating system (OS) environments. Deploying AI at the edge, for instance, is not feasible on low-memory PLCs or old SCADA systems. In such cases, organisations might need to invest in industrial-grade AI accelerators for localised inferencing.

Moreover, many organisations lack a framework for assessing technology maturity. Jumping directly into advanced technologies like digital twins without first implementing foundational technologies, such as sensors, real-time monitoring and automation, would lead to failed pilots. A technology readiness level assessment or digital maturity model must be used to plan the incremental roadmap.

Version control and update management also present technical issues. Many smart devices rely on firmware and software updates to patch vulnerabilities or improve functionality.

Without proper orchestration tools, such as Over-The-Air (OTA) update platforms, companies risk exposing themselves to cyber threats or functionality mismatches between devices.

Another challenge is the lack of explainability in complex AI models. Deep learning systems used for defect detection or anomaly classification are often perceived as 'black boxes', making it challenging for operators to trust their outputs without the aid of visualisation tools or human-in-the-loop systems.

Lastly, high-fidelity data simulations and real-time digital twins require ultra-reliable latency, which is typically under 10 ms. These precision requirements are challenging to meet without 5G connectivity and dedicated edge computing units.

14.3.7 Financial Considerations

Cost remains one of the most significant barriers to Industry 4.0 adoption, particularly for small and medium-sized enterprises (SMEs). The financial challenge is multifaceted, comprising capital expenditure (CAPEX) on new equipment, operational expenditure (OPEX) on software and services, hidden costs related to downtime and training and uncertainty in return on investment (ROI) timelines.

Initial setup costs can be overwhelming. Deploying a smart factory setup may involve purchasing robots, sensor suites for predictive maintenance, edge computing hardware and cloud-based platforms, all of which are very costly. This does not even include ancillary costs, such as redesigning the factory layout, power backup systems, or cybersecurity infrastructure.

Another aspect is the recurring cost of licensing software tools, such as MES, SCADA, CAD/CAM integration tools and AI/ML analytics platforms. While SaaS models allow easier adoption, they can become expensive at scale. For example, advanced analytics tools from AWS or Azure may charge per computation, per GB of data storage, or per user.

Skilling and training costs are often underestimated. A workforce unfamiliar with PLC programming, robotics maintenance, data analytics, or cybersecurity needs substantial onboarding time and structured skill development programs, which come at an added cost.

Compounding this is the uncertainty in ROI. Most digital transformation benefits, such as predictive maintenance, energy efficiency, or quality improvements, take 1 to 3 years to impact fully. This timeline is often too lengthy for SMEs operating under tight cash flows or those in competitive markets with low-profit margins.

Inadequate access to low-interest loans, government subsidies, or public-private partnerships further limits adoption. While several Industry 4.0 incentive programs exist globally (e.g., Germany's Industrie 4.0 and India's SAMARTH Udyog Bharat 4.0), access is often limited to larger segments of the industries.

To mitigate these challenges, some companies are turning to 'smart factory as a service' (SFaaS) models, where third-party providers manage the entire transformation on a subscription basis. This reduces upfront investment but requires trust and the sharing of data with the service provider.

Lastly, the risk of technology obsolescence deters investment. The rapid pace of change in robotics, AI and industrial software means that systems purchased today may need to be upgraded or replaced within a few years. Without modular and future-proof architecture, these upgrades result in sunk costs and repeated investment cycles.

14.4 THE WAY FORWARD

To navigate the challenges of Industry 4.0 adoption, a phased, strategic and collaborative approach is essential. Companies should begin by identifying high-impact, low-complexity areas where digital transformation can yield immediate benefits. Pilot implementations involving AI-based quality control, predictive maintenance, or inventory optimisation can demonstrate ROI and help build organisational confidence.

Skill development must be a central component of this transformation. A digitally enabled workforce will require continuous training across data literacy, cloud usage, robotics programming, cybersecurity awareness and systems thinking. Industry-academia collaboration can accelerate this by introducing practical, modular certification programs aligned with real-world use cases.

On the regulatory front, governments and international bodies must evolve frameworks that encourage innovation while ensuring accountability and compliance. Harmonised standards, cybersecurity certifications and cloud governance models must be implemented through industry-wide cooperation.

Most importantly, organisations must gear up for the next phase of industrial evolution, Industry 5.0. This new paradigm places the human back at the centre of manufacturing, emphasising collaborative robotics, sustainability and mass personalisation. Industry 5.0 envisions a future where intelligent machines work in tandem with humans, augmenting their capabilities rather than replacing them. It focuses on ethical AI, eco-friendly production and resilient supply chains that are adaptable to crises like pandemics or climate disruptions. Apart from the collaboration concept, as seen in Industry 4.0, there will be a focus on compassion, where machines will sense human preferences, needs and emotions and humans will care for the robots with empathy. Additionally, there will be co-evolution, where humans and machines interact and enhance their capabilities together.

With Industry 5.0, the focus shifts from automation alone to human-machine symbiosis, where intuition and creativity are complemented by speed and precision. Companies that embrace this evolution early will not only unlock higher productivity but also differentiate themselves through innovation, sustainability and customer intimacy.

The journey from Industry 4.0 to Industry 5.0 is not linear. Still, organisations that invest today in building modular, secure and human-centric digital ecosystems will be best positioned to lead in tomorrow's intelligent industrial economy.

SUMMARY

- Industry 4.0 represents the convergence of advanced technologies, including IIoT, AI, Big Data and Analytics, Cloud Computing, Cybersecurity, Additive Manufacturing, AR/VR, Blockchain, Digital Twins and Automation and Robotics, all working together to create smart, autonomous and data-driven industrial systems.
- The digital transformation enabled by Industry 4.0 is not limited to production. It spans the entire product lifecycle, from R&D and design to supply chain, manufacturing, maintenance and customer support, enabling a holistic digital enterprise.

- 5G plays a transformative role in enabling ultra-fast, low-latency connectivity, especially for applications that require real-time synchronisation, such as collaborative robotics, real-time video-based AR guidance and decentralised control systems.
- Despite the availability of mature technologies, adoption remains limited. Many industries, particularly small and medium-sized enterprises (SMEs), are struggling to implement Industry 4.0 due to challenges in infrastructure, finances, change management, interoperability and skills development.
- Organisational readiness is a major barrier. Many companies lack clear digital strategies, leadership alignment, or internal coordination between IT and operations. Change management is often overlooked, resulting in cultural resistance and fragmented efforts. Surrounding industry ecosystems, including suppliers, logistics providers and service partners must also be digitally ready for a full-scale transformation.
- Technological complexity is another hurdle. Integrating legacy systems with new technologies, ensuring data interoperability, managing multiple vendors and building cybersecurity into every layer demands deep technical capability.
- The lack of a skilled workforce trained in digital technologies, such as AI, robotics programming, data analytics, cybersecurity and digital design tools, is a critical bottleneck. Traditional job roles need to be redefined and upskilled.
- The Way Forward advocates for a multi-pronged strategy, including robust government policy support, global and national standards, targeted financial incentives, educational reforms and industry-academia collaborations.
- Features of Industry 5.0 emerge as the next phase, emphasising human-centricity, sustainability and resilience. It complements Industry 4.0 by reintroducing the human as a creative collaborator with smart machines. Compassion and co-evolution will be added to collaboration.

EXERCISE

Multiple Choice Questions (1 Mark)

1. Which platform processes and stores the massive data generated from IIoT devices?
   (a) SCADA
   (b) Cloud
   (c) PLC
   (d) RFID
2. How does Big Data Analytics contribute to Industry 4.0?
   (a) By laying composite fibres
   (b) By simulating manufacturing layouts
   (c) By identifying patterns and predicting anomalies
   (d) By replacing human operators

3. What is the primary role of Digital Twins in a smart factory setup?
   (a) Tracking inventory
   (b) Manufacturing goods
   (c) Virtual prototyping and performance simulation
   (d) Machine calibration
4. Blockchain ensures:
   (a) Visual display of factory layouts
   (b) Secure and tamper-proof data exchange
   (c) Reduction in AI model size
   (d) Predictive maintenance
5. How is cybersecurity enforced across the Industry 4.0 setup?
   (a) With legacy firewalls
   (b) Through zero-trust architecture and secure access protocols
   (c) Only with antivirus
   (d) Via physical security
6. What does 5G enable in Industry 4.0?
   (a) Component painting
   (b) Physical layout changes
   (c) High-speed, low-latency communication
   (d) Packaging
7. Which of the following are key factors in the slow adoption of Industry 4.0?
   (a) Lack of mobile phones
   (b) Organisational and regulatory factors
   (c) Technical, security and financial issues
   (d) Both B and C
8. Why is data leverage still a challenge in smart factories?
   (a) Too much use of IoT
   (b) Only structured data is collected
   (c) Data is often siloed, unstructured, or lacks context
   (d) Robots refuse to transmit data
9. What are the primary challenges related to technology readiness in smart factories?
   (a) Overuse of AI
   (b) High cost, Incompatibility of protocols,
   (c) Lack of explainability and upgrade issues
   (d) Both B and C

10. Why do financial aspects become a major concern for SMEs in adopting Industry 4.0?
    (a) Free electricity
    (b) High CAPEX and long ROI timelines
    (c) Scarcity of hardware
    (d) Too many engineers
11. Additive Manufacturing is particularly helpful for:
    (a) Heavy metal forging
    (b) On-demand part customisation
    (c) Large volume injection moulding
    (d) Conveyor belt maintenance
12. What does the Digital Thread facilitate in a manufacturing environment?
    (a) Real-time network redundancy
    (b) Human-robot collaboration
    (c) Seamless information flow across lifecycle stages
    (d) Secure payment transactions
13. Cybersecurity in Industry 4.0 helps protect:
    (a) Supply Chain only
    (b) Only cloud systems
    (c) Only sensors
    (d) Entire digital ecosystem from threats
14. What is considered the foundational enabler of Industry 4.0?
    (a) Cybersecurity
    (b) Blockchain
    (c) Connectivity and Data
    (d) Additive Manufacturing
15. A technician in a smart factory needs to visualise wiring paths in a tight space. Which technology should they use?
    (a) Blockchain
    (b) Digital Twin
    (c) AR Smart Glasses
    (d) Cloud Dashboard
16. A production line is constantly misaligned. The QA team cannot inspect every product manually. Which solution will best help?
    (a) Blockchain
    (b) Cobots
    (c) Vision Systems integrated with AI
    (d) SCADA

17. A factory is experiencing frequent machine downtimes due to unplanned failures. What Industry 4.0 solution should be implemented first to address this?

(a) Blockchain-based contracts
(b) Predictive maintenance with AI and Big Data
(c) Use of collaborative robots
(d) Additive Manufacturing for tooling

18. A firm wants to track the origin of parts and ensure quality throughout its global supply chain. What would be the most suitable solution?

(a) AR with visual tracking
(b) 3D Printing with SLM
(c) Blockchain ledger with component tagging
(d) Local server-based SCADA

19. A company aims to train new workers more efficiently while minimising errors in complex wiring processes. Which technology is most appropriate?

(a) Vision systems
(b) Additive Manufacturing
(c) Augmented Reality
(d) Cybersecurity protocols

20. What is Industry 5.0 focused on?

(a) Full machine control and no human involvement
(b) Only cybersecurity improvements
(c) Human-machine collaboration, sustainability and personalisation
(d) Large-scale manual production

Short Answer Type Questions (2 Marks)

1. What is the significance of interconnectivity in Industry 4.0?
2. How does the Industrial Internet of Things (IIoT) enhance factory performance?
3. What is the function of the digital thread in Industry 4.0?
4. Define the role of 5G in smart manufacturing environments.
5. Why is edge computing important in Industry 4.0?
6. What is meant by horizontal and vertical integration in Industry 4.0?
7. Why is change management a critical factor in Industry 4.0 transformation?
8. Explain how predictive analytics supports the development of smart factories.
9. How does cloud computing support Industry 4.0?
10. How does Industry 5.0 build upon the offerings of Industry 4.0?

Long Answer Type Questions (5 Marks)

1. Explain how Industry 4.0 technologies interact to enable smart manufacturing.
2. Discuss the challenges in data leverage in Industry 4.0 and their implications.
3. What are the major organisational readiness issues affecting Industry 4.0 transformation?
4. Elaborate on the infrastructural limitations that hinder Industry 4.0 adoption.
5. How do financial constraints impact the pace of Industry 4.0 adoption, particularly for SMEs?
6. Explain how human factors affect Industry 4.0 adoption.
7. How does regulatory uncertainty delay Industry 4.0 implementation?
8. Discuss the significance of 5G in enabling autonomous systems in smart factories.
9. Why is integration across legacy systems a technical challenge in Industry 4.0?
10. In what ways does a robust data strategy enable sustainable Industry 4.0 adoption?

Index

3D modelling, 161, 201
3D modelling software, 176
3D printing, 6, 10, 157–182
  applications of, 179–182
  benefits of, 178–179
  history of, 159–161
  jet technologies, 169
  metal printing, 172
  overview of 3d modelling software, 176–178
  technologies, 163–176
  working concept of, 161–163
3D scanning, 177
3D slicers, 162
5C architecture of CPS, 44
5G compatibility, 17
5G use cases in industry 4.0, 16–17
5S workplace organisation, 64
5th generation (5G), 15

A

Account take over (ATO), 435
Acrylonitrile butadiene styrene (ABS), 166
Additive manufacturing (AM), 10, 157, 160
Advanced message queueing protocol (AMQP), 86, 94
Advanced research projects agency (ARPA), 431
Adware, 435
Agile manufacturing approach, 67
AI-based predictive maintenance, 118
AI-driven robotics, 118
Amazon managed blockchain, 410
Amazon simple storage service (S3), 410
Analytics, 12, 309–345
  benefits of, industry 4.0, 325
  business use cases, 342–343
  descriptive, 312
  evolution of, 310
  implementation strategies, 331–342
  key types of, 312
  need for, in industry 4.0, 311
Andon system, 64
Anomalies, 322
  collective, 322
  contextual, 322
  point, 322
Anomaly detection, 322
Anti-fragile operations, 72
Anti-lock braking systems (ABS), 6
Anti-money laundering (AML), 420
Application level security, 437
Application security, 446
Articulated (6 axis) robots, 128
Artificial intelligence (AI), 2, 80
  chatbots and, 113
Association rule mining, 313
Atomicity, consistency, isolation, and durability (ACID), 279
Augmented reality (AR), 2, 10, 16, 189
  application areas of, 209
  generic digital augmentation, 198
  history of, 194
  in smart factories, 205
  key technologies behind, 200
  manipulation and interaction with objects, 202
  superimposition-based augmentation, 197
  trigger-based augmentation, 195
  types of, 195
  view-based augmentation, 197
  working of, 195
Augmented workforce and human-machine collaboration, 69
Automated guided vehicles (AGVs), 16, 130

Automated teller machines (ATMs), 113
Automation, 64, 121
  fixed, 121
  flexible, 122
  integrated, 122
  programmable, 121
Automation and robotics, 10
Autonomous mobile robots (AMRs), 16, 69, 71, 130
Autoregressive integrated moving average (ARIMA) model, 320

B

Batch processing, 276
Batch processing using mapReduce, 291
Big data, 11, 267–304
  characteristics, 274
  evolution of, 268
  implementation approach, 296–299
  management techniques, 285
  processing, 288–296
  storage solutions, 277
  use of, across industries, 299
Big data analytics, 2, 80
Big data and analytics, 6
Big data processing, 288–296
  batch processing, 291
  distributed data processing, 290
  parallel data processing, 289
  real-time (Stream) processing, 291
Binary classification, 317
Binder jetting, 169
Bitcoin whitepaper, 391
Block(s), 391, 393, 397
  body of, 402
  hash, 401, 402
  header, 400
  height, 402
  structure of, 400
  timestamp, 401
  transaction, 402
Blockchain, 6, 389–423
  benefits for industry 4.0, 415
  challenges and considerations for, 419–422
  components of, 393
  consensus mechanism, 397
  consortium, 403
  event, 396
  federated, 403
  hashing, 397
  implementation strategy, 415–419
  key concepts of, 391–411
  ledger, 397
  merkle root, 402
  networks, 403
  nonce, 401
  principles of, 393
  protocols relevant for industry 4.0, 403
  smart contracts, 400
  transaction, 396, 402
  transaction workflow, 392
  use cases in industry 4.0, 411–415
  wallets, 399
Blockchain as a service (BaaS), 406
  advantages, 408
  identity and access management (IAM) implementation, 418
  importance of, 407
  model, 407
  monitoring and maintenance, 419
  network construction, 417
  network launch, 418
  network testing, 418
  potential issues with, 409
  protocol selection, 417
  provider selection, 416
  scaling and optimisation, 419
  solution providers, 410
  systems integration, 417
Blockchain networks, 403
  consortium blockchains, 403
  federated blockchain, 403
  permissioned blockchain networks, 403
  private blockchain networks, 403
  public/permissionless blockchain networks, 403
  types of, 403
Botnet, 436
Bound powder extrusion, 174
Broadband IoT, 16
Brownfield integration, 16
Building information modelling (BIM), 373
Business case evaluation, 296
Business email compromise (BEC), 435
Business to business (B2B), 80
Business to consumer (B2C), 80
Byzantine fault tolerance (BFT), 398
Byzantine generals' problem (BGP), 399

C

California consumer privacy act (CCPA), 13
Capital expenditure (CAPEX), 471
Cardano (ADA), 405
Cartesian robots, 126
Case studies
  A digital transformation journey in aircraft manufacturing, 460–467
  Agile manufacturing in industry 4.0, 70–71
  Digital twin implementation for a smart hospital building, 378–382
  Lean manufacturing in industry 4.0, 66–67
  Real-life implementation of automation and robotics in Amazon warehouse, 143–147
Centralised and decentralised systems, 389
Centroid-based clustering, 314
Certificate signing request (CSR), 258
Chatbots, 113
Chlorinated polyethylene (CPE), 166
Classification, 317
Cloud computing, 2, 6, 11, 221–261
  cloud deployment models, 229–232
  cloud security, 256–259
  cloud service models, 225–228
  definition, 222
  evolution of, 223–225
  for industry 4.0, 247–256
  key components of cloud architecture, 232–247
  manufacturing execution systems (MES) in cloud, 250
  manufacturing through cloud ERP, 248
  programmable logic controllers (PlCs) in cloud, 253
  supervisory control and data acquisition (SCaDa) in cloud, 250
Cloud computing in IIoT architecture, 87
Cloud gateway, 83
Cloud security, 256, 438
Cloud security services for IoT applications, 257
Clustering, 314
  centroid-based, 314
  density-based, 314
  grid-based, 314
Cognitive digital twin (CDT), 363–364
Cold path analytics, 95
Collaborative robots (COBOTs), 129
Collective anomalies, 322
Column-family stores, 281
Commercial off-the-shelf (COTS), 253
Common vulnerability scoring system (CVSS), 448
Compliance assembly robot arm (SCARA) robots, 127
Computational simulation model, 353
Computer numerically controlled (CNC) machines, 67
Computer security, 446
Computer-aided design (CAD), 6, 161, 176
Computer-aided manufacturing (CAM), 6
Computer-integrated manufacturing (CIM), 91
Computing technologies, 44
Consensus, 393
Consortium blockchains, 403
Constrained application protocol (CoAP), 86
Content delivery networks (CDN), 445
Contextual anomalies, 322
Continuous improvement (Kaizen), 64
Convex programming, 323
Convolutional neural networks (CNNs), 318
Corda, 405
Cox proportional hazards model, 321
Critical IoT, 16
Cross-functional teams, 68
Cross-industry standard process for data mining (CRISP-DM), 338
Cryptography, 399
Customer relationship management (CRM), 96
Cyber threats, 433
  account take over (ATO), 435
  adware, 435
  data breach, 436
  denial of service (DoS), 435
  distributed denial of service (DDoS), 436
  insider threat, 436
  malware, 433
  phishing, 435
  ransomware, 434
  smishing, 435
  spear phishing, 435
  spyware, 435
  trojan, 434
  virus, 434
  vishing, 435
  whaling, 435
  worms, 434

Cyber-physical systems (CPS), 32
  5C architecture of, 44
  communication and networking technologies, 44
  key technology layers of, 43
Cyber-physical systems (CPS), 9, 36, 457
Cybersecurity, 6, 13, 429–452, 469
  balancing the elements for, 433
  challenges in implementing, 450–451
  confidentiality, 432
  defence in depth model, 442–447
  demilitarised zone (DMZ) in, 443
  detect function, 449
  evolution of, 431
  framework, 447
  implementation framework, 447–450
  in industry 4.0, 438–442
  integrity, 432
  layered security approach, 445
  principles, 432
  protect function, 449
  recover function, 450
  respond function, 449
  security at different layers, 437–439
  types of cyber threats, 433

D

Data
  acquisition, 297
  breach, 436
  column-family stores, 281
  document stores, 281
  identification, 297
  ingestion, 297
  lakes, 284
  leverage, 467
  marts, 283
  partitioning, 289
  processing and transformation, 298
  semi-structured, 272
  storage, 297
  structured, 272
  types of, 270
  unstructured, 272
  variety, 276
  veracity, 276
  visualisation, 298
  warehouse, 283
Data acquisition, 94
Data breach, 436
Data ingestion, 297
  batch ingestion, 297
  streaming data ingestion, 297
Data lakes, 85, 94
Data storage, 94
Data visualisation and analysis, 85
Data visualisation techniques, 312
Data warehouses, 85
Data warehousing, 293
Database as a service (DBaaS), 85
Database management system (DBMS), 277
Decentralisation, 393
Decentralised autonomous organisations (DAO), 422
Decision review system (DRS), 189
Decision tree, 318
Deformable mirror device (DMD), 165
Demilitarised zone (DMZ) in cybersecurity, 443
Denial of service (DoS), 435
Density-based clustering, 314
Density-based spatial clustering of applications with noise (DBSCAN), 314
Dependent variable, 316
Depth APIs, 201
Depth tracking mechanisms, 201
Descriptive analytics and data visualisation, 298
Descriptive analytics, 312
Device management systems, 86
Device security, 446
Device technologies, 43
Device to dashboard data flow, 340
Device-to-device (D-to-D), 88
Digital kaizen, 65
Digital light processing (DLP), 165
Digital poka-Yoke, 65
Digital supply chain, 441
Digital supply chain and on-demand logistics, 69
Digital thread and its relationship with digital twins, 359
Digital transformation, 34
Digital twin aggregates (DTAs), 357
Digital twin instance (DTI), 357
Digital twin prototype (DTP), 356
Digital twin technology, 140
Digital twins, 12, 119, 351–384
  and smart vehicles design, 366
  cognitive, 363–364

definition, 352
digital thread and its relationship with, 359–360
evolution of, 353–356
implementing a, 376–378
in aerospace, 371
in automotive, 372
in construction, 373
in health care, 374
in infrastructure, 369
in manufacturing, 369
in oil and gas, 373
in product life cycle, 365–368
in the operational phase, 368
key implementation considerations, 375–376
types of, 356–358
use cases of, 369–375
vs augmented reality, 361–363
vs simulation, 360–361
Digital value stream mapping (DVSM), 65
Digital wallets, 399
Digitalisation, 33
Digitisation, 33
Direct metal laser sintering (DMLS), 169
Direct modelling approach, 176
Discrete event simulation, 360
Disruptive testing, 141
Distributed control systems (DCS), 16, 440
Distributed denial of service (DDoS), 436, 445
DMZ layers in ISA-95 framework, 443
Document stores, 281
Domain name system (DNS), 440
Double spend problem, 391
Dynamic simulation, 360
Dynamic trigger, 196

E

Edge computing, 6, 83
Edge computing in automation, 119
Electron beam melting (EBM), 172
Electronic data interchange (EDI), 273
Electronic fuel injection (EFI), 6
End-of-arm tools (EOAT), 117, 145
End-to-end (e2e) communication, 13
Enterprise DMZ, 444
Enterprise resource planning (ERP), 248
Ethereum, 404
Exploratory data analysis (EDA), 312
Extended reality (XR), 212
eXtensible markup language (XML), 269
Extract load transform (ELT), 287
Extract, transform, load (ETL), 286

F

Federated blockchain, 403
Field gateway, 82
Field of view (FOV), 203
Fixed automation, 121
Flexible automation, 122
Flexible manufacturing systems (FMS), 67, 68
Fog computing, 83
Fourth-generation (4G), 14
Frame rate, 203
Frames per second (FPS), 203
Full nodes, 395
Fully immersive VR, 200
Fused deposition modelling (FDM), 159, 166
Fused filament fabrication (FFF), 174

G

G-code, 161
General data protection regulation (GDPR), 98, 469
Generic digital augmentation, 198
Genetic algorithm, 323
Geometric programming, 323
Global data protection regulation (GDPR), 13
Graph databases, 281
Graphical user interface (GUI), 118
Grid-based clustering, 314

H

Hadoop, 294
Hadoop distributed file system (HDFS), 270, 294
Hazard analysis and risk assessment (HARA), 139
HBase, 270
Head mounted devices (HMDs), 202
Heuristic search methods, 323
History of AR/VR, 194
Hive, 269
Horizontal integration, 47, 48–51, 249
Hot path analytics, 95

Human-machine collaboration in future workplaces, 23
Human-machine interface (HMI), 118
Hyperledger, 404
  burrow, 405
  fabric, 404
  indy, 405
  iroha, 404
  sawtooth, 404
Hypertext transfer protocol secure (HTTP), 94

I

IBM blockchain platform, 410
Identity and access management (IAM), 410
Identity and access management (IAM) implementation, 418
Image processing and projection, 201
Immutability, 390, 393
Independent variables, 316
Industrial automation and robotics, 111–151
  actuators for, 116
  controllers for, 116
  cybersecurity for, 119
  definition, 112
  evolution of, 114
  key building blocks of, 116
  sensors for, 116
  types of, 120
Industrial automation IoT, 16
Industrial control systems (ICS), 13
Industrial control system (ICS) networks, 439
Industrial DMZ, 444
Industrial internet of things (IIoT), 6, 81
  application domains of, 99
  benefits of, 96
  challenges in implementing, 97
  cloud computing and, 119
  cloud computing in, 87
  gateways, 82
  implementation strategy, 89–95
  organisational setup for implementing, 89
  platforms, 88
  protocols used in, 86
  reference architecture of, 82
  technological setup for implementing, 90
Industrial internet of things (IIoT), 9, 79–105
Industrial internet reference architecture (IIRA), 44
Industrial IoT (IIoT), 40
Industrial manufacturing system architecture (IMSA), 45
Industrial revolutions and their impact, evolution of, 7
Industry 1.0, 3–4
Industry 2.0, 4–5
Industry 3.0, 5–6
Industry 4.0, 2, 6–7, 457
  10 steps to implement analytics for, 331–336
  5G use cases in, 16–17
  and sustainable development, 22
  benefits of, 18
  benefits of analytics in, 325–330
  blockchain benefits for, 415
  blockchain protocols relevant for, 403–406
  blockchain use cases in, 411–415
  building blocks of, 8
  cloud computing for, 247–256
  connectivity for, 13
  cybersecurity in, 438–442
  digital transformation in, 35
  enabled agile manufacturing process, 68
  enabled lean manufacturing process, 65
  evolution of, 3
  factors affecting the adoption of, 467–471
  financial considerations, 471
  foundational principles of, 8
  horizontal and vertical integration, 9
  implementing analytics in, 331–342
  in healthcare and medical technologies, 20
  key technologies of, 9
  key tenets of, 31–56
  legal and regulatory factors, 468
  on workforce and education, 23
  operational methodologies, 9
  role of higher education in preparing students for, 24
  systematic adoption of automation and robotics in, 133
  technology integration, 470
  use cases of digital twin applications in, 369–375
  use cases of, 18
Information security, 430
Information technology (IT), 52, 80, 429, 467
Infrastructure as a service (IaaS), 87
Ingress protection (IP) rating, 120

Insider threats, 436
compromised insiders, 436
malicious insiders, 437
negligent insiders, 436
Integrated automation, 122
Integrated circuits (ICs), 43
Intellectual property (IP), 469
Intelligent production systems, 19
Internet of things (IoT), 2, 80
Inter-quartile range (IQR), 322
In-vehicle infotainment (IVI) systems, 372
ISA 95 framework, 52–53

J

Jidoka, 64
Just in time (JIT), 51, 61
Just in-time (JIT) inventory management, 9
Just-in-sequence (JIS), 71

K

K nearest neighbour, 318
Kaizen, 63
Kaplan-Meier estimator, 321
Key performance indicators (KPIs), 95, 309, 376
Knowledge discovery in databases (KDD), 336

L

Laser triangulation, 177
Latency, 203
Late-stage customisation, 67
Lean 4.0, 65
Lean manufacturing, 62
Light detection and ranging (LiDAR), 117
Light nodes, 395
Linear programming (LP), 323
Liquid crystal display (LCD), 165
Location-based trigger, 196
Lockout-tagout (LOTO), 140
Logistic regression, 318
Long short-term memory (LSTM) networks, 320
LTE-Advanced (4G+), 14

M

Machine learning (ML), 2, 80
Machine-to-machine (M2M) communication, 13
Malicious industrial add-in, 441
Malware, 433
Manufacturing execution systems (MES), 11, 118, 250
Manufacturing execution systems (MES) in cloud, 250
Map function, 292
MapReduce framework, 289
Marker-based trigger, 196
Massive IoT (MIoT), 16
Material jetting, 171
Mathematical optimisation, 323
Maximum a posteriori (MAP) estimation, 200
Mean time between failures (MTBF), 327, 333
Merkle root, 402
Mesh modelling, 177
Message queuing telemetry transport (MQTT), 94, 334
Metaverse, 214
Microsoft azure blockchain service, 410
Microsoft SQL server, 279
Miner nodes, 396
Mirrored spaces model' (MSM), 354
Mixed reality (MR), 10, 192
Mobile human machine interface (HMI), 441
Mobile security, 438
Motion control systems, 117
MQ telemetry transport (MQTT), 86
Muda, 62
Multi-factor authentication (MFA), 445
Mutability, 390
MySQL, 279

N

Naïve bayes, 318
National institute of standards and technology (NIST), 36, 222
National vulnerability database (NVD), 448
Natural language processing (NLP), 113, 467
Network intrusion detection system (NIDS), 442
Network level security, 437
Network security, 446
Networking and communication protocols, 119
Neural networks, 318
NIST's cybersecurity framework, 447

Node, 395
  full, 395
  light, 395
  miner, 396
Nonce, 401
Non-immersive VR, 199
NoSQL databases, 280

O

Online analytical processing (OLAP), 286
Online transaction processing (OLTP), 285
Open connectivity foundation (OCF), 88
Operational performance indicators (OPIs), 376
Operational technology (OT), 80, 467
Operational technology (OT) networks, 13
Oracle blockchain service, 410
Oracle database, 279
Organisational setup for implementing IIoT, 89
Overall equipment effectiveness (OEE), 66, 327, 333, 335
Over-the-air (OTA) software, 6

P

Parallel data processing, 289
Parametric modelling approach, 176
Peer-to-peer (P2P) network, 394
Peer-to-peer transactions, 391
Permissioned blockchain networks, 403
Personally identifiable information (PII), 98
Phishing, 435
Photogrammetry, 177
Physical security, 446
Pig, 269
Pilot deployment and operator training, 141
Point anomalies, 322
Point-of-sale (POS) system, 271
Poka-Yoke, 64
Polycarbonate (PC), 166
Polylactic acid (PLA), 166
Polynomial regression, 316
PostgreSQL, 279
Postponement or late-stage customisation, 67
Power-by-the-hour (PBH) approach, 329
Practical byzantine fault tolerance (PBFT), 399
Predictive analytics, 314
Predictive maintenance and equipment monitoring, 319
Prescriptive analytics, 323
Private blockchain networks, 403
Process simulation, 360
Product lifecycle management (PLM), 354
Product-as-a-service, 72
Product-as-a-service (PaaS), 329
Production monitoring dashboards, 312
Programmable automation, 121
Programmable logic controllers (PLCs), 14, 114, 253
Programmable logic controllers (PLCs) in cloud, 253
Proof of authority (PoA), 398
Proof of capacity (PoC), 398
Proof of stake (PoS), 396, 398
Proof of work (PoW), 396, 398
Proportional-integral-derivative (PID), 116
Public/permissionless blockchain networks, 403
Purdue model for control hierarchy, 443

Q

Quality control and defect detection, 319
Quick-change tooling and SMED, 68

R

Radio frequency identification (RFID), 99
Random forest, 318
Ransomware, 434
Rapid manufacturing (RM), 160
Rapid prototyping (RP) technique, 159
Really simple syndication (RSS), 273
Real-time andon and performance monitoring, 66
Real-time configuration management, 258
Real-time production scheduling, 68
Recurrent neural networks (RNNs), 318
Reduce function, 292
Reference architectural model industries 4.0 (RAMI4.0), 45
Regression analysis, 315
Reinforcement learning (RL), 324
Relational database management system (RDBMS), 94, 269
RepRap model, 159
Representational model, 353
Resilient manufacturing, 72
Reskilling, 23
Return on investment (ROI), 32

Ridge and lasso regression, 316
Robot operating system (ROS), 146
Robot safety standards, 119
Robotic process automation (RPA), 113
Robotics, 113
Robots, 113
 articulated (6 axis), 128
 automated guided vehicles , 130
 autonomous mobile, 130
 cartesian, 126
 collaborative, 129
 key parameters of, 125–126
 SCARA, 127
 types of, 126

S

SAMARTH Udyog Bharat 4.0, 471
Sample explore modify model assess (SEMMA), 336
SCARA robots, 127
Second-order cone programming (SOCP), 323
Secure sockets layer (SSL), 438
Security
 application, 446
 application level, 437
 cloud, 438
 computer, 446
 device, 446
 mobile, 438
 network, 446
 network level, 437
 physical, 446
 system level, 437
 transmission level, 437
Security information and event management (SIEM), 445, 469
Security management system, 86
Selective laser melting (SLM), 169
Selective laser sintering (SLS), 168
Semidefinite programming (SDP), 323
Semi-immersive VR, 200
Semi-structured data, 272
Sensors in cloud, 253
SHA-256, 402
Simple storage service (S3), 410
Simulated annealing, 323
Simulation, 12, 119
Simulation-based optimisation, 324
Simultaneous localisation and mapping (SLAM), 130
Simultaneous location and mapping (SLAM), 200
Single-minute exchange of die (SMED), 64
Slicer software, 161
Small and medium-sized enterprises (SMEs), 471
Smart cities and infrastructure development, 21
Smart contracts on blockchain, 400
Smart factory, 440
Smart JIT and inventory control, 65
Smart manufacturing system, 19
Smart six sigma (Quality 4.0), 71
Smishing, 435
Software-as-a-service (SaaS), 250
Solana (SOL), 406
Solid modelling, 176
Spatial audio, 204
Spear phishing, 435
Spyware, 435
Stereolithography (SLA), 163
Stream processing, 276
Stress testing, 141
Structured data, 272
Structured light, 177
Structured query language (SQL), 269, 278–279
Superimposition-based augmentation, 197
Supervisory control and data acquisition (SCADA), 14, 118, 250, 440
Supervisory control and data acquisition (SCaDa) in cloud, 250
Supply chain and logistics optimisation, 19
Supply chain optimisation, 319
Support vector machine (SVM), 318, 322
Surface modelling, 177
Survival analysis, 321
Synchronisation, 353
System integration testing (SIT), 418
System level security, 437
System of systems' (SoS), 38

T

Technical and process changes required for 5G compatibility, 17
Technological setup for implementing, 90
Theory of constraints (TOC), 71
Thermoplastic polyurethane (TPU), 166
Third-generation (3G), 14
Time series database (TSDB), 94

Time series forecasting, 319
Time-sensitive networking (TSN), 15
Timestamp, 401
Total productive maintenance (TPM), 64
Transmission level security, 437
Transparency, 393
Trigger
  dynamic, 196
  location-based, 196
  marker-based, 196
Trigger-based augmentation, 195
Trojan, 434

U

Ultra-reliable low-latency communication (URLLC), 15, 16
Unstructured data, 272
Update paradox, 440
Upskilling, 23
Use cases of digital twin applications in industry 4.0, 369–375
Use management systems, 86
User acceptance testing (UAT), 418

V

Value stream mapping (VSM), 64
Vat photopolymerisation, 163
Vertical integration, 47, 51–54, 249
View-based augmentation, 197
Virtual PLC, 253
Virtual private networks (VPNs), 438, 445
Virtual reality (VR), 10, 16, 189, 191
  application areas of, 209
  controllers, 204
  fully immersive, 200
  headsets, 202
  history of, 194
  in smart factories, 205
  key technologies behind, 200, 202
  non-immersive, 199
  semi-immersive, 200
  software, 204
  types of, 199
  working of, 199
Virtual representation, 353
Virtual simulation, 140
Virus, 434
Vishing, 435
Vision systems, 118
Voice assistants, 113
VR headsets, 202

W

Web application firewall (WAF), 445
Web indexing, 293
Whaling, 435
Wide area network (WAN), 252, 341
Wireless personal area network (WPAN), 341
Wireless sensor networks (WSNs), 253
Worms, 434

Y

Yet another resource negotiator (YARN), 294
Yield, energy and throughput (YET) analysis, 328

Z

Zero-knowledge proofs (ZKPs), 420
Zookeeper, 270